Cellular Microbiology

Bacteria–Host Interactions in Health and Disease

Contents

Abbreviations Used in this Book

AA	arachidonic acid
ABC (proteins)	ATP-binding cassette
AD	transactivation domain
ADP	adenosine diphosphate
A/E (lesion)	attaching and effacing lesion
AHL	acyl homoserine lactone
AIDS	acquired immunodeficiency syndrome
AIL (*ail*)	attachment-invasion locus
AMP	adenosine monophosphate
AP	alkaline phosphatase
APC	antigen-presenting cell
ARF	ADP-ribosylation factor
ATFI	activator protein 1 (AP1) transcription factor
ATP	adenosine triphosphate
BALT	bronchial-associated lymphoid tissue
Bcl-2	B cell lymphoma
Bcr	break point cluster region
BFP	bundle-forming pili
BMEC	brain microvascular endothelial cells
BMP	bone morphogenetic protein
Boss	bride of sevenless
bp	base pairs
BSA	bovine serum albumin
BSE	bovine spongiform encephalopathy
CaM	calmodulin
cAMP	cyclic AMP (3', 5' cyclic monophosphate)
CAK	Cdk-activating kinase
CALLA	common acute lymphoblastic leukaemia antigen
CAP	catabolite activator protein
CAP	cationic antimicrobial protein
CBP	CREB binding proteins
CD	cluster of differentiation

cdc	cell division cycle
CDR	complementarity-determining region
CDT	cytolethal distending toxin
CETP	cholesterol ester transfer protein
CFTR	cystic fibrosis transmembrane conductance regulator
CFU	colony-forming unit
CGD	chronic granulomatous disease
cGMP	cyclic guanosine monophosphate
CGRP	calcitonin gene-related peptide
CJD	Creutzfeldt-Jacob disease
CKI	Cdk inhibitor
CNF	cytotoxic necrotizing factor
CO I-III	cytochrome oxidase complex
ConA	concanavalin A
COP	coat protein
COX (I)(II)	cyclooxygenase
Cpn	chaperonin
CRE(B)(M)	cAMP response element
crmA	cytokine response modifier (A)
CRP	C-reactive protein
CRP	cyclic AMP receptor protein
CSF	colony-stimulating factor
CT	cholera toxin
DAF	decay-accelerating factor
DAG	diacylglycerol
DBD	DNA-binding domain
DD-PCR	differential display-polymerase chain reaction
DH	Db1 homology (domain)
DIC	disseminated intravascular coagulation
DNT	dermonecrotic toxin
DT	diphtheria toxin
DTH	delayed-type hypersensitivity
EAP	endotoxin-associated protein
EB	elementary body
EBV	Epstein-Barr virus
ECM	extracellular matrix
EDIN	epidermal differentiation inhibitory factor
EF	(o)edema factor (of anthrax)
EF(2)	elongation factor
EGF(R)	epidermal growth factor (receptor)
EHV2	equine herpesvirus 2
EP	endogenous pyrogen (an early name for IL-1)
EPEC	enteropathogenic *E. coli*
EPO	erythropoietin
ER	endoplasmic reticulum
ERK	extracellular signal-regulated kinase
ES	embryonal stem (cells)

EST	expressed sequence tag
ET(-1)	endothelin
ETA	exotoxin A
FACS	fluorescence-activated cell sorter
FADD	Fas-associated death domain
FAF	fibroblast-activating factor
FAK	focal adhesion kinase
FGF(R)	fibroblast growth factor (receptor)
FHA	filamentous haemagglutinin (of *Bord. pertussis*)
FNR	regulator for fumarate and nitrate reduction
GABA	γ-aminobutyric acid
GAG	glycosaminoglycan
GALT	gut-associated lymphoid tissue
GAP	GTPase-activating protein
GAPD	glyceraldehyde 3-phosphate dehydrogenase
GDI	guanine nucleotide dissociation inhibitor
GEF	guanine nucleotide exchange factor
GI	gastrointestinal
GM-CSF	granulocyte-macrophage colony-stimulating factor
GOD	generation of diversity
GPI	glycosylphosphatidylinositol
GSH	reduced glutathione
GSP	general secretory pathway (of bacteria)
GTP	guanosine triphosphate
hBD-1	human β-defensin 1
HCMV	human cytomegalovirus
HDL	high-density lipoprotein
HEL	hen egg lysozyme
HEV	high endothelial venule
HFF	human foreskin fibroblast
HGF	human gingival fibroblast
HIV	human immunodeficiency virus
HLA	human leukocyte antigens
Hsp	heat shock protein
HSV	herpes simplex virus
HUC	human uroepithelial cell
HUS	haemolytic uraemic syndrome
HUVEC	human umbilical vascular endothelial cell
HV	hypervariable (region – as applied to the antibody molecule)
ICAM(-1, 2)	intercellular adhesion molecule
ICE	pro-IL-1β-converting enzyme
IEL	intraepithelial lymphocyte
IFN(α, β, γ)	interferon
IGF	insulin-like growth factor
IL (e.g. IL-1)	interleukin
IL-1ra	interleukin 1 receptor antagonist
IL-χR	interleukin χ receptor (χ = 1, 2, . . .)

$IP_3(R)$	inositol 1,4,5-triphosphate (receptor)
IRAK-1 and -2	IL-1 receptor-associated serine/threonine kinase
IS	insertion sequence
ITAM	immunoreceptor tyrosine-based activation motif
IVET	*in vivo* expression technology
IVOC	*in vitro* organ culture
Jak (kinase)	Janus kinase
JNK (1)(2)	Jun N-terminal kinases
KC	murine form of the human chemokine – melanoma growth stimulatory activity (MGSA)
LAD	leukocyte adhesion deficiency
LAF	lymphocyte-activating factor
LAMP(-1, 2)	lysosome-associated membrane protein
LAP	lipid A-associated protein
LBP	lipopolysaccharide-binding protein
LDL	low-density lipoprotein
LEE	locus of enterocyte effacement
LF	lethal factor (of anthrax)
LFA(-1, 2, 3)	lymphocyte function associated molecule
LGL	large granular lymphocyte
LGP	lysosomal membrane glycoprotein
LIF	leukocyte inhibitory factor
LINEs	long interspersed elements
LPS	lipopolysaccharide
LT	labile toxin
LTA	lipoteichoic acid
MAC	membrane attack complex (of complement pathway)
MALT	mucosa-associated lymphoid tissue
MAP (kinase)	mitogen-activated protein (kinase)
MARCKS	myristoylated alanine-rich C kinase substrate
MASP	MBP-associated serine protease
MBP	mannose-binding protein
MCP	membrane co-factor protein
M-CSF	macrophage colony-stimulating factor
MCP-1	monocyte chemoattractant protein 1
MEK	MAPK/ERK kinase
MHC	major histocompatibility complex
MIP-1α/β	macrophage inflammatory protein
MIS	Mullerian-inhibiting substance
MLC	myosin light chain
MM	α-methyl-D-mannoside
MMTV	murine mammary tumour virus
MOMP	major outer membrane protein
MPF	maturation-promoting factor
MRSA	methicillin-resistant *Staphylococcus aureus*
MTB	main terminal branch (of bacterial protein secretion)
NAD	nicotinamide adenine dinucleotide

NDH1–5	NADH dehydrogenase complex
NF-κB	nuclear factor κB
NMR	nuclear magnetic resonance
NO	nitric oxide
NOS	nitric oxide synthase (inducible (i)NOS, endothelial (e)NOS, neuronal (n)NOS)
Nramp(1, 2 etc.)	natural resistance-associated macrophage protein
NSAID	non-steroidal anti-inflammatory drug
NSF	*N*-ethylmaleimide-sensitive fusion
ODFR	oxygen-derived free radical
OMP	outer membrane protein
ORF	open reading frame
oriC	origin of replication
2'5'OS	2'–5' oligoadenylate synthetase
PA	protective antigen
PAF	platelet-activating factor
PAI	pathogenicity island
PAI	plasminogen activator inhibitor
PAMP	pathogen-associated molecular pattern
PAK	P21 activated kinase
PCR	polymerase chain reaction
PDGF	platelet-derived growth factor
PG	prostaglandin
PH	pleckstrin homology (domain)
PI3K	phosphoinositide 3-kinase
PI5K	phosphoinositide 5-kinase
PID	phosphotyrosine interaction domain
PIP_2	phosphatidylinositol 4,5-bisphosphate
PIP_3	phosphatidylinositol 3,4,5-triphosphate
PKA	protein kinase A
PKB	protein kinase B
PKC	protein kinase C
PKG	protein kinase G
PKR	RNA-activated protein kinase
PLA_2	phospholipase A_2
PLC	phospholipase C
PLD	phospholipase D
PLTP	phospholipid transfer protein
PMN	polymorphonuclear leukocyte
PMT	*Pasteurella multocida* toxin
Pol II	RNA polymerase II
PRR	pattern recognition receptor
PT	pertussis toxin
PTB	phosphotyrosine binding (domain)
PTP	phosphotyrosine phosphatase
RAG-1,2	recombination activating genes
ralGDS	guanidine nucleotide dissociation stimulator

RANTES	regulated on activation, normal T expressed and secreted
RAP-PCR	RNA arbitrarily primed PCR
RB	reticulate body
RCA	regulator of complement activation
RFLP	restriction fragment length polymorphism
Rho	Ras homology
RhoGAP	Rho GTPase activating protein
RIP	receptor interacting protein
RNA	ribonucleic acid
RSV	Rous sarcoma virus
RTK	receptor tyrosine kinase
RT-PCR	reverse transcriptase-polymerase chain reaction
SAA	serum amyloid A
SAg	superantigen
SALT	skin-associated lymphoid tissue
SAP	serum amyloid P
SAPK (1, 2)	stress-activated protein kinases
SCID	severe combined immunodeficiency
SE (A, B, C, D, E)	staphylococcal enterotoxin
SERCA	smooth endoplasmic reticulum Ca^{2+} ATPase
SH2/SH3	Src homology domain (2, 3)
Shc	SH2 domain-containing protein
Sifs	salmonella-induced filaments
SINEs	short interspersed elements
SLT	Shiga-like toxin
SNAPs	soluble NSF attachment proteins
SNAREs	SNAP receptors
snRNP	small nuclear ribonucleoprotein
Sos	son of sevenless
SPe	streptococcal pyrogenic exotoxin
SPI-1(2)	*Salmonella* pathogenicity island 1 (2)
SR-A	scavenger receptor (class) A
SRP	signal recognition particle
ST	stable toxin
STAT	signal transducers and activators of transcription
STI	*Salmonella typhimurium*-derived inhibitor of T cell proliferation
STM	signature-tagged mutagenesis
T3/T4	thyroid hormones
TAF	TBP-associated factors
TAP-1(2)	transporter associated with antigen processing
TCP	toxin co-regulated pilus (e.g. of *V. cholerae*)
TCR	T cell receptor
TD (antigen)	thymus-dependent
TFR	transferrin receptor
TGF(R)	transforming growth factor (receptor)
TI (antigen)	thymus-independent
Tiam 1	tumour invasion and metastasis factor

Tir	translocated intimin receptor
TK	tyrosine kinase
TMV	tobacco mosaic virus
TNF(R)	tumour necrosis factor (receptor)
tPA	tissue plasminogen activator
TRADD	TNF receptor-associated death domain
TRAF	TNF receptor-associated factor
TSE	transmissible spongiform encephalopathy
TSH	thyroid-stimulating hormone
TSST	toxic shock syndrome toxin
TUNEL	TdT-mediated dUTP-biotin nick end labelling
Tus	termination utilization substance
UTI	urinary tract infection
VAMP	vesicle-associated membrane protein
VAP	virulence-associated protein
VCAM	vascular cell adhesion molecule
VEC	vascular endothelial cell
VCP	vaccinia complement control protein
VPI	*V. cholerae* pathogenicity island
VT	Vero toxin

Bacterial Taxonomy and Nomenclature

Many of the readers of this book will not be microbiologists and are likely to be unfamiliar with bacterial taxonomy and nomenclature. Taxonomy is the scientific discipline involved with biological classification and can be subdivided into identification and nomenclature. In microbiology the basic taxonomic unit is the *species*. A species can be operationally defined as a collection of similar strains that differ sufficiently from all other groups of strains to warrant recognition as a basic taxonomic unit. Speciation is increasingly using ribosomal RNA gene sequences as a basis and it has been proposed that two bacteria whose 16S rRNA sequences are greater than 97% identical are likely to be of the same species.

A bacterial species is normally defined from the characteristics of several strains. Groups of species are collected into *genera* (singular: genus). A genus can be defined as a collection of different species, each sharing some major property or properties but differing in other characteristics. Groups of genera are collected into *families* which in turn are collected into *orders* and thus into *divisions*. The highest-level taxon is the *domain*.

The system used to name bacteria is the binomial system. All bacteria are given genus and species names. The genus name of an organism can be abbreviated to a single (capital) letter but the species name is never abbreviated. Normally, the first time an organism is mentioned in print it is denoted in full. Thereafter the genus name is abbreviated. Thus *Escherichia coli* is written in full and thereafter as *E. coli*. Genus and species names are normally of Latin or Greek derivation and are printed in *italics*.

Unfortunately, a number of the bacteria referred to in this book have genus names starting with a common letter, e.g. staphylococci, streptococci, salmonella. To allow the reader to understand which organisms are being referred to we have slightly modified the basic convention and have either used the full (unabbreviated) nomenclature for bacteria or have used non-standard abbreviations. The following list provides the full and abbreviated names of the bacteria described in this volume.

Actinobacillus actinomycetemcomitans	*Act. actinomycetemcomitans*
Actinobacillus pleuropneumoniae	*Act. pleuropneumoniae*
Actinobacillus suis	*Act. suis*
Actinomyces naeslundii	*A. naeslundii*
Actinomyces viscosus	*A. viscosus*
Aeromonas hydrophila	*Aer. hydrophila*

Bacillus alvei	*B. alvei*
Bacillus anthracis	*B. anthracis*
Bacillus cereus	*B. cereus*
Bacillus laterosporus	*B. laterosporus*
Bacillus thuringiensis	*B. thuringiensis*
Bordetella pertussis	*Bord. pertussis*
Borrelia burgdorferi	*Borr. burgdorferi*
Brucella melitensis	*Br. melitensis*
Brucella suis	*Br. suis*
Burkholderia cepacia	*Burk. cepacia*
Campylobacter jejuni	*C. jejuni*
Chlamydia psittaci	*Chl. psittaci*
Chlamydia trachomatis	*Chl. trachomatis*
Citrobacter freundii	*Cit. freundii*
Clostridium bifermentans	*Cl. bifermentans*
Clostridium botulinum	*Cl. botulinum*
Clostridium caproicum	*Cl. caproicum*
Clostridium chauvoei	*Cl. chauvoei*
Clostridium difficile	*Cl. difficile*
Clostridium histolyticum	*Cl. histolyticum*
Clostridium novyi	*Cl. novyi*
Clostridiuim oedematiens	*Cl. oedematiens*
Clostridium perfringens	*Cl. perfringens*
Clostridium septicum	*Cl. septicum*
Clostridium sordellii	*Cl. sordellii*
Clostridium spiroforme	*Cl. spiroforme*
Clostridium tetani	*Cl. tetani*
Coxiella burnetii	*Cox. burnetii*
Dichelobacter nodosus	*D. nodosus*
Escherichia coli	*E. coli*
Haemophilus ducreyi	*Haem. ducreyi*
Haemophilus influenzae	*Haem. influenzae*
Helicobacter hepaticus	*Hel. hepaticus*
Helicobacter mustelae	*Hel. mustelae*
Helicobacter pylori	*Hel. pylori*
Klebsiella pneumoniae	*K. pneumoniae*
Lactobacillus acidophilus	*L. acidophilus*
Legionella pneumophila	*Leg. pneumophila*
Listeria ivanovii	*Lis. ivanovii*
Listeria monocytogenes	*Lis. monocytogenes*
Listeria seeligeri	*Lis. seeligeri*
Mycobacterium avium	*M. avium*
Mycobacterium leprae	*M. leprae*
Mycobacterium tuberculosis	*M. tuberculosis*
Mycoplasma genitalium	*Myco. genitalium*
Neisseria gonorrhoeae	*N. gonorrhoeae*
Neisseria meningitidis	*N. meningitidis*

Pasteurella multocida	*P. multocida*
Pneumocystis carinii	*Pn. carinii*
Porphyromonas gingivalis	*Por. gingivalis*
Pseudomonas aeruginosa	*Ps. aeruginosa*
Rickettsia prowazekii	*Rick. prowazekii*
Rickettsia rickettsii	*Rick. rickettsii*
Salmonella dublin	*Sal. dublin*
Salmonella typhi	*Sal. typhi*
Salmonella typhimurium	*Sal. typhimurium*
Shigella dysenteriae	*Sh. dysenteriae*
Shigella flexneri	*Sh. flexneri*
Shigella sonnei	*Sh. sonnei*
Staphylococcus aureus	*Staph. aureus*
Staphylococcus epidermidis	*Staph. epidermidis*
Staphylococcus saprophyticus	*Staph. saprophyticus*
Streptococcus mutans	*Strep. mutans*
Streptococcus pneumoniae	*Strep. pneumoniae*
Streptococcus pyogenes	*Strep. pyogenes*
Streptococcus sanguis	*Strep. sanguis*
Streptococcus sobrinus	*Strep. sobrinus*
Treponema denticola	*Tr. denticola*
Treponema pallidum	*Tr. pallidum*
Vibrio cholerae	*V. cholerae*
Vibrio fischeri	*V. fischeri*
Vibrio harveyi	*V. harveyi*
Yersinia enterocolitica	*Y. enterocolitica*
Yersinia pestis	*Y. pestis*
Yersinia pseudotuberculosis	*Y. pseudotuberculosis*

Introduction

The introduction of antibiotics into clinical practice in the 1950s and 1960s heralded a revolution in medicine and suggested that bacterial infectious diseases were a 'thing of the past'. One consequence of this 'clinical miracle' was the decline in research into the mechanisms by which bacteria caused disease. Many scientists interested in the mechanisms of infectious diseases decamped into the discipline of virology and tremendous advances have been made since the 1960s in our understanding of the molecular and cellular processes which occur in virally infected cells. However, relatively little work was done during this period on the mechanisms used by bacteria to infect multicellular hosts or on the pathologies induced by such infections.

In the 1990s it is clear that our belief that the introduction of antibiotics had defeated bacterial infection was premature. Bacterial diseases such as tuberculosis (which now kills an estimated 3 million individuals each year), that were thought to have been defeated, are returning worldwide, even to the developed world. Diarrhoeal diseases are still major health care problems worldwide, killing an estimated 3–4 million people each year, and similar numbers are estimated to die from bacterial respiratory infections. Not only are established bacterial diseases returning to haunt us but there has been the discovery, during the past three decades, of new bacterial diseases such as Legionnaire's disease and Lyme disease and the surprising discovery of the role of *Helicobacter pylori* as the causative agent of gastric ulceration and some gastric cancers. There is growing evidence that certain cardiovascular diseases may be caused by bacteria. In addition, new forms of well-known bacteria, such as *Escherichia coli* 0157, now recognized as one of seven distinct strains of this bacterium capable of causing diarrhoea, have shot to international prominence.

In consequence, there has been an enormous resurgence of interest in bacterial infections during the last decade, largely fuelled by the realization that many species of bacteria are becoming resistant to all known antibiotics. Indeed, while this book was being written it was reported that *Staphylococcus aureus*, a major cause of hospital infections, had become insensitive to vancomycin, the last effective antibiotic against this organism. No new antibiotic classes have been introduced into clinical practice in the last twenty years. Thus, we face an uncertain future with regard to our ability to cope with bacterial infections.

The renewed interest in bacterial infections, stimulated largely by the growing resistance of bacteria to antibiotics, has occurred at a time when advances in microbiology, molecular biology and eukaryotic cellular biology have enabled more profound questions to be asked of the infectious process. The synthesis of these three related disciplines has been termed cellular microbiology (a term introduced by Cossart and colleagues, 1996) and, although only a very new branch of science, has already produced striking discoveries that are revolutionizing our view of the microbial world and of how bacteria and eukaryotic cells interact during the process of infection, and to maintain the enormous numbers of bacteria (the normal microflora) that colonize all multicellular organisms.

Much of the advance that has been made is due to the development of the techniques of molecular biology. Analysis of 16S ribosomal RNA has, for example, shown that all life on our planet is based on three basic cell types which can be classified into three great domains: Bacteria, Archaea and Eukarya (the last containing *Homo sapiens*). The first two are unicellular life forms and thus the most diverse and numerous life forms on our planet are single-celled creatures. Given that there are so many bacteria and single-celled Archaea (formerly Archaebacteria) living on our planet it is surprising how few bacteria cause disease. So far no Archaeal species is known to cause disease. New techniques in microbial and eukaryotic molecular biology such as *in vivo* expression technology (IVET), signature-tagged mutagenesis, differential display–polymerase chain reaction (DD-PCR), yeast two-hybrid analysis and phage display are being used to probe the interactions that occur between bacteria and eukaryotic host cells.

What is emerging from the study of cellular microbiology is just how intimate are the interactions which occur between bacteria and host cells. Bacteria such as enteropathogenic *E. coli*, *Salmonella typhimurium*, *Shigella flexneri* and *Yersinia* spp. have been shown to be able to hijack many of the complex molecular functions of eukaryotic cells in order to enter such cells as part of the infectious process. Bacteria also produce a wide range of exported proteins, known as exotoxins, which have profound effects on eukaryotic cells. Indeed such exotoxins have been widely used by cell biologists to investigate cell function. Advances in cell biology have also fed back into microbiology and it is now established that many bacterial species produce intercellular signalling molecules to determine bacterial density. This cell-to-cell communication is known as quorum sensing and may be an important virulence mechanism.

Bacterial infection, and the ability of the host to survive it, is now seen as an evolutionary battle between bacteria and the widespread immune responses of the host. There has been a resurgence of interest by immunologists in how the immune system recognizes and deals with bacteria. This has revealed that all creatures on our planet produce antibacterial peptides. It has also overturned immunological dogma with respect to how T lymphocytes recognize antigens. It is now clear that there is an evolutionary battle *par excellence* between bacteria, which are trying to avoid immunological defeat, and the multifaceted immune systems of vertebrates.

Surprisingly, despite the considerable interplay between bacteria and the cells of higher organisms, the academic disciplines of microbiology and cell biology are taught as separate subjects. No textbook of microbiology or cell biology pays more than lip-service to the other discipline. This is in spite of the fact that both disciplines require an understanding of each other. Many universities are now recognizing this and are reorganizing their courses to provide greater flexibility. This textbook is designed to bridge the gap between cell biology and microbiology and provide a current synthesis of the relevant science. The first half of the book introduces certain key concepts of prokaryotic and eukaryotic cell biology, cell signalling mechanisms and current molecular biological techniques used in cellular microbiology. The second half of the book describes how bacteria interact with host eukaryotic cells during infections and explains the interactions with the immune system which enable the infected individual to recover from infections.

This textbook is written for final-year graduates and for postgraduate students studying microbiology, medical microbiology, pathology, immunology, pharmacology, cell biology, molecular biology or biochemistry. It will also be of interest to scientists in these disciplines.

REFERENCE

Cossart P, Boquet P, Normark S, Rappuoli R (1996) Cellular microbiology emerging. *Science* 271: 315–316.

PART 1

Background to Cellular Microbiology

CHAPTER 1

An Introduction to Cellular Microbiology

Ring-a-ring of roses
A pocketful of posies
Atishoo atishoo
We all fall down

Traditional children's chant

Introduction

Twenty years ago it would have been difficult to comprehend the foundation of this traditional children's chant as society, or at least Western society, had lost its fear of bacterial infection. Of course, outside this privileged zone, death due to bacterial infections was, and indeed still is, relatively common, particularly if you are either very young or very old (Table 1.1). For example, it is estimated that 50 000 000 individuals die, worldwide, each year from all causes and of these 20 000 000 die from infectious diseases. Antibiotics, which were developed in the period between the 1940s and the 1970s, have brought unimaginable changes to our lives. Their use swept away the isolation hospitals, where those suffering from communicable bacterial diseases would be nursed. Gone from our towns are the sanitariums with their populations of chronic sufferers from tuberculosis. These institutions were replaced by the antibiotics prescribed by your general practitioner.

One consequence of the development and introduction of antibiotics into clinical practice was the decline in research into the mechanisms by which bacteria cause disease. Funding such research probably seemed pointless at the time as the prevailing belief was that antibiotics had defeated bacterial infections and that these diseases were 'a thing of the past'. This view was given its most profound support in 1969 by the then United States Surgeon General in his address to Congress when he stated 'We can close the book on infectious diseases.' Unfortunately, it is now clear that as soon as antibiotics were introduced bacteria began to develop resistance (Figure 1.1). For example, in the late 1940s the first penicillin-resistant

Table 1.1 Infectious diseases: estimated yearly death statistics

Cause of death	Estimated numbers	Infectious agents
Acute respiratory diseases	7–10 000 000	Bacteria, viruses, protozoa, fungi
Diarrhoeal diseases	4 300 000	Bacteria, viruses
Tuberculosis	3 300 000	Bacteria
AIDS	1–2 000 000	Virus
Malaria	1–2 000 000	Protozoa
Hepatitis	1–2 000 000	Virus
Measles	220 000	Virus
Bacterial meningitis	200 000	Bacteria
Schistosomiasis	200 000	Parasitic worm
Pertussis (whooping cough)	100 000	Bacteria
Amoebiasis	100 000	Protozoa
Hookworm	50 000	Parasitic worm
Rabies	35 000	Virus
Yellow fever	30 000	Virus
Trypanosomiasis (sleeping sickness)	20 000	Protozoan

Source of information: World Health Organization.

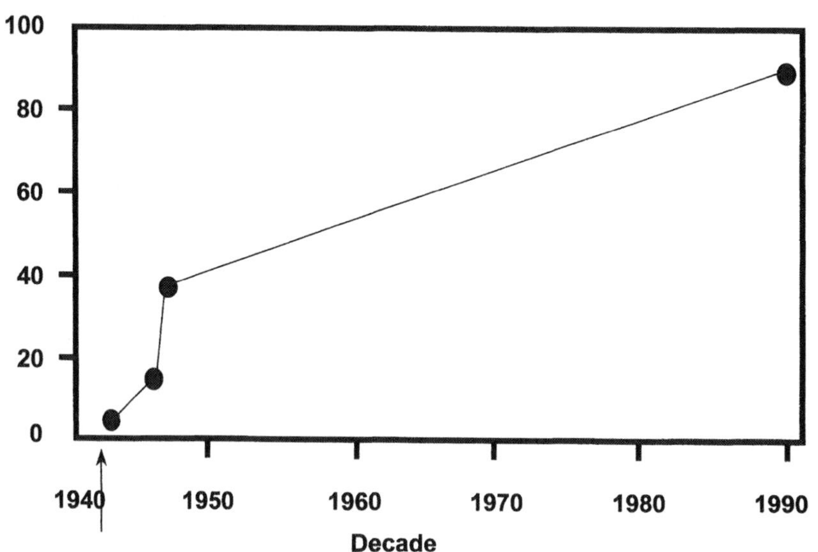

Figure 1.1 The rise in resistance to penicillin by *Staphylococcus aureus*. Almost as soon as this 'wonder drug' was introduced, resistant strains were identified. The incidence of such strains increased steadily until the 1990s, by which time the majority of strains were resistant to all forms of penicillin

strains of the commensal bacterium, *Staphylococcus aureus*, were discovered. The percentage of resistant strains increased over the years and now this bacterium is resistant to all forms of penicillin and in consequence is known as methicillin (sometimes multiple)-resistant *Staphylococcus aureus* (MRSA) and is a scourge of hospitals in the Western world. More worrying still are the recent reports of strains of *S. aureus* which are tolerant to vancomycin – the last remaining effective antibiotic against this bacterium. Other bacteria (e.g. *Shigella* spp., *Salmonella* spp., *Neisseria gonorrhoeae*, *Haemophilus influenzae*, *Mycobacterium tuberculosis*, *Enterococcus* spp., *Streptococcus pneumoniae*) have also begun to develop antibiotic resistance. Most worrying of all is the fact that no new antibiotic classes have been introduced into clinical use for more than 20 years. These worries have stimulated the major pharmaceutical companies to begin searching for novel aproaches to develop new antibacterial compounds.

Thus the 'golden age of antibiotics' looks jaded after only two generations. This was highlighted by a recent report from the House of Lords of the United Kingdom. Tuberculosis, which we thought had been defeated, has returned both to the Third World and to North America and Western Europe. MRSA is a major scourge of our hospitals. It is also clear from epidemiological studies that there is a continuing emergence of new infectious diseases. Historical examples of this are the plague (caused by *Yersinia pestis*) and syphilis (caused by *Treponema pallidum*). Emerging diseases in the readers' lifetime include legionellosis (*Legionella pneumophila*), Lyme disease (*Borrelia burgdorferi*) and enteritis (*Campylobacter jejuni*) (Table 1.2). It is therefore clear that we must face the problem of bacterial diseases anew and seek further solutions.

Table 1.2 Infectious agents and diseases recognized since 1950

Agent	Disease
Hantavirus	Korean haemorrhagic fever
Flavivirus	Dengue haemorrhagic fever
Arenavirus	Argentine haemorrhagic fever, Lassa fever
Kyasanur virus	Kyasanur forest disease
Chikungunya virus	Chikungunya
O'nyong-nyong virus	O'nyong-nyong fever
Haemorrhagic virus	Oropouche
Encephalitis virus	LaCross encephalitis
Marburg virus	Marburg disease
Legionella micdadei	Pontiac fever
Toxoplasma gondii	Human toxoplasmosis
Borrelia burgdorferi	Lyme disease
Ebola virus	Ebola disease
Campylobacter jejuni	Enteritis
Clostridium difficile	Pseudomembranous colitis
Cryptosporidium parvum	Diarrhoea
Staphylococcus aureus	Septic shock
Staphylococcus aureus	Toxic shock syndrome
Gardnerella vaginalis	Bacterial vaginosis
Helicobacter pylori	Gastritis (cancer?)
Legionella pneumophila	Legionnaires' disease
Mobiluncus spp.	Bacterial vaginosis
Escherichia coli 0157:H7	Haemolytic uraemic syndrome (HUS)
Hepatitis C virus	Hepatitis
Human herpesvirus 6	Exanthum subitum
Human immunodeficiency virus	AIDS
Parvovirus	Fifth disease
Rotavirus	Infantile diarrhoea
'Small round' virus	Gastroenteritis
BSE prion	Bovine spongiform encephalitis (BSE)
Chlamydia pneumoniae	Coronary artery disease?

These concerns have been the driving forces behind a renaissance in the study of bacteria and their role, both as causative agents of disease and as members of the normal microflora which all multicellular organisms maintain. This resurgence in interest in bacteriology has been aided by major advances in molecular biology and cell biology and these techniques are enabling us to comprehend that understanding the mechanisms of bacterial infection requires that the mutual interactions between bacteria and eukaryotic cells are understood in molecular and cellular detail. This realization has given rise to what can be termed the 'interface science' of cellular microbiology, which is a blend of classical microbiology, prokaryotic and eukaryotic molecular biology and cell biology. Cellular microbiology attempts to explain how bacteria interact both with themselves and with eukaryotic cells (mainly the epithelial cells, which are barriers to bacterial entry, and cells of the immune systems of the body) to produce infectious diseases.

We start our discussion with a brief historical overview, to enable the reader to see the genesis of cellular microbiology.

A BRIEF HISTORY OF BACTERIOLOGY

Except under abnormal circumstances, soft-tissued animals, and the soft tissues of animals with skeletons, fail to appear in the fossil record and the pages of this record are written in bone. It is fortunate then that bone is prey to bacterial infections and that the signs of such infection are so clearly visible. The earliest example of osteomyelitis (bone infection) is in a 200-million-year-old skeleton of *Dimetrodon* from the Permian period. We also know that the dinosaurs suffered from osteomyelitis as did the mastodon, the sabre-tooth tiger and even Neanderthal man. The unwritten history of bacterial infections can therefore be traced back for hundreds of millions of years.

The evolution of early man from wandering hunter–gatherer to geographically established farmer with domesticated animals, and the subsequent development of established settlements, are all believed to have had major impacts on the ecology of the infectious diseases of *Homo sapiens*. The writings of the earliest civilizations, such as those of the Egyptians, Sumerians and Chinese, all attest to their knowledge of infections and of the devastating effects of epidemics. Many readers will be familiar with the description in Exodus (written around 1500 BC) of the plague which beset the Egyptians. This epidemic was characterized by 'sores that break into pustules on man and beast'. The Hippocratic texts of ancient Greece describe the symptoms of infections such as tuberculosis in terms clearly recognizable to modern physicians. However, the causation of the epidemics that periodically swept the various ancient human populations remained a mystery. Philip Ziegler, in his eminently readable history of the Black Death, reveals that even in the fourteenth century, with Europe being devastated by the plague, its causation was still unknown and 'corruption of the atmosphere' was regarded as the most likely cause.

It was not until the seventeenth century that the concepts were developed which would eventually link disease to infection with living unicellular organisms. Like so many other aspects of scientific thought the impetus was the development of new technology, in this case the microscope. In 1665 Robert Hooke, also known for his studies of elasticity, published his book *Micrographica*, which introduced the concept of the cell and was based upon his microscopical examination of plant and animal material. It was the Dutchman, Antoni van

Leeuwenhoek, however, who discovered the universe of microscopic life, including the bacteria. Using his skilfully hand-ground lenses, Leeuwenhoek's single-lens microscopes could produce magnifications in the range 50 to approximately 300 diameters and using these instruments he discovered the existence of spermatozoa and of red blood cells. Leeuwenhoek's greatest claim to fame was his discovery of a world which had not been dreamed of in 5000 years of recorded history, the microbial world, or the world of 'animalcules' (little animals), the term used by Leeuwenhoek and his contemporaries. One of his most surprising findings was the abundance of these microscopical life forms and there is no better description of this abundance than in a letter from Leeuwenhoek describing, for the first time, the characteristic bacteria present in the human mouth.

I have had several gentlewomen in my house, who were keen on seeing the little eels in vinegar; but some of them were so disgusted by the spectacle, that they vowed they'd never use vinegar again. But what if I should tell such people in future that there are more animals living in the scum on the teeth in a man's mouth, than there are men in the whole kingdom?

Leeuwenhoek's sentiments are shared by dentists in the twentieth century who are still trying to teach us the correct manner in which brushing can remove these 'animalcules' from our teeth and so prevent dental caries and the periodontal (gum) diseases – two of mankind's most common chronic (bacterial) diseases.

The discovery of unicellar microorganisms, including bacteria, led to the obvious question – where did they come from? Two schools of thought existed: (i) they arose spontaneously or (ii) they formed from the 'seeds' or 'germs' of the animalcules. Spontaneous generation or abiogenesis was an ancient doctrine and held that many plants, animals or birds could arise spontaneously under given conditions. For example, the yearly appearance of migrating swifts was believed to be due to the generation of these birds in the mud of ponds. This idea probably arose because swifts build their nest of pond mud. Abiogenesis was finally laid to rest by the work of Redi, who showed that maggots failed to develop spontaneously on meat if adult flies were excluded, and Spallanzani, who demonstrated that sterile infusions of organic matter in water failed to develop animalcules if air was excluded. These experiments were later refined by Louis Pasteur, who used his famous swan neck-shaped flasks (now called Pasteur flasks) in which nutrient media were sterilized, but left exposed to the air, to demonstrate that it was not the air itself that gave rise to the microorganisms but that the microorganisms were already present in the air.

Pasteur, a chemist by training, was instrumental in demonstrating that bacteria have chemical actions and used this information to show that contamination of distillers' yeast cultures resulted in spoilage of wine and beer. Indeed Pasteur classified these spoilages as 'diseases' of beer and wine. It was during the period of the mid- to late nineteenth century that the pioneering work of clinicians such as Semmelweiss, surgeons like Lister and the microbiologists – Pasteur and Koch in particular – led to the discovery that certain diseases were due to the actions of bacteria. For example, Koch in 1876 conclusively demonstrated the bacterial causation of anthrax and in doing so introduced a set of criteria which were needed to establish the causal relationship between an infectious organism and the disease it causes. These are now known as Koch's postulates and in a generic manner are as follows: (i) the microorganism must be present in every case of the disease, and the pattern of isolation must follow the pattern of the disease; (ii) the microorganism must be isolated from the diseased host and grown in pure culture; (iii) the specific disease must be reproduced when a pure culture of the microorganism is inoculated into a healthy, susceptible host; and (iv) the microorganism must, again, be recoverable from the experimentally infected host. These concepts played a valuable role in convincing a sceptical public that bacteria can cause disease.

The pioneering work of Pasteur and Koch led, within 25 years, to the discovery of many of the major bacterial agents of human disease (Table 1.3). Looking back one hundred years later, the pace of these discoveries is truly staggering. Between 1877 and 1900 many of the major bacterial killers were identified. These studies also provided methods for the prevention of many of these diseases by the application of hygienic measures and immunization. It led to the discovery of mechanisms by which bacteria could cause disease and of the nature of the immune responses of the host to such bacteria. For example, in 1888 Roux and Yersin demonstrated that the diphtheria organism produced a soluble toxin that could mimic all of the symptoms of diphtheria. This led to the finding by von Behring and Kitasato in 1890 that injection of animals with bacterial toxins would result in the production of a substance in the serum capable of preventing disease. Such animal sera could cure children infected with diphtheria or tetanus, thus giving rise to the term antitoxin, subsequently antibody. This idea was further developed to produce vaccines, based on toxins, which could be used prophylactically (see Chapter 7 for a detailed description of bacterial toxins). The term antigen was given to the agents eliciting the synthesis of antibody. At the turn of the nineteenth century Buchner and Bordet found that serum had the capacity to kill certain organisms and the complex multi-protein system we now know as complement was discovered (the pathways of complement activation are described in detail in Chapter 8). The discovery and investigation of these soluble antibacterial factors (antibody and complement) was paralleled by the work of the Russian zoologist, Metchnikoff, whose pioneering studies showed the role that phagocytic cells (known as macrophages – Greek *makros*, large; *phagein*, to eat) play in dealing with infections. We now know that macrophages, antibodies and complement interact in order to deal with infectious bacteria (see Chapters 8 and 9). Another bacterial toxin that plays a major role in Gram-negative bacterial infections, and is the major agent stimulating macrophages, is endotoxin – a part of the cell envelope of bacteria which can be released when bacteria die. The biological actions of endotoxin – also known (incorrectly) as lipopolysaccharide (LPS) – were first described in the 1890s.

While much of the pioneering work on antibodies or humoral immunity occurred in France, the beginnings of cellular immunity were pioneered by Robert Koch while attempting to produce a vaccine for tuberculosis. He found that injection of the supernatants of *M. tuberculosis* into the skin of animals infected with this organism resulted in a severe local inflammatory reaction 24 hours later. This reaction is known today as the delayed-type hypersensitivity (DTH) reaction but it was not until 1942 that Merrill Chase showed that this response could be transferred, not by serum, but by lymphoid cells. We now know that the DTH response is due to the actions of specialized lymphocytes known as Th_1 cells. The role of these cells in antibacterial immunity is described in detail in Chapters 8 and 9.

Thus from the pioneering work of Koch, Pasteur and colleagues up until the 1940s, bacteriology and immunology developed hand in hand. The discovery in the 1930s and 1940s of antibiotics and the rapid introduction of these chemical therapeutics into clinical practice in the 1950s and 1960s had a major impact on research into the interactions between bacteria and the eukaryotic cells of the host. Many immunologists, for example, deserted the infectious diseases and became fascinated by the concept of self and non-self in immunity and of autoimmunity and autoimmune diseases – conditions where the immune system appears to be recognizing and attacking 'us' rather than 'them'. Ironically, it now appears that

Table 1.3 The discovery of the major bacterial diseases

Disease	Causative bacterium	Year of discovery	Discoverers
Anthrax	*Bacillus anthracis*	1877	R. Koch
Pus formation	*Staphylococci*	1878	R. Koch
Gonorrhoea	*Neisseria gonorrhoeae*	1879	A.L.S. Neisser
Typhoid fever	*Salmonella typhi*	1880	C.J. Eberth
Pus formation	Streptococci	1881	A. Ogston
Tuberculosis	*Mycobacterium tuberculosis*	1882	R. Koch
Cholera	*Vibrio cholerae*	1883	R. Koch
Diphtheria	*Corynebacterium diphtheriae*	1883	T.A.E. Klebs
Tetanus	*Clostridium tetani*	1884	A. Nicolaier
Diarrhoea	*Escherichia coli*	1885	T. Escherich
Pneumonia	*Streptococcus pneumoniae*	1886	A. Fraenkel
Meningitis	*Neisseria menigitidis*	1887	A. Weischselbaum
Food poisoning	*Salmonella enteritidis*	1888	A.A.H. Gaertner
Gas gangrene	*Clostridium perfringens*	1892	W.H. Welch
Plague	*Yersinia pestis*	1894	S. Kitasato/A.J.E. Yersin
Botulism	*Clostridium botulinum*	1896	E.M.F. van Ermengem
Dysentry	*Shigella dysenteriae*	1898	K. Shiga
Paratyphoid	*Salmonella paratyphi*	1900	H. Schottmüller
Syphilis	*Treponema pallidum*	1903	F.R. Schaudinn and E. Hoffman
Whooping cough	*Bordetella pertussis*	1906	J. Bordet and O. Gengou

autoimmune diseases – conditions like rheumatoid arthritis affecting the joints and myasthenia gravis which affects muscle – are triggered by infection. Likewise pathologists, biochemists and even microbiologists deserted bacteriology for richer pastures in other apparently more fruitful scientific disciplines, such as virology or protozoology. The most active arena of microbiology, and one which was taken over largely by the biochemist, molecular biologist and cell biologist, was the biology of bacterial exotoxins (reviewed in Chapter 7). The study of bacterial exotoxins has had profound implications for our understanding of the ways in which eukaryotic cells control signals arriving at their external surfaces (see Chapters 2 and 3).

BACTERIAL DISEASES

As will be described below, our world teems with microbial life forms including an enormous number and diversity of bacterial species. Our own bodies are populated with large numbers of bacteria: the normal or commensal microflora. It is estimated that up to 1000 different species of bacteria live on the epithelial surfaces of the human body and this is probably an underestimate. This should be contrasted by the rather small numbers of bacteria which can colonize *Homo sapiens* to cause disease (see Tables 1.2 and 1.3). In spite of the small number, these organisms, as described by Arno Karlen in his book, *Plague's Progress*, have helped shaped world history. Epidemic typhus, a disease borne by lice, has been a major participant in European history since its introduction from the East in the fifteenth century. In this century, from 1917 to 1921, typhus infected 20 million Russians and killed 3 million. This led Lenin to declare at the height of the epidemic 'Either socialism will defeat the louse or the louse will defeat socialism'.

Mutualism, commensalism and parasitism

Jonathan Swift, the satirist and author of Gulliver's Travels wrote:

> So, naturalists observe, a flea
> Hath smaller fleas that on him prey;
> And those have smaller fleas to bite 'em,
> and so proceed *ad infinitum*

This satirical broadside provides a particular, unpleasant, view of parasitism which is probably shared by most of the readers of this book. It is likely that all life forms are parasitized. Bacteria parasitize humans and in turn are parasitized by bacteriophages. In turn modern man could be described as the worst parasite, living off the life-blood of the world and destroying everything for his own benefit. However, the association between different species is much more complex and dynamic than the vision conjured up by the word parasitism. Associations between living organisms in which one species lives on or within another is termed symbiosis, which means 'living together'. Symbiosis can be divided into three broad and, importantly, overlapping categories: mutualism, commensalism and parasitism. The former state is one in which both benefit from the association. Good examples are the bacteria and protozoa which live within the stomachs of domestic ruminants. These are necessary for the digestion of the cellulose present in the plant material eaten by the ruminant. The microorganisms have a sheltered and controlled environment and a constant supply of nutrients. In turn, the ruminant host is able to extract nutrients from the vegetable matter that it eats. Commensalism is a state in which one species uses another as its physical environment, normally existing within the larger species. The commensal microbial flora of *Homo sapiens* is a good example. The interactions between commensal bacteria and man is highly specialized, resulting in only certain bacterial species being found in particular locations in the human body. Such interactions are normally harmless although they may benefit the host. The best example of this is the role played by the commensal microflora in preventing infectious bacteria colonizing mucosal surfaces. In contrast to mutualism and commensalism, the term parasitism describes a mechanism of association in which the parasite gains most of the advantage from the interaction. However, it should be appreciated that the terms mutualism, commensalism and parasitism describe interactions which are dynamic and dependent on environmental conditions. For example, certain members of the commensal microflora of humans are known as opportunistic pathogens. Examples include *Staph. aureus, Neisseria meningitidis, Haemophilus influenzae* and *Strep. pneumoniae*. In addition, almost any member of the commensal microflora can cause symptoms of disease if it is released into the body from its normal site of colonization.

How do bacteria cause disease?

While most of the bacteria causing the major infections of man were discovered in the last century we still do not fully understand how they cause disease and this is the major subject of cellular microbiology. Nevertheless, various principles have been established.

All infectious diseases have several stages of cell – cell interaction from the inception of the contact between the infectious bacterium and the host until the induction of physical signs of the infection (Figure 1.2). The first, and probably most important, is the adhesion of the bacterium to some host epithelial cell population. This can be followed by invasion of the epithelium and the passage of the bacteria through these cells and into the underlying submucosa. Multiplication of the infectious organism can then ensue. This will normally invoke a host immune response and evasion of such protective host responses is vital for survival of the bacterium. Much of the host immune responsiveness is driven by small proteins known as cytokines and it is increasingly being perceived that such cytokines are targets for bacteria. Much of the pathology caused by bacterial infections is due to the actions of factors released by the bacteria (e.g. exotoxins) or to the overproduction of host factors, mainly cytokines, produced to defend against the bacterium.

The ability of a bacterium to adhere, invade, evade host defences and cause tissue damage is largely due to its ability to produce a variety of molecules which have been termed virulence factors. These have been traditionally grouped into the following classes: adhesins (responsible for adhesion of the organism to host tissues); invasins (responsible for tissue invasion); impedins (allowing the bacterium to overome host defence mechanisms); and aggressins (bacterial factors able to damage host cells or tissues). Two of the authors (BH and MW) have proposed a further class of bacterial virulence factor, termed modulins, that are bacterial components which stimulate cytokine synthesis. The induction of cytokines modulates the behaviour of the cell producing the cytokines. These various virulence factors will be described in more detail throughout this book.

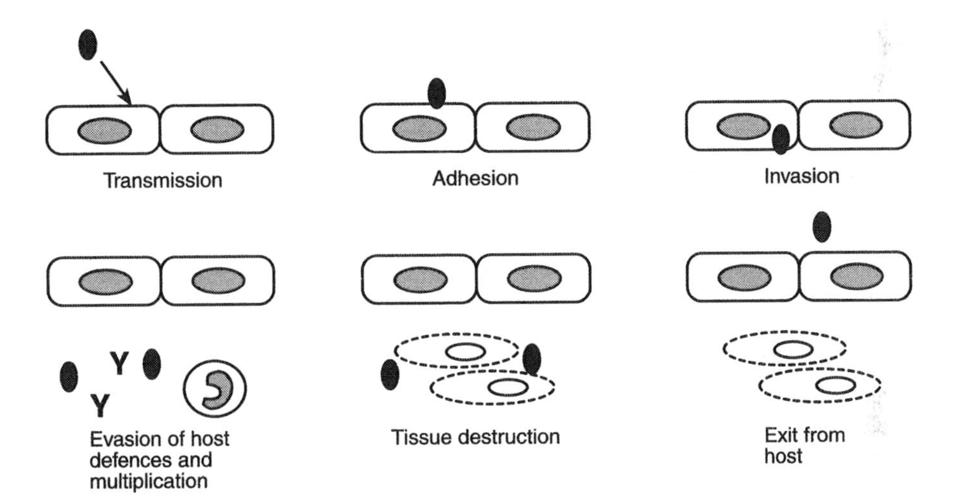

Figure 1.2 The various stages of bacteria–eukaryotic cell interaction during the infection of the host. Going from left to right and from top to bottom we begin with transmission of the bacterium and its adhesion to the epithelial (mucosal) surface. Adhesion is a key process in infection. This is followed by the invasion of the organism. This can occur by passage of the bacterium through the epithelial cells or other cell populations in the epithelia (e.g. M cells in the gastrointestinal tract) or by damage to the epithelial layer. Within the host, the bacteria can multiply and in so doing will encounter the immune responses of the host. Although not shown, there are also powerful antibacterial systems on the external epithelial surface which the bacteria must cope with (antibacterial peptides, metal-chelating compounds, destructive enzymes – see Chapter 8). To survive, the bacterium must overcome or evade these various responses. As part of the infectious process host cells may be killed or damaged, either by direct actions of the bacterium or, more likely, by the actions of exotoxins. Finally, as the infectious bacterium only survives by spreading from host to host it must be able to exit from the host and encounter another uninfected host

Bacteria and Idiopathic Diseases

In addition to the classic infectious diseases caused by bacteria, there is increasing speculation that bacteria may be involved in the pathology of idiopathic diseases such as gastric ulceration, gastric cancer, arthritis, psoriasis and even coronary artery disease which includes conditions such as atherosclerosis. *Helicobacter pylori* is now firmly linked to the pathogenesis of stomach ulcers, a condition which only a decade or so ago was believed to be due solely to stressful living and which was treated by histamine receptor (H_2) antagonists or by surgery. Indeed, for a time the most commercially successful drug was Zantac, an H_2 antagonist used solely for the treatment of gastric ulceration. This condition is now largely treated by antibiotics to kill *Hel. pylori.* Psoriasis, a relatively common skin condition in which there is hyperproliferation of the keratinocytes of the skin, may be caused by superantigens released from skin-associated bacteria. Indeed, superantigens might be responsible for a range of idiopathic diseases, including a number of the conditions known as autoimmune diseases. Scientists working on coronary artery disease, which kills many tens of millions of individuals each year, are now awakening to the possibility that bacteria may contribute to tissue pathology. There is indirect evidence that *Chlamydia pneumoniae* and *Hel. pylori* may contribute to pathology. More recently, bacteria associated with dental diseases have been postulated to increase susceptibility to coronary heart disease. In addition, it has been revealed that *Bartonella henselae,* the causative organism of cat scratch fever, can interact directly with blood vessel endothelial cells and promote new blood vessel formation (angiogenesis). This is an exciting area of cellular microbiology research and one that awaits detailed exploration.

EMERGENCE OF THE NEW DISCIPLINE OF CELLULAR MICROBIOLOGY

By the 1980s it was clear that bacterial infections continued to be a major cause of morbidity and mortality, that antibiotic resistance was increasing at an alarming rate and that new bacterial diseases were emerging at a regular rate. The massive increase in global travel over the past three decades has contributed to the worldwide dissemination of pathogens previously confined to geographically defined regions. Even bacteria which were well known, and apparently well understood, such as *Escherichia coli,* were beginning to appear in new guises as causes of diarrhoeal diseases with the pathology being caused by direct cell-to-cell interactions or by the production of a range of specific toxins some of which are newly discovered (Table 1.4). The *E. coli* strain 0157, also known as the 'hamburger bug', is now recognized by the general public because of its propensity for causing outbreaks of 'food poisoning' (actually haemolytic uraemic syndrome) which are often lethal to the elderly.

There has also been an appreciation, during the past decade, that the enormous numbers of bacteria that colonize multicellular organisms (the normal or commensal microflora described previously) must interact with host eukaryotic cells without provoking an immune response . How this occurs is likely to be important for our understanding of the mechanisms of infection. For example, the average human body is composed of 10^{13} eukaryotic cells but supports 10^{14} bacteria on its epithelial (mucosal) surfaces. Thus purely on the basis of cell numbers humans are 90% bacterial. We know very little about our normal microflora. However, there have been attempts over many years to try and modify it for therapeutic

Table 1.4 Diarrhoeagenic strains of *Escherichia coli*

Strain	Clinical symptoms	Mechanism
Enteropathogenic (EPEC)	Watery diarrhoea	Pili, type III secretion
Enterohaemorrhagic (EHEC)[a]	Bloody diarrhoea, HUS[b]	Shiga-like toxin
Enteroinvasive (EIEC)	Dysentry	Cellular invasion and cell–cell spread
Enterotoxigenic (ETEC)	Watery diarrhoea	Colonization factors, heat-labile/-stable toxins
Enteroaggregative (EAggEC)	Watery diarrhoea, persistent disease	Fimbriae, heat-stable toxin
Diffusely adherent (DAEC)	Watery diarrhoea, persistent disease	Toxins?
Cytolethal toxin (CDT)-producing	Diarrhoea	Cytolethal distending toxins

[a] *E. coli* 0157.
[b] HUS – haemolytic uraemic syndrome.

benefit. The best-known current example of this is the use of fermented milk products (Yakult® being the product most people would recognize) to change the microflora of the colon. The concept of using dietary methods to change the microflora of the digestive tract has expanded in recent years to using microfloral bacteria or modifying such bacteria as a means of increasing resistance to infections or to boost host nutrition. Agents which contain such microfloral bacteria are referred to as probiotics and are being looked at as potential therapies in a range of conditions. One ancient and extreme form of the use of probiotics is the practice of coating the umbilicus of new borns with manure. The modern explanation for the effectiveness (if any) of this 'treatment' would be that the bowel bacteria of the domestic animal is non-infectious and would prevent colonization of the cut cord tissue with infectious organisms. The role of the commensal microflora and its mysterious ability to exist without provoking an immune response will be discussed in Chapters 8–10.

During the 1980s major advances were being made in molecular biology and cell biology and it has been the application of these methodologies to microbiology that has enabled cellular microbiology to emerge during the past decade. There has been significant cross-fertilization between these disciplines. Thus the paradigm of the bacterium as a very simple version of the 'evolutionarily superior' eukaryotic cell, acting as no more than a unicellular automaton, has begun to crumble. This view was largely due to the fact that bacteriologists grew bacteria in pure (axenic) culture on artificial substrates and in the absence of 'appropriate signals'. Bacteria are increasingly being recognized as complex, beautifully adapted, and adaptable, organisms which can respond to their environment in many and varied ways. Multicellular organisms are now recognized to be wholly dependent on intercellular communication for their survival. Although such communication was not thought to be important for bacteria it is now established that bacteria do signal to one another. The best-understood bacterial cell-to-cell signalling mechanism is *quorum sensing*. This denotes the ability of certain bacteria to determine their cell density and, in consequence, switch on (or off) particular genes (Figure 1.3). Quorum sensing was discovered in the marine bacterium *Vibrio fischeri*, which is also a symbiont found in the light organs of certain marine fish and

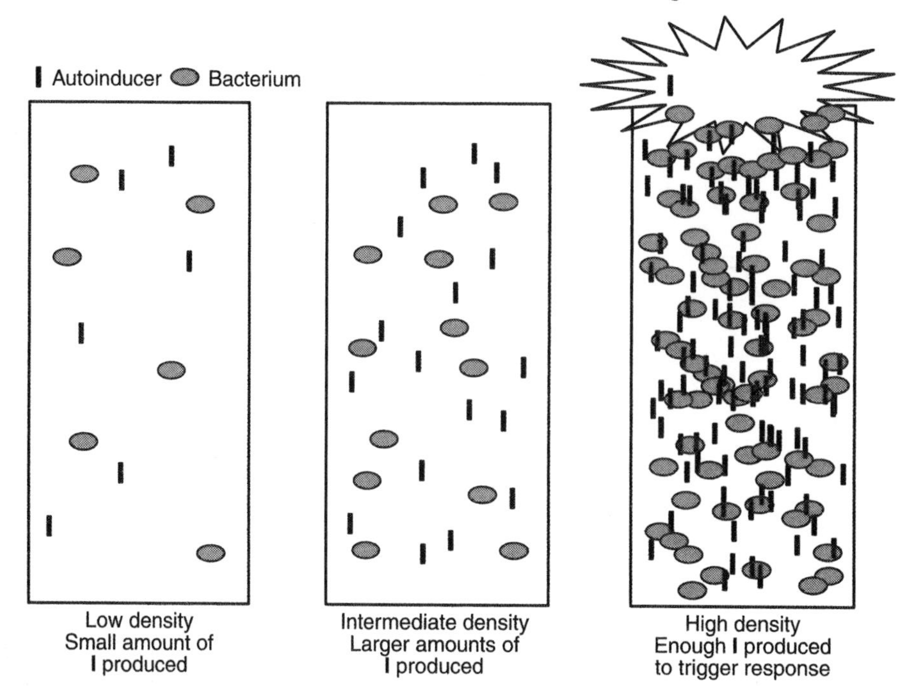

Figure 1.3 Schematic diagram showing the mechanism of quorum sensing. Each bacterium produces the autoinducer (I) which is involved in cell-to-cell signalling. At low and intermediate bacterial densities there is not sufficient autoinducer to stimulate the bacteria. However, at high cell density autoinducer concentration becomes large enough to stimulate bacteria. The nature of the stimulation is the switching on of particular genes or sets of genes (operons) which are involved in promoting bacterial survival (see Chapter 3)

squids and is described in Chapter 3. Quorum sensing is not limited to marine bacteria and it is suggested that this census taking by bacteria may be a virulence mechanism allowing them to increase their numbers without switching on the virulence genes that would alert the host to their presence. In addition, there is recent evidence that the compounds involved in quorum sensing can inhibit the function of both macrophages and T lymphocytes.

Bacterial cell-to-cell signalling is simply one example of the complex interplay that bacteria have with their environment. There is growing evidence that bacteria can respond to signals emanating from eukaryotic host cells and can even react to direct cell-to-cell contact with host cells. For example, bacteria have been found to be responsive to a variety of mammalian mediator molecules such as serotonin, catecholamines, insulin and cytokines such as interleukin 1 (IL-1), IL-2, tumour necrosis factor (TNF) and transforming growth factor (TGF)α. Kapreylants and Kell have suggested the term 'microendocrinology' to describe this prokaryotic – eukaryotic communication. In turn, bacteria produce a wide range of molecules which can regulate the production of the local hormones, known as cytokines, which are key local cell-to-cell regulatory molecules in multicellular organisms. The cytokines are also the molecules which control the innate and acquired immune responses in vertebrates and are described in detail in Chapters 3, 8 and 9.

Another set of molecules which are important in the interaction of bacteria with the host are antibacterial peptides. These are small peptides produced by most cells in multicellular organisms and which have the capacity to kill Gram-negative and Gram-positive bacteria by binding to the bacterial cell wall and forming pore-like structures. Bacteria are therefore killed by the loss of their intracellular fluids. Well over a hundred such peptides have now been discovered and they are produced by all living creatures, including bacteria. The finding that virtually all cells in vertebrates produce these potent antibacterial peptides reveals that the constant exposure to infectious bacteria has been a major evolutionary driving force. The role of such peptides in our defence against bacterial infection will be reviewed in Chapter 8.

Cellular microbiology emphasizes just how intimate are the contacts between bacteria and the cells of the host (Figure 1.4). The infectious process starts with the adherence of the bacterium to the target cell. Bacteria produce a range of molecules called adhesins which selectively bind them to the appropriate cell. A wide range of cell surface molecules including lipids, glycolipids, carbohydrates and proteins are ligands for bacterial adhesins. The host cell is not an inactive partner in this process as adhesion can activate the host cell, which can be of advantage either to the eukaryotic cell or the bacterium. For example, adhesion often triggers the production of cytokines and thus can start the inflammatory process which has evolved to defeat infectious agents. With a number of bacteria, adhesion results in the release by bacteria of specific proteins by a specialized mechanism called a type III secretion pathway (described in more detail in Chapters 2 and 6). This mechanism results in the release of bacterial gene products directly into the cytoplasm of the host cell. This is an extremely efficient method of influencing eukaryotic cell behaviour and is used by organisms such as *Sh. flexneri* and *Salmonella typhimurium* to hijack the cytoskeletal machinery of the host cell to enable the bacterium to enter the (in these cases, epithelial) cell. Proteins injected into host eukaryotic cells include kinases and phosphatases and it is clear that bacteria talk the same language as their eukaryotic 'prey'. The ability of bacteria to bind to and to enter cells and the molecular cross-talk between the bacterium and its eukaryotic host will be described in detail in Chapters 5 and 6. Exported proteins with potent actions on eukaryotic cells termed exotoxins are also major weapons in the armamentarium of the infecting bacterium. A very large number of exotoxins have now been discovered and they have a wide range of actions on mammalian cells. Exotoxins have been used extensively by eukaryotic cell biologists as probes of cell function and much of our understanding of cell signalling has been determined by the use of toxins. These molecules are also being used as therapeutic agents.

Much of the new understanding about bacteria–host cell interactions has been derived from the development of novel molecular biological techniques which allow the identification and analysis of the genes that are activated during infection, in both bacteria and eukaryotic cells. These various techniques are defined in brief in Table 1.5. The determination of the complete DNA sequence of the genomes of a number of bacteria including *Haem. influenzae, Hel. pylori* and *E. coli* K12 have been reported and a growing number of the genomes of other bacteria are being sequenced or their sequencing is being planned. Such genomic databases enable candidate virulence genes to be identified and compared to those in different bacteria in order to identify differences in pathogenic lifestyles. A key question is – what bacterial genes are switched on/off when the infectious organism makes contact with the host or with the specific host cell? A number of techniques have been developed in the past few years to address this vital question. These include *in vivo* expression technology (IVET) which, as the name suggests, can identify genes which are switched on by the bacterium within the experimental animal infected with the organism of interest. Signature-

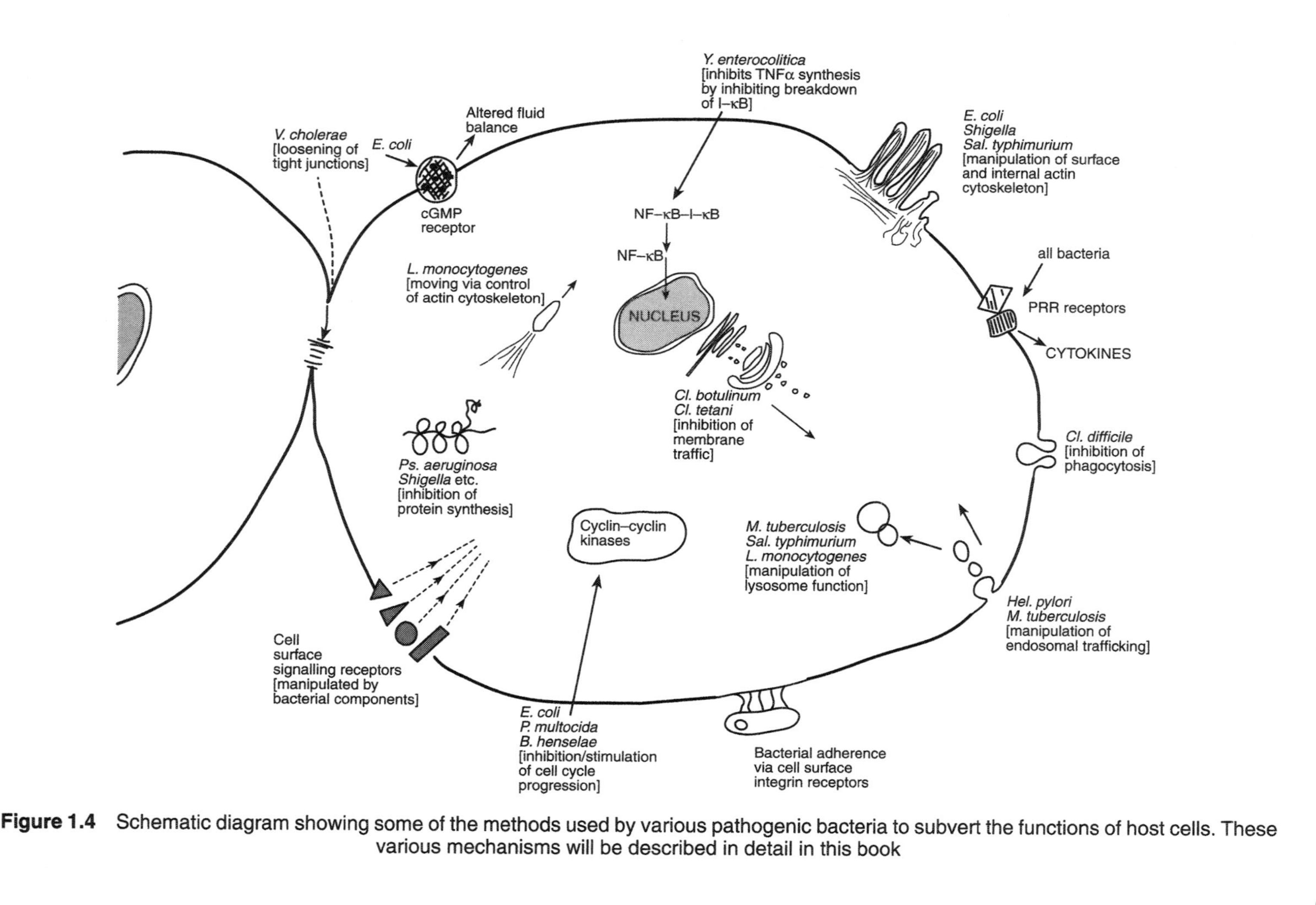

Figure 1.4 Schematic diagram showing some of the methods used by various pathogenic bacteria to subvert the functions of host cells. These various mechanisms will be described in detail in this book

tag mutagenesis is another technique which provides a bank of individually 'labelled' mutants that can be screened in batches to detect which do not survive in animal infection studies, again to define genes important in the infectious process. Arbitrarily primed and differential display-polymerase chain reaction are techniques which allow a glimpse of the global expression of genes within bacteria maintained under different conditions (e.g. under culture conditions or within cells). The use of these molecular techniques is very important in our growing understanding of bacterial interactions with host cells These molecular biological techniques are described in Chapter 4.

Table 1. 5 Molecular biological techniques for defining genes involved in bacterial infection

Technique	Applications
Genome mapping	Identification of homologous genes
Proteomics	Identification of multiple protein sequences
PhoA fusion cloning	Cloning of exported (virulence?) proteins
Mutagenesis (site-directed or transposon)	Identification of gene function by gene disruption
Allelic exchange	Replacement of genes
In vivo gene expression technology (IVET)	Identification of genes involved in infection
Signature-tag mutagenesis	Identification of genes involved in infection
Subtractive hybridization DD-PCR RNA-arbitrarily-primed PCR	Identification of genes expressed under different conditions
Yeast two-hybrid analysis	Identification of protein–protein interactions within cells
Phage display	Identification of interacting gene products
Homologous recombination	Production of animals lacking particular gene products

A discovery of monumental proportion which has arisen from the application of the techniques of molecular biology is described in the final section of this chapter.

AN EVOLUTIONARY HISTORY OF PROKARYOTIC AND EUKARYOTIC CELLS

By measurement of the decay products of uranium 238 which decays to lead 206 (half-life 4.5 billion years) and thorium 232 which decays to lead 208 (half-life 14.0 billion years) it is possible to estimate the age of the earth. The figure derived is 4.5 – 4.6 billion years, which is similar to estimates made of the age of moon rocks. When first formed, the earth was molten, but it is estimated to have cooled sufficiently by 4 billion years ago to allow the formation of the oceans. The oldest rocks on earth have been found in the Isua Formation in Greenland and are estimated to be 3.8 billion years old. The first record of life in the fossil record, and believed to be Cyanobacteria, is found in rocks 3.6 billion years old. Thus life, in the form of what appears to be complex single-celled organisms, sprang into being within 500 million years or so of the appearance of liquid water on the planet. There has been enormous speculation about the mechanisms which could account for the development of the molecules of life, never mind life itself. The pioneering work of Miller and Urey applied an electrical discharge to mixtures of gases believed to be present in the earth's early atmosphere (methane, carbon dioxide, water vapour etc.) and found that this resulted in the generation of amino acids, the building blocks of proteins. This has resulted in much paper being sacrificed in attempts to explain what happened next to form the unimaginably complex organization that we call a cell. The answer is that we simply don't know and have little idea of how to find the answer. The most appealing idea (at least to the authors) is that proposed by Stuart Kauffman in a beautifully written and stylish book entitled: *At Home in the Universe: The Search for Laws of Complexity*. Kauffman argues that self-organization is an inherent property of complex systems and that the laws of complexity, which we are just beginning to glimpse, will explain how the enormously complex systems such as cells could have evolved in the geological winking of an eye.

In order to understand life in all its complexity, it is vital that it be catalogued so that relationships between organisms can be recognized. Biologists have been classifiying life forms since the pioneering studies of Linnaeus in the eighteenth Century. The division of living things into two categories – plants and animals – was only superseded in the late nineteenth century with the recognition of microbial forms of life, and up until the 1980s textbooks were subdividing life into five kingdoms: Monera (bacteria), Protoctista (single-celled organisms with nuclei), Fungi, Animalia and Plantae. The study of the evolutionary relationships of living creatures is called phylogeny (Greek *phylon*, race; *genesis*, origin). During the past two decades there has been increasing application of the techniques of molecular biology to the study of phylogeny. These studies have revolutionized our concept of phylogeny and have introduced a completely unexpected view of the relationships and diversity of living creatures – one that reveals that single-celled creatures are the major and most diverse life forms.

Our new understanding of molecular phylogenetics is based on the idea that molecules can be used to determine how different one organism is from another in terms of DNA or protein sequence. By choosing a particular molecule, and comparing its sequence between different living forms, one can use the sequence data as a molecular clock. In other words, the evolutionary distance between two species can be measured by differences in the nucleotide or amino acid sequences between homologous molecules. The molecule of choice for such studies turns out to be genes encoding the RNA within the ribosomes (ribosomal or rRNA). Most studies have been conducted using 16S rRNA genes from prokaryotic cells and 18S rRNA genes from eukaryotic cells. The method used to sequence the 16S rRNA gene is explained in Figure 1.5. Computer analysis of the rRNA gene sequences (termed molecular phylogeny) has suggested that there are not five kingdoms of life, but that cellular life has evolved along three major lineages (Figure 1.6). Two of these lineages – Bacteria and Archaea – are solely microbial and are composed of prokaryotic cells. Archaeal species have been called archaebacteria in the past, but this is now clearly not correct. The third lineage – Eukarya – is also largely composed of single-celled organisms but also contains all the multicellular creatures which we recognize as the denizens of our planet. This application of molecular biology to phylogeny restructured our view of cellular evolution. As can be seen from Figure 1.6 it is also now established that the domain Eukarya is not of recent origin but is as ancient as the prokaryotic domains. Early eukaryotes are believed to be like modern-day microsporidia and diplomonads, which are cells that lack mitochondria. *Giardia lamblia*, a parasitic single-celled diplomonad, has been suggested to be the coelacanth of cells – that is, a living fossil – and is the cause of an acute form of gastroenteritis called giardiasis. This organism has not one, but two nuclei, no endoplasmic reticulum or mitochondria, although it does have a cytoskeleton. *Giardia* has the accolade of being the first intestinal microorganism to be observed under the microscope, by the great microscopist Antoni van Leeuwenhoek, and is the most commonly diagnosed intestinal parasite in the United States.

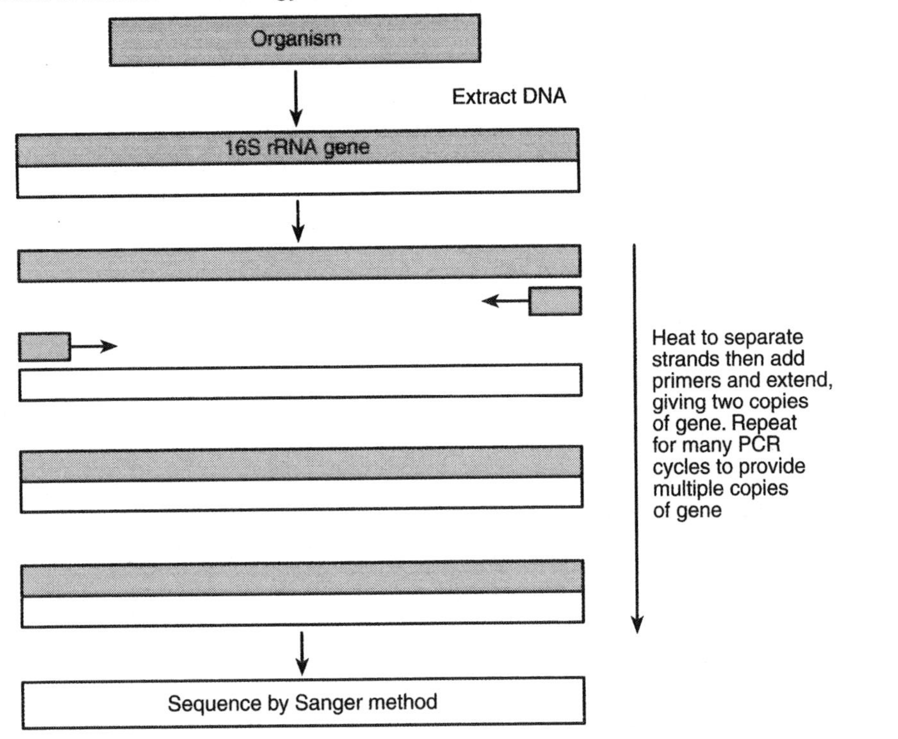

Figure 1.5 Ribosomal RNA sequencing using the polymerase chain reaction (PCR). Individual organisms or collections of organisms (for example, ecological samples) have their DNA isolated and the 16S ribosomal RNA gene or genes can then be amplified by PCR. For single organisms the amplified DNA can then be sequenced by the Sanger method, normally using an automated DNA sequencer. If ecological samples containing many different microbial species are being used the individual 16S genes are cloned and sequenced individually

Evolution is measured by diversification of life forms. Our own general perspective of the diversity of life is based on our observations and what we may read. Thus many people would probably regard insects as the most diverse life form. However, molecular phylogenetic studies conducted in the past two decades have established that the most diverse life form on our planet is microbial and is distributed within the three domains of life. This means that the greatest eukaryotic diversity is not to be found among the insects or fish, but among single-celled organisms such as protozoa, amoebae, algae, single-celled fungi etc. Our knowledge of the biology of life forms is normally achieved by 'capturing' and breeding or growing them and then examining their forms and function. Surprisingly, this strategy has only identified and catalogued about 5000 non-eukayotic organisms. This contrasts with the 500 000 or so insect species that have been classified. Studies of a number of environments have suggested that only a fraction of 1% of organisms seen microscopically can be cultured. Thus we appear to grossly underestimate the diversity of the microbial world. Fortunately, by using molecular phylogenetic methods it is not necessary to cultivate organisms in order to classify them and this technology is opening up a whole new dimension to our appreciation of species diversity. Another exciting development from the analysis of 16S RNA has been the discovery of diverse microbial life forms in the oceans, and the appreciation that microorganisms are the major life forms in this environment. Even more spectacular is the

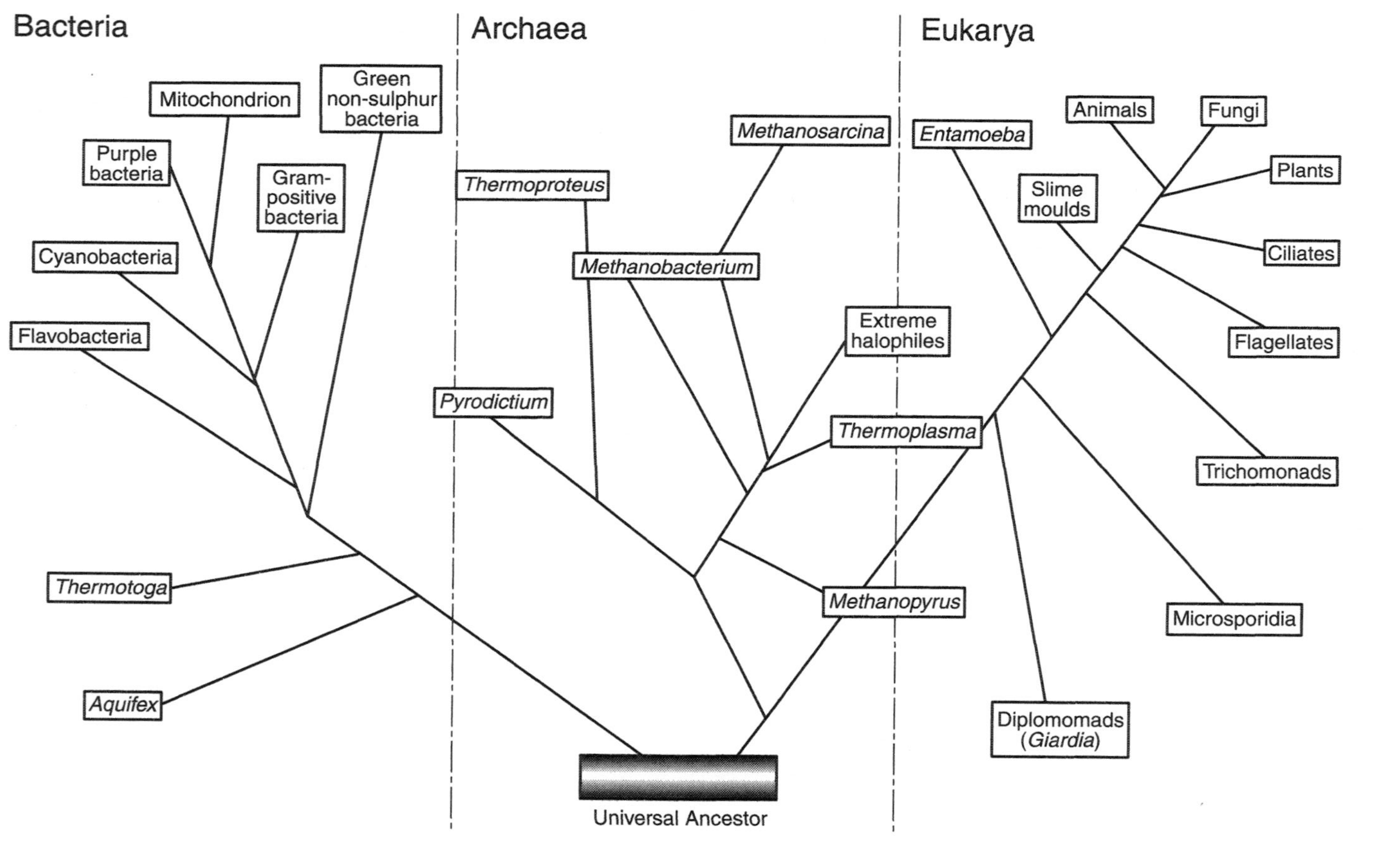

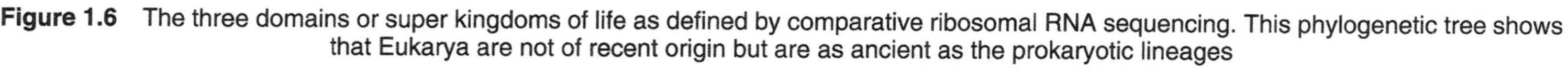

Figure 1.6 The three domains or super kingdoms of life as defined by comparative ribosomal RNA sequencing. This phylogenetic tree shows that Eukarya are not of recent origin but are as ancient as the prokaryotic lineages

finding that microorganisms live in the very rocks of the earth itself. Recent estimates have suggested that the earth supports 4–6x10^{30} prokaryotic cells and that this represents 350–550 Pg of carbon, where 1 Pg equals 10^{15} g. This prokaryotic biomass contains as much carbon and 10 times as much nitrogen and phosphorous as do the earth's plants, making it the largest pool of these nutrients in living organisms. It is likely that this is an underestimate as it is difficult to estimate the numbers of prokaryotes which live in the subsurface (i.e. habitats > 8 m below the surface of the earth and 10 cm below marine sediments).

CONCLUSIONS

We live in a world which contains enormous numbers of single-celled life forms based on the three great domains of cellular evolution. *Homo sapiens* as a species has made peace with a great many examples of the domain Bacteria. However, this peace is shattered from time to time by the entry and colonization of our bodies by a small number of bacterial species which can cause pathology. The evolution of these infectious agents appears to occur with some rapidity as a large number of novel infectious organisms are continually being recognized (Table 1.2). How these bacteria cause disease is the main focus of this textbook. We are also subject to colonization by a range of single-celled members of the domain Eukarya. The major eukaryotic microbial parasites are fungi (moulds and yeast) and protozoa (a large group of infecting organisms which include trypanosomes, amoebae, *Plasmodium* etc.). Of the third great domain of cellular life – Archaea – we know very little. Humans do contain some Archaeal species. For example, *Methanobrevibacter smithii* is found in the gut. Methanogenic Archaeal species are found in the stomachs of cows and of termites. However, we know very little about the distibution of this cell form in multicellular eukaryotes. Perhaps the most fascinating aspects of the Archaea is that it is believed that there are no Archaeal species which cause disease. Other 'life forms' also cause disease in multicellular eukaryotes. Viruses are a fascinating and successful parasitic life form able to colonize and hijack the function of eukaryotic cells and bacteria with diverse pathological sequelae. The mechanisms by which viruses evade the immune responses of the host will be described in this book as it is presumed that bacteria will employ similar mechanisms. The transmissible spongiform encephalopathies (TSEs), named by some as prions, are the latest infective agent to be described and have yet to be fully defined. These cause fatal diseases such as bovine spongiform encephalopathy (BSE) and Creutzfeld–Jacob disease (CJD). The mechanisms by which eukaryotic microorganisms, viruses and TSEs cause disease will not be discussed in any detail except where such mechanisms are shared with, or highlight, those of infectious bacteria.

This book will concentrate on the cellular microbiology of humans and will largely concentrate on bacteria which cause infections. We will also consider the largely unexplored mystery of how we cope with the enormous numbers of commensal bacteria which populate all multicellular creatures. In doing this we will largely ignore the growing body of experiments which are revealing the wonderful ways in which bacterial plant pathogens produce pathology. This includes the ability of one well-known plant pathogen, *Agrobacterium tumefaciens*, to genetically transform plant cells by transfer of a piece of single-stranded DNA.

REFERENCES

Books

de Bruijn FJ, Lupski JR, Weinstock GM (1998) *Bacterial Genomes: Physical Structure and Analysis*. Chapman & Hall, London.

Dorman CJ (1994) *Genetics of Bacterial Virulence*. Blackwell Oxford.

Ewald PW (1994) *Evolution of Infectious Disease*. Oxford University Press, Oxford.

Fortey R (1997) *Life: An Unauthorized Biography*. Flamingo London.

Henderson B, Poole S, Wilson M (1998) *Bacteria/Cytokine Interactions in Health and Disease*. Portland Press, London.

Karlen A (1996) *Plague's Progress: A Social History of Man and Disease*. Gollancz, London.

Kauffman S, (1995) *At Home in the Universe: The Search for Laws of Complexity*. Viking Press, London.

Majno G (1991) *The Healing Hand: Man and Wound in the Ancient World*. Harvard University Press, Cambridge, MA.

Rappuoli R, Montecucco C (1997) *Guidebook to Protein Toxins and their use in Cell Biology*. Oxford University Press, Oxford.

Roberts DMcL, Sharp P, Alderson G, Collins M.(1996) *Evolution of Microbial Life*. Society for General Microbiology Symposium 54. Cambridge University Press, Cambridge, UK.

Stanley SM (1987) *Earth and Life Through Time* (2nd edn). Freeman, New York.

Wills C (1996) *Plagues: Their Origin, History and Future*. HarperCollins London.

Ziegler P (1970) *The Black Death*. Pelican, London.

Reviews

Baker B, Zambryski P, Staskawicz B, Kumar-Dinesh SP (1997) Signalling in plant–microbe interactions. *Science* 276: 726–733.

Finlay BB, Cossart P (1997) Exploitation of mammalian host cell functions by bacterial pathogens. *Science* 276: 718–725.

Fuqua C, Winans SC, Greenberg EP (1996) Census and concensus in bacterial ecosystems: the LuxR-LuxI family of quorum-sensing transcriptional regulators. *Annu Rev Microbiol* 50: 727–751.

Kaprelyants AS, Kell DB (1996) Do bacteria need to communicate with each other for growth? *Trends Microbiol* 237: 237–242.

Lartey R, Ghoshroy S, Sheng J, Citovsky V (1997) Transport through plasmodesmata and nuclear pores: cell-to-cell movement of plant viruses and nuclear import of *Agrobacterium* T-DNA. In *Molecular Aspects of Host-Pathogen Interactions* (eds McCrae MA, Saunders JR, Smyth CJ, Stow ND), pp. 253–280. Cambridge University Press, Cambridge, UK.

Lee CA (1997) Type III secretion systems: machines to deliver bacterial proteins into eukaryotic cells. *Trends Microbiol* 5: 148–156.

Mekalanos JJ (1992) Environmental signals controlling virulence determinants in bacteria. *J Bacteriol* 174: 1–7.

Pace NR (1997) A molecular view of microbial diversity and the biosphere. *Science* 276: 734–740.

Quinn FD, Newman GW, King CH (1997) In search of virulence factors of human bacterial diseases. *Trends Microbiol* 5: 20–26.

Strauss EJ, Falkow S (1997) Microbial pathogenesis: genomics and beyond. *Science* 276: 707–712.

Woese CR (1987) Bacterial evolution. *Microbiol Rev* 51: 221–271.

Papers

Mukamolova G, Kaprelyants AS, Young DI, Young M, Kell DB (1998) A bacterial cytokine. *Pro Natl Acad Sci USA* 95: 8916–8921.

Whitman WB, Coleman DC, Wiebe WJ (1998) Prokaryotes: the unseen majority. *Proc Natl Acad Sci USA* 95: 6578–6583.

CHAPTER 2

The Cellular Biology Underlying Prokaryotic–Eukaryotic Interactions

INTRODUCTION

Cellular microbiology is a scientific discipline which encompasses cell biology and microbiology. Being a composite scientific discipline cellular microbiology will be of interest to students in many subject areas and therefore it is important to furnish certain background information which will be necessary to understand the later chapters. To this end Chapters 2 and 3 provide a brief overview of key topics in microbiology and cell biology which the authors believe are required to understand the contents of this book. These will be set in the context of the bacterial interactions with the host. Chapter 4 presents a compendium of the molecular biological techniques used in current cellular microbiological research.

BACTERIAL ULTRASTRUCTURE

Bacteria, the smallest organisms capable of independent existence, vary in size with diameters ranging from 0.1 μm to 50 μm – the diameter of the 'average' bacterium being approximately 1.0 μm, but larger cells do exist as will be described later in the chapter. Apart from their overall shape – spherical, rod-like or spiral – and their cellular arrangement (pairs, clusters, chains etc.), little can be discerned of their structure by light microscopy. Examination of the internal stucture of these prokaryotic microorganisms by electron microscopy generally reveals only a DNA-containing region and ribosomes. Membrane-bound organelles and a nuclear membrane are notably absent from all bacteria. In contrast, the cell envelope of bacteria is far more complex than that of eukaryotic cells, consisting of a cytoplasmic membrane and a rigid cell wall which, in some species (Gram-negative bacteria), is multi-layered. Many, if not most, bacteria also have an additional layer (capsule or glycocalyx) external to their cell wall and may have one or more of a variety of surface appendages. The only exceptions to this generalized description of a bacterial cell are the mycoplasmas, which do not have a cell wall.

The bacterial cell envelope

Cytoplasmic membrane

The innermost layer of the envelope in all bacteria is the cytoplasmic membrane and this has the classical unit membrane structure comprising a phospholipid bilayer in which proteins are embedded. However, it differs from the cytoplasmic membrane of eukaryotes in that it does not contain sterols. It is approximately 8 nm thick and, as in the cells of higher organisms, it functions mainly as an osmotic barrier and in the transport of nutrients and waste products. However, in bacteria it is also involved in ATP production, the synthesis and export of cell wall components and the secretion of extracellular hydrolytic enzymes and toxins. The cytoplasmic membrane also constitutes the outermost structure of *Mycoplasma* spp. The composition of the cytoplasmic membrane of *Mycoplasma* spp. that lack a cell wall differs from that of other prokaryotes in that it may contain cholesterol.

The cell wall

Unlike the homeostatically controlled environment of cells of higher organisms, free-living bacteria are subject to the vagaries of their external environment. Protection against such fluctuations is provided by the cell wall, which prevents osmotic lysis and also functions as a molecular sieve. On the basis of a simple staining procedure developed more than 100 years ago, bacteria can be divided into two major groups, Gram-positive and Gram-negative, and this reflects major differences in the structure of the cell wall of these two bacterial groups. In the case of disease-causing bacteria, it is important to realize that virtually every component of the bacterial cell wall has some role to play in the interaction between the bacterium and its host and in the induction of pathology.

Gram-positive cell wall

Electron microscopic examination of the walls of Gram-positive bacteria shows a simple structure consisting of a single layer with a thickness of between 20 and 50 nm. Chemical and immunological analysis, however, has revealed greater complexity (Figure 2.1a). The main structural component of the cell walls of Gram-positive bacteria is a heteropolymer (known as peptidoglycan) which consists of chains of alternating *N*-acetylglucosamine and *N*-acetyl muramic acid residues cross-linked by short peptide chains containing several of a small group of amino acids: L-alanine, D-alanine, D-glutamic acid, lysine and diaminopimelic acid. The cross-links may be formed directly between these peptide chains or may involve short peptide 'cross-bridges'. The exact composition of the peptides and the cross-bridges is species-dependent and is unusual in nature in that D-amino acids are present. It is also interesting to note that *N*-acetylmuramic acid and diaminopimelic acid are found only in prokaryotes. Cross-linking occurs in all three planes resulting in one macromolecule (between 20 and 40 layers thick) enclosing the whole bacterial cell. Peptidoglycan is the main component of the Gram-positive cell wall (comprising at least 40% of its mass) and is responsible for its mechanical strength and also acts as a permeability barrier.

The other major components of the wall are anionic polymers such as teichoic and lipoteichoic acids (the latter abbreviated LTA). Teichoic acids consist of chains of glycerol, ribitol, mannitol or sugars linked by phosphodiester bonds and are attached to muramic acid

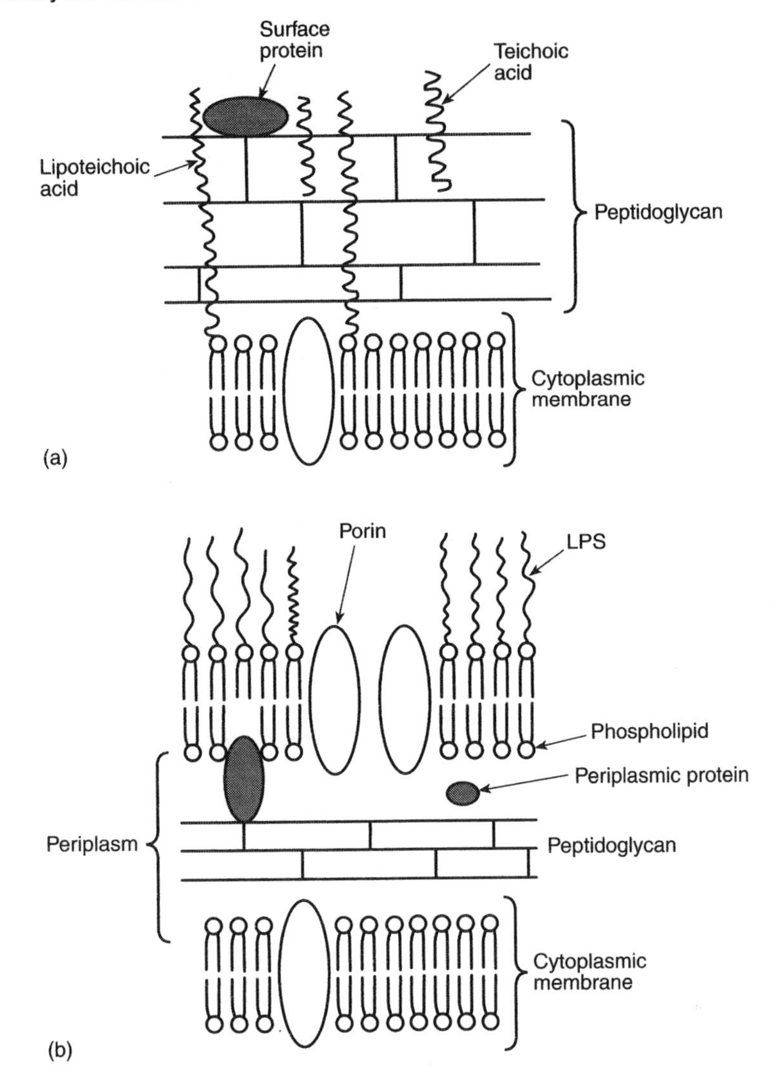

Figure 2.1 Structure of the cell walls of (a) Gram-positive and (b) Gram-negative bacteria. The cell walls of Gram-positive bacteria contain large amounts of peptidoglycan with associated teichoic and lipoteichoic acids. These components have pro-inflammatory properties. Likewise the outer component of the Gram-negative bacterium, lipopolysaccharide, is a well-known potent pro-inflammatory mediator. Not shown are proteins loosely associated with the cell wall and which, with secreted carbohydrates, form the bacterial capsule

residues in the peptidoglycan. D-Alanine or L-lysine are common substituents of the chains, which usually contain approximately 40 residues. LTAs consist of chains of glycerol phosphate, with D-alanine and sugar substituents, attached to a glycolipid (or diglyceride) in the cytoplasmic membrane. Both teichoic acids and LTAs are antigenic and often constitute the major somatic antigens of Gram-positive bacteria. In some respects, LTAs can be regarded as the Gram-positive equivalent of lipopolysaccharide (LPS), although the latter has a much greater potency and range of biological activities. Both the LTAs and peptidoglycan are able to induce cytokine synthesis and can contribute to the inflammation found in infections with Gram-positive bacteria (see Chapters 8 and 9).

Although disregarded until comparatively recently, proteins covalently linked to peptidoglycan are now recognized as being important cell wall components, particularly in the case of protein A in *Staphylococcus aureus* and the M proteins of streptococci. In some streptococci M proteins can account for 16% of the dry mass of the cell wall. Many of these proteins function as adhesins, are involved in evading host defence systems, are important mediators of host tissue destruction and are known to induce cytokine release from host cells.

Gram-negative cell wall

In Gram-negative bacteria, the cell wall is a more complex multi-layered structure consisting of a thin peptidoglycan layer (consisting, possibly, of only one sheet of peptidoglycan) and a lipid bilayer forming an outer membrane external to this. The outer membrane is linked to the peptidoglycan by a lipoprotein which spans a gelatinous region (containing proteins, enzymes and highly hydrated peptidoglycan with a low degree of cross-linking) known as the periplasm (Figure 2.1b).

The innermost leaflet of the outer membrane consists principally of phospholipids together with some proteins, while the outer leaflet is composed of LPS molecules of which there are approximately 3.5×10^6 per cell in *Escherichia coli*. LPS is an amphiphilic molecule consisting of three regions: (a) a glycolipid (lipid A) embedded in the outer layer of the membrane; this is responsible for most of the biological activities characteristic of LPS; (b) a 'core' oligosaccharide containing a characteristic sugar acid, 2-keto-3-deoxyoctulonic acid (KDO) and a heptose; and (c) an antigenic polysaccharide (O-antigenic side chain) composed of a chain of repeating oligosaccharide units. This projects away from the cell interior and can render the bacterial surface extremely hydrophilic. LPS is the main somatic antigen of Gram-negative bacteria and its structure is very much strain-dependent. LPS can provoke a wide range of responses in animals and many of these are attributable to its ability (at extremely low concentrations, e.g. pg/ml) to induce the release of a variety of cytokines from a range of host cells. Approximately 50% of the dry mass of the outer membrane is protein and more than 20 immunochemically distinct proteins (termed outer membrane proteins, OMPs) have been identified in *E. coli*. Some of these proteins (porins) form trimers which span the outer membrane and contain a central pore with a diameter of about 1 nm. These porins (e.g. OmpC and OmpF of *E. coli*) are permeable to molecules with molecular mass up to approximately 500–600 Da. They are also potent inducers of cytokine release from monocytes and lymphocytes. Some methods of extracting LPS from bacteria result in an LPS/OMP complex (to which the term 'endotoxin' should be applied). The protein(s) associated with the LPS is known as lipid A-associated protein (LAP) or endotoxin-associated protein (EAP) and is known to have biological activities distinct from those of LPS.

Cell wall of* Mycobacterium *species

In this genus the peptidoglycan layer is covered by lipid-rich layers so that up to 60% of the dry mass of the cell wall may consist of lipids, rendering it extremely hydrophobic. A variety of lipids, glycolipids and lipoproteins have been isolated from mycobacterial cell walls and several of these contain mycolic acids, which are unique to the mycobacteria, nocardiae and corynebacteria. One of the major constituents of the cell wall is lipoarabinomannan which, as well as playing an important role in enabling survival within the host, is also a potent inducer of cytokine release from monocytes and macrophages.

The periplasm

This is usually defined as the region between the cytoplasmic membrane and the outer membrane of the wall of a Gram-negative bacterium and it can occupy as much as 40% of the cell volume. Gram-positive bacteria do not have a comparable region. The periplasm is a fluid-filled region with a width of between approximately 7 nm and 15 nm although its size varies between species and also depends on the growth phase of the organism. It consists of a gel comprising a large number of proteins, oligosaccharides and uncross-linked peptidoglycan but is interrupted at many points by Bayer bridges which link the cytoplasmic membrane to the outer membrane. The functions of the periplasm are best described in terms of the major constituents of this region: the periplasmic proteins and oligosaccharides. The main functions of the proteinaceous constituents of the periplasm are: electron transport, transport of low molecular mass organic and inorganic nutrients into the cell, degradation of polymers to enable their import into the cell, biosynthesis of cell wall components and detoxification of toxic molecules present in the external environment. The oligosaccharides are thought to act as osmotic stabilizers of the periplasm.

Cell surface appendages

Some, but not all, bacteria (both Gram-positive and Gram-negative) have one or more of a number of appendages anchored in the cell wall and/or cytoplasmic membrane. These include flagella, fimbriae, conjugative pili and fibrils.

Flagella are organelles responsible for bacterial motility and confer on the organism the ability to move towards or away from beneficial or adverse environmental conditions respectively. Bacterial species exist which display phototaxis, chemotaxis, aerotaxis and magnetotaxis. Flagella consist of a globular protein (flagellin) which aggregates in a helical arrangement to form a long, hollow filament (usually many times the length of the bacterial cell) terminating in a complex 'basal body' embedded in the cell wall and cytoplasmic membrane. Flagellins are highly antigenic and constitute the H antigens of motile bacteria. Important pathogenic bacteria which are motile include *Clostridium* spp., *Salmonella* spp., *Vibrio cholerae*, *E. coli*, *Pseudomonas* spp., *Campylobacter* spp. and *Helicobacter pylori*. The mechanisms by which bacteria can sense and respond to chemical gradients are described in Chapter 3.

Fimbriae (also termed pili) are rod-shaped structures originating in the cytoplasmic membrane and are composed of proteinaceous subunits. They are shorter (0.5–10 μm) and thinner (1–11 nm) than flagella and their main function is to enable adhesion of the organism to host cells or to other bacteria. Each fimbria is composed of approximately 1000 identical proteinaceous major subunits and a much smaller number (1–10) of minor subunits – either, or both, of these may function as an adhesin. In addition, some fimbriae (e.g. those of *Neisseria* spp.) have a polysaccharide or glycoprotein at their tip which constitutes the major adhesin. A bacterium may be able to elaborate a number of fimbriae each with a different adhesin, so enabling adhesion to different host receptors and hence, possibly, different types of host cells. They are found mainly on Gram-negative bacteria although they have been detected on streptococci and actinomycetes. A detailed description of the role of pili in bacteria adhesion is provided in Chapter 5.

Conjugative pili are similar in structure to fimbriae (although much thicker and longer) but are involved in the transfer of DNA during bacterial conjugation.

Fibrils are shorter and much thinner than fimbriae or pili and may cover the whole cell surface. They are found on the surfaces of oral streptococci, *Bacteroides* spp. and other bacteria and probably have an adhesive function.

Cell surface-associated components

Most bacteria, especially when first isolated from their natural habitat, have additional layers of material on their surface which are generally only loosely associated with the cell wall (Figure 2.2). Examples of such components include capsules, slime layers and S layers. These readily extractable surface-associated components comprise a complex mixture of molecules (Figure 2.3) many of which are capable of interacting with and inducing changes in host cells.

The capsule of a bacterium is a gelatinous, highly hydrated matrix composed of polysaccharide(s) and/or protein(s) which is loosely associated with the cell wall and is often shed into the environment. Although generally regarded as being amorphous, there is evidence to suggest a highly ordered secondary structure in the capsule of *E. coli*. There is also evidence implying that capsular material may be linked covalently to the cytoplasmic membrane by means of phospholipid substituents. Capsules are highly antigenic, constituting the K antigens of bacteria, and function as molecular sieves, as adhesive structures, as nutrient sequestering/storage systems and in protecting the organism against phagocytosis (see Chapter 8). Organisms in which the capsule is known to be a major virulence factor include *Streptococcus pneumoniae, Haemophilus influenzae, Neisseria* spp., *Bacteroides* spp., *Pseudomonas*

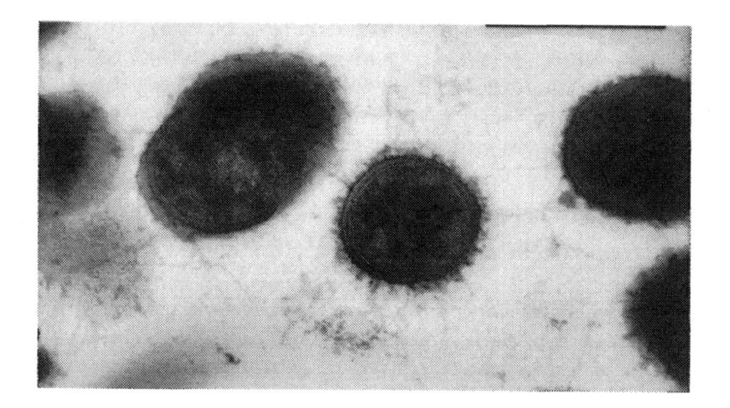

Figure 2.2 Electron micrograph showing abundant surface-associated material on the Gram positive bacterium *Streptococcus sanguis.* This material, which is stained with the electron-dense compound ruthenium red, can also be seen to be shed from cells and exists between the bacteria

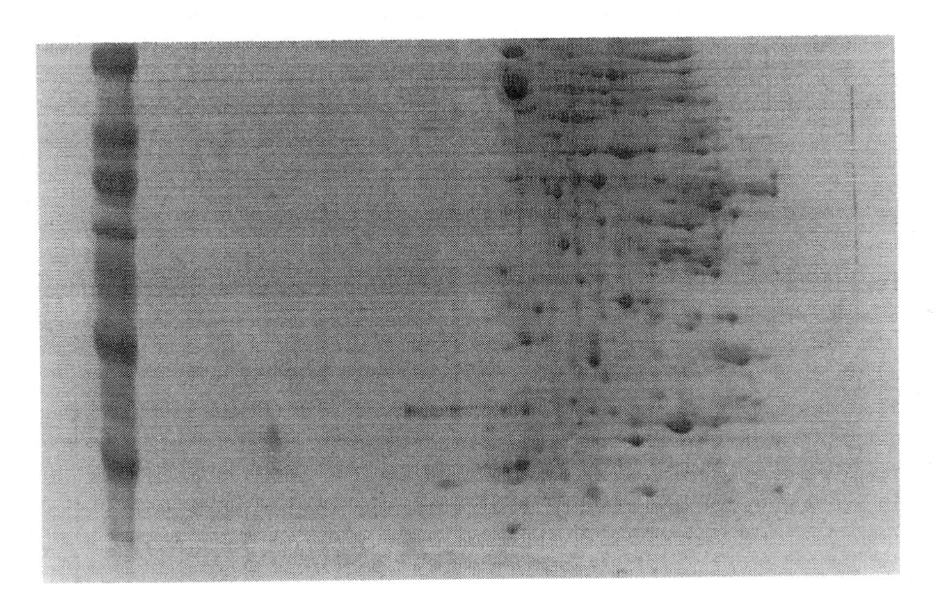

Figure 2.3 In this figure the surface-associated material from the Gram-negative bacterium *Actinobacillus actinomycetemcomitans* has been isolated by washing the bacteria gently in saline. This removes the components associated with the bacterial surface without killing or damaging the bacteria. To determine how many proteins are present in this material it has been separated by two-dimensional SDS-PAGE and the individual isolated proteins have been stained. Each spot represents a distinct protein and at least 100 proteins are present

spp. and *Klebsiella* spp. Slime layers, produced notably by bacteria of the genus *Staphylococcus*, are similar to capsules except that they have a higher water content and are even more loosely attached to the cell surface than capsules.

The surfaces of many species of bacteria (e.g. *Campylobacter* spp., *Actinobacillus* spp., *Eubacterium* spp.) are completely covered in a crystalline array of identical protein, or glycoprotein, self-assembling units. Such layers (termed S layers) have protective and adhesive functions similar to capsules.

In addition to these 'structures' which are external to the bacterium it is increasingly being noted that a variety of proteins are to be found loosely associated with the surfaces of bacteria. Some of these proteins, such as the heat shock protein, chaperonin 60, are normally thought of as intracellular proteins. Be that as it may, there are increasing numbers of examples of bacteria which demonstrate the presence of this protein external to the cell.

The nuclear region

Bacteria have only one chromosome, which consists of a singular, and in most cases circular, molecule of DNA. Some proteins are associated with the molecule and are thought to have a structural role, while the negative charge on the DNA is partially neutralized by polyamines and Mg^{2+} ions. Although not separated from the cytoplasm by a nuclear membrane, the DNA tends to aggregate and is visible by electron and light microscopy as a distinct structure termed a nucleoid. The DNA molecule varies in size depending on the particular species – the DNA of *E. coli*, for example, contains 4700 kilobase pairs. The molecule, if linearized, would be approximately 1 mm in length, that is, 1000 times the length of the cell; therefore in the cell it exists as a tightly-coiled ('supercoiled') structure in which the regions of supercoiling are stabilized by proteins. In many species genes are also carried on plasmids in addition to the chromosome. These are generally small circular molecules of DNA, although linear plasmids are found in some species such as *Borrelia burgdorferi*. These plasmids often encode proteins responsible for virulence factors and antibiotic resistance.

Cytoplasm

In contrast to the cytoplasm of eukaryotic cells, the bacterial cytoplasm is devoid of organelles but contains large numbers of ribosomes. It has a gel-like consistency due to its high protein content and also contains imported nutrients and waste materials that are subsequently excreted. The cytoplasm of some bacteria may also contains granules composed of poly-β-hydroxybutyric acid, glycogen or polyphosphate which function as storage polymers for carbon and phosphate respectively.

BACTERIAL GENE TRANSCRIPTION, OPERONS, MOVABLE GENETIC ELEMENTS

Adapt or perish

The absence of a separate nuclear compartment bounded by a membrane system has some important consequences for bacterial gene expression, not least of which is the rapid adaptive response bacteria are capable of mounting when environmental conditions change. This is particularly important in the context of bacterial pathogenesis and disease. For example, consider the shift in environmental conditions a *Salmonella* bacterium must endure following ingestion in contaminated water or food. Firstly, they must survive the acid conditions of the stomach. Following this they invade the cells lining the intestinal epithelium, and to cause systemic disease they must also survive in blood and grow and replicate in macrophages. At each stage of this journey the environmental conditions for growth are very different, and if the bacterium cannot (a) recognize where it is, or (b) adapt to the conditions of that location, then it will not survive.

Gene transcription

The synthesis of RNA from a DNA template (transcription) is performed by RNA polymerase, a multi-subunit enzyme. The core enzyme which catalyses RNA synthesis comprises four subunits known as α_2,β,β'. The sigma factor (δ) which associates with the core enzyme has no catalytic activity but instead is necessary for core enzyme recognition of gene promoters. Sigma factor specificity is the first of many levels of regulation of gene expression open to bacteria. Bacteria possess multiple sigma factor genes, and each sigma factor allows the RNA polymerase core enzyme to recognize a specific set of promoters and transcribe only those genes. Thus, by changing the sigma factor that is expressed in response to altered environmental conditions, a fundamental shift in gene expression can occur to provide proteins or metabolic functions appropriate to the new conditions. Stress in its many forms (e.g. temperature, pH, osmolarity) often results in altered sigma factor expression and consequent expression of a new set of proteins that are necessary to protect the cell from such environmental insults.

Another means for co-regulation of genes encoding proteins of related or integrated function is the organization of genes within operons. An operon is the tandem arrangement of genes to form a single transcriptional unit. For example, the locus encoding capsule biosynthesis in the human pathogen *Strep. pneumoniae* comprises 15 genes (for polysaccharide synthesis, export and attachment to the cell surface) that are transcribed as a single mRNA of some 15 kb. Regulons, on the other hand, are groups of (frequently unlinked) genes or operons that are subject to control by a single transcriptional regulator. An example is the *vir* regulon of *Strep. pyogenes*. The transcriptional activator Mga, which demonstrates homology with response regulators of two-component signal-transducing systems, controls the expression of multiple virulence-associated genes including those encoding M and M-related proteins, secreted cysteine protease and oligopeptide permease.

Induction and repression

The regulation of gene expression by induction or repression of transcription is an important means by which bacteria maintain fitness. The transcriptional factors involved are DNA-binding proteins that interact with sequences in the vicinity of the gene promoter to influence (enhance or block) binding of RNA polymerase. In turn, the affinity of these transcriptional factors for so-called operator sequences near the gene promoter is determined by the binding of small inducer or co-repressor molecules.

Another very important means of transcriptional activation for bacteria is the two-component signal transduction system, also known as the histidyl-aspartyl phosphorelay which is described in detail in Chapter 3. As the name suggests, this system is comprised of two proteins. The histidine protein kinase sensor is a transmembrane receptor and the response regulator is a cytoplasmic protein, and together they mediate reversible phosphorylation events that control gene expression in response to environmental signals. The histidine protein kinase sensor catalyses the autophosphorylation of a conserved histidine residue within the cytoplasmic domain of the protein in response to a specific external signal. Activated histidine kinase then serves as a phospho-donor for the response regulator, which once phosphorylated mediates changes in gene expression at the level of transcription.

Two-component signal transduction systems were once thought to be restricted to prokaryotes. They have, however, now been identified also in eukaryotes including plants and fungi (this system will be described in detail in Chapter 3).

Mobile genetic elements

Plasmids are generally circular DNA molecules that can exist independently of the chromosome and are present in many bacteria. Plasmids vary both in size and in the numbers within the bacterium. Many plasmids are not essential and can be cured from bacteria without adverse effects. However, many impart advantageous phenotypes on the host bacterium. Plasmids are often characterized on the basis of the phenotype they confer on the host bacterium. Resistance factors, or R plasmids, are widespread amongst both Gram-positive and Gram-negative bacteria, and carry (often multiple) antibiotic resistance genes. Often, too, the resistance genes are carried by transposons on the R plasmid, and this allows strains to rapidly develop multiple resistance plasmids. Virulence plasmids encode functions that may enhance the survival or ability of bacteria to cause disease within the host. Some virulence plasmids encode toxins or iron uptake systems, for example. Others may impart a complex virulence phenotype. Thus a large virulence plasmid of the enteropathogen *Shigella flexneri* carries multiple genes enabling *Shigella* to invade non-phagocytic human cells.

A group of plasmids also encode genes for their conjugal transfer between bacteria, and these too promote rapid dissemination of virulence factors such as antibiotic resistance amongst many and varied bacteria. Episomes are plasmids that can exist separately or integrated onto the bacterial chromosome, and conjugal transfer of integrated episomes can result in mobilization also of the donor chromosomal DNA.

Transposable elements are distinct DNA segments that can move around the bacterial genome. The event does not require extensive areas of homology between the transposon and the insertion site, but instead is more or less random and requires the products of genes carried by the transposable element.

The simplest transposable elements are the insertion sequences or IS elements. These are small (typically 0.75–1.5 kb) that carry only gene(s) encoding the protein(s) necessary for the transposition event, and are defined at either end by short (15–25 bp) inverted repeats. IS element insertion results in duplication of the chromosomal target sequence creating, generally, small direct repeats at either end of the IS elements.

Composite transposons consist of a central DNA region of variable size, and containing extra genes, which are flanked on either side by IS elements that are identical or very similar. Composite transposons may have originated by an IS element transposing close to the original integration site. The accessory genes carried by composite transposons may include genes encoding antibiotic resistance, toxin production and metabolic function. Insertion sequences and composite transposons move by non-replicative (conservative) transposition that requires just a transposase activity and results in release of the element from the donor DNA during transfer, in a cut-and-paste type reaction.

Type II transposons are not bound by IS elements but comprise distinct units that carry genes for transposition and additional accessory genes encoding, for example, antibiotic resistance. These transposons are flanked by short inverted repeats and do generate short direct repeats of the transposition target sequence, but they move by replicative transposition. This mechanism involves duplication of the element such that the transposing molecule is a copy of the original element. Replicative transposition requires a transposase enzyme and, in addition, a resolvase activity. A third group of elements, the conjugative transposons, are more closely related to phage. They transpose by a non-replicative method but are not bounded by IS elements, nor do they duplicate the target sequence in transposition. Conjugative transposons, as the name suggests, can move between the genetic elements of two different bacteria in close contact. Transfer is independent of plasmids or bacteriophages and conjugatve transposons can be relatively promiscuous. For example, the host range of the Tn916 family of conjugative transposons covers both Gram-positive and Gram-negative organisms and includes in excess of 40 species. Some conjugative transposons have a very broad host range and carry multiple antibiotic resistance genes, and these elements, too, contribute to the rapid dissemination of antibiotic resistance genes.

PATHOGENICITY ISLANDS

Mobile genetic elements such as plasmids and transposons are known to have contributed significantly to bacterial pathogenicity, and indeed continue to do so, through the inter-species transfer of virulence factors. It is becoming increasingly apparent also that interspersed within the genomes of many pathogenic bacteria are large distinct chromosomal elements encoding virulence-associated genes. These chromosomal loci have been termed 'pathogenicity islands' and represent another important mechanism contributing to microbial evolution.

Pathogenicity islands (PAIs) were first described in isolates of uropathogenic *E. coli*. The genes required for α-haemolysin production were shown to be present on large chromosomal regions which could be lost spontaneously, resulting in an avirulent phenotype. A more detailed description of the *E. coli* haemolysin-associated PAIs will serve to provide a general definition of PAIs overall (Table 2.1). Firstly, they are large; Pai I is 70 kb while Pai II is 190 kb and carries, additionally, genes encoding adherence-mediating P fimbriae. Secondly, PAIs are found in pathogenic isolates but not, or only infrequently, in non-pathogenic isolates or strains. Pai I and Pai II are flanked by short direct repeats (16 bp for Pai I, 18bp for Pai II), one of which remains on the chromosome following spontaneous deletion of the element. The site of insertion of Pai I and Pai II is the 3' end tRNA genes, which act also as sites for the integration of many bacteriophages into the bacterial genome. Thus Pai I of uropathogenic *E. coli* is flanked by the *selC* gene for selenocysteine tRNA, and Pai II by *leuX* encoding a minor leucine tRNA. Perhaps the most striking characteristic of Pai I and II of *E. coli*, and of many other PAIs, is that these chromosomal elements have a lower G + C content compared to the host bacterial chromosome. Pai I and II have a G + C content of 41% compared to 51% for the *E. coli* host. This information suggests that PAIs were acquired by horizontal gene transfer from other bacterial species (Table 2.2).

Pai I and Pai II of uropathogenic *E. coli* represent chromosomal elements that encode a single (haemolysin; Pai I) or multiple (haemolysin and P fimbriae; Pai II) distinct virulence determinants. A second class of PAI is represented by LEE (locus of enterocyte effacement; Pai III) of enteropathogenic *E. coli* strains. LEE is a 35 kb chromosomal element containing multiple genes that define a complex virulence property. In the case of LEE, this property is the induction of attaching and effacing lesions on enterocytes. LEE encodes a type III secretion system (see 'Bacterial protein secretion systems' below and Chapter 6) and a number of secreted proteins that induce a signal transduction cascade within enterocytes that results in

Table 2.1 Common features of pathogenicity islands

1. Carriage of virulence-associated genes
2. Pathogen-specific
3. Large (frequently > 30 kb) distinct chromosomal units often flanked by direct repeats
4. Different G + C content compared to host chromosome
5. Associated with tRNA genes and/or IS elements

Table 2.2 PAIs of selected bacteria

Pathogen	PAI designation	Size (kb) (PAI/host)	G + C	Boundary sequences	tRNA	Phenotype
E. coli	Pai I	70	40/51	16bp DR[a]	*selC*	Haemolysin production
	Pai II	190	40/51	18bp DR	leuX	Haemolysin, P-fimbriae production
	LEE (Pai III)	35	39/51	–	*selC*	Induction of attaching and effacing lesions on enterocytes
Sal. typhimurium	SPI-1	40	42/52	–	–	Invasion of non-phagocytic cells
	SPI-2	40	45/52	–	*valV*	Survival in macrophages
	SPI-3	17	?[b]	–	*selC*	Survival in macrophages
V. cholerae	VPI	39.5	35/46	13 bp DR	*ssrA*[c]	Colonization, expression of phage CTXΦ receptor
Yersinia pestis	HPI	102	46–50/46–50	IS*100*	–	Haemin storage, iron uptake
D. nodosus	VAP1	12	52/45	19bp DR		Acquisition of other virulence factors?

[a]DR, direct repeat; [b]nucleotide values not yet availble; [c]a tRNA-like gene.

major cell cytoskeleton rearrangements and the formation of a pedestal upon which the bacterium sits (see Chapter 5). LEE (Pai III) is situated at exactly the same site in the *selC* tRNA locus as Pai I of uropathogenic *E. coli*. In contrast to Pai I, however, LEE is not flanked by direct repeats and appears to be stable. Nevertheless, the distinct G + C content (39% versus 51% for the *E. coli* chromosome) and the absence of this element from non-enteropathogenic *E. coli* clearly mark LEE as having been acquired from another species by horizontal gene expression. Introduction of LEE into a non-pathogenic laboratory strain of *E. coli* results in the single-step acquisition of the attaching and effacing phenotype.

PAIs are not restricted to *E. coli*, and many more pathogens demonstrate chromosomal loci that fulfil the criteria of PAIs – that is, large distinct chromosomal elements that differ in G + C content and that encode defined virulence functions. Many, but not all, of these are unstable, being flanked by direct repeats or associated with insertion sequences, as shall be discussed, and Table 2.2 lists the relative characteristics of selected PAIs.

Pathogenic *Salmonella* species have a complex lifestyle, and this is reflected by the large (>100) number of genes required for virulence as identified using some of the techniques detailed in Chapter 4. Many of these virulence-associated genes are encoded within pathogenicity islands, four of which have been defined to date. *Salmonella* pathogenicity island (SPI)-1 of 40 kb encodes some 25 genes including those for secreted proteins and an associated type III secretion system. This locus is essential for invasion of non-phagocytic cells by *Salmonella*, and mutation of genes within SPI-1 attenuates the virulence of bacteria when inoculated orally in mice but not when inoculated intraperitoneally. Additionally, SPI-1 gene products can induce apoptosis of *Salmonella*-infected macrophages. SPI-1 does not have flanking direct repeats nor is it associated with a tRNA gene, and in contrast to the *E. coli* PAIs, SPI-1 is species-specific rather than strain-specific. However, the size, orientation and organization of the invasion genes on SPI-1 are broadly similar to those present on a *Shigella* virulence plasmid, and the G + C content is significantly different from that of the *Salmonella* chromosome (42% compared with 52%) – clues which point to horizontal acquisition also for SPI-1. It is thought that SPI-1 may have been stably acquired early in the evolution of the genus. A second *Salmonella* PAI was identified in the vicinity of a valine tRNA gene. The 40 kb SPI-2 element comprises some 17 genes that encode a distinct type III secretion system and a two-component regulatory system. SP1–2, which has a G + C content of 45%, is required for survival of *Salmonella* within macrophages.

Since the *selC* tRNA gene is the site in *E. coli* of Pai I and LEE, in addition to being the target for integration of bacteriophage DNA, it was reasoned that the *Salmonella selC* locus may harbor a pathogenicity island. This was indeed the case, and DNA sequencing in the vicinity of this locus indicated the presence of a 17 kb element that was absent from *E. coli* and additionally demonstrated a somewhat reduced G + C content compared with the *Salmonella* chromosome. A gene within the SPI-3 element was shown to be essential for bacterial growth at low magnesium concentrations and for survival within macrophages. SPI-3, like LEE, is not flanked by direct repeats. The *Salmonella* chromosome appears to be dotted additionally with several smaller elements, both single genes and operons, that have been termed pathogenicity islets. These are required for full virulence and demonstrate evidence of having been acquired by horizontal gene transfer: different G + C content, flanking direct repeats. Collectively, these regions can be considered a pathogenicity archipelago within the *Salmonella* chromosome.

The bacterial species *V. cholerae* includes harmless aquatic strains as well as strains capable of causing cholera epidemics. The virulence of *V. cholera* is associated with the synthesis and secretion of cholera toxin encoded by *ctxA* and *ctxB* genes that are carried by the filamentous bacteriophage CTXΦ. The bacterial receptor for phage infection is the toxin co-regulated pilus (TCP) that serves also as an important adherence determinant for the bacteria. The TCP gene cluster, which includes an accessory colonization factor, is present on a 39.5 kb pathogenicity island, termed VPI. Acquisition of VPI by aquatic strains of *V. cholerae* allows them to colonize the intestines of humans and animals and is a prerequisite (through expression of TCP, the CTXΦ phage receptor) for conversion of colonization-proficient *V. cholerae* to epidemic and pandemic strains. VPI has a low G + C content compared to the *Vibrio* chromosome, is flanked by short direct repeats and is inserted adjacent to a tRNA-like gene, suggesting an island status.

The property of instability demonstrated by many PAIs was used as a molecular handle to identify a novel PAI of enteropathogenic *Sh. flexneri*. The chromosomal *she* locus of *Sh. flexneri* comprises overlapping genes *set1A* and *set1B* that encode *S. flexneri* ShET1 enterotoxin. These genes are contained entirely within the oppositely oriented *she* open reading frame (ORF) that encodes a predicted protein with putative haemagglutinin and mucinase activity. The limited distribution of this unusual chromosomal locus amongst *Sh. flexneri* strains raised the possibility that this virulence-associated element was part of a larger PAI. To investigate this possibility, a tetracycline resistance determinant was inserted within the *she* locus on the *Shigella* chromosome. Tetracycline sensitive mutants arose at a frequency of 10^{-5}–10^{-6} and six independent isolates were selected and demonstrated to have lost a 51kb chromosomal fragment containing the *she* locus, and additionally several IS-like elements and a gene encoding an IgA protease-like protein.

Chromosomal elements which correspond fully with the definition of PAIs (Table 2.1) have not been described, as yet, in Gram-positive bacteria. Nevertheless, there are a number of examples of specific clusters of virulence factors in Gram-positive bacterial pathogens that can be considered PAIs in the broadest sense. Chief amongst these is a 19 kb Locus, designated PaLoc, of toxigenic *Clostridium difficile* isolates that cause antibiotic-associated diarrhoea and colitis. PaLoc comprises five genes, *tcdA–E*, including those encoding the *Cl. difficile* enterotoxin (*tcdA*) and cytotoxin (*tcdB*). The PaLoc is absent from non-toxigenic *Cl. difficile* isolates, but where present demonstrates no direct repeats nor is it associated with a tRNA gene. Instead, the target site for integration is probably a 115 bp sequence that is present in non-toxigenic strains, but absent from isolates with the PaLoc, and that is predicted to form a stem loop structure.

A 10 kb chromosomal element encoding a cluster of six virulence-associated genes is present in *Listeria* species pathogenic to humans and animals (including *Lis. monocytogenes* and *Lis. ivanovii*) but absent from non-pathogenic species. Included in this cluster are genes necessary for escape of *Listeria* from phagosomes (*hly*, encoding listeriolysin O) and for intra- and intercellular movement (*actA* and *plcB*). As for PaLoc of *Cl. difficile*, this locus demonstrates no direct repeats nor is it associated with a tRNA gene. The non-pathogenic *Lis. seeligeri* does possess this locus, but it is not properly expressed.

As has been related, there is strong evidence for horizontal acquisition, by Gram-negative bacteria at least, of PAIs. This raises several issues, not least of which are: how were they acquired and where did they come from?

It is not fully understood how PAIs might be transferred from one organism and integrated into the chromosome of a second unrelated bacterium, although some clues are provided in the form of genetic 'footprints' within the PAIs themselves. The transfer of DNA, including plasmids and transposons, is known to occur *in vivo* by several mechanisms. These are transformation (uptake of naked DNA from surroundings) and conjugation (requiring direct cell–cell contact). Chromosomal DNA transfer is mediated also by bacteriophage transduction.

The target for prophage integration onto the bacterial chromosome is commonly the 3' end of tRNA genes, a site also frequently used for the integration of PAIs such as Pai I–III of pathogenic *E. coli* species. Indeed, non-expressed genes are found on Pai I and II of uropathogenic *E. coli* that resemble the integrase genes of bacteriophages, and it has been speculated that some PAIs originated from integrated prophages. Putative transposase and integrase genes are present on some PAIs including VPI of *V. cholerae*. Frequently, also, IS elements are associated with PAIs and there is a probable role for transposition events in the acquisition of some PAIs.

The virulence of *Dichelobacter nodosus*, which causes footrot in sheep, is associated with the presence of multiple chromosomal regions encoding virulence-associated proteins. One of these so-called VAP regions (12 kb) is located next to a tRNA gene, contains a bacteriophage-like integrase gene, and is flanked by direct repeats, as is characteristic of many PAIs. Interestingly, one pathogenic *D. nodosus* strain possesses a plasmid that carries a modified VAP cluster. This plasmid is able to integrate into the *D. nodosus* chromosome resulting in PAI formation.

Following acquisition and integration of PAIs, the bacterium is faced with two difficulties: those of retention and integration of function. The new host must ensure that it can retain these previously mobile elements that clearly provide beneficial phenotypes and improved fitness within the animal host. Mutations within mobility gene(s) associated with an integrated element will render them non-functional and will effectively immobilize the element. Many pathogenicity islands possess molecular footprints of such mobility genes. Natural genetic competence and/or a high rate of mutation and genetic recombination are thought to be powerful forces that drive microbial evolution. These factors may contribute significantly also to the acquisition and stable maintenance of PAIs.

Even if PAIs encode specific regulatory elements for controlling expression of associated virulence factors, as is the case for SPI-1 and 2 for example, the incorporated sequences must integrate their expression within the regulatory networks of the host bacterium to ensure full and correct (in terms of temporal and spatial cues) expression of virulence functions. The PhoP/PhoQ regulatory system, which is present in pathogenic and non-pathogenic *Salmonella* species, governs bacterial response to low Mg^{2+} environments. A number of horizontally acquired genes of *Salmonella*, including some encoded on PAIs, are regulated by the PhoP/PhoQ system. The importance of correct regulation of expression for virulence is demonstrated by the non-virulent *Lis. seeligeri* which, as was mentioned earlier, possesses a *Listeria* chromosomal element encoding a cluster of six virulence-associated genes. In contrast to *Lis. monocytogenes*, however, which also possesses this element, *Lis. seeligeri* is non-virulent due to incorrect expression of PAI-associated genes. Transformation of *Lis. seeligeri* with a plasmid encoding the *Lis. monocytogenes* transcriptional activator of *hly* (listeriolysin gene) imparts a virulent phenotype on *Lis. seeligeri*.

Finally, where did the PAIs originate? Analysis of the stable PAIs of *Salmonella* indicates that these elements were acquired early in the evolution of the genus *Salmonella*. Consequently, the donor organism may no longer be extant nor, indeed, recognizable as the donor. Molecular analysis of emerging pathogenic strains of bacteria combined with microbial genome sequencing (Chapter 4) may be useful in identifying recent PAI acquisition events and both donor and recipient bacteria.

Catching bacteria in the act would also provide considerable detail regarding the mechanisms of transfer, integration, retention and regulation of PAIs. Large horizontally acquired chromosomal elements encoding discrete functions are not restricted to pathogenic bacteria. Rather, the quantum evolution of bacterial species by means similar to the PAI-mediated evolution of pathogens may be widespread. A recent study has described the horizontal transmission of a large (500 kb) chromosomal element amongst soil bacteria. The chromosomal element encodes nodulation and nitrogen fixation genes that allow *Mesorhizobium loti* species to form a symbiotic relationship with the roots of certain plants. This 'symbiosis island' is transmissible in laboratory matings to at least three species of non-symbiotic mesorhizobia. The element integrates into a tRNA gene, is flanked by 17 bp direct repeats, encodes a bacteriophage-like integrase just within its left end, and offers a tremendous opportunity to witness evolution in action.

BACTERIAL PROTEIN SECRETION SYSTEMS

It has been estimated that approximately 20% of the polypeptides and proteins synthesized by a bacterium are destined for a location external (either partially or completely) to the cytoplasmic membrane. Such proteins include outer membrane and periplasmic proteins of Gram-negative bacteria, cell surface-associated proteins of Gram-negative and Gram-positive bacteria, flagella and fimbrial proteins and a whole range of extracellular enzymes and toxins. There appear to be at least five pathways available for the export of such proteins: the general secretory pathway and the Types I, II, III and IV secretion systems. The general features of these systems are compared in Figure 2.4.

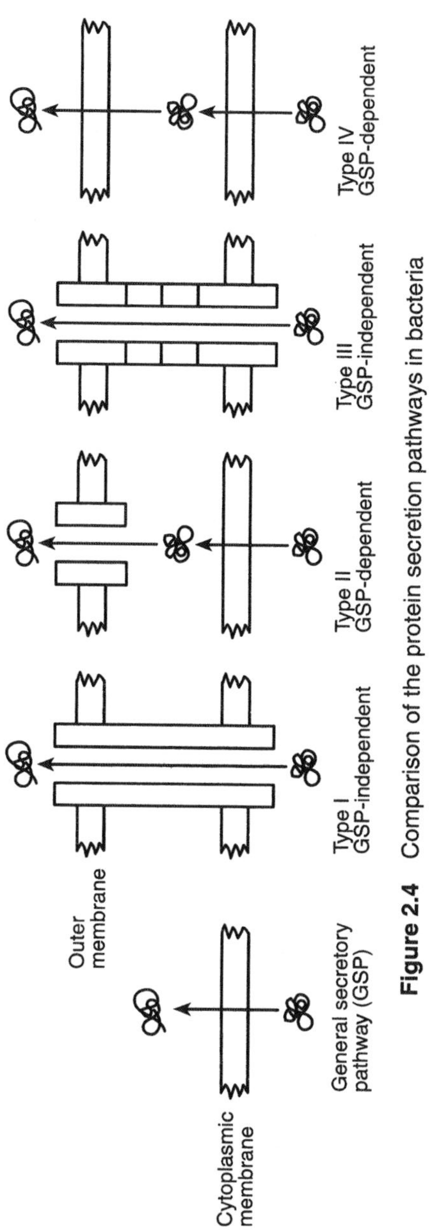

Figure 2.4 Comparison of the protein secretion pathways in bacteria

The general secretory pathway

Most of the proteins whose ultimate destination lies beyond the cytoplasmic membrane are secreted via this pathway. The first step in the general secretory pathway (GSP) involves the translocation of the protein across the cytoplasmic membrane and, in the case of Gram-positive organisms, this results in the export of proteins to the cell surface or the external environment. In Gram-negative bacteria, however, this would merely deliver proteins to the periplasm. Proteins destined for export from Gram-negative bacteria need to be transported across the periplasm and the outer membrane by means of what are known as the terminal branches of the GSP.

Proteins which are transported by the GSP are known as secretory proteins or, if they are in the form of precursors, presecretory proteins. All such proteins have a characteristic 'secretory signal sequence' at the amino terminal which consists of a region with at least 10 hydrophobic amino acids (H domain) preceded by a short hydrophilic domain (N domain) containing at least one positively charged residue. The main function of this signal sequence is to direct the protein to the cytoplasmic membrane. The early stage of the GSP is a common pathway concerned with translocation of the presecretory protein across the cytoplasmic membrane; the pathway then has a number of branches ('terminal branches') depending on the ultimate destination of the protein (Figure 2.5).

The GSP in *E. coli* (the organism whose GSP has been most intensively studied) consists of seven proteins: SecA, B, D, E, F, G and Y (Figure 2.6). SecA is found mainly in the cytosol although it can also associate with the cytoplasmic membrane. It exists as a dimer and has ATP-hydrolysing activity. SecB exists as a tetramer in the cytosol and binds to presecretory proteins but not to most cytoplasmic proteins; it probably functions as a GSP-specific molecular chaperone. SecY, SecE and SecG proteins have long stretches of hydrophobic amino

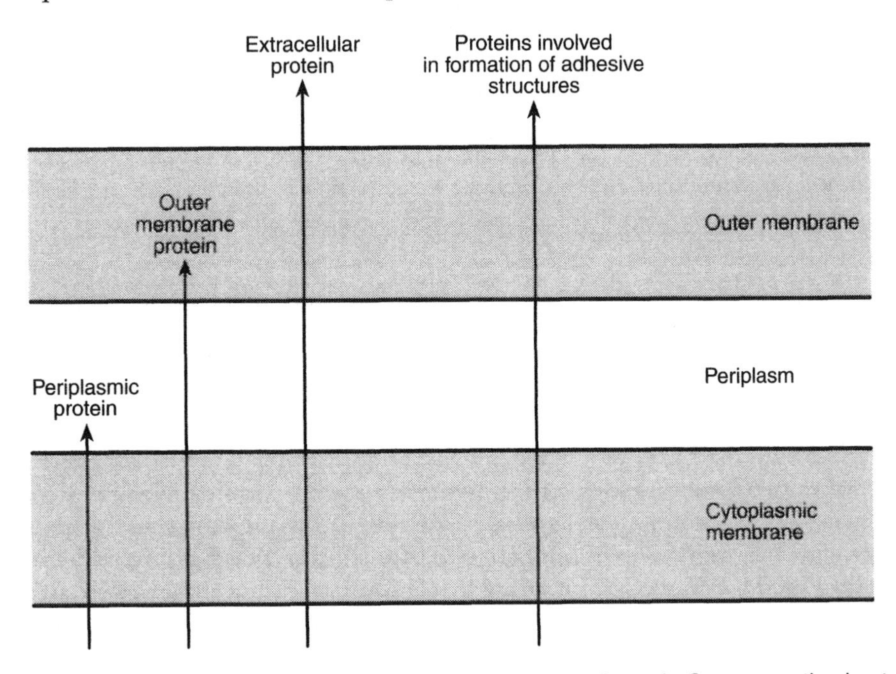

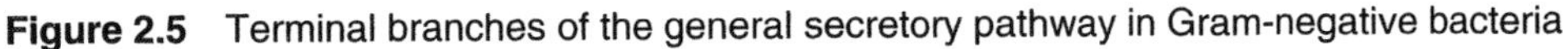

Figure 2.5 Terminal branches of the general secretory pathway in Gram-negative bacteria

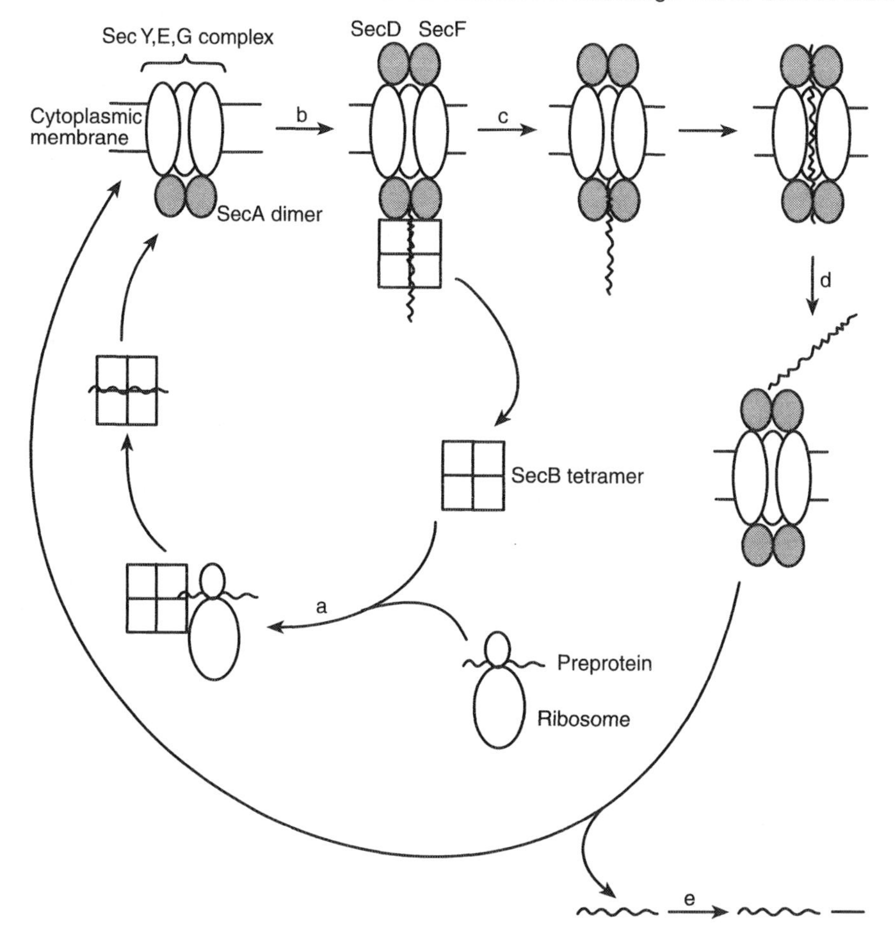

Figure 2.6 Operation of the general secretory system: (a) SecB tetramer binds to preprotein as it is synthesized by the ribosome; (b) secB delivers the preprotein to the SecA dimer; (c) preprotein begins its translocation through the SecY, E, G translocase; (d) preprotein is shed from the translocase system; (e) signal sequence is cleaved by a signal peptidase

acids which anchor them in the cytoplasmic membrane and, together with SecA, constitute the translocase system. SecD and SecF are also involved in translocation as described below.

The first stage in translocation of the secretory protein across the cytoplasmic membrane involves binding of SecB to the unfolded secretory protein as it is synthesized at the ribosome. SecB both prevents folding of the nascent protein and targets it to the translocase at the cytoplasmic membrane. A negatively charged region of the SecB molecule binds to a positively charged region on the SecA located at the last 22 carboxy terminal residues. The signal sequence of the nascent protein then interacts with SecA resulting in the transfer of the mature protein to SecA. ATP then binds to one of the nucleotide binding sites of SecA and induces a profound conformational change in the SecA, enabling the insertion of the signal sequence of the protein into the other part of the translocase system. This consists of a complex of SecY, SecE and SecG which form a pore in the cytoplasmic membrane. At the same time, SecB is released from the complex and is free to bind to a new nascent protein.

Hydrolysis of the ATP bound to SecA reverses the conformational change in the molecule thereby releasing the protein. Binding of another molecule of ATP again enables binding of SecA to the protein but this occurs 2.5 kDa further along the protein chain. The cycle of ATP binding and hydrolysis is repeated many times, so shunting the protein through the translocation channel of the membrane 2.5 kDa at a time. SecD and SecF are thought to prevent backward sliding of the protein in the translocation channel and may also be involved in the release of the protein from the translocation channel. The signal sequence of the protein is cleaved by one of several signal peptidases, most of which are anchored in the cytoplasmic membrane and have their active sites on the periplasmic face of the membrane.

The part of the GSP described so far is sufficient to ensure the export of proteins from Gram-positive bacteria and/or their incorporation into the cytoplasmic membrane and cell wall of such organisms. However, in the case of Gram-negative bacteria, this section of the GSP merely delivers the protein into the periplasm. Within the periplasm, proteins undergo extensive conformational changes regardless of their ultimate destination. While some of these proteins remain in the periplasm, others are incorporated into the outer membrane and secreted proteins will additionally need to be transported across the outer membrane; this is achieved by various terminal branches of the GSP. The most widely used of the terminal branches of the GSP is known as the main terminal branch (MTB), which is used to secrete a variety of toxins and hydrolytic enzymes. It is also sometimes referred to as the type II secretion pathway. The secretion of pullulanase from *Klebsiella oxytoca* is the most extensively studied example of the MTB. This involves at least 14 proteins, some of which form a complex spanning the periplasm while others form a pore in the outer membrane through which the pullulanase is transported. Details of the way in which the system functions remain to be established. Another important terminal branch is the chaperone/usher pathway which is used for the assembly of pili and other adhesive structures.

Other secretion pathways

A number of important virulence factors do not have a classical N-terminal signal sequence and their secretion does not involve the GSP. Examples of such proteins include the haemolysin of *E. coli*, the adenylate cyclase of *Bordetella pertussis*, the alkaline protease of *Pseudomonas aeruginosa* and the leukotoxin of *Pasteurella haemolytica*. Such proteins are exported via the Type I secretion pathway, which involves the transfer of the protein from the cytoplasm to the outside of the cell without an intermediate stop in the periplasm. They have a C-terminal secretion signal in the last 60 amino acids which generally contains a negatively charged amino acid followed by several hydrophobic residues. The system consists of two cytoplasmic membrane proteins, an ATPase, a dimeric protein which spans the inner and outer membranes (known as a membrane fusion protein) and an outer membrane polypeptide.

The type III secretion pathway is a recently identified pathway which can not only export proteins directly to the cell surface (as in type I secretion) but also often is involved in the delivery of bacterial virulence proteins directly into the cytoplasm of a host cell. It is triggered by contact with the target cell and so is often referred to as 'contact-dependent secretion'. The system has many features in common with that used to export flagellar proteins in both Gram-positive and Gram-negative species. Indeed, many of the genes encoding components of the type III system are homologous to those encoding proteins involved in flagellar secretion. The presence of this pathway has been demonstrated in a number of important pathogens including *Yersinia* spp., *Salmonella* spp., *Shigella* spp., enteropathogenic *E. coli* and, possibly, *Chlamydia* spp. The components of the system include an energy-generating system to drive secretion, a regulatory system, cytoplasmic chaperones to stabilize the effector molecules and proteins located in the cytoplasmic and outer membranes which constitute the secretory apparatus itself. The components of the system are highly conserved among different species and, in some cases, have been shown to be interchangeable. Hence, effector molecules involved in the uptake of *Salmonella* spp. by epithelial cells also enable uptake of *Shigella* spp.

A number of proteins are translocated across the cytoplasmic membrane by the GSP but then appear to form a pore in the outer membrane to enable their secretion to the external environment – this is described as a type IV secretion pathway. Such proteins are termed 'autotransporters' and include the IgA proteases of *Neisseria gonorrhoeae* and *Haem. influenzae*, the vacuolating cytotoxin of *Hel. pylori* and pertussis toxin.

THE BACTERIAL CELL CYCLE

Given adequate nutrients and suitable environmental conditions, a bacterial cell will grow in size and eventually undergo binary fission giving rise to identical daughter cells. The whole sequence of events bears all the hallmarks of a tightly controlled system. Hence, all daughter cells are the same size, DNA replication is initiated at the same specific cell mass (known as the initiation mass), cell division takes place when the cell mass has doubled and takes place exactly in the centre of the cell. Four main phases can be recognized during the bacterial growth cycle: cell growth, DNA replication, DNA segregation and cell division. The length of the growth cycle depends on many factors including the particular species involved, the availability of nutrients and environmental factors. In the case of *E. coli*, it can be as short as 20 minutes. DNA replication, DNA segregation and cell division together constitute what is known as the 'bacterial cell cycle' – a subject which has been studied extensively in only a few species including, naturally, *E. coli*. These three processes are not simply three phases of one complex operation but are three independent processes that are tightly coordinated.

DNA replication is a complex process involving more than 30 different proteins. Initiation of replication occurs at a specific region on the chromosome known as *oriC* (origin of replication) which is 245 bp in length and is located at 84 minutes of the *E. coli* chromosome. At this point, the complementary strands of the DNA molecule are separated to allow the entry and assembly of the proteins involved in replication. The key protein involved in initiation is DnaA and its concentration is critical in determining when the process commences. Hence, overproduction of the protein stimulates DNA synthesis and decreases the cell mass at which initiation takes place. The protein binds to four 9 bp sequences at *oriC* (known as *dna* boxes) forming a complex containing up to 40 molecules of the protein as well as five molecules of a histone-like protein. This induces the opening of the DNA helix at AT-rich sequences in *oriC*, so allowing the ingress of other replication proteins including DnaB helicase (to unwind the DNA in front of the replication fork) and DNA polymerases. The DNA polymerases move away from the fork in both directions, synthesizing DNA at a rate of 1000 nucleotides per second. Immediately after initiation there is a period of time during which reinitiation from the newly formed *oriC*s is prevented. The explanation of this is as follows. The adenines in GATC sites in mature DNA of *E. coli* are all methylated. However, in newly replicated DNA only the old strand will be methylated – methylation of the new strand takes a few minutes. This hemimethylated DNA is unable to replicate. At *oriC*, however, the methylation process is delayed considerably and it remains in the hemimethylated state for approximately 40% of the cell cycle. This delay in methylation is probably due to binding of the *oriC* to the outer membrane – a hypothesis supported by the isolation of DNA–outer membrane complexes and the finding that outer membrane material can bind specifically to hemimethylated *oriC*. Once this refractory period has passed, initiation of the next round of replication is controlled by two main factors: the concentration of DnaA and the activation of *oriC*. It would appear that once DnaA has been used to initiate one round of replication, it cannot be used for another – fresh DnaA must be synthesized. This loss of activity could be the result of some conformational change in the molecule triggered by its involvement in initiation or it may be inactivated by binding to the membrane. There is evidence in support of both of these suggestions. If initiation is dependent, in part, on the synthesis of new DnaA, what controls this process? It has been shown that guanosine tetraphosphate, the concentration of which is inversely proportional to growth rate, inhibits one of the promoters of the *dnaA* operon. This means that at high growth rates more DnaA will be produced. Activation of *oriC* appears to be under the control of the promoters of two neighbouring genes: *gid* and *mioC*. Transcription from both P*gid* and P*mioC* is inhibited by guanosine tetraphosphate. These findings enable a link to be established between the growth rate and the frequency of initiation of replication. At fast growth rates the guanosine tetraphosphate concentration will be low, hence expression of *dnaA*, *mioC* and *gid* genes will be increased and the initiation frequency will also increase. Replication terminates at a 450 kb region (*ter*) directly opposite *ori* which contains 'traps' for the replication forks can enter but not leave this region. One of these traps, T1, prevents anticlockwise-moving forks from passing, while another, T2, inhibits the passing of clockwise-moving forks. Both require the presence of the 36 kDa Tus (termination utilization substance) protein which inhibits unwinding of the DNA.

Once the replication forks meet at the *ter* site in the chromosome, terminus region decatenation of the daughter chromosome occurs. This is mediated by DNA gyrase and topoisomerase IV and takes place within a few minutes of the replication of the terminus region. There is, therefore, essentially no G_2 phase in the bacterial cell cycle (see section on the mammalian cell cycle). The DNA molecules are then partitioned into those regions of the cell destined to become daughter cells. This partitioning is mediated by MukA (from the Japanese word *mukaku* meaning anucleate) and MukB. MukB is a 177 kDa protein (the largest protein found in *E. coli*) with a secondary structure similar to that of the myosin heavy chain of eukaryotic cells. It can bind to DNA, has ATP/GTP binding activity and the structural features appropriate for a molecule designed to move molecules/structures around inside a cell, implying that it could be part of the chromosome partitioning apparatus.

Cell division involves the formation of a structure known as a septum at the mid-point of the cell. The septum is an invagination of the cell wall and cytoplasmic membrane which ultimately divides the parent cell into the two daughter cells, and its formation is directed by a cytoskeletal element known as the Z (or FtsZ) ring. The FtsZ protein is central to the assembly of this division apparatus and this is a tubulin-like protein capable of polymerizing to form filaments. An oligomer of FtsZ binds to a membrane protein, ZipA, which localizes at the mid-point of the cell and then polymerizes to form an annular structure on the internal face of the cytoplasmic membrane. A whole series of proteins necessary for division then accumulate at the Z ring including FtsA, FtsI, FtsQ, FtsL, FtsW, FtsK and FtsN. The function of many of these has not been determined although FtsI is known to be involved in peptidoglycan synthesis. Following assembly of these division proteins, ingrowth of the septum commences; the Z ring remaining associated with its leading edge. The complete septum is formed within approximately 10 minutes of the segregation of the daughter chromosomes to opposite ends of the cell. As *E. coli* has three potential septation sites – one central and one at each pole – there has to be a system to ensure that septation occurs only at the mid-point. This involves products of the *minC*, *minD* and *minE* genes which function as inhibitors of septation. MinE directs the septation-inhibiting complex (comprising MinC and MinD) to the polar sites, so preventing septation at these regions. Septum formation results in the production of a double layer of peptidoglycan across the middle of the dividing cell and this is split apart by the EnvA protein. This is the end of the process for Gram-positive bacteria as the cell envelope is comprised of the cytoplasmic membrane and the peptidoglycan. In contrast, Gram-negative bacteria have an additional layer, the outer membrane, in their cell wall. Invagination of this structure occurs as the separation of the double peptidogycan layer takes place.

EUKARYOTIC CELL STRUCTURE

Having reviewed certain key aspects of the cell and molecular biology of bacteria attention will now turn to the eukaryotic cells. The most obvious difference between bacteria and eukaryotic cells is size. Bacteria are normally in the size range 0.1 to 1μm in diameter and 1–5 μm in length. However, some extremely large bacteria are found, such as *Epulopiscium fishelsoni*, related to the Gram-positive genus *Clostridium*, which can reach a size of 80x600 μm. This is much larger than the average cell in a multicellular organism. However, certain eukaryotic cells have evolved much larger sizes and, for example, the amoeba *Amoeba proteus* has 100 000 times the volume of *E. coli* and can, when fully extended, be more than 1 mm in length. The physical size of the eukaryotic cell is also reflected in its DNA content (Table 2.3). The range of genome size (measured in numbers of nucleotide base pairs) in bacteria is from 0.6 to 10 million base pairs (Mbp). In eukaryotic cells the range is from 6 to >3000 Mbp. However, much of the DNA in eukaryotic cells is non-coding and certain flowering plants, molluscs, amphibians, insects and cartilaginous fishes have larger genomes than mammals. The relationship between genome size and cell behaviour is not clear. It should also be appreciated that 5–10% of the genome of eukaryotic cells is composed of mobile retrovirus-like elements known as retrotransposons and which has been called 'parasitic DNA'.

The principal feature which distinguishes a eukaryotic cell from a prokaryotic cell is the presence of a nucleus with its associated nuclear membrane. It is not clear when the nucleus evolved and what it evolved from and this problem will be briefly described in the section on the nucleus. Other major differences which distinguish eukaryotic cells from bacteria probably evolved in response to the size constraints of the former. With increasing size comes problems with the transport of molecules by simple diffusion. This presumably led to the evolution of compartmentalization within the eukaryotic cell. The nucleus is one example of this compartmentalization of cellular function. Other organelles in eukaryotic cells include the mitochondria, chloroplast, peroxisome and the lysosome. The mitochondrion will be discussed in the next section as there is now good evidence that it originated as an

Table 2.3 DNA content of bacteria and Eukarya

Organism	Haploid DNA content (million base pairs)
Bacteria	
Mycoplasma genitalium	0.6
Escherichia coli	4.6
Myxobacteria	10
Archaea	
Methanococcus jannaschii	1.8
Archaeoglobus fulgidus	2.2
Eukarya	
Spraguea lophii (protozoan)	6.2
Saccharomyces cerevisiae (unicellular)	14
Dictyostelium discoideum (unicellular)	70
Euglena (unicellular)	3000
Caenorhabditis elegans (multicellular)	100
Drosophila melanogaster	165
Mouse	3000
Human	3000

intracellular bacterial symbiont (an endosymbiont). The peroxisome has also been suggested to have been derived from an endosymbiotic bacterium, and the evidence will be briefly aired. The lysosome is part of an extensive intracellular vesicular apparatus, part of whose function is to kill pathogens. The role of the lysosomal apparatus in the killing, and in the nurturing, of intracellular bacteria is described in detail in Chapter 6. With the increased size of the eukaryotic cell has come the problem of targetting the proteins to the correct destination. Proteins can either be for intracellular use of they can be for export. If they are required within the cell then they may be destined for the nucleus, the mitochondria, the lysosomes etc. Two cytoplasmic organelles, the endoplasmic reticulum (ER) and the Golgi apparatus (named after its discoverer), have evolved to transport proteins to specific addresses within the cell. The ER is an extensive network of intracellular membranes stretching from the nucleus to the plasma membrane on which proteins are translated and lipids are synthesized. Protein synthesis is a key target of certain bacterial toxins and many toxins utilize the protein-sorting apparatus of the cell to get to their destinations. Examples of toxins targetting protein synthesis are diphtheria toxin, Shiga toxins and *Ps. aeruginosa* exotoxin A. The cellular mechanisms of bacterial exotoxins are described in detail in Chapter 7.

While bacteria have flagella to allow them to move, eukaryotic cells have evolved a complex arrangement of intracellular proteins, some of which are reversibly polymerizing proteins, which form a cytoskeleton and can be used by the cell to provide a motive apparatus. More importantly, this complex system of protein filaments and fibres, which extends through the cytoplasm, provides a structural framework for the cell, determining the shape and organization of the cytoplasm. It is also responsible for intracellular transport and positioning of the cellular organelles and is an integral part of the mechanism of the movement of the chromosomes during cell division. The cytoskeleton is now recognized to be a key target for bacteria in their attack on eukaryotic cells and both type III secretion systems and some of the most powerful bacterial exotoxins have evolved to manipulate this complex central element of eukaryotic cell biology.

A particularly fascinating aspect of modern cell biology is the concept that certain of the intracellular structures in eukaryotes are derived from endosymbiotic bacteria. This has been termed the serial endosymbiosis theory and proposes that mitochondria and chloroplasts were originally bacteria. In addition, peroxisomes and the eukaryotic flagellum/cilium, which is also termed the undulipodium, are also thought to have evolved from bacteria. If this can be verified then it provides a completely different perspective about the past evolutionary relationships between bacteria and nucleated cells and may have important consequences for our future understanding of the interactions between these two great domains of life. Thus, at the very heart of the topic of cell biology lies the most intimate interaction possible between bacteria and eukaryotes.

PROKARYOTES WITHIN EUKARYOTES: MITOCHONDRIA, PEROXISOMES and CENTRIOLES

The concept that cell organelles arose from symbiotic associations with bacteria can be traced back to the end of the nineteenth century. However, it has been the American biologist, Lynn Margulis, who has pioneered and driven this field forward in the face of scepticism and even derision.

Mitochondria

Strong evidence now exists to support the hypothesis that mitochondria and chloroplasts arose from free-living bacteria. The static structures of mitochondria as they appear under the electron microscope is a common image reproduced in textbooks. However, time-lapse photography reveals the ability of mitochondria to move through the cell, to change shape and even to fuse with other mitochondria and undergo division. Mitochondria are never produced *de novo* but always arise from the division of pre-existing mitochondria. This division process resembles that seen in bacteria. Within the cytoplasm they are often found associated with microtubules which play a role in determining their distribution and orientation.

Mitochondria resemble bacteria and are surrounded by a double membrane system, consisting of inner and outer mitochondrial membranes, separated by an intermembrane space and containing some 300–400 proteins. The inner membrane, with its multiple folds or cristae, is the site of the electron transport chain central to aerobic respiration. Mitochondria, as predicted from the endosymbiosis hypothesis, contain circular DNA molecules like those found in bacteria. The size of these circular DNAs in animals is small – around 16–18 kb – although larger mitochondrial genomes are found in yeast (80 kb) and plants (200–2000 kb). The size range of mitochondrial genomes is from 6 kb in *Plasmodium* spp to 2.5 Mb in the flowering plant *Cucumis melo* – the musk melon. In the case of these larger genomes, much of the DNA is non-coding. For comparison, the genome size of the smallest known free-living pathogen *Mycoplasma genitalium* is 580 kb. In human mitochondria the genome has been sequenced and contains 16 569 bp, which form 37 genes encoding 13 proteins, 12S and 16S ribosomal RNAs and 22 transfer RNA molecules (Figure 2.7). These tRNAs are required for the translation of the mRNAs encoded by the mitochondrial genome. The bacterial-like behaviour of the translation of mitochondrial mRNA into protein is seen by its susceptibility to antibiotics that affect bacterial protein synthesis and by the fact that polypeptides start with formylmethionine. Mitochondrial genomes contain some intriguing molecular surprises. For example, prokaryotic and eukaryotic genomes encode 32 or more distinct tRNA molecules while mitochondria encode only 22. The reader will be aware that in DNA/RNA there are 64 possible triplet codons of which 61 encode the 20 different amino acids which produce proteins. Many tRNAs in prokaryotes and eukaryotes are able to recognize more than one single codon in the mRNA because of a process called 'wobble' which allows some mispairing between the tRNA anticodon and the third position of certain complementary codons. This 'wobble' means that at least 30 tRNA molecules are necessary to translate the universal code. Mitochondria, having only 22 tRNAs, utilize an extreme form of the 'wobble' rules in which uracil (U) in the anticodon of the tRNA can pair with any of the four bases in the third codon position of the mRNA, allowing four codons to be recognized by a single tRNA. Moreover, some codons specify amino acids in mitochondria that differ from the accepted universal code.

Mitochondrial DNA has a high mutation rate, estimated to be 5–10 times that of nuclear DNA. It also displays maternal inheritance. The latter is the case because the mitochondria of

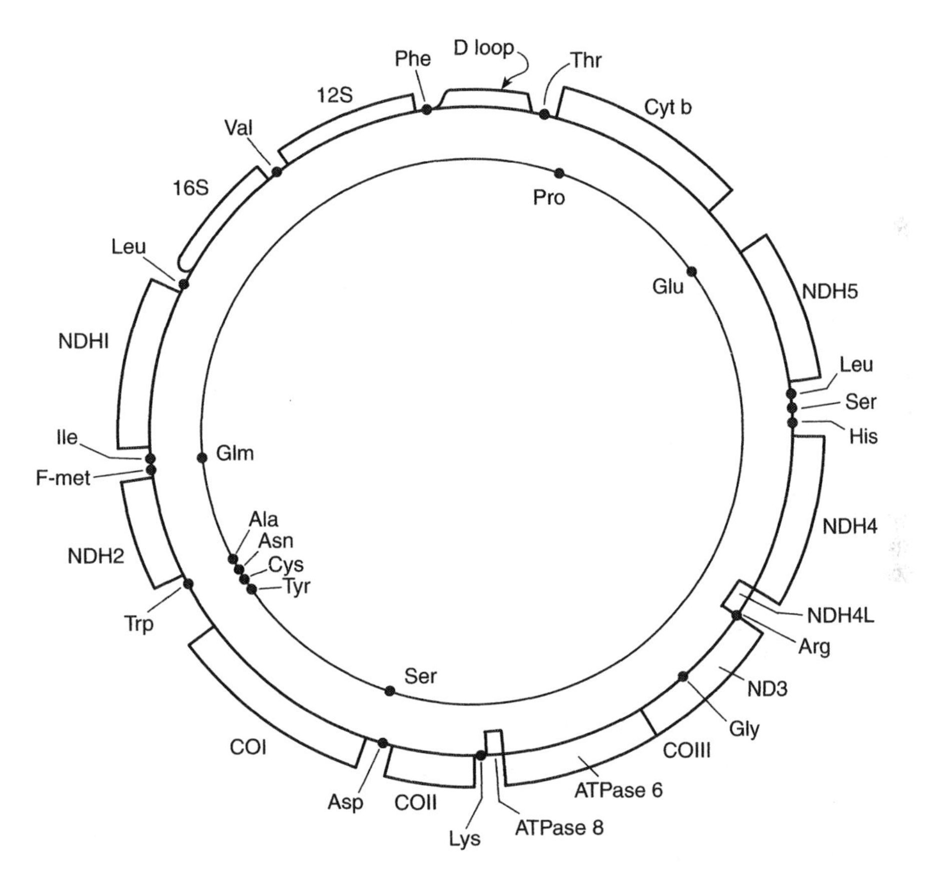

Figure 2.7 Genetic map of the circular chromosome of the human mitochondrial genome which contains 16 569 base pairs NDH1 5. The genome contains the genes for 12S and 16 S rRNAs and the filled circles indicate the positions of the tRNA genes. The remaining genome encodes a variety of proteins (NDH1 5 are components of the NADH dehydrogenase complex and COI–III are subunits of the cytochrome oxidase complex

the sperm are not retained after fertilization. The study of mitochondrial DNA has recently revealed that mutations in the mitochondrial genome can be responsible for certain diseases. These conditions can only be inherited from the individual's mother. A number of such conditions have now been recognized which affect tissues requiring large amounts of mitochondrial activity. These include Leber hereditary optic neuropathy and Kearns–Sayre syndrome, which also involves the optic nerve and includes ataxia, heart disease and hearing loss. The inheritance of these conditions do not show the regular Mendelian patterns and there can be a wide variation in the type and severity of symptoms in different offspring from the same mother.

It has also been hypothesized that the high mutation rate in mitochondrial DNA, which reflects a combination of a mutagenic environment (due to free radicals and other noxious factors produced by mitochondria), an inadequate DNA repair system and the consequence of the paucity of non-coding segments (such that chance mutations will involve coding DNA), could implicate the mitochondrion in the normal ageing process. This would particularly be true for the mitochondria in neurones. Mitochondrial DNA has leapt to prominence in the last decade because it is now being used by anthropologists to trace the origins of *Homo sapiens*. This has given rise to the 'Out of Africa' hypothesis and the concept of the African Eve – the mother of us all.

The majority of the genes required for mitochondrial function, including those required for DNA replication, transcription and translation, are contained within the nucleus of the cell. These presumptive bacterial genes are transcribed and translated on free cytosolic ribosomes and the proteins are imported into the mitochondria as intact polypeptide chains. Thus, at some time in the past the majority of the mitochondrial genome must have become incorporated into the genome of the host. It is not known how many of these bacterial genes may have been lost in this process. However, the number of mitochondrial proteins can be compared with the genome of *Myco. genitalium*, an intracellular pathogen capable of being cultivated outside of the cell, which has recently been shown to have a genome of only 580 kb with 470 predicted coding sequences. This is very similar to the number of genes believed to encode the mitochondrion. The evolutionary pressure which retains the 37 genes in the human mitochondrial genome is unclear given that this only represents some 10% of the genes required to produce a mitochondrion. As the mitochondrion is the product of both mitochondrial- and a large number of nuclear-encoded genes – there must be a mechanism for coordinating the transcription of these multiple genes. If this were to be controlled by the mitochondrion then all nucleated cells would be, to a limited extent, constantly under the direct control of a prokaryote.

What was the bacterium (or bacteria) that was taken up by eukaryotic cells and eventually turned into the mitochondrion? It was proposed in the 1970s, from biochemical evidence, that non-sulphur purple bacteria (Proteobacteria) are the closest contemporary bacterial relatives of mitochondria. This proposal has received support from ribosomal RNA sequencing and the identification has been narrowed to one of the two major subdivisions of the α-proteobacteria – which contain the obligate intracellular parasitic genera *Rickettsia*, *Ehrlichia* and *Anaplasma*.

Peroxisomes

The peroxisome is another interesting intracellular organelle. Peroxisomes are single membrane-bound organelles which do not contain DNA or ribosomes but do contain at least

50 different enzymes. These organelles are so named because they contain one or more enzymes which utilize molecular oxygen to remove hydrogen from specific organic substrates and in doing so produce hydrogen peroxide. The enzyme catalase then uses this hydrogen peroxide to oxidize a variety of other substrates by a 'peroxidative' reaction. In the liver and kidneys the peroxisomes are involved in detoxifying toxic molecules that enter the bloodstream. Another major function of the oxidative reactions carried out by peroxisomes is the β-oxidation of fatty acids, a process which produces acetyl CoA for biosynthetic reactions.

Peroxisomes are morphologically similar to lysosomes. However, one aspect of the behaviour of peroxisomes has suggested that they may, like the mitochondrion and chloroplast, have come from some prokaryotic ancestor. Peroxisomes are assembled, like mitochondria and chloroplasts, from proteins that are synthesized on free ribosomes and then imported into the peroxisome as full-length proteins. Another similarity to mitochondria and chloroplasts is that peroxisomes are not formed *de novo* but replicate by division. The Nobel Prize winner Christian DeDuve, who discovered the lysosome, has proposed that peroxisomes were acquired as microbial symbionts. It is of interest that *Giardia lamblia,* one of the most primitive eukaryotes, has neither mitochondria nor peroxisomes.

Centrioles

The centriole, which is found in the centrosome adjacent to the nucleus, is the major centre for microtubule organization in cells. During mitosis, the centrosome is duplicated and becomes the centre for the microtubular organization responsible for chromosomal movement. The centrosomes of most animal cells, but not of plant cells, contain a pair of centrioles, oriented perpendicular to each other. The centrioles are cylindrical structures consisting of nine triplets of microtubules, similar to the basal bodies of cilia and flagella (undulipodia). The role of the centrioles, which used to be thought of as organizing the microtubular apparatus in mitosis, is currently uncertain. Like mitochondria and peroxisomes the centrioles arise from the duplication of pre-existing centrioles.

Lynn Margulis has argued that the centriole is derived from bacteria known as spirochaetes. Spirochaetes are widespread, both in aquatic environments and as commensal and pathogenic organisms. Perhaps the best-known species is *Treponema pallidum,* the causative agent of syphilis. Spirochaetes are typically slender, sinuous, helical-shaped bacteria which have a unique method of locomotion in which the cell itself is the engine, rather than utilizing flagella. Fibrils, referred to as axial fibrils or filaments, are attached to either pole of the cell and are wrapped around the coiled protoplasmic cylinder. Both the axial fibrils and the protoplasmic cylinder are surrounded by a three-layered membrane called the outer sheath of the cell envelope. These axial fibres, in terms of protein structure and ultrastructure, are similar to those of bacterial flagella. Each fibre is anchored at one end and extends for about two-thirds of the length of the cell. The axial fibrils rotate rigidly, as do bacterial flagella. Because the protoplasmic cylinder is also rigid, whereas the outer sheath is flexible, when both axial fibrils rotate in the same direction the protoplasmic cylinder rotates in the opposite direction. This mechanism enables spirochaetes to move both through fluids and on solid substrata (Figure 2.8).

Spirochaetes in aquatic environments and in commensal niches (e.g. in the gut of termites) are found attached to the surface of protozoa and can act to move these eukaryotic cells.

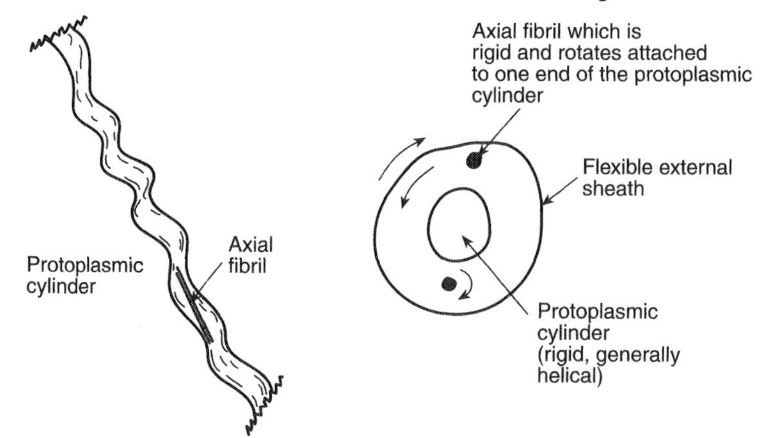

Figure 2.8 Diagram showing, on the left, the general shape of a spirochaete indicating the position of the axial fibril. The cross-section of the spirochaete on the right shows the arrangement of the protoplasmic cylinder, axial fibrils and external sheath and the relationship between them, which results in rotation of the protoplasmic cylinder and the opposite rotation of the external sheath

Margulis and colleagues have argued from this so-called motility symbiosis of spirochaetes that centrioles are the evolutionary result of an early symbiosis between a eukaryote and a spirochaete. Obviously, much more experimentation will be required before this hypothesis can be properly tested.

We have little information on the evolutionary development of the eukaryotic cell and it is only within the last 20 years that the revolutionary concepts of Margulis and coworkers on the endosymbiotic source of mitochondria and chloroplasts have received experimental backing and general support from the scientific community. How many more of the complex systems employed by eukaryotes had their derivation in the domains Bacteria or Archaea is unknown. What do these findings mean for our understanding of bacteria–eukaryote interactions? It is too early to say. Can extracellular or intracellular bacteria communicate with mitochondria, or the mitochondral genes in the nucleus, or have any effect on their functions, or vice versa? The mitochondria, being the 'powerhouse' of the cell, could be an excellent target for pathogens. However, as yet, little information is available to answer this question.

THE PLASMA MEMBRANE

The cytoplasmic and outer membrane of the bacterium has been briefly described. In eukaryotic cells the plasma membrane serves similar functions to the bacterial outer membrane, being the container of the cells contents, the barrier to ingress and egress of molecules and the site of the reception of incoming signals. Eukaryotic cells also have, unlike bacteria, many internal membranes in the form of the nuclear membrane, endoplasmic reticulum (ER), Golgi apparatus etc., which will be described in later sections of this chapter. The plasma membrane and other internal membranes take the form of a lipid bilayer in which are embedded a variety of proteins. The composition of the lipid bilayer controls its fluidity and flexibility and this can be altered depending on diet and the cell's requirements. There are four major phospholipids in the plasma membranes of the mammalian cell: phosphatidylcholine, phosphatidylethanolamine, phosphatidylserine and sphingomyelin. In addition to these phospholipids the plasma membranes also contain glycolipids and cholesterol. These lipid components are not distributed symmetrically within the plasma membrane and it is established that the outer layer of the lipid bilayer contains mainly phosphatidylcholine and sphingomyelin with the glycolipids (whose sugar residues are exposed on the cell surface). Some of these glycolipids act as receptors for bacteria or for bacterial exotoxins. The inner layer of the membrane consists predominantly of phosphatidylethanolamine and phosphatidylserine. An additional phospholipid, phosphatidylinositol (which, as will be described in Chapter 3, is involved in cell signalling events) is also localized to the inner half of the membrane. Cholesterol is found in both halves of the membrane.

Only about half the mass of the typical mammalian cell plasma membrane is lipid. The remainder is protein with a small component (5–10%) being the carbohydrate components of glycolipids and glycoproteins and other carbohydrate-containing molecules. On a molar basis it is estimated that there are about 50–100 lipid molecules for every molecule of protein. Our current view of the biological membrane is that it is a two-dimensional fluid in which are inserted protein molecules. This is the the so-called *fluid mosaic model* which was proposed by Singer and Nicolson in 1972. In this model two classes of membrane-associated proteins were defined: peripheral and integral membrane proteins. The former can be isolated without disruption of the lipid bilayer (e.g. by use of high salt or low pH) and are not inserted into the hydrophobic interior of the membrane. In contrast, integral membrane proteins are inserted within the hydrophobic interior of the lipid bilayer and can only be released by detergents. Many integral proteins completely span the lipid bilayer and have domains on both sides of the membrane. These are known as transmembrane proteins and such proteins are important in, for example, cell signalling.

Membrane proteins can be subdivided into four major functional classes, although it is obvious that there are overlaps in this classification. Many of these proteins are involved in the processes of cell signalling and will be reviewed in detail in Chapter 3. The first functional class are proteins involved in transport. A good example of this are transmembrane ion transporters which are involved in, for example, pumping sodium out of and potassium into cells. One such protein is the cystic fibrosis transmembrane conductance regulator (CFTR), a member of the large ABC (ATP-binding cassette) family which regulates chloride ion secretion across the plasma membrane. In cystic fibrosis mutation of the gene encoding this protein results in its failure to insert correctly into the plasma membrane. Sufferers have defective Na^+/Cl^- transport with elevated levels of Na^+ in extracellular secretions. In the lungs this results in inhibition of antibacterial defence mechanisms and recurrent infections of the lungs with bacteria such as *Ps. aeruginosa* (see Chapter 8 for more details). Some ion channels function as receptors. Such receptors are normally referred to as transmitter-gated ion channels, as binding of the ligand (e.g. acetylcholine) to the receptor (in this case the acetylcholine receptor) causes a conformational change that opens or closes the ion channel.

A second functional class of plasma membrane proteins are enzymes. A good example of a membrane-associated enzyme which is also involved in cell signalling is adenylate cyclase, which converts ATP to the signalling molecule cyclic adenosine monophosphate (cAMP). Another example are the group of receptors which also have tyrosine kinase activity and are consequently known as receptor tyrosine kinases (RTKs). In both of these cases the enzyme activity is expressed on the inside of the cell. However, there are membrane-associated enymes in which the catalytic site is on the exterior of the cell. An interesting example is the zinc-dependent metalloproteinase known variously as neutral endopeptidase, neprilysin, enkephalinase or common acute lymphoblastic leukaemia antigen (CALLA, also CD10 – see Chapter 9 for information on CD nomenclature). This enzyme cleaves peptides on the cell surface and is believed to downregulate cellular response to peptide signals. Inhibition of this enzyme on polymorphonuclear leukocytes (PMNs) results in the ability of these cells to detect a chemotactic gradient of the bacterial chemoattractant, F-met-leu-phe, being inhibited. Knockout of this enzyme in mice results in animals which are extremely sensitive to the effects of Gram-negative bacterial lipopolysaccharide (LPS). This suggests that this enzyme plays a role in desensitizing mice to the effects of bacterial LPS. A number of other zinc-dependent metalloproteinases are also found on the plasma membrane.

A third group of plasma membrane-associated proteins are the various receptors for hormones, cytokines, lipid mediators etc., which are responsible for the complex intracellular signalling pathways that will be described in the next chapter. Such molecules can either function to bind their ligand and then the signalling can be produced by some additional protein (or proteins) or, as in the case of the RTKs, the receptor has inherent enzymatic (kinase or phosphatase) activity.

The final functional class of plasma membrane-associated proteins are also receptors but those that function to link cells together. One example is the group of transmembrane proteins called the β_1-integrins which function to bind cells to the extracellular matrix and which are also used as receptors by various microorganisms. At cell–matrix junctions, focal adhesions and hemidesmosomes (see next section), the integrins also make contact with components of the cytoskeleton. This provides a stable linkage between the cell and its external connective tissue matrix. In addition to providing a molecular glue, the integrins can also function to carry signals into the cell via a non-receptor tyrosine kinase called FAK (focal adhesion kinase) which eventually leads to activation of Ras and the MAP kinase cascade (see Chapter 3 for more details). Other plasma membrane receptors for extracellular matrix components include CD44 which binds to hyaluronan, the β_2-integrins which can bind to fibrinogen and CD29. A large group of variegated receptors which are grouped together through use of the CD nomenclature also function as cell–cell receptors acting to link cells together. In Chapter 9 the role of cell surface proteins such as CD4, CD8, CD80 and MHC class I in immune responsiveness will be described. To give the reader an idea of how many such cell surface receptors there are, the recently published book entitled *The Leukocyte Antigens Facts Book* provides potted details on the proteins CD1 to CD166 and more have been discovered and named since this book was published in 1997.

In addition to proteins, the plasma membrane also contains a variety of carbohydrates. These are associated with glycolipids and glycoproteins and generally the chains of the carbohydrates are short and are termed oligosaccharides. Other proteins contain one or more longer carbohydrate chains (e.g. syndecan and betaglycan) and are called proteoglycans. Other heavily glycosylated plasma membrane proteins are termed mucins. Episialin is one such mucin and like the proteoglycan, syndecan, is involved in the process of tumorigenesis. All of these carbohydrates are found on the exterior side of the plasma membrane and form a 'sugar coating' on cells which has been termed the glycocalyx. As will become clear in later chapters, this glycocalyx is important in the recognition of animal cells by bacteria and in the adherence of bacteria to such cells. Recognition of cell-bound carbohydrate also occurs in the inflammatory process when PMNs are recognized by activated blood vessels which bind to particular carbohydrate residues on the plasma membrane of these leukocytes (see Chapter 8).

The plasma membrane is only two molecules thick, lacks mechanical rigidity, and therefore has to be strengthened. This is done in most eukaryotic cells by a framework of proteins attached to the membrane via transmembrane proteins. These proteins form a meshwork of fibres which is called the cell cortex and can allow cells to change their shape. The nature of this cell cortex will be the subject of the next section.

THE CYTOSKELETON

During the last decade a significant proportion of the literature which now forms the basis of cellular microbiology has been concerned with the effects of bacteria (and viruses) on the eukaryotic cytoskeletal apparatus. A number of bacterial toxins (see Chapter 7) have been found to target Rho, Rac or Cdc42 – members of the Ras family of small guanine triphosphatases (GTPases) and intimately involved in the control of the actin cytoskeleton (see Chapter 3). These toxins can have a wide range of effects including the inhibition of macrophage phagocytosis, cell cycle progression and, of course, cytotoxicity. Bacterial pathogens such as *Listeria, Salmonella, Shigella* and *Yersinia* are able to enter non-phagocytic cells by rearrangement of the host cell's actin cytoskeleton. Moreover, bacteria such as *Lis. monocytogenes, Sh. flexneri* and some Rickettsiae have the ability to manipulate the actin cytoskeleton to move within host cells. Thus it is clear that bacteria have evolved to modulate the cell cytoskeleton as a major pathogenic mechanism.

The composition and nature of the cytoskeleton

Time-lapse photography of cells reveals that they have significant plasticity, undergoing constant change of shape and with their plasma membranes being able to produce a wide variety of protrusions (called lamellipodia, filopodia, pseudopods, ruffles etc.). In addition, most cells in culture have some capacity to move. These visible changes, and the less visible movements of organelles and other structures within cells, are due to a dynamic set of intracellular proteins which are described as the cytoskeleton. This consists of a network of protein filaments extending throughout the cytoplasm of all eukaryotic cells and serving as a structural framework or scaffold that determines cell shape, cell movement and the general organization of the cytoplasm. The cytoskeleton is also involved in the complex processes of mitosis and cytokinesis. It is important to remember that, unlike the skeleton of the body, which is rigid, the cytoskeleton is a dynamic structure which is continually reorganizing to meet the ever-changing needs of the cell. The cytoskeleton is a composite of three major and distinct protein filaments – intermediate filaments, actin filaments (or microfilaments) and microtubules – which may be linked together or joined to subcellular organelles or to the plasma membrane by a variety of accessory proteins (Figure 2.9). Although these three distinct filaments will be discussed in turn, within the cell these fibril-forming proteins interact closely. Microbes (bacteria, viruses, protozoa) are known to interact with the actin filaments or microfilaments of the cell in many and varied ways. Less is known about microbial interactions with other cytoskeletal proteins, although it is likely that some may turn out to be targets for microbial action.

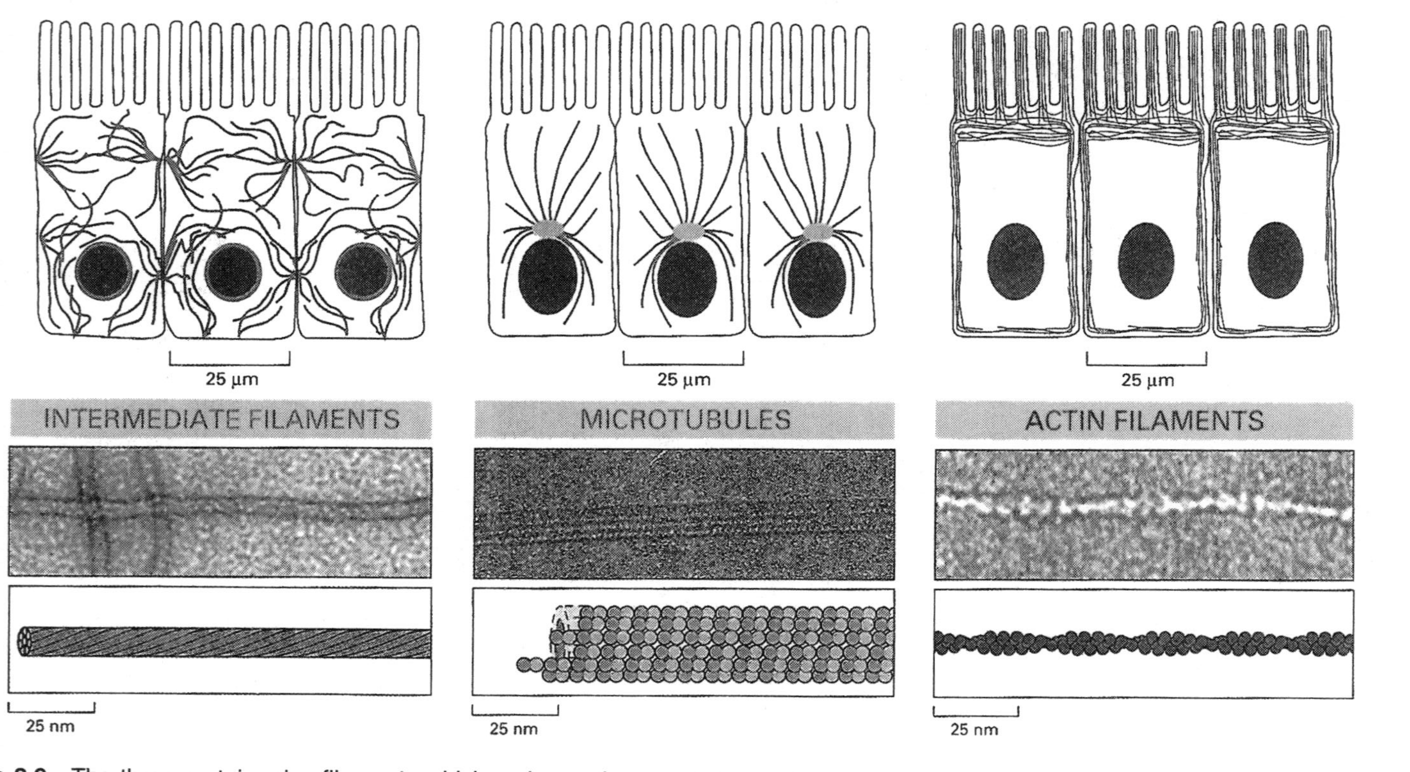

Figure 2.9 The three protein microfilaments which make up the cytoskeleton of eukaryotic cells. As described in the text, the intermediate filaments are around 10 nm in diameter and form into rope like fibres that function to cope with the mechanical stresses that cells are exposed to. In contast, microtubules, which form from the protein tubulin, are long straight fibres which function in organellar movement and in cytokinesis. As shown in this diagram microtubules generally have one end attached to a single microtubule-organizing centre known as the centrosome. The third form of microfilament is the actin filament or microfilament – helical polymers composed of the protein actin. Interactions between actin filaments and actin-binding proteins results in these microfilaments performing different functions in the cell. In this diagram the actin filaments are shown forming microvilli. (Reproduced from Alberts *et al.* (1998) *Essential Cell Biology*. Garland, New York, with permission)

Intermediate filaments

Intermediate filaments are so called because their 10 nm diameter is intermediate between that of microfilaments (actin filaments), which are 6 nm in diameter, and microtubules, which have a diameter of 23 nm. These protein filaments exhibit great tensile strength and their main function is to enable cells to withstand the mechanical stress that occurs when they are stretched. One immediately thinks of muscle cells stretching. However, cells in epithelia, in mucosal layers, blood vessels, lungs etc. are constantly being exposed to mechanical stress. In *Homo sapiens*, for example, there are more than 50 different genes coding for intermediate filaments. These proteins can be subdivided into a number of groupings: keratins, vimentin and vimentin-related proteins, neurofilaments and nuclear lamins. Intermediate filaments form an elaborate network in the cytoplasm of most cells, extending from the nucleus, where they form a ring, all the way to the plasma membrane. In epithelial cells, keratin filaments are tightly anchored to the plasma membrane at two areas of specialized cell contact termed desmosomes and hemidesmosomes. Desmosomes are junctions between adjacent cells, at which cell–cell contacts are due to proteins termed cadherins. Hemidesomsomes link epithelial cells to their underlying connective tissue base. At the present time the evidence implicating intermediate filament as having a role in infectious diseases (such as will be described for the actin filaments) is minimal. The major effect is providing epithelial cells with a mechanism for linking themselves together and thereby excluding bacteria from the body's interior. The oral bacterium *Treponema denticola* has been found to cause a decrease in the keratin content of cultured epithelial cells and the proportion of the cells expressing desmoplakin II, a protein involved in maintaining intercellular junctions, was also decreased. Zonula occludens toxin from *V. cholerae* has the ability to 'loosen' the tight junction complex in intestinal epithelial cells but appears to do so by targetting the actin cytoskeleton and the zonula occludens proteins. However, we would speculate that bacteria will have found ways of modifying the intermediate filaments.

Mutations in the genes encoding these intermediate filament proteins are now recognized to produce a number of blistering conditions of the skin and mucosa such as epidermolysis bullosa simplex which are due to the loss of mechanical strength of the keratinocytes in skin and to a decrease in the binding to the underlying connective tissue. These blistering conditions are obviously accentuated by concurrent infections. However, no mutations have yet been described which make sufferers directly more susceptible to infections.

Microtubules

In contrast to the numbers of intermediate filament proteins, microtubules are composed of a single form of globular protein called tubulin, which is a dimer consisting of two closely related 55 kDa proteins known as α-tubulin and β-tubulin. A third form, γ-tubulin, is specifically localized to the centromere of the cell. These tubulin dimers polymerize to form relatively long and stiff tubes which in cross-section generally consist of 13 linear protofilaments assembled round a hollow core; in other words these microtubules look very much like the scaffolding on building sites (Figure 2.10). Microtubules, like actin filaments (as will be described) are polar structures with two distinct ends. There is the fast-growing plus end and the slow-growing minus end. This polarity determines the direction of movement along microtubules. Microtubules are extremely dynamic structures undergoing rapid polymerization and depolymerization, depending on the cells' needs.

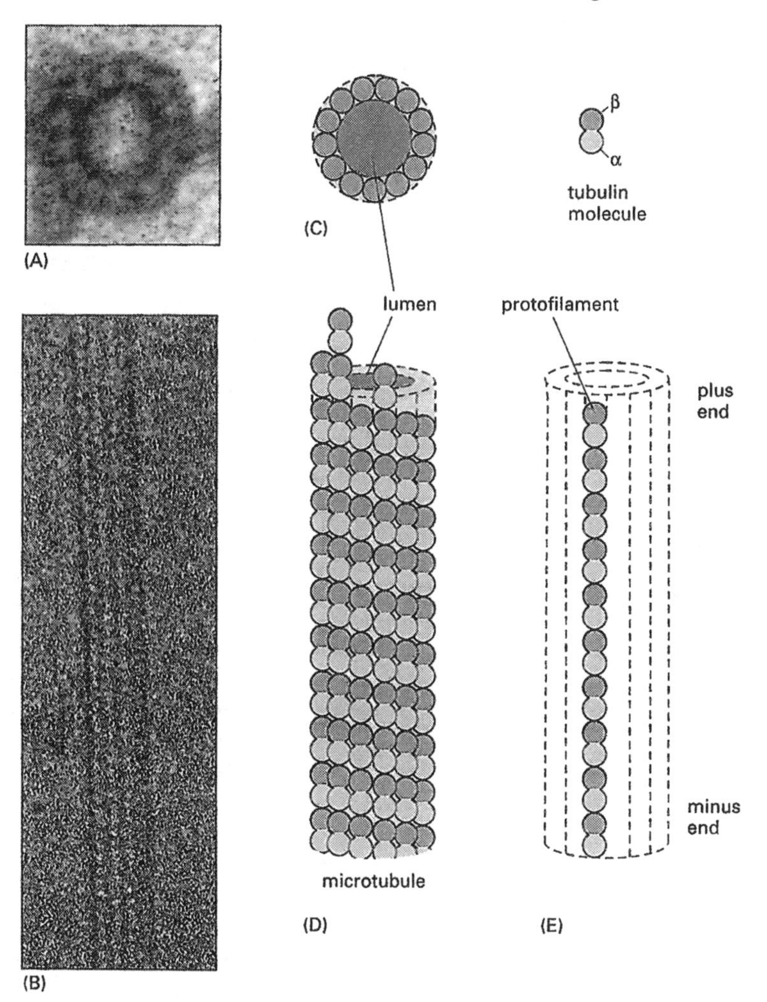

Figure 2.10 Structure of a microtubule. (A) An electron microscopic image of a microtubule showing the distinctive ring of 13 subunits, each of which is a separate tubulin dimer. (B) In this electron micrograph the microtubule is being viewed lengthwise. (C, D) In schematic form the packing of the tubulin molecules is indicated. (E) This shows one tubulin molecule and how it is formed into a polymer (protofilament). It should be noted that the tubulin molecules are all arranged in the same orientation, thus providing the microtubule with a structural polarity (plus end/minus end). (Reproduced from Alberts *et al.* (1998) *Essential Cell Biology*. Garland, New York, with permission)

Microtubules have less of a structural role than intermediate filaments, although stabilized microtubules are used to build cilia and flagella in eukaryotic cells. The major role played by microtubules is in the intracellular movement of cellular components and the best-known example is the movement of chromosomes through the cell at mitosis in order to form two daughter nuclei. Microtubules also transport and position membrane vesicles and organelles. In this process the best analogy is that the microtubules act like rails and the motive power or engine is produced by large ATP-hydrolysing proteins called kinesins and dyneins which, collectively, are called motor proteins. These are complex multi-subunit proteins which have an ATP binding domain which interacts with the microtubules and a tail which makes a stable attachment to cell components including vesicles, or organelles such as mitochondria. Kinesins generally move toward the plus end of microtubules (i.e. away from the centrosome, which is the major microtubule organizer in cells) and dyneins move toward the minus end. Microtubules and their associated motor proteins act in the control of the positioning of membrane-bound organelles within eukaryotic cells. This includes the positioning and shape of the ER and the Golgi apparatus.

Microtubules are highly labile and dynamic. To date there is no evidence to suggest that bacteria can interfere with microtubules, although plant microtubules are known to interact with the tobacco mosaic virus (TMV) protein, P30. Plants produce a variety of non-peptidic compounds which can interfere with microtubules. These include colchicine, an alkaloid from the meadow saffron which binds strongly to tubulin monomers and prevents their polymerization. Taxol, extracted from the bark of the yew tree, has the opposite effect. It binds tightly to microtubules and stabilizes them. These compounds, which interefere with microtubules, are used in cancer therapy as they are able to inhibit cell division.

The actin cytoskeleton

Actin filaments, and a growing number of associated structural and cell signalling proteins, are involved in many of the activities of cells, particularly those involving the cell surface. Actin filaments are necessary for the processes of cell locomotion, phagocytosis and cell division. As the former two activities are utilized by leukocytes to find and kill bacteria it may not be surprising that bacteria have evolved systems for manipulating the actin cytoskeleton.

Actin

Actin, which was first isolated from muscle in 1942, is now known to be a ubiquitous eukaryotic proteins that constitutes 5–10% of the total protein content of most cells. Yeasts encode only one actin protein but higher eukaryotes have several actin genes. For example, mammals have at least six actin genes, with four genes being expressed in different forms of muscle and the other two in non-muscle cells. Actins are highly conserved proteins, with yeast actin being 90% identical in amino acid sequence to the actins of mammalian cells. Actin monomers are clam shell-shaped globular proteins consisting of 375 amino acids (molecular mass 43 kDa) with binding sites that enable head-to-tail interactions with two other actin monomers allowing the formation of filaments. Polymerization of actin monomers requires ATP, which is trapped within the crevice made up by the two halves of the clam shell. The first step in actin polymerization (a process called nucleation) is the formation of a small aggregate consisting of three actin monomers. Within an actin filament, all actin monomers are oriented in the same direction, actin filaments having a distinct polarity, and their ends, called the 'plus' and 'minus' ends, are distinguishable from each other. Actin monomers are able to grow by the reversible addition of monomers at both ends of the filament. Like microtubules, the plus end elongates 5–10 times faster than the minus end. Each monomer is rotated by 166° in the filaments giving the appearance of a double-stranded helix. This polarity of actin filaments is important both in their assembly and in establishing a unique direction of movement of a second muscle protein, myosin, relative to the actin mirofilament. Hydrolysis of the bound ATP in actin filaments weakens the bonds between monomers and promotes depolymerization. A number of fungal toxins have been used to study actin polymerization. Cytochalasins bind to the plus end of actin filaments and prevent polymerization. Phalloidins are toxins from the *Amanita* mushroom which bind to and stabilize actin filaments. The mechanism of action of these toxins is similar to those of the microtubule 'toxins' colchicine and taxol.

Actin-associated proteins

Actin filaments are organized within cells into linear bundles, two-dimensional networks and three-dimensional gels. The ability of actin filaments to form such structures is dependent on a growing number of accessory proteins, some of which are shown in Table 2.4. The concentration of actin monomers within the cytosol is high (50–200μM), which is very much greater than the concentration at which monomers normally polymerize (around 1μM). There has to be some mechanism for keeping these actin monomers from spontaneously polymerizing. Two proteins have been found which carry out this inhibitory function. The first is thymosin, a 5 kDa peptide which appears to be the major protein responsible for binding ADP-bound actin monomers and preventing their polymerization. The second molecule is profilin, which has a similar mechanism. However, in addition, profilin can enhance monomer polymerization by promoting the exchange of bound ADP for ATP. The production of actin filaments can also be regulated by so-called capping proteins that bind to one

Table 2.4 Some of the actin accessory and control proteins

Protein	Function
Actin	Forms fibrils
Tropomyosin	Strengthens actin microfilaments
Thymosin	Binds actin monomers/prevents polymerization
Cofilin	Regulates actin polymerization
Profilin	As thymosin, also enhances actin monomer ADP/ATP exchange
Gelsolin	Fragments and caps actin filaments
Fimbrin	Actin-bundling protein forming closely packed bundles
α-Actinin	Actin-bundling protein forming more loose structure than fimbrin
Filamin	Cross-links actin filaments into a gel
Villin	Actin-bundling protein in microvilli
Fascin	Actin-bundling protein
Spectrin	Attaches actin microfilaments to plasma membrane
Myosin-I	Moves intracellular vesicles on actin filaments
Myosin-II	Moves actin filaments
Erzin	Links actin filaments to the plasma membrane
Radixin	Links actin filaments to plasma membrane
Nuclear actin-binding protein (NAB)	
Ponticulin	Actin-binding plasma membrane glycoprotein
Tenuin	Protein located at adherens junctions and microfilament bundles
Rho A, B, C	Intracellular signalling proteins
Rac 1, 2	Intracellular signalling proteins
Cdc42	Intracellular signalling protein
RhoG	Intracellular signalling protein
TC10	Intracellular signalling protein

end of the actin filament to prevent the addition (or loss) of actin monomers. The best-known example of such proteins is gelsolin, which is also able to sever filaments and bind to the plus end.

In addition to these proteins which act to control the generation of actin microfilaments three other proteins control the form or topology of intracellular actin filaments. Individual actin filaments are assembled into two basic types of structure – the actin bundle and the actin network, both of which play distinct roles in the cell. In bundles, the actin filaments are cross-linked into closely packed parallel arrays. In networks, the actin filaments are loosely cross-linked in orthogonal arrays forming three-dimensional meshworks with the properties of semi-solid gels (Figure 2.11). Two proteins have been identified which cross-link actin filaments into bundles. The first is fimbrin, a 68 kDa protein containing two adjacent actin-binding domains. It binds to actin filaments as a monomer holding two parallel filaments close together. In such bundles, all the filaments have the same polarity, with their plus ends adjacent to the plasma membrane. Fimbrin was first isolated from intestinal microvilli and was later found in the surface projections of a variety of cells. A second type of actin bundle is composed of more loosely spaced filaments which are capable of contraction and which are found in the contractile ring that divides cells following mitosis or in stress fibres. The protein involved here is α-actinin and in contrast to fimbrin it binds to actin filaments as a dimer with each subunit being a 102 kDa protein with one actin-binding site. Filaments cross-linked by α-actinin are separated by a distance of 40 nm compared with 14 nm when cross-linked by fimbrin. This increased separation of the filaments by α-actinin allows the motor protein myosin to interact with the actin filaments in these bundles and enables them to contract. A third protein cross-links actin microfilaments. Filamin is a dimer of two identical 280 kDa subunits which forms a flexible V-shaped molecule that cross-links actin filaments into orthogonal networks. Such networks of actin filaments underlie the plasma membrane and support the surfaces of cells.

Cellular functions of the actin cytoskeleton

Actin filaments are found to be highly concentrated at the periphery of the cell and to form a three-dimensional network beneath the plasma membrane. This network of actin filaments and its associated actin-binding proteins is called the cell cortex by some cell biologists and it determines the shape that cells can take, and certain key aspects of cell behaviour including the movement of cells (Figure 2.12). Actin microfilaments can produce the following cell structures:

Cell protrusions

Many cells have protrusions which stick out from the main body of the cell and these are largely due to the presence of actin filaments. Perhaps the best characterized of these cell surface protrusions are the microvilli, which are finger-like extensions of the plasma membrane. Microvilli are abundant on the surfaces of cells involved in absorption – the best example being the epithelial cells lining the intestine, where the microvilli on the apical surface are known as a brush border. There are approximately one thousand microvilli per intestinal epithelial cell and this strategy increases the surface area for absorption by up to 20-fold. The intestinal epithelial cell microvillus consists of a protrusion of the plasma membrane formed by 20–30 closely packed parallel bundles of actin microfilaments. The filaments

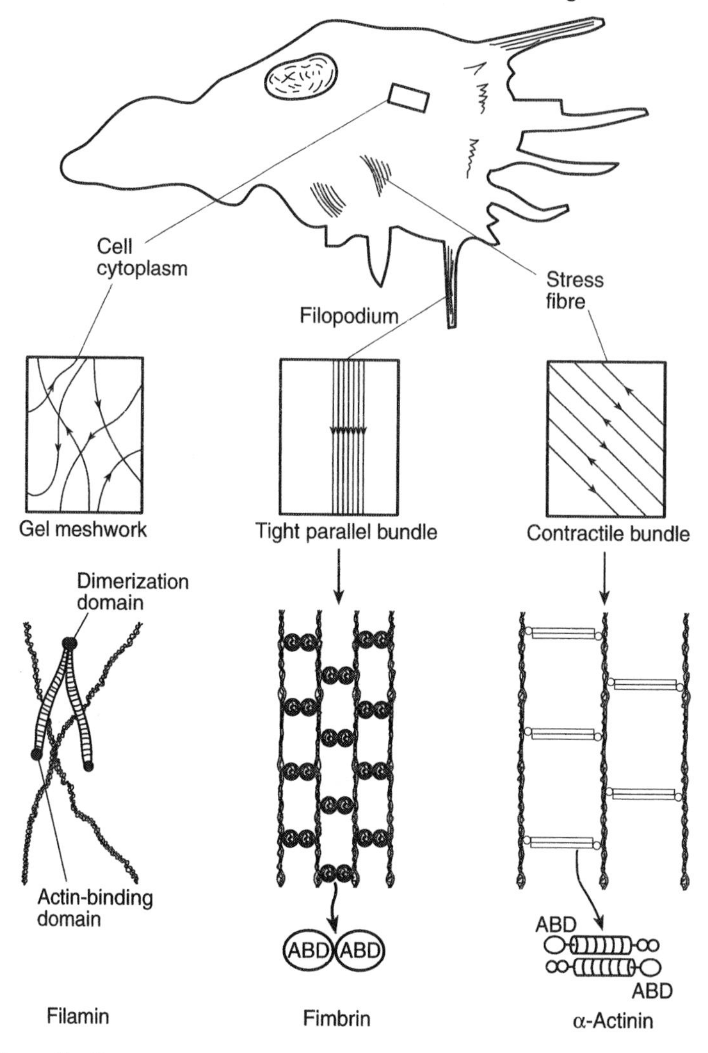

Figure 2.11 Schematic diagram showing the presence of actin microfilaments in a cell and the three types of filaments formed by the association of the actin polymers with the actin cross-linking proteins: filamin, fimbrin and α-actinin

in these bundles are cross-linked partly by fimbrin but also by villin, a 95 kDa protein present in the microvilli of only a few specialized cell types. The actin bundles are attached to the plasma membrane by the calcium-binding protein calmodulin associated with myosin I. At their base the actin bundles are anchored to the part of the cell cortex called the terminal web which cross-links and stabilizes the microvilli (Figure 2.13). The microvilli are targets for bacteria which produce attaching/effacing lesions on intestinal epithelium. The best example of this is the enteropathogenic strain of *E. coli* which causes massive actin cytoskeletal rearrangement leading to the formation of pedestals on which the bacteria bind strongly. The mechanism of pedestal formation and the result on cell behaviour are described in more detail in Chapters 5 and 6.

Other surface protrusions are transient, for example the pseudopodia produced by phagocytes. These are produced by three-dimensional networks of actin microfilaments and

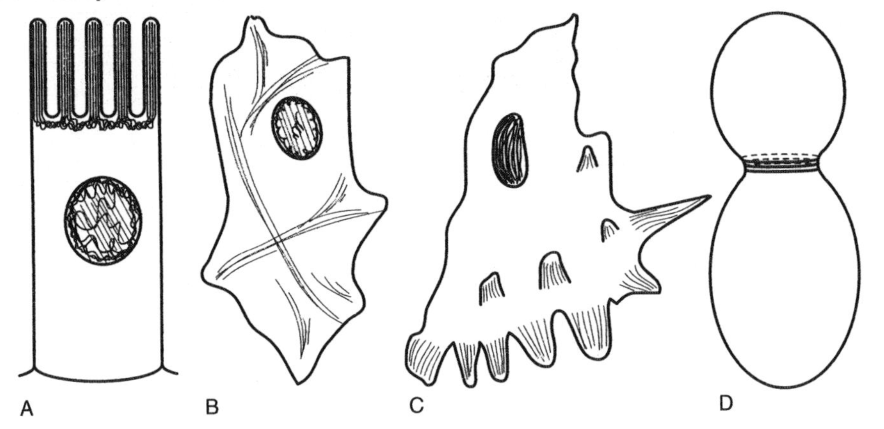

Figure 2.12 The roles played by the actin cytoskeleton in the cell. In (A) the actin filaments form the microvilli which are present in cells in a number of organs including the enterocyte of the intestine. In (B) the actin is in the form of contractile bundles which define cell shape and the contact of the cell with the substratum. In (C) the actin is forming cell protrusions known as lamellipodia (sheet-like structures) and filopodia (finger-like cell protrusions) which form at the leading edge of the cell. In (D) the actin filaments are forming a contractile ring which is involved in the division of cells at the end of the cell cycle/mitosis

can engulf bacteria to form an endosome. Lamellipodia (ruffles) are broad, sheet-like extensions at the leading edges of cells, such as fibroblasts, which also contain a network of actin filaments. Filopodia or microspikes are another transient cell extension – in this case thin projections of the plasma membrane.

Actin cytoskeleton and cell adhesion

In epithelial cell layers the actin cytoskeleton is anchored to regions in which the cells make contact, known as the adherens junctions. In sheets of epithelial cells these junctions form a continuous ribbon-like structure termed an adhesion belt. Contact between cells is mediated by cadherins, which are transmembrane proteins linked on the cytosolic side of the cells to three cytoplasmic proteins called catenins (α, β, γ). These catenins are in turn linked to the actin cytoskeleton. *Listeria* uses cadherin as a cell surface binding receptor to enter into cells.

Readers with experience of tissue culture will know how difficult it can be to remove fibroblasts from the plastic surfaces of culture dishes. This is because these cells secrete extracellular matrix proteins which form a contact layer on the plastic. This allows the fibroblasts to attach to the culture dish by utilizing transmembrane proteins called β_1-integrins. Cells do not bind over their whole surface but rather at discrete sites called focal adhesions. These focal adhesions also serve as attachment sites for large bundles of actin filaments called stress fibres which interact with myosin and it is this interaction which enables the fibres to be 'stressed' (Figure 2.12). These focal adhesions are an important part of the normal signalling machinery of the cell and indicate that the cell is in contact with appropriate neighbours (see Chapter 3).

Cell movement

Many cells have the capacity for locomotion on solid substrata. Good examples are fibroblasts moving to sites of wound healing and leukocytes migrating to sites of infection. The

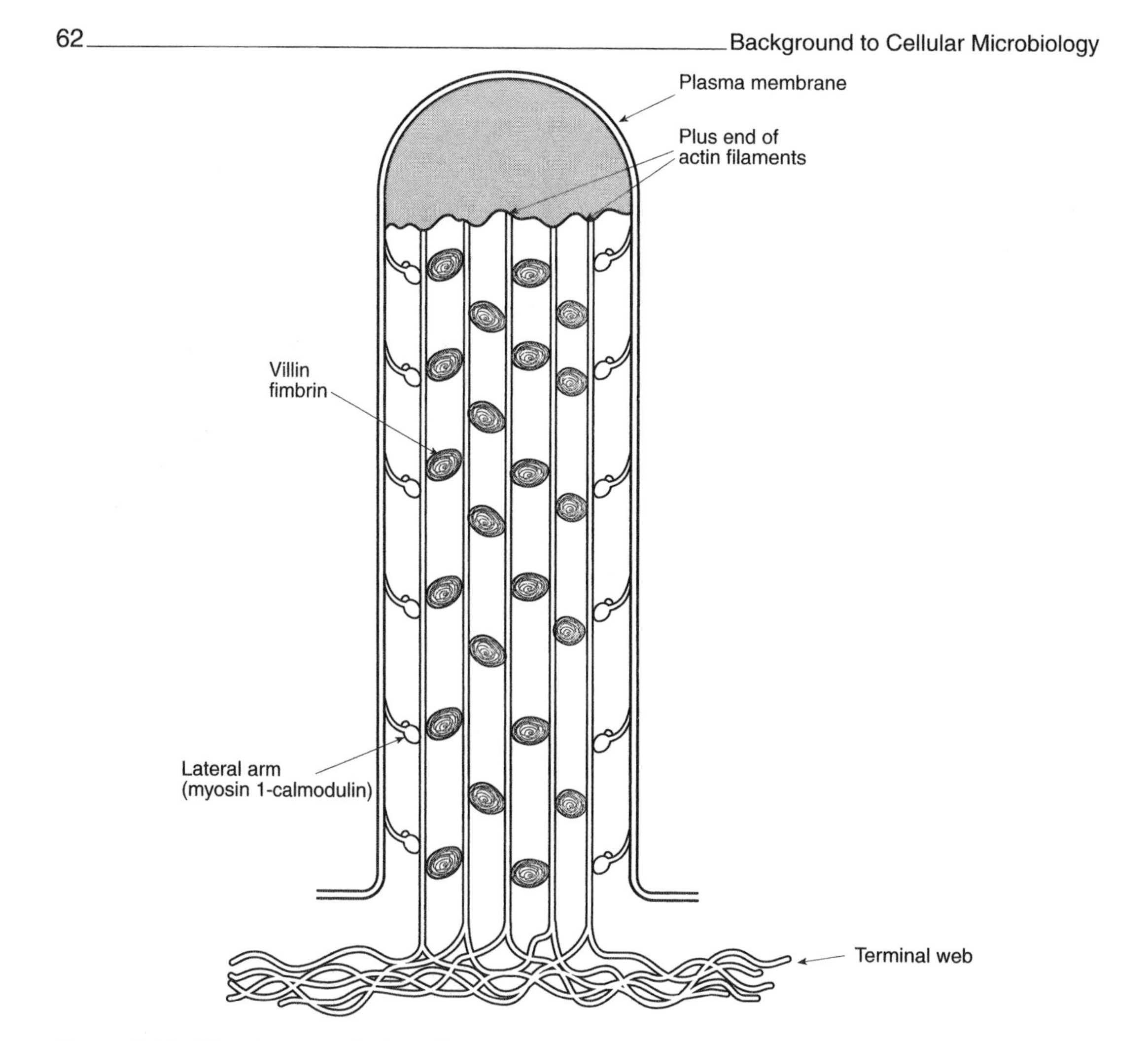

Figure 2.13 The structure of microvilli. Actin filaments form a core structure within the microvilli which are cross-linked into closely packed bundles by fimbrin and another actin-binding protein called villin. The actin filaments are attached to the plasma membrane by structures called lateral arms which consist of myosin I and calmodulin. The actin filaments have one orientation with their plus ends at the tip of the microvillus embedded in a cap of, as yet, unidentified proteins

mechanism of cell movement is not understood in molecular detail but it is believed to involve both actin and mysosin. Cell movement involves a sequential set of events. First, cells produce extensions from their leading edge. Second, these extensions need to attach to the substrate on which the cells are crawling. Finally, the trailing edge of the cell must dissociate from its substratum and retract into the main cell body.

Cell signalling and the actin cytoskeleton

The dynamic nature of the cytoskeleton and its involvement in many key cellular processes such as cell shape, organelle movement, mitosis, endocytosis and exocytosis suggested that it must be integrated and tightly controlled. It is only within the last decade that the proteins responsible for controlling this complex system have begun to be elucidated. It is now emerging that the control of the actin cytoskeleton involves low molecular mass GTP-binding proteins which are a subfamily of the well-known Ras superfamily. The proteins identified to date are members of the Rho subfamily and are: RhoA, B, C; Rac1, 2; Cdc42, Rho G and TC10. It is fascinating that our understanding of these signal transduction proteins is being elucidated by the use of growing numbers of bacterial exotoxins which target the Rho subfamily and its interactions with the cytoskeleton. Several of these proteins have been shown to be involved in different aspects of cytoskeletal organization (see Chapter 3). We now understand that activation of Rho leads to the assembly of stress fibres and local focal adhesion complexes. Activation of Rac leads to the production of lamellipodia and membrane ruffles. The role of Cdc42 appears to be the formation of filopodia. It has also been found that Rho and Rac are required for the production of cadherin-based adherens junctions. These Rho GTPases thus seem to link cell surface receptors to the organization of the actin cytoskeleton. The bacterial exotoxins which affect the cytoskeleton are shown in Table 2.5 and are discussed in detail in Chapter 7.

Table 2.5 Bacterial exotoxins acting on the actin cytoskeleton

Toxin	Bacterium	Mechanism and effect
C2 toxin	*Cl. botulinum*	ADP ribosylation of actin/inhibition of actin polymerization
Iota toxin	*Cl. perfringens*	As above
Toxins A and B	*Cl. difficile*	Glucosylation of Rho/Rac/Cdc42/depolymerization of F-actin
C3 toxin	*Cl. botulinum*	ADP ribosylation of Rho/depolymerization of F-actin
Cytotoxic necrotizing factors	*E. coli*	Modification of Rho/inhibition of actin polymerization
ZOT	*V. cholerae*	Unknown/increase in permeability of tight junctions

Integrins associated with the actin cytoskeleton can also induce cell signalling events. At sites of focal adhesions the cytoplasmic domains of the integrins can interact with a non-receptor tyrosine kinase known as focal adhesion kinase (FAK). This kinase is localized to focal adhesions and rapidly undergoes autocatalytic tyrosine phosphorylation when the integrin molecules bind to an extracellular ligand (e.g. fibronectin). This initial phosphorylation leads to the activation of a range of other signalling proteins which will not be dealt with in this chapter but which will be described in detail in Chapter 3.

Hijacking the actin cytoskeleton

As we have seen, bacteria can interact with the actin cytoskeleton as part of their virulence mechanisms. Perhaps the most audacious aspect of this behaviour is the use that bacteria such as *Listeria* makes of the process of actin microfilament formation – using it to propel them through the cell. This will be discussed in Chapter 6. Many viruses also abuse the actin cytoskeleton for their own purposes. For example, vaccinia virus utilizes the process of actin polymerization to propel it into neighbouring cells. Many of these viruses appear able to induce actin polymerization and formation of cell projections which aids the process of infection and transmission. Further discussion of the actin cytoskeleton will be found in Chapters 5–7.

While this section has focused on the eukaryotic cytoskeleton recent studies have established that bacteria also have cytoskeletal proteins, with functions similar to tubulin and actin, that play roles in chromosome segregation and cytokinesis.

VESICULAR TRANSPORT PATHWAYS: EXOCYTOSIS AND ENDOCYTOSIS

Eukaryotic cells have evolved mechanisms for the transport of membrane vesicles to selected destinations within the cell. Material, such as nutrients or bacteria, entering the cell in a process called endocytosis (which encompasses phagocytosis and pinocytosis) is sequestered in vesicles which can then be transported to appropriate destinations. Similarly, proteins produced in the ER and modified by the Golgi apparatus are packaged into vesicles and addressed to either the lysosomes or to the plasma membrane for membrane incorporation or for cell secretion (exocytosis) respectively.

Vesicle transport is another facet of cell biology which has become a target for bacteria in their evolutionary struggle with their hosts. Many bacteria have evolved an intracellular lifestyle, with a number living within the vesicular apparatus of host cells (*Mycobacterium spp., Legionella spp., Chlamydia spp.*), particularly the macrophage. Survival depends on controlling the functions of the endosome and its fusion with lysosomes. In addition, a number of potent bacterial toxins such as tetanus neurotoxin, botulinum neurotoxins and the vacuolating toxin from *Hel. pylori* target the cell's vesicular apparatus. In the case of the neurotoxins, their effect on vesicular transport results in the inhibition of release of neurotransmitters, causing spastic or flaccid paralysis (see Chapter 7).

Role of the ER in protein synthesis and sorting

The volume : cell surface ratio in eukaryotic cells has presumably been the evolutionary driving force behind the development of membrane-defined compartmentalization. The main membrane-bounded compartments in the typical eukaryotic cell are defined in Table 2.6. Proteins can traffic into these various membrane-bound organelles by three distinct mechanisms (Figure 2.14). Proteins destined for the nucleus or for the mitochondria and peroxisomes are produced by free ribosomes in the cytoplasm. Nuclear proteins are taken up through the nuclear pores, whose structure and function are briefly described in the next section. Proteins destined for the mitochondria and peroxisomes are unfolded in the cytoplasm, transported across the organelle membrane and refolded on the other side. This unfolding/refolding is aided by proteins known as molecular chaperones. The third mechanism by which proteins can be 'posted' to specific organelles in the cell is vesicular transport and relies on the concerted action of the endoplasmic reticulum (ER) and the Golgi apparatus. The discovery of vesicular transport mechanisms began with George Palade in the 1960s, who studied the mechanism of protein secretion in pancreatic acinar cells – cells that secrete digestive enzymes into the small intestine. These studies defined the pathway taken by secreted proteins and now known as the secretory pathway as

$$\text{rough ER} \rightarrow \text{Golgi} \rightarrow \text{secretory vesicles} \rightarrow \text{cell exterior}$$

It was later shown that proteins found in the plasma membrane and lysosomes also traversed this pathway. The rough ER is simply the description given to the ER when it is studded with ribosomes.

Table 2.6 The main membrane-bounded cell compartments

Compartment	Major function	% Total cell volume in hepatocyte
Cytosol	Site of metabolic pathways and protein synthesis	54
Endoplasmic reticulum	Synthesis of proteins and lipid distribution system for cell components	12
Golgi apparatus	Modification of proteins and lipids and their sorting and packaging for intracellular delivery or secretion	3
Nucleus	Contains major genome, site of DNA and RNA synthesis	6
Mitochondria	ATP synthesis	22
Peroxisomes	Oxidation of toxic molecules	1
Lysosomes	Intracellular degradation	1
Endosomes	Sorting of endocytosed material	1

The ER serves as a point of entry for proteins destined for other organelles and for the ER itself. Two types of protein are transferred from the cytosol to the ER: (i) soluble proteins translocated into the lumen of the ER and destined for secretion or for insertion into some other organelle and (ii) prospective transmembrane proteins which become embedded in the ER membrane. In contrast to the proteins destined for the nucleus, mitochondria and peroxisomes, and which are synthesized on ribosomes found in the cytosol, proteins which enter the ER are produced on ribosomes attached to the ER. These two populations of ribosomes are indistinguishable and their topology within the cell depends on the proteins that they are synthesizing. Proteins which are made on ER-bound ribosomes contain what is known as an ER signal sequence. This is typically a sequence of hydrophobic amino acids at the N-terminus of the protein. When proteins containing a signal sequence begin to be translated, the signal sequence is recognized and bound by a complex containing six polypeptides and a small RNA molecule called a signal recognition particle (SRP). This binding halts further translation and targets the complex (ribosome, polypeptide chain and SRP) to the ER, where it binds to an SRP receptor. This results in the release of the SRP from the ribosome and the ribosome then binds to a so-called protein translocation complex in the ER membrane. Protein translation then commences but now the newly formed polypeptide chain is threaded through the lumen of the ER via what is termed a translocation channel (Figure 2.15). Once the protein has been synthesized, the translocation channel, which binds to the

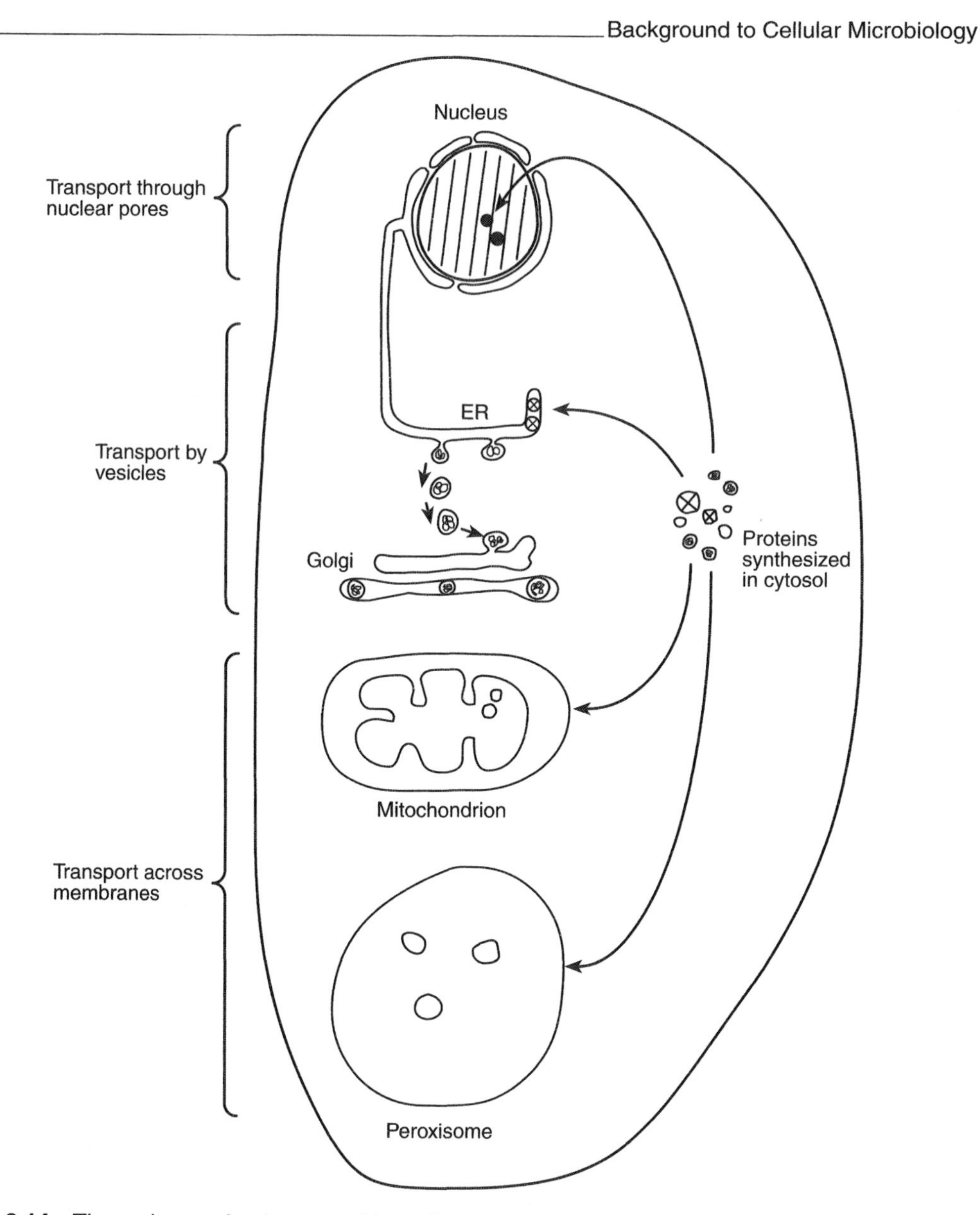

Figure 2.14 The major mechanisms used by cells to transport proteins into organelles. Protein transport into the nucleus is through pores. Protein transport into the endoplasmic reticulum (ER) generally requires that the protein be synthesized within the ER lumen. Proteins in the ER can then traffic within vesicles to other sites in the cell, mainly the Golgi apparatus. To transport proteins into mitochondria or peroxisomes the proteins need to be unfolded in the cytoplasmic side and refolded within the organelle. These processes are 'catalysed' by proteins known as molecular chaperones

signal sequence, opens sideways and releases the signal sequence into the lipid bilayer, where it is cleaved by a protease called a signal peptidase.

Some proteins produced on the ER are not released into the lumen but are destined to become membrane proteins. In the case of a transmembrane protein with just one

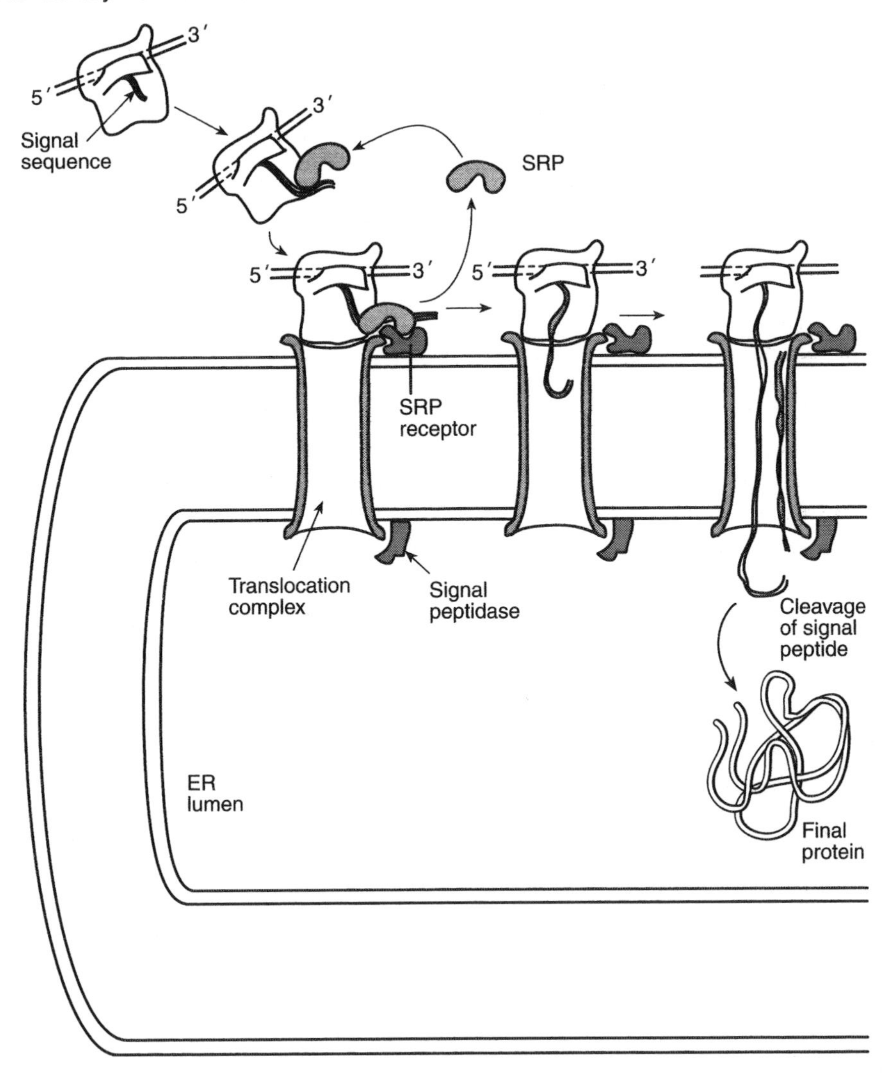

Figure 2.15 Protein targetting in the ER. A ribosome translating a secretory protein produces the signal peptide which, as it emerges from the ribosome, is recognized and bound by the signal recognition particle (SRP). This recognition events leads to the ribosome being bound to the SRP receptor on the ER membrane. The SRP is then released and the ribosome binds to a membrane structure called a translocation complex. The signal sequence is inserted into this translocation complex, which is a channel into the lumen of the ER. Protein translation then continues and the growing polypeptide chain is effectively translocated across the membrane. Cleavage of the signal sequence by a protease termed a signal peptidase releases the polypeptide chain into the lumen of the ER, where it can undergo further transport to its final destination

membrane-spanning segment the initiation of protein synthesis is identical to that described. However, the threading of the newly formed protein chain is halted by an additional sequence of hydrophobic amino acids called a stop-transfer sequence encoded by the mRNA. This second sequence is released sideways into the lipid bilayer from the translocation channel but is not cut by the signal peptidase, leaving an α-helical membrane-spanning segment. This process can be repeated to produce proteins, such as the many seven membrane-spanning receptor proteins, which couple to G proteins and are key elements in cell signalling.

Vesicular transport

Proteins exported into the lumen or inserted into the membrane of the ER are on the first stage of a complex process of vesicular trafficking which is a dynamic endomembrane-driven system in which there is continual budding and fusion of transport vesicles. The movement of these vesicles is controlled by elements of the cytoskeleton. The eukaryotic cell exhibits a major outward secretory pathway leading from the synthesis of proteins on or in the ER, through the Golgi apparatus (where they are modified) to the cell surface. From the Golgi, a side branch of the pathway leads to the endosomes and lysosomes. There is also a major inward flow of vesicles through the endocytic pathway, which is responsible for the ingestion and degradation of extracellular molecules and leads from the plasma membrane of the cell through the endosomes to the lysosomes (Figure 2.16). Indeed, cells constantly sample their environments by the mechanism of endocytosis.

An enormous amount of specificity has had to be built into this vesicular transport system. Vesicles budding off from a particular compartment must carry only those proteins appropriate to the vesicle's final destination and must be able find and specifically fuse only with the appropriate target membrane. While much is known about the mechanisms of vesicular transport, a complete description of the process has still to be attained, and some of the descriptions which will be given are still speculative. Vesicle formation is a dynamic process which is controlled by a number of proteins with specific functions. One group of proteins are found on the cytoplasmic face of the vesicles and their function appears to be that of distorting membrane conformation to cause vesicle budding. Because these proteins appear to coat the vesicles, the vesicles are termed 'coated vesicles' and currently three distinct types, which appear to function in different types of vesicular transport, have been identified. The first to be described were the clathrin-coated vesicles which are involved in the uptake of extracellular molecules from the plasma membrane by endocytosis. They are also responsible for the transport of molecules from the Golgi to lysosomes. The exterior or 'coat' of the clathrin-coated vesicles are composed on two types of proteins. Clathrins contain three protein chains which assemble into a basket-like lattice structure which is stabilized by a second class of proteins called adaptins. The specificity of the material (cargo) taken into the nascent vesicle is determined by the adaptins. These proteins bind the clathrin coat to the vesicle and bind to another set of molecules called cargo receptors which, in turn, bind to the specific proteins that the vesicle is destined to carry. Once the vesicle is loaded with its selected protein cargo a small GTP-binding protein called dynamin assembles a ring around the neck of the vesicle and, in a GTP-dependent process, pinches off the vesicle from the membrane. After budding the coat proteins dissociate and the naked vesicle can move and fuse with its intended target (Figure 2.17). If the vesicle has only a short distance to travel it does so by simple diffusion. However, if the vesicle has to move from the Golgi to the plasma membrane then it requires the help of the cytoskeletal system with its motor proteins for moving intracellular constituents.

Two other types of coated vesicles have been identified. These do not contain clathrin and in consequence have been denoted non-clathrin-coated, coatomer or COP vesicles (COP: *co*at

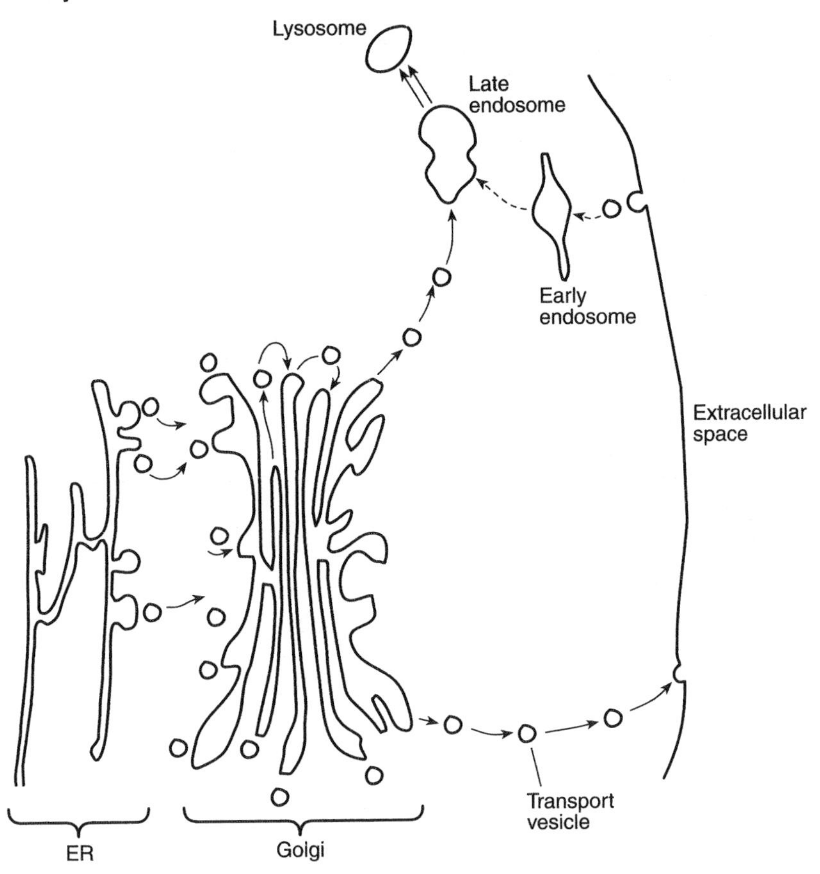

Figure 2.16 Simple schematic diagram outlining the pathways of vesicular movement within cells. In the outward secretory pathway (solid arrows) proteins are transported from the ER, through the Golgi apparatus to the plasma membrane or, via late endosomes, to the lysosome. There is an inward endocytic pathway (broken arrows) in which extracellular molecules are ingested in vesicles derived from the plasma membrane and are delivered to early endosomes and then, via late endosomes, to the lysosomes

protein). COPI-coated vesicles bud from the Golgi but their final destination has not been fully delineated. COPII-coated vesicles bud from the ER and are destined for the Golgi. These vesicles are transported along microtubules. The budding of both clathrin- and COPI-coated vesicles from the Golgi requires the participation of an additional GTP-binding protein called ADP-ribosylation factor (ARF), which is related to the Ras superfamily (described in detail in Chapter 3). ARF alternates between a GTP- and a GDP-bound state. When bound to GTP, ARF associates with the membrane of the Golgi and promotes the binding of the COP protein, resulting in vesicle budding. Hydrolysis of the bound GTP converts ARF to the GDP-bound state, leading to disassembly of the vesicle coat prior to fusion with the target membrane. This cycle can then repeat. Several other Ras-related GTP-binding proteins have also been found to be involved in vesicle-mediated secretion. At the present time more than 30 Ras-related proteins (called Rab proteins) have been implicated in vesicular transport in mammalian cells (see Chapters 3 and 7). The plasma membranes of most cells contain

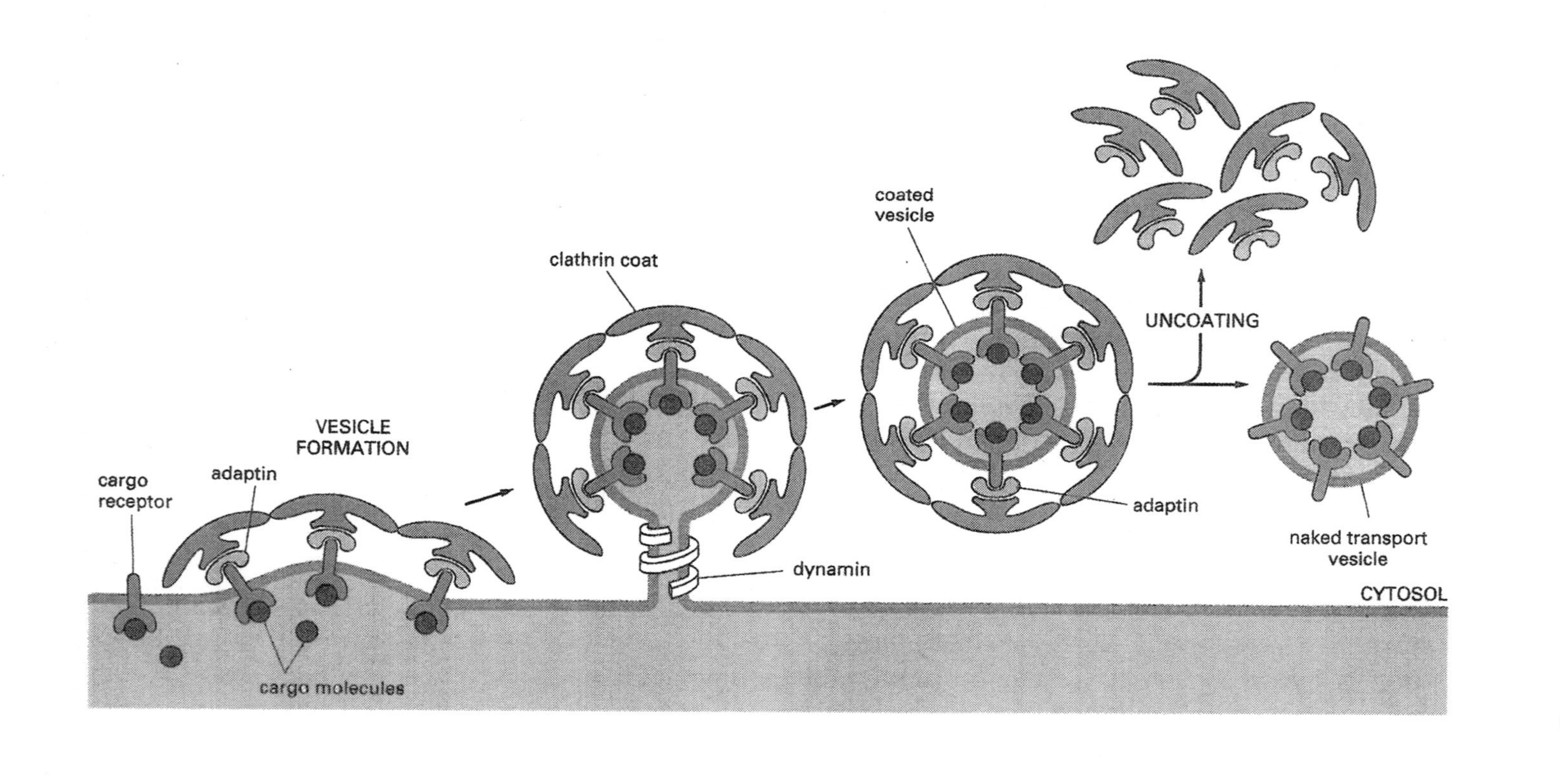

Figure 2.17 Formation of clathrin-coated vesicles. Cargo receptors which have bound to cognate proteins are 'captured' by adaptins which, in turn, bind clathrin molecules to the cytosolic surface of the nascent vesicle, To cause the vesicle to bud, dynamin proteins assemble round the neck of the vesicle and in a GTP-dependent process pinch off the vesicle. After the vesicle has budded the coat proteins are removed and the naked membrane can fuse with its target membrane. Reproduced from Alberts *et al* (1998) *Essential Cell Biology*, Garland, New York, with permission)

morphologically and biochemically distinct invaginations called caveolae which are also believed to form coated vesicles, although less is known about them than those described above.

Vesicle recognition and fusion

Having formed a transport vesicle how does the cell ensure that it reaches its correct destination where it can fuse? For example, a vesicle carrying lysosomal enzymes must fuse with a lysosome and not with the plasma membrane. The process involves two separate events. Firstly, the transport vesicle must specifically recognize the correct target membrane. Secondly, the vesicle and the target membrane must fuse, thus delivering the contents of the vesicle to the target organelle. Specific recognition between a vesicle and its target is now believed to be due to selective ligand/receptor interactions. The first protein identified in vesicle recognition was NSF (short for *N*-ethylmaleimide-sensitive fusion). This is a soluble cytoplasmic protein that complexes with another group of proteins called SNAPs (*s*oluble *N*SF *a*ttachment *p*roteins). The major hypothesis currently proposed to account for specific vesicle trafficking involves a further group of proteins called SNAP receptors or SNAREs. According to this so-called SNARE hypothesis, the interactions between specific SNAREs on vesicles (called v-SNAREs) and on the target membranes (t-SNARES) are responsible for the specific docking of vesicles with the correct target membranes (Figure 2.18). Once this receptor–ligand binding has been satisfied the SNARE complex recruits NSF and SNAPs and this then leads to fusion of the vesicle and the target membrane. Interactions between the SNARE proteins are believed to be regulated by the Rab proteins described earlier.

Endocytosis and exocytosis: bacterial involvement

Phagocytosis by leukocytes is a major part of the antibacterial defences of multicellular organisms and one which has been subverted by a number of organisms (as described in Chapters 6 and 8). In this section a brief description will be given of bacterial phagocytosis and of the role that certain bacterial exotoxins play in inhibiting neuronal exocytosis (neuroexocytosis).

Endocytosis and the phagocytosis of bacteria

During phagocytosis, binding of bacteria to receptors (e.g. C3b receptors) on the cell surface triggers the extension of pseudopodia, which are actin microfilament-driven extensions of the plasma membrane described in an earlier section. These pseudopodia eventually surround the bacterium and the membranes fuse to form an intracellular vesicle called a phagosome of around 0.25 μm diameter. The phagosomes then fuse with lysosomes to produce what is termed a phagolysosome. Lysosomes are membrane-enclosed organelles that contain an array of enzymes capable of breaking down all types of biological polymers. The best analogy for the lysosome is that it is the digestive apparatus of the cell. These organelles display considerable variation in size and shape as a result of the diversity of the materials that are taken up for digestion. The enzymes within lysosomes, which include proteases, glycosidases and lipases, all have acidic pH optima. This is because the pH of the interior of the lysosome is maintained at around pH 5 by an ATP-driven pump system which imports

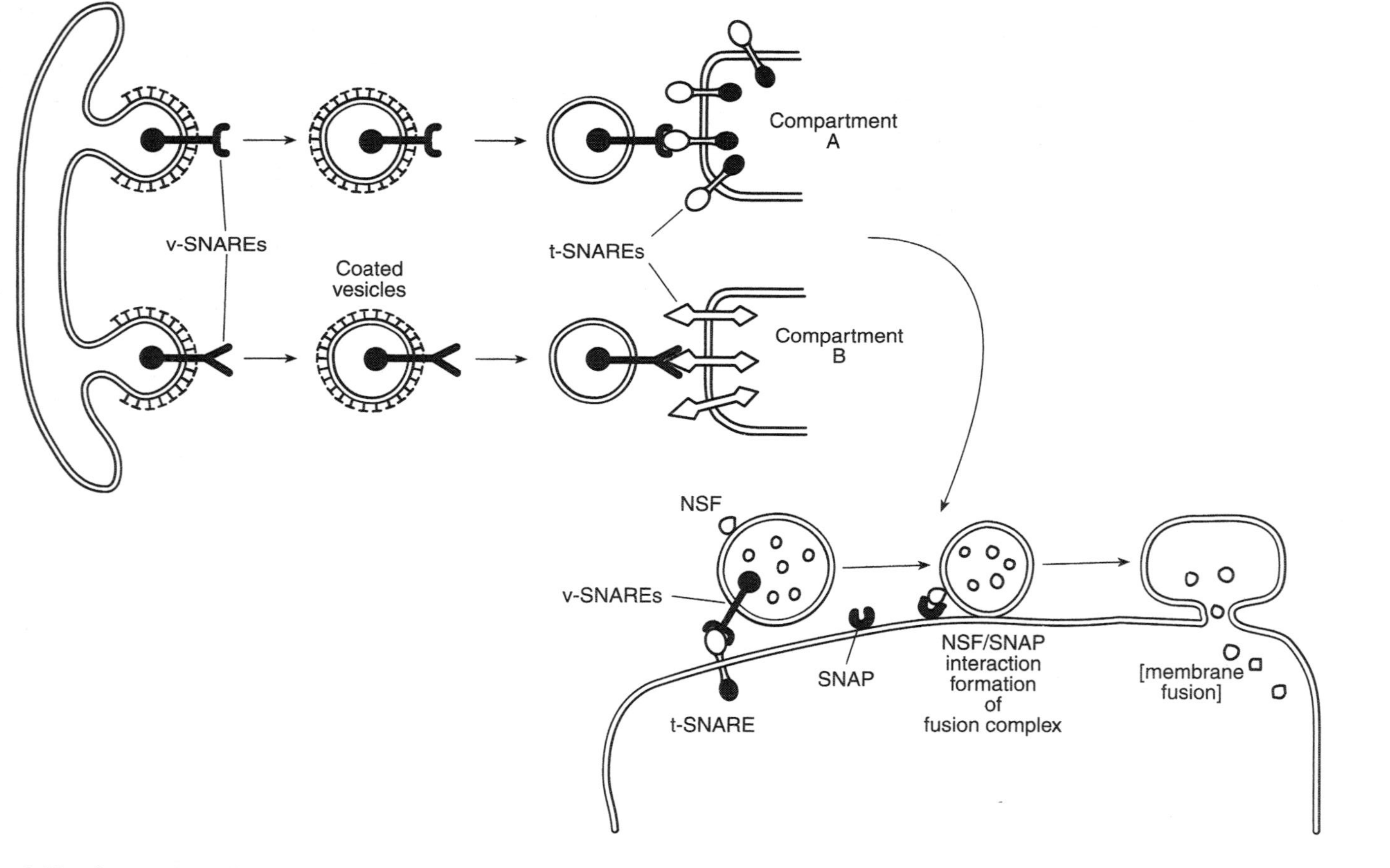

Figure 2.18 Current hypothesis to account for specificity of membrane trafficking. Vesicles budding off from donor organelles (e.g. Golgi apparatus) express specific marker proteins known as vesicle SNAREs (v-SNAREs) on their surface. Vesicles, after budding, diffuse, or are moved, until they recognize their complementary target SNAREs (t-SNAREs) on the membranes of target organelles (plasma membrane, endosome etc.). There are thought to be many different pairs of complementary v- and t-SNAREs. Once the vesicle has bound (docked) to the correct target membrane proteins such as NSF and SNAP catalyse the fusion of the vesicle with the target membrane and allow the vesicle contents to enter the interior of the target organelle

protons from the cytosol. Lysosomal membranes contain two highly glycosylated proteins – LAMP-1 and -2 (lysosome-associated membrane protein) – which are thought to protect the lysosomal membrane from degradation and which are used as markers for lysosomes. Within the lysosomes, bacteria are generally killed by exposure to the lytic enzymes, antibacterial peptides and free radicals that are either present, or which can be generated. Another protein family which has been found to be important in killing intracellular bacteria has been termed the natural resistance-associated macrophage protein (Nramp). These proteins are involved in ion and proton transport and their role in resistance to intracellular bacterial infections is described in Chapter 8. However, certain bacteria can defend themselves against these lysosomal defences by manipulating the vesicular transport system for their own ends. Perhaps the most interesting method of defeating the vesicular transport system is that of *Toxoplasma gondii,* which causes toxoplasmosis. This parasite enters cells and encloses itself in a vacuole that is a composite of the host cell plasma membrane and parasite organelle components produced during the invasion process. Thus *T. gondii* overcomes the defence mechanisms of the lysosomal apparatus by not fusing with the endosomal vesicular system. Curiously, the interaction of this parasite with the host cell results in the formation of a new organelle which is not present in either of the participants. In Chapter 6 the methods intracellular bacteria use to subvert the vesicular apparatus of the cell are described in detail.

Exocytosis and bacterial exotoxins

A number of bacterial exotoxins have been shown to interefere with vesicular traffic. The best-known neurotoxins are from bacteria of the genus *Clostridium.* Neuronal signalling depends on structures called synapses. These structures represent the sites where signals are transmitted from one neurone, called the presynaptic neurone to another neurone, the postsynaptic neurone. The postsynaptic neurone can also be on an effector cell such as a muscle. Two types of synapses exist – electrical and chemical synapses – and we will only be concerned with the latter. The structure of a chemical synapse consists of a synaptic knob, a synaptic cleft and the plasma membrane of the postsynaptic neurone. The synaptic knob is a small bulge at the end of a terminal branch of a presynaptic neurone and contains a neurotransmitter packaged into synaptic vesicles which are a specialized part of the exocytic vesicular apparatus. When an action potential reaches a synaptic knob, voltage-gated calcium channels in its membrane open and allow calcium levels to rise. This increase in intracellular calcium triggers the movement of neurotransmitter vesicles to the plasma membrane of the synaptic knob. Once there, they fuse with the membrane and release their neurotransmitter via exocytosis. Of the proteins involved in vesicle–membrane fusion three – VAMP/synaptobrevin, SNAP-25 and syntaxin – are the specific targets of eight neurotoxins produced by clostridia (see Chapter 7). SNAP-25 is a member of the SNAPs described earlier in this section. Syntaxin is a t-SNARE. All of the neurotoxins have zinc-dependent proteinase activity and it is the specific cleavage of one or other of the three proteins described that result in all of these toxins blocking neuroexocytosis.

Hel. pylori produces a vacuolating toxin which causes massive formation of endosomal vacuoles in epithelial cells. The mechanism of vacuole formation is now known and it is believed that this toxin will provide a molecular probe for defining more clearly the mechanism underlying endosome formation.

NUCLEAR STRUCTURE AND GENE ORGANIZATION

The nucleus is the defining organelle of eukaryotic cells and contains the DNA. In multicellular organisms every cell contains a complete complement of genes. Each cell is totipotent and it is this totipotency that has allowed animals (such as Dolly the sheep) to be cloned from somatic cells. The eukaryotic nucleus is a very complex organelle and the organization of the DNA in eukaryotes, and the manner in which it is transcribed, is also much more complicated that the mechanisms used by bacteria. Moreover, many bacterial products can induce eukaryotic cells to transcribe genes, particularly those involved in the immune and inflammatory responses, and there is growing evidence that microorganisms can manipulate the transcriptional control machinery of eukaryotic cells for their own ends.

The human genome is estimated to contain 60–100 000 genes. *Myxococcus xanthus*, which has the largest bacterial genome thus discovered, has an estimated 8000 genes. The numbers of genes are, however, not reflected in the size of the genomes of these two organisms – 9.4 Mb for the bacterium compared with 3000 Mb for the human genome. The yeast genome contains 14 Mb and is believed to contain 5–10 000 genes, Thus there is at most a log order increase in complexity (in terms of number of genes) between bacteria and man but a 2–3 log order increase in the number of base pairs. Indeed cells from certain plants (lily) and amphibians (the lungfish) have up to 50 000 Mb of DNA in their nuclei. Even the onion has 15 000 Mb in its genome. Much of the extra DNA found in eukaryotic cells does not encode proteins but it is not clear what evolutionary pressure resulted in the accumulation of this extra non-coding DNA in eukaryotes. The amount of DNA contained in the haploid genome of a species is called the C value and the very large amounts of DNA found in eukaryotes has been termed the C value paradox.

Structure of the nucleus

The chromosome of prokaryotes is typically a non-membrane enclosed covalently closed circular molecule which enables bacterial genes to be transcribed and translated simultaneously. In contrast, eukaryotic genes are separated from the translational apparatus by the nuclear membrane and it is now established that eukaryotic mRNA undergoes post-transcriptional processing before being exported from the nucleus. In addition, with many genes that are stimulated by signals arriving at the cell's surface, the factor causing transcription is a DNA-binding protein which is produced in the cytoplasm and has to enter the nucleus to deliver its signal. This facet of gene control will be described in Chapter 3.

The nuclear envelope is composed of two concentric membranes called the inner and outer nuclear membrane. The outer nuclear membrane is continuous with the ER. The inner and outer nuclear membranes are joined at specific sites termed nuclear pore complexes and these are the only channels for the passage of molecules into and out of the nucleus. The nuclear pore complex is a large multimolecular structure (composed of possibly more than 100 different proteins) of 120 nm diameter with an estimated molecular mass of 120–130 million daltons. Molecules less that 20 kDa can freely traverse these pores in either direction. Larger molecules of protein and RNA are too large to cross by free diffusion and instead utilize an active process which involves recognition and selective transport in one or other direction. Proteins entering the nucleus from the cytoplasm, for example histones, polymerases and transcription factors, are targetted to the nucleus by specific amino acid sequences called nuclear localisation signals. The first such localization signal to be recognised was in the simian virus 40 (SV40) T antigen that initiates viral DNA replication in infected cells. The HIV-1 Rev protein is another viral protein recognized as a nuclear-targetting signal. These are good examples of infectious agents utilising eukaryotic cell rules for their own ends. Another more recent example has been the discovery that bacteria can interfere with nuclear transport. For example, leptomycin B, a secondary metabolite of *Streptomyces* spp., inhibits nucleocytoplasmic transport and, in consequence, blocks the cell cycle in G_1 and G_2.

In dividing cells the nuclear envelop breaks down in mitosis during prophase and is re-formed in the daughter cells at the end of mitosis.

Underlying the inner nuclear membrane is the nuclear lamina, a fibrous meshwork composed of proteins called lamins which are part of the intermediate filaments of the cell. The lamins, and the nuclear lamina they produce, provide structural support to the nucleus and are believed to act as sites for chromatin attachment. The lamins are also believed to contribute to the nuclear matrix which remains following removal of the nuclear DNA by DNase and the histones and other nuclear proteins (see below) by high salt buffers. This residual framework of lamins and other proteins maintain the shape of the nucleus. The presence of the nuclear matrix is still controversial but is claimed by some scientists to be the framework on which the DNA-protein complex (chromatin) of the eukaryotic cell is hung. The structure of the nucleus is shown in Figure 2.19.

The most prominent structure within the eukaryotic nucleus is the nucleolus. This non-membranous structure is organized around the chromosomal regions that contain the genes for the ribosomal RNA molecules and is involved in the production of the ribosomes on which proteins are translated.

Although the nucleus of each human cell contains 60–100 000 genes, no one cell will express all of these genes. Muscle cells will produce a different pattern of gene expression from liver parenchymal cells and both will be different from neurones. Thus cells have the capacity only to utilize a certain proportion of their genes. This is being clearly seen from the Human Genome Project studies in which shotgun cloning is used to randomly clone short sections of genomic (termed sequence-tagged sites – STSs) or cDNA (termed expressed sequence tags – ESTs). Each short clone can then be used to design PCR primers to amplify that particular piece of DNA wherever it is found. Use of the EST primers shows that different patterns of genes are clearly expressed in different tissues. In all cells a proportion of the nuclear DNA is highly condensed and transcriptionally inactive. This highly condensed DNA can be recognized in microscopical preparation of cells and is termed heterochromatin. The remainder of the chromatin is termed euchromatin, and is decondensed and distributed throughout the nucleus. The different expression pattern of genes in individual cells and tissues of a multicellular organism appears to be related to the pattern of methylation of DNA. Cytosine residues in DNA can be methylated at the C-5 position and such methylation appears to be correlated with a reduction in the transcriptional activity of genes.

Organization and transcription of eukaryotic DNA

The genome in eukaryotes is composed of multiple chromosomes, each one composed of a linear stretch of DNA which forms complexes with proteins to allow the DNA to be

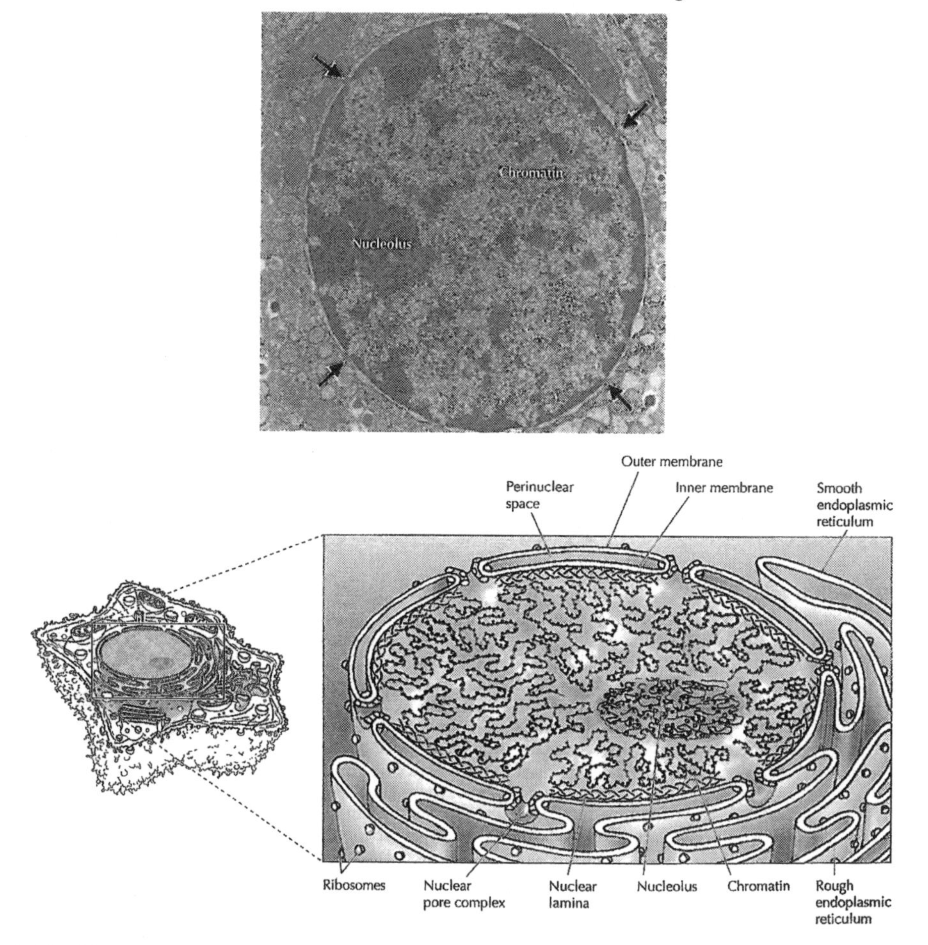

Figure 2.19 The nucleus of the eukaryotic cell. In (A) the structure of the nucleus is seen at electron microscopic magnification with the nucleolus and chromatin and the nuclear pore complexes (arrowed). In (B) the structure of the nucleus is shown diagrammatically with the various constituent parts denoted. (Reproduced from Cooper (1997) *The Cell*, ASM Press,Washington, DC, with permission)

packaged within the nucleus. To give an idea of how complex this organization is – the DNA in a human cell is packaged into a nucleus with a diameter of only 5–10 μm, yet if it were to be fully extended it would be 2 m long.

The term chromatin describes the complex between DNA and protein. The major proteins are the histones, which are small proteins containing a large proportion of basic amino acids (arginine and lysine) that promote their binding to the negatively charged DNA molecule. Five major histones, H1, H2A, H2B, H3 and H4, have been identified. A wide variety of so-called nonhistone chromosomal proteins are also associated with the DNA. The basic structural unit of chromatin is called the nucleosome. This consists of 146bp of DNA wrapped 1.75 times round a histone core consisting of two molecules of each of H2A, H2B, H3 and H4 and one molecule of H1 that acts to bind the DNA onto the nucleosome. The non-histone proteins are found in between the nucleosomes. The packaging of DNA into nucleosomes produces 10 nm diameter fibres and shortens the length by approximately six-fold. The chromatin is further condensed by coiling into 30 nm fibres. It is through this condensation mechanism that all of the DNA in the cell can be packaged into the nucleus. Further coiling of the DNA occurs at mitosis when the chromatin fibres condense into chromosomes. Of course, the disadvantage of this arrangement is that in order to transcribe genes the DNA needs to be freed in order to interact with the RNA polymerases which produce RNA from the DNA template. Before describing this key process, it is important to realize that, unlike bacteria, 90–95% of the DNA in the eukaryotic nucleus is non-coding. Over the past two decades the nature of this non-coding DNA has been defined although its function is not completely understood. The DNA in eukaryotic cells can be subdivided into non-repetitive, moderately repetitive and highly repetitive DNA. The non-repetitive DNA constitutes the genes. Several types of highly repeated sequences exist. One class, called simple-sequence DNA, contains a tandem array of thousands of copies of short sequences ranging from 5 to 200 nucleotides. Because of their base composition, many simple sequence DNAs can be separated from the rest of the genome and because of the method use to isolate them they have been termed satellites and this class of highly repetitive sequence DNA is now called satellite DNA. These sequences can be repeated millions of times per genome and can account for 10–20% of the DNA of most higher eukaryotes. Satellite DNA regions are not transcribed nor involved in gene control but may have a role to play in the determination of chromosome structure. Moderately repetitive DNA sequences are scattered throughout the genome as either interspersed or tandemly repeated sequences. The interspersed sequences can be either short or long. Short interspersed elements, or SINEs, are less than 500 bp long and as many as 500 000 copies are present in the human genome. The best-characterized human SINE is a set of closely related sequences called the *Alu* family, so-called because of the presence of a restriction site for the restriction endonuclease *AluI*. Members of this family are 200–300 bp long and comprise almost 10% of the human genome. *Alu* are thought to have arisen from an RNA element whose DNA complement was interspersed throughout the genome as a result of the activity of a reverse transcriptase. Retroviruses are the best-known source of reverse transcriptase, suggesting that the *Alu* family may have a viral origin. Indeed, it is now established that retroviral DNA is incorporated into human cells and that some of these genes are transcribed and the proteins found in human tissues. It has been proposed that these proteins may play some role in the pathology of autoimmune diseases. It is also thought that the enormous repetition of the *Alu* sequences occurred within the past few tens of millions of years. Long interspersed elements (LINEs) are another category of moderately repetitive DNA. In humans the most prominent example is a family of LINEs designated L1 which are about 6400 bp long and are present up to 40 000 times in the genome.

Introns and exons

Some 20 years ago it was found that most eukaryotic genes also contained non-coding segments of DNA within their genes. Archaea can also contain such 'intervening' DNA. Thus eukaryotic genes have a segmented structure in which coding sequences (called exons, or *ex*pressed) are separated by non-coding sequences also known as intervening sequences or introns (Figure 2.20). Some genes can have many introns. The 'winner' at present is the gene for type I collagen, which contains 50 introns. The entire gene is transcribed to yield a long mRNA molecule and regions encoded by the introns are removed within the nucleus by a process known as splicing. Interestingly, introns were first discovered in adenovirus DNA. The initial RNA transcript contains signal sequences at the intron–exon boundaries which can be recognized by a large nuclear-associated structure that has been termed the spliceosome. This nuclear 'machine' utilizes a number of small ribonucleoproteins (snRNP or 'snurps') which are involved with the cutting and religation of the RNA. Some genes have the ability to use different combinations of exons and introns to produce alternative forms of the encoded protein. This process is termed alternative splicing and is used by bacteriophages and by mammalian cells. The first example of alternative splicing was the identification of two related peptides: calcitonin, which is involved in the control of calcium metabolism; and calcitonin gene-related peptide (CGRP), which has a number of roles particularly in inflammation and control of blood pressure. Alternative splicing therefore allows the production of multiple homologous proteins without the need to evolve a multigene family. This may be the reason that alternative splicing evolved.

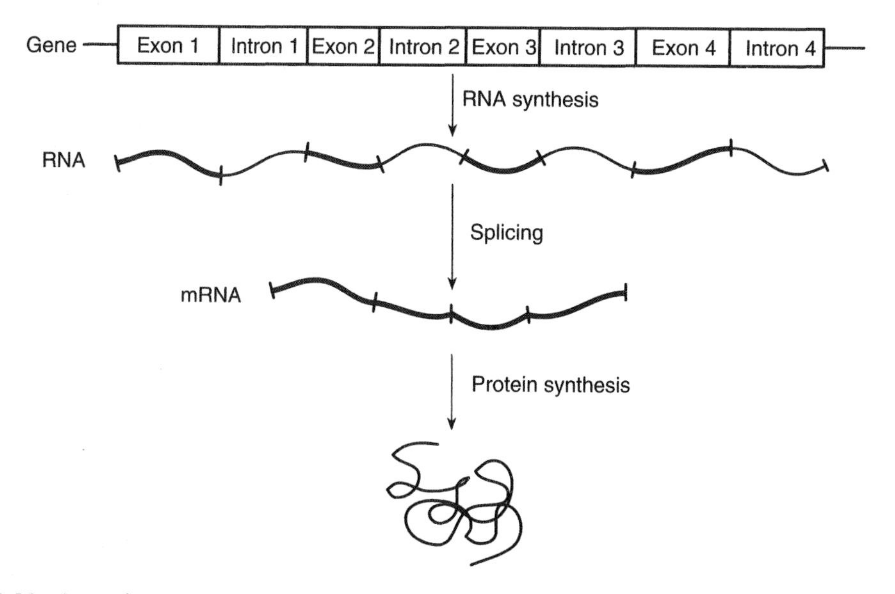

Figure 2.20 Intron/exon structures of eukaryotic genes. Many eukaryotic genes contain intervening sequences called introns which are not found in the final encoded protein. When a gene containing such intervening DNA is transcribed all the sequence, including exons and introns, is transcribed into RNA. The RNA is then processed and the intervening sequences are removed, leaving an mRNA molecule only encoding the exons. Translation of this mRNA results in the production of the exon-encoded protein

Control of transcription

The role of RNA polymerase in bacterial gene transcription has been described under 'Gene transcription', above. In contrast to bacteria, eukaryotic cells contain three RNA polymerases which transcribe: (1) ribosomal (r) RNA (RNA polymerase I); (2) messenger (m) RNA (RNA polymerase II); and (3) transfer (t) RNA (RNA polymerase III). These are multi-subunit enzymes which form complexes of 500–700 kDa. Each polymerase requires additional proteins, termed transcription factors, that enable the enzyme to bind to the DNA template and initiate and maintain transcription. We will only consider the mechanism of RNA polymerase II (pol II) as its modulation by bacterial factors is of recognized importance. This enzyme is composed of 12 subunits.

Gene transcription is an extremely complex process, requiring recognition of the gene to be transcribed, binding of a complex series of proteins which then unwind the DNA, transcribe the RNA polymer, and rewind the DNA. Pol II by itself is incapable of initiating transcription and must bind to additional proteins which are termed transcription factors. These transcription factors can, in turn, be separated into generic (also basal) and specific types. The secret of the transcription factors is their ability to bind to specific sequences of DNA. In eukaryotic cells, individual genes have significant amounts of upstream DNA (sequences 5' to the gene) associated with them which are involved in the control of transcription. These upstream sequences are known as enhancers and promoter regions. Most promoters have a sequence known as the TATA box which is approximately 25 bp upstream of the initiation site for the gene. It is at the TATA box that the complex of proteins making up the pol II transcription complex is formed. This involves binding of a factor termed TFIID to a region that extends upstream from the TATA sequence. The specific binding to the TATA box is conferred by the presence of the TATA-binding protein (TBP) and other accessory proteins called TAFs (TBP-associated factors). Additional TFII proteins (TFIIA, B, E etc.) bind to the TFIID complex, forming a very large multimolecular complex, to which the pol II eventually binds.

The RNA pol II transcription complex (or initiation complex) formed at the transcription startpoint is dependent, for its efficiency and specificity, upon short DNA sequences further upstream which bind to specific transcription factors. Over the past decade a large number of transcription factors have been discovered which bind to specific DNA sequences and facilitate gene transcription. Transcription factors generally consist of two domains: one which binds to the DNA and the other which interacts with the RNA polymerase complex. The known transcription factors fall into four structurally defined families based on the DNA binding motif. The *zinc-finger* motif contains repeats of cysteine and histidine residues that bind zinc ions and form loop structures – the so-called 'fingers' that bind to the DNA. Steroid hormone receptors are examples of transcription factors containing zinc fingers. The *helix-turn-helix* motif was originally described in prokaryotic DNA-binding proteins and one helix binds to DNA while the other helices stabilize this interaction. The *leucine zipper* and *helix-loop-helix* families of DNA- binding proteins form DNA-binding domains by dimerization of polypeptide chains. In the case of the former the two chains are held together by hydrophobic interactions between leucine side chains. While the nature of the DNA-binding domains of these transcription factors is understood, less is known about the activation domains of these proteins. The function of these activation domains is to interact with basal transcription factors, such as TFIIB or TFIID, to facilitate the assembly of a transcription complex (Figure 2.21).

This brief review of the eukaryotic nucleus and its DNA reveals a complex and far from understood system for regulating gene expression. Many questions arise from the current knowledge we have of eukaryotic gene expression and remain to be answered. In Chapter 3 the mechanism by which external signals arriving at the eukaryotic cell surface can induce the transcription of specific genes will be discussed. In the penultimate section of this chapter the mechanism by which cells regulate their growth and division will be briefly reviewed.

THE CELL CYCLE AND APOPTOSIS

The ability of some viruses to transform cells and produce cancer has been known for many decades, and it is established that one host antiviral defence mechanism is programmed cell

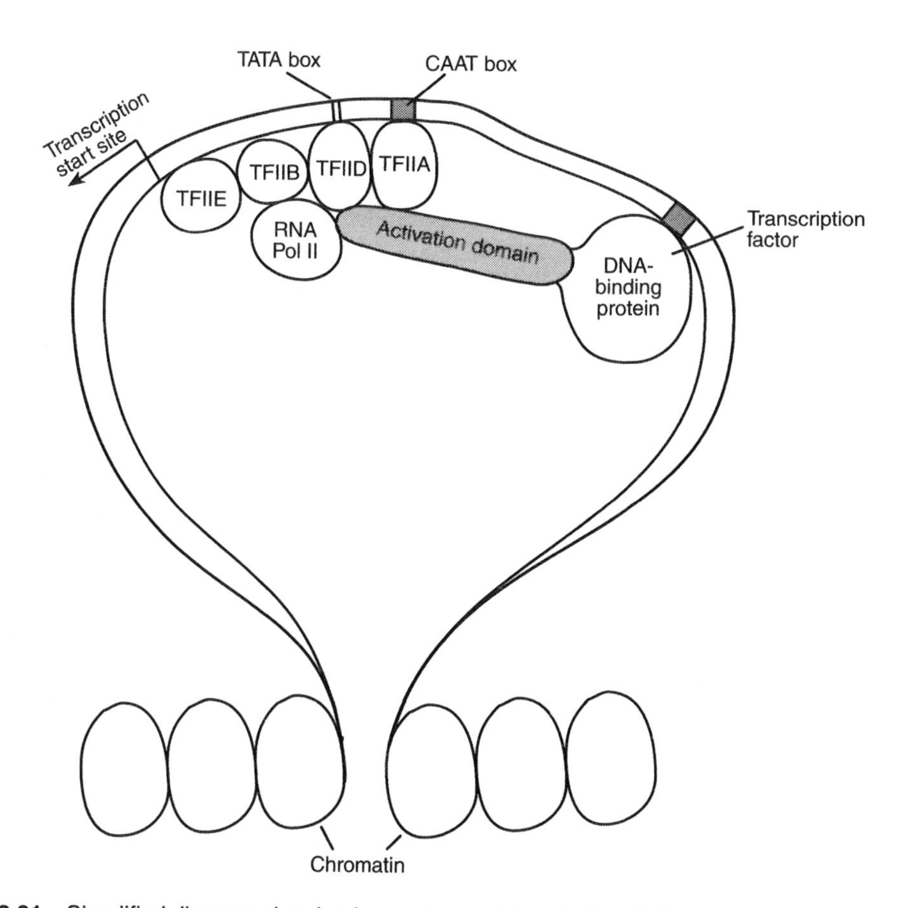

Figure 2.21 Simplified diagram showing how a transcription factor binding some way upstream of the transcription start site for a particular gene can influence gene transcription by interacting with the transcriptional complex

death or apoptosis. Within recent years it has become apparent that some bacteria can stimulate or inhibit the cell cycle and interfere with apoptosis. The action of bacteria on the cell cycle and on apoptosis is described in Chapter 10.

Historically, cell cycle analysis had its origins in the 1950s with the pioneering work of Mitchison in Edinburgh on the fission yeast *Schizosaccharomyces pombe*. Genetic analysis of these yeasts began at about this time but it was not until the 1970s that Nurse and coworkers applied genetic methods to cell cycle analysis and isolated the first yeast cdc (cell division cycle) mutants. Cell cycle genetics of both *Schizosaccharomyces pombe* and *Saccharomyces cerevisiae* advanced markedly in the late 1970s and early 1980s and, with the ability to clone, express and characterize cell cycle genes and their products, the last 20 years has seen the mechanisms of the eukaryotic cell cycle largely elucidated. Many of the terms utilized in the cell cycle literature have come from yeast genetics, which tends to make the nomenclature confusing. Every effort has been taken to simplify the description of the cell cycle mechanism and keep abbreviations to a minimum.

The cell cycle

For a cell to replicate itself it must: (i) faithfully reproduce its DNA; (ii) manufacture sufficient cellular organelles, membranes, soluble proteins etc. to enable the daughter cells to survive; and (iii) partition the DNA and the cytoplasm (containing organelles) equally to form two daughter cells. These processes must be done in order, and requires significant amounts of feedback control to ensure that the process is correct. Failure to control the cell cycle process carries with it a high price – diseases such as cancer. Microscopic examination of cells maintained in culture reveals the two major visible phases of the cell cycle: mitosis and interphase (Figure 2.22). Mitosis, first described in the nineteenth century, is the most visually spectacular phase of the cell cycle, involving the condensation of the chromosomes and their partitioning within the cell followed by the process of cytokinesis, an actin microfilament-dependent event resulting in the formation of the two daughter cells. These events last no more than an hour or so, and so for the majority of the time a dividing cell exists in interphase – the period between mitoses. In dividing cell populations the cell grows at a steady rate during interphase. However, the DNA is synthesized only during a certain portion of interphase. The timing of DNA synthesis thus divides the cycle of eukaryotic cells into four distinct phases. The M phase is the period of mitosis. This is followed by a period which was termed the first gap, and is now known as the G_1 phase, which corresponds to the time from the end of mitosis to the beginning of DNA synthesis. The phase of cellular DNA synthesis is termed the *synthesis* or S phase. The time period from the cessation of DNA synthesis to the initiation of mitosis was the second gap in our understanding of the cell cycle and was termed gap $(G)_2$. The duration of these cell cycle phases varies considerably in

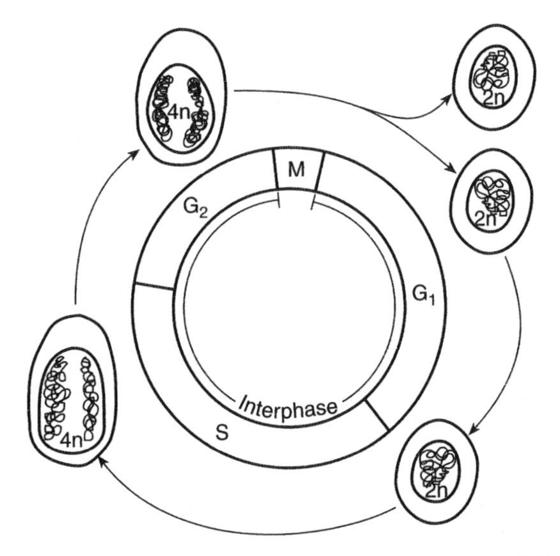

Figure 2.22 Simple diagram showing the four phases of the cell cycle. Mitosis, which is followed by cytokinesis, is the most visible sign of the complex process known as the cell cycle. This process produces two daughter cells which contain the $2n$ complement of cellular DNA. In continually cycling cells the phase following mitosis is called G_1 and lasts approximately 10–24 hours. This is followed by S phase during which the cell's DNA is replicated resulting in cells having a $4n$ DNA content. Following on from S phase cells enter G_2, which allows them to check that they have correctly replicated their DNA. At the end of G_2 cells enter mitosis and then divide

various cell types. For a rapidly proliferating human cell with a cell cycle duration of 24 hours G_1 may last 11 hours, S phase around 8 hours and G_2 3–5 hours, with M lasting 1 hour. In comparison, budding yeast can complete all four phases of the cell cycle in 90 minutes. A further phase of the cell cycle has been defined. When cells such as neurones stop dividing they are believed to exit G_1 and enter a quiescent state which has been called G_0.

Control of the cell cycle

Cell cycle control is regulated by external signals, such as growth factors, acting in concert with internal mechanisms which centre around three groups of proteins: cyclins, protein kinases and protein phosphatases. The cell cycle also presents cells with a number of critical decision points called checkpoints which determine whether a cell should proceed through the cell cycle, or not. A major checkpoint in most cells occurs in late G_1 and controls progression to S. The best-understood example is in the budding yeast *S. cerevisiae*, where it was first discovered and termed START (Figure 2.23). Passage through START is subject to a complex control network and the decision to proceed depends upon the integration of nutritional, hormonal and cell size parameters. For example, if cells are in a nutrient-poor environment they arrest their cell cycle at START. Yeast mating hormones can also arrest cells at START. The proliferation of most animal cells is regulated at G_1 and this decision point is called the restriction point. Of course, for most animal cells nutritional availability is not a control factor – the major cell cycle progression factors being extracellular growth factors. Once cells enter the S phase they replicate their DNA. Cells then enter the second gap phase (G_2) which represents an important second checkpoint for the cell cycle. It is important that a cell does not enter mitosis without having a completely replicated genome. If it did, it would lead to unequal distribution of the genome and non-functioning cells. Thus G_2 is the period in the cell cycle during which the cell can detect unreplicated DNA or DNA which has been damaged (by mutagens, radiation etc.) (Figure 2.22). The ability to detect DNA damage is vital in order to prevent the induction of cancer. DNA damage can also arrest the cell cycle in G_1 to allow repair to be made. This G_1 arrest is controlled by a protein known as p53 which is induced by the presence of DNA damage. Of great interest to cancer biologists is the finding that p53 is mutated in many human cancers leading to damaged DNA being replicated rather than repaired (see Chapter 3).

Over the past 10–15 years the proteins which control the progression of the cell cycle through the various checkpoints described have been identified using a variety of cells and animals. The first such factor to be identified came from experiments in which frog oocytes were stimulated by progesterone, which triggers their progression from G_2 into M phase, a process known as maturation. Such cells produced a factor which stimulated oocytes into the M phase in the absence of progesterone. This factor was termed maturation-promoting factor (MPF). Other workers studying yeast genetics discovered cell division cycle (cdc) mutants which were defective in their ability to complete cell cycle progression. Such mutants were arrested at specific points in the cell cycle and this suggested that specific gene products were involved in the progression through the various cell cycle checkpoints. Mutants such as *cdc*28, which caused cell cycle arrest at START, and *cdc*2, which blocked the G_1 to S transition and the G_2/M transition, were discovered. As it turned out the products of *cdc*28 and *cdc*2 were the same protein, which is now known as Cdc2. It was later discovered that Cdc2 was a protein kinase (see Chapter 3 for definition and description of the actions of protein kinases). Thus we have the discovery of MPF, which stimulates the G_2 to M transition, and Cdc2,

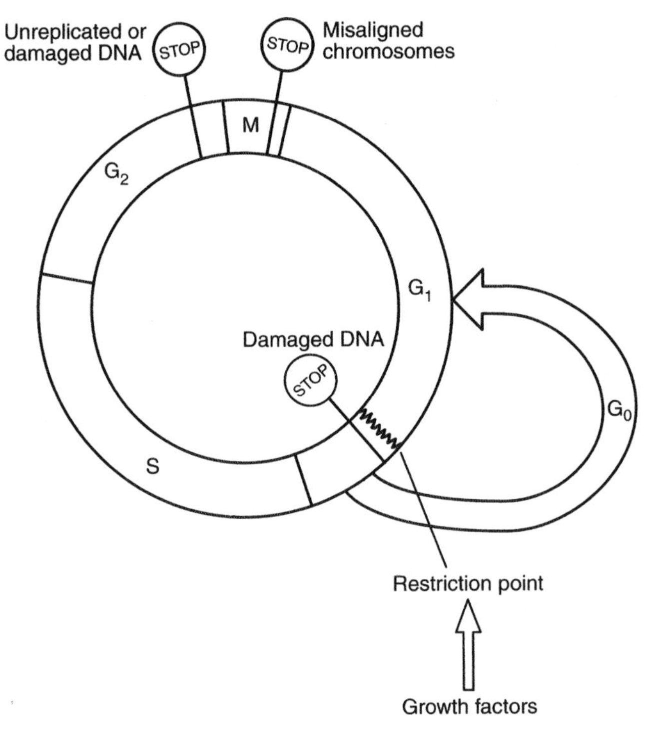

Figure 2.23 Restriction and checkpoints in the mammalian cell cycle. The availability of growth factors controls the cell cycle of animal cells at a point late in G_1 called the restriction point (START in yeast). If growth factors are not available during G_1, the cell enters a quiescent stage of the cell cycle known as G_0. The cell cycle also contains a number of checkpoints which are there to ensure that the cell is faithfully replicating its genome. Damage to DNA results in cell cycle arrest in either G_1 or G_2. In G_2 it is also possible to arrest the cell cycle if there is unreplicated DNA. Failure to align daughter chromosomes is another reason for cell cyle arrest, this time in mitosis

which can control the G_1/S and G_2/M transition. A third discovery made the link which eventually provided a satisfactory hypothesis to explain the control of cell cycle progression. This arose from studies of early sea urchin embryos which undergo a series of rapid cell division cycles following their fertilization. Using protein synthesis inhibitors it was shown that these embryos required new proteins in order to enter M phase. In the early 1980s two proteins were identified which displayed a periodicity, accumulating during interphase and disappearing at mitosis. These proteins were called cyclins (in fact cyclin A and cyclin B) and it was suggested that their build-up during the cell cycle might function to induce mitosis. This hypothesis was subsequently tested by microinjection of cyclin A into oocytes, a procedure which indeed was able to stimulate cell progression from G_2 into mitosis.

These various and diverse experimental approaches were eventually fused when MPF was isolated and shown to be composed of Cdc2 and cyclin B. The molecular ballet which Cdc2 and cyclin B undergo has now been elucidated (Figure 2.24). In mammalian cells, cyclin B synthesis is initiated in S phase and this molecule accumulates and forms complexes with Cdc2 (which is at a relatively stable concentration throughout the cell cycle) throughout S and G_2. An important facet of the control of this complex is the phosphorylation of Cdc2. This occurs at two key positions. One is at threonine 161 and is required for the kinase activity of Cdc2. The second is a dual phosphorylation at threonine 14 and tyrosine 15. Addition of these phosphate groups inactivates Cdc2 and thus the cell accumulates an inactive Cdc2/cyclin B complex during S and G_2. The key event enabling cells to transit from G_2 to mitosis is the dephosphorylation of Cdc2 at positions threonine 14 and tyrosine 15 by a protein phosphatase called Cdc25. Once activated, Cdc2 phosphorylates a variety of key proteins which initiates the process of mitosis and also triggers the degradation of cyclin B by a ubiquitin-dependent proteolysis mechanism, thus accounting for the precipitous fall in cyclin B levels found by earlier investigators. One of the functions of cyclin B is to activate Cdc2. Loss of cyclin B therefore shuts off the activity of Cdc2 and cells then exit mitosis and return to interphase. As will be described in Chapter 10 bacterial proteins have recently been discovered which inhibit the synthesis of cyclin B or inhibit the kinase activity of Cdc2. The consequence of these bacterial proteins that cells are blocked in G_2. These findings have only been made within the last year of writing and it is not known how many other mechanisms bacteria may use to block cell cycle progression.

Cyclin B is not the only cyclin and other cyclin/cyclin kinase complexes appear to have roles in ensuring that cells progress through other cell cycle checkpoints. It is now established that in yeast Cdc2 controls passage from G_2 to M in association with mitotic B-type cyclins (Clb1, Clb2, Clb3 and Clb4). Cdc2 is also involved in passage through START.

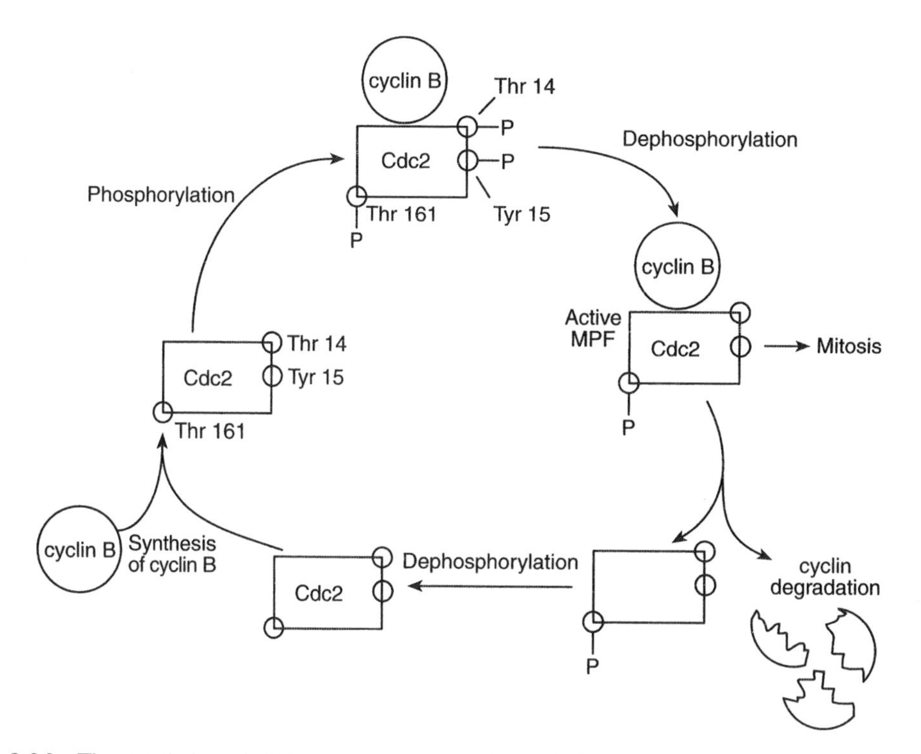

Figure 2.24 The regulation of MPF during the cell cycle. In S phase and G_2 Cdc2 forms a complex with cyclin B. This complex is phosphorylated at Thr14, Tyr15 and Thr161. The latter phosphorylated residue (Thr161) is required for activation of the MPF. However, phosphorylation of the other two sites renders the complex inactive. Dephosphorylation of Thr14 and Tyr15 by Cdc25, a protein phosphatase, renders the complex active and drives cells into mitosis. MPF activity is terminated by the proteolytic degradation of cyclin B

However, in order to drive cells through this checkpoint it complexes with different B-type cyclins (Clb5 and Clb6). The cell cycle of higher eukaryotes is also controlled by multiple cyclin kinases which are now termed cyclin-dependent kinases (Cdks). To confuse matters the original MPF kinase Cdc2 is now known as Cdk1, with the other identified kinases being numbered sequentially. These various Cdks associate with specific cyclins to drive cells through the various stages of the cell cycle. For example, progression from G_1 to S is primarily regulated by Cdk2 and Cdk4 in association with cyclins D and E.

To summarize, the cell cycle control system is dependent on the activity of specific kinases, the Cdks. The activity of these Cdks is, in turn, controlled by four distinct mechanisms: (1) association with specific cyclins which activates the kinases and this, in turn, depends on the synthesis and degradation of the cyclins; (2) activation of the Cdk/cyclin complexes requires phosphorylation at a conserved residue around position 160. This phosphorylation is catalysed by a Cdk-activating kinase (CAK) which is probably composed of a Cdk complexed with cyclin H; (3) inhibitory phosphorylation at positions 14 and 15 allows the G_2 phase to be completed before the cell enters mitosis. The importance of these phosphorylations is shown by mutation of tyrosine 15 in Cdk1 for phenylalanine which results in premature entry into mitosis; (4) a recently discovered fourth level of control is a family of proteins known as Cdk inhibitors (CKIs) which bind to cyclin/Cdk complexes and inhibit complex formation or complex activity. These additional levels of control represent targets for parasites including bacteria. It is fully expected that bacteria will show the capacity to manipulate the complex cell cycle mechanism at multiple points as part of novel virulence mechanisms.

Apoptosis

We are the only species to know that we will die and, in consequence, we tend to have a fear of death. However, death is necessary in biology. At the species level we are currently experiencing what happens if one species (namely ourselves) lives longer and multiplies more than should occur with the normal ecological checks and balances. Perhaps it is our natural fear of death that blinded us to the fact that in all organisms, death, in this case cell death, is absolutely essential for survival. Take the average human body, composed as it is of 10^{13} cells. Every day around 5×10^{11} blood cells are produced. If this happened without some form of compensatory loss of an equal number of old blood cells the results are obvious: a complete clogging up of the circulatory system.

It was only about a quarter of a century ago that it was recognized that cell death was a normal homeostatic process. Pathologists had long recognized cells which were dying and had described this process by a variety of terms including zeiosis, popcorn-type cytolysis, necrobiosis, pyknosis and karyorrhexis. However, it is only since the late 1980s that we have begun to appreciate the role that normal cell death – a process termed apoptosis, programmed cell death or sometimes 'cell suicide' – plays in a wide variety of normal and pathological processes. For example, during embryonic development, apoptosis (the term now most widely used by biologists) controls cells numbers in mesenchymal, neural and epithelial cell populations and is the central process in the deletion of tissues during organ folding and rotation. It is responsible for the loss of the web of soft tissues between the developing fingers. It is also a major mechanism for the negative selection of B and T lymphocytes and is thought to play a role in the pathology of autoimmune diseases and AIDS and also cancers. In AIDS, activation of apoptosis in CD4 T lymphocytes results in immunosuppression. Epstein–Barr virus, the causative agent of Burkitt's lymphoma, encodes a viral oncogene homologous to an oncogene known as *bcl-2*, which is capable of inhibiting apoptosis and contributing to malignancy. Indeed, many virally infected cells eliminate themselves by apoptosis, thus minimizing damage to uninfected neighbouring cells. As described in Chapter 9 and 10, viruses have evolved genes encoding proteins which inhibit apoptosis, and recent studies have shown that bacteria can induce apoptosis in eukaryotic cells.

Apoptosis has evolved such that cells can die and their contents, many of which may be noxious to other cells, can be disposed of in a way that does not damage bystander cells. This discriminates apoptosis from necrosis, another mechanism by which cells die. Necrotic cell death leads to the production and release of noxious stimuli which can cause extensive tissue damage. During apoptosis, chromosomal DNA is fragmented as a result of cleavage between nucleosomes. This produces a large number of pieces of DNA about 200 nucleotides long. These fragments can be detected on polyacrylamide gel electrophoresis. A more informative technique is TUNEL (TdT-mediated dUTP-biotin nick end labelling), which can be used on tissue sections or with cultured cells. This and related methods work on the basis that the single- or double-stranded breaks introduced into the DNA as part of the apoptotic process can be labelled and therefore the cells undergoing apoptosis can be recognized. The chromatin condenses and the nucleus breaks up into small fragments. The cell as a whole shrinks and breaks up into membrane-enclosed fragments which have been termed apoptotic bodies. Cells undergoing apoptosis and the fragments they produce are recognized by macrophages and by neighbouring cells and are phagocytosed and removed.

Mechanism of apoptosis

Apoptosis can be induced by a wide variety of factors, including radiation, hyperthermia, withdrawal of growth factors, glucocorticoids, certain cytokines and various pharmacological agents such as inhibitors of kinases. However, the common pattern of morphological and biochemical changes in cells undergoing apoptosis suggested that there was a single pathway of mortality which was independent of the apoptotic stimulus. This pathway involves a group of related enzymes, which are called caspases, that are present in the cell as inactive pro-forms which can be activated by a variety of stimuli. These proteases, either as part of a cascade or as a collection of overlapping enzymes, cleave key cellular proteins and result in cells dying without releasing components which could cause further tissue damage. It is not yet clear what constituents of cells are key to this process of proteolytic cleavage.

Much of our understanding of apoptosis came from the study of the nematode *Caenorhabditis elegans*. It was in this organism that it was demonstrated that mutation of the gene *ced-3*, which encodes for one of the caspaces, blocks apoptosis. Mutation studies showed that another gene in this worm, *ced-9*, was responsible for inhibiting apoptosis. Indeed, mutations that inactivate *ced-9* are lethal, as they cause the death of cells that should survive. In vertebrates the counterpart of *ced-9* is known as *bcl-2*. This was originally discovered as a proto-oncogene that is activated in lymphomas and the protein is overexpressed. It was subsequently shown that when added to cells Bcl-2 protected against growth factor withdrawal-induced apoptosis. Bcl-2 belongs to a family of proteins that can undergo homo- or hetero-dimerization. One of these proteins is termed Bax and counteracts the ability of Bcl-2 to protect against apoptosis. It is likely that this Bcl-2/Bax interplay is a key controlling element in apoptosis.

CONCLUSIONS

The key to cellular microbiology is the understanding of the interactions which occur between bacteria and host eukaryotic cells. In this chapter attention has focused on the structure of bacteria and eukaryotic cells and on the key cellular systems required for the functioning and replication of these cell types. In the next chapter attention switches to the mechanisms by which bacteria and eukaryotic cells recognize and respond to signals.

FURTHER READING

Books

Alberts B, Bray D, Johnson A, Lewis J, Raff M, Roberts K, Walter P. (1998) *Essential Cell Biology: An Introduction to the Molecular Biology of the Cell*. Garland Publishing, New York.
Cann AJ. (1997) *Principles of Modern Virology* (2nd edn.) Academic Press, London.
Cooper GM. (1997) *The Cell*. ASM Press, Washington, DC.
Gerhart J, Kirschner M. (1997) *Cells, Embryos and Evolution*. Blackwell Press, Malden, USA.
Latchman D (1998) *Gene Regulation : A Eukaryotic Perspective*. Stanley Thornes, Cheltenham.
Lewin B. (1997) *Genes VI*. Oxford University Press, Oxford.
McCrae MA, Saunders JR, Smyth CJ, Stow ND (1997) *Molecular Aspects of Host–Pathogen Interactions*. Cambridge University Press, Cambridge, UK.
Murray A, Hunt T (1993) *The Cell Cycle: An Introduction*. Freeman, New York.
Rappuoli R, Montecucco C (1997) *Guidebook to Protein Toxins and their Use in Cell Biology*. Oxford University Press, Oxford.

Reviews

Brennan PJ, Nikaido H (1995) The envelope of mycobacteria. *Annu Rev Biochem* 64:29–63
Collazo CM, Gallan JE (1997) The invasion-associated type-III protein secretion system in Salmonella: a review. *Gene* 192: 51–59
Cudmore S, Reckmann I, Way M (1997) Viral manipulation of the actin cytoskeleton. *Trends Microbiol* 5: 142–148.
Donachie WD (1993) The cell cycle of *Escherichia coli*. *Annu Rev of Microbiol* 47: 199–230.
Driessen AJM, Fekkes O, van der Wolk JPW (1998) The Sec system. *Curr Opin Microbiol* 1: 216–222.
Duong F, Eichler J, Price A, Leonard MR, Wickner W (1997) Biogenesis of the gram-negative bacterial envelope. *Cell* 91: 567–573.
Fuchs E, Cleveland DW (1998) A structural scaffolding of intermediate filaments in health and disease. *Science* 279: 514–519.
Groisman EA, Ochman H (1997) Pathogenicity islands: bacterial evolution in quantum leaps. *Cell* 87: 791–794.
Groisman EA, Ochman H (1997) How *Salmonella* became a pathogen. *Trends Microbiol* 5: 343–349.
Hall A (1998) Rho GTPases and the actin cytoskeleton. *Science* 279: 509–514.
Hancock IC (1997) Bacterial cell surface carbohydrates: structure and assembly. *Biochem Soc Trans* 25: 183–187.
Hirokawa N (1998) Kinesin and dynein superfamily proteins and the mechanism of organelle transport. *Science* 279: 519–526.
Holtje JV (1998) Growth of the stress-bearing and shape-maintaining murein sacculus of *Escherichia coli*. *Microbiol Mol Biol Rev* 62: 181–203.
Lee CA (1996) Pathogenicity islands and the evolution of bacterial pathogens. *Infect Agents Dis* 5: 1–7.
Lutkenhaus J (1998) The regulation of bacterial cell division: a time and a place for it. *Curr Opin in Microbiol* 1: 210–215.

Mermall V, Post PL, Mooseker MS (1998) Unconventional myosins in cell movement, membrane traffic, and signal transduction. *Science* 279: 527–533.
Moens S, Vanderleyden J (1996) Functions of bacterial flagella. *Crit Rev Microbiol* 22: 67–100.
Pugsley AP (1993) The complete general secretory pathway in Gram-negative bacteria. *Microbiol Rev* 57: 50–108.
Pugsley AP, Francetic O, Possot OM, Sauvonnet N, Hardie KR (1997) Recent progress and future directions in studies of the main terminal branch of the general secretory pathway in Gram-negative bacteria: a review. *Gene* 192: 13–19.
Rothfield LI, Justice SS (1997) Bacterial cell division: the cycle of the ring. *Cell* 88: 581–584.
Salvesen GS, Dixit VM (1997) Caspases: intracellular signalling by proteolysis. *Cell* 91: 443–446.
Shasham S, Shuman MA, Herkowitz, I (1998) Death-defying yeast identify novel apoptosis genes. *Cell* 92: 425–427.
Sleytr UB, Sjara M (1997) Bacterial and archaeal S-layer proteins: structure–function relationships and their biotechnological applications. *Trends in Biotechnol* 15: 20–26.
Smith GA, Portnoy DA (1997) How the *Listeria monocytogenes* ActA protein converts actin polymerization into a motile force. *Trends Microbiol* 5: 272–276.
Staley JP, Guthrie C (1998) Mechanical devices of the spliceosome: motors, clock, springs and things. *Cell* 92: 315–326.
Stephens C, Shapiro L (1996) Delivering the payload: bacterial pathogenesis. *Curr Biol* 6: 927–930.

Papers

Singer SJ, Nicolson GL (1972) The fluid mosaic model of the structure of cell membranes. *Science* 175: 720–721.
Sullivan JT, Ronson CW (1996) Evolution of rhizobia by acquisition of a 500-kb symbiosis island that integrates into a phe-tRNA gene. *Proc Nat Acad Sci USA* 95: 5145–5149.

CHAPTER 3

Prokaryotic and Eukaryotic Signalling Mechanisms

INTRODUCTION

The human body is composed of 10^{13} cells which are organized into a number of specialized tissues. It is self-evident that the activities of this multitude of cells have to be integrated to produce a healthy organism. The corollary to this statement is that tissue pathology may be caused by interfering with the integration of biological signals. In a world saturated with the transmission of information, the importance of information transfer and the consequences of defective transfer should be obvious to the reader.

The ability of microorganisms to manipulate host cellular messages for their own ends is becoming a common theme in cellular microbiology and in this chapter the nature of cell-to-cell signalling in both eukaryotes and prokaryotes will be described and the possible interactions between these systems considered. Cell-to-cell signalling is produced by a wide variety of molecules. In eukaryotes many of these signals are proteins and one class of protein, the cytokines, are clearly seen to be important in the response to bacteria. Cytokines both control the innate and acquired immune response and are the major mediators of tissue pathology. The role of cytokines in the organisms' response to bacteria will be discussed in more detail in Chapters 8–10. In this chapter the intercellular signalling systems used by both eukaryotes and prokaryotes will be described, followed by a discussion of the intracellular signalling systems used both by prokaryotes and eukaryotes. The systems of intercellular and intracellular signalling are intimately connected, with the molecules involved in the former specifically acting on cells to induce particular patterns of intracellular signalling activity. It is now clear that there is much that is shared between bacteria and eukaryotic cells in terms of signalling between cells and, in particular, signalling within cells.

One purpose of this chapter is to provide basic information to enable readers to understand later chapters which will discuss intercellular and intracellular signalling and its relationship to cellular microbiology. The three signalling systems used in multicellular organisms are neuronal, endocrine and cytokine signalling (Figure 3.1). This chapter will not

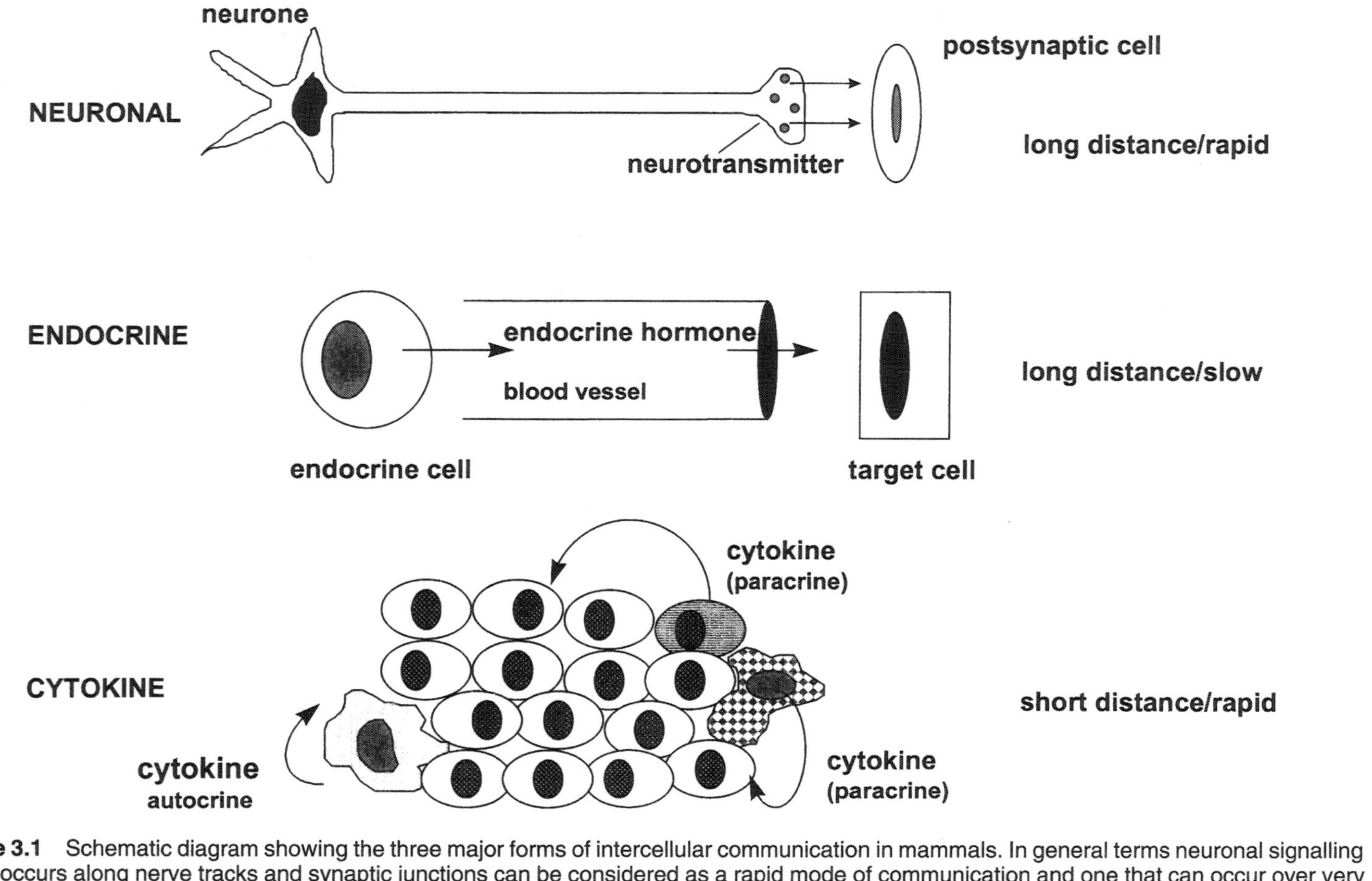

Figure 3.1 Schematic diagram showing the three major forms of intercellular communication in mammals. In general terms neuronal signalling which occurs along nerve tracks and synaptic junctions can be considered as a rapid mode of communication and one that can occur over very long distances (e.g. brain to big toe). It is now appreciated that many chemicals are involved in signalling at synapses and also have roles in inflammation. A good example is substance P, mentioned in the text. Endocrine signalling involves the release of a hormone from its glandular tissue and the transport of this molecule in the blood (possibly on a carrier protein) to a limited number of cells in the target tissue. Such signalling occurs over a long distance and is limited by the rate of blood flow and diffusion. The most recently discovered intercellular signalling system is that of the cytokines. Much of this signalling occurs over short distances either cell to nearby cell (paracrine signalling) or by stimulation of the cell producing the cytokine (autocrine signalling). Some cytokines (EPO, IL-6) have systemic actions. Indeed, there is a growing appreciation of the overlap between homone and cytokine signalling systems (reproduced from Henderson *et al* Bacteria-Cytokine Interactions in Health and Disease (1998), Portland Press, London, with permission)

discuss the first, and possibly most important, system of cell-to-cell comunication, namely neuronal signalling, although certain of the endocrine hormones which will be briefly mentioned act as neurotransmitters. As we learn more about the organization of complex multicellular organisms we see just how interrelated are the cell-to-cell signalling systems and how they can be targets for attack by infectious agents. Chapters 2 and 7 briefly mention that certain bacterial exotoxins target neural signalling. These are the neurotoxins produced by *Clostridium tetani* and *Cl. botulinum*, which have metalloproteinase activity and cleave specific intracellular proteins, thus preventing the release of neurotransmitters. It is this blockade of neurotransmitter release which produces the classic pathology of these infections. A more recent discovery also points to the interaction of a bacterial toxin with neuroendocrine signalling and cytokine synthesis. In this case the example is toxin A from *Cl. difficile*. This toxin causes diarrhoea and intestinal inflammation in man and the mouse. In mice, in which the gene for the neurokinin-1 receptor (which binds the neurohormone, substance P) has been knocked out by homologous recombination, there is no pathological response to toxin A and this protective effect correlates with a diminution in intestinal levels of the proinflammatory cytokine TNFα. This is a fascinating example of the interactions between signalling systems, and brings home the point of the connectedness of all cellular systems.

EUKARYOTIC CELL-TO-CELL SIGNALLING

Our understanding of the integrative nature of biological systems owes much to the pioneering work of the Frenchman Claude Bernard (1813–1878), who created the concept of the maintenance of the *milieu interieur* and who suggested that the system of ductless glands, which we now know as the endocrine system, performed an integrating function to maintain homeostasis. However, it was not until 1902 that Bayliss and Starling demonstrated that an acid extract of the duodenum, when injected into dogs, induced a marked flow of pancreatic juice. The substance responsible was named *secretin* and Starling subsequently coined the term hormone (from the Greek word meaning 'I excite') for such intercellular messengers. From this study, the concept developed of hormones as molecules produced by specific glands and secreted directly into the blood, in which they are conveyed to selected organs or tissues where they exert their effect (Figure 3.1). These ideas were further developed by the American physiologist Walter Cannon, who coined the term homeostasis (Greek: *homoios* 'the same' and *stasis* 'standing') which literally means 'standing or staying the same'. However, Cannon was aware of the dynamic nature of physiological systems and defined homeostasis as 'a condition that may vary, but which is relatively constant'. At the end of the Second World War advances in mathematics and the development of the computer had given rise to the concepts of cybernetics, information theory and games theory. Cybernetics, its offshoot General System Theory, and also non-equilibrium thermodynamics have had marked influences on our understanding of the mechanisms used in the homeostatic control of biological systems. It was during this period that the pioneering work was begun on endotoxin-induced fever, an aberration in the control system of temperature homeostasis caused by lipopolysaccharide (LPS) from infecting Gram-negative bacteria, and led to the discovery of protein factors which we now know as cytokines.

Endocrine hormone signalling

We now recognize three major groups of endocrine hormones. A number of specialized tissues such as the hypothalamus, pituitary and parathyroid glands and tissues such as the pancreas and intestine produce a range of polypeptide hormones. A number of the peptide hormones, particularly those produced by the intestines, have neurotransmitter-like activity. The second group are the steroid hormones produced by the adrenal cortex, gonads and skin. The latter tissue produces a very interesting 'pro-hormone' which is hydroxylated in the liver and kidney to 1,25-dihydroxy-vitamin D_3 (commonly termed vitamin D or calcitriol). This sterol hormone was first recognized for its importance in the maintenance of the skeleton, and deficiency results in rickets in children and osteomalacia (demineralization and softening of bone) in adults. In recent years it has been found that human macrophages, if activated by the cytokine gamma (γ)-interferon (IFN), produce a hydroxylase which also converts the circulating inactive vitamin D metabolite to the active 1,25-dihydroxy-vitamin D_3. This vitamin D metabolite is a potent activator of macrophages and enhances their antibacterial activities, particularly against *Mycobacterium tuberculosis*, the causative organism of tuberculosis. This example shows how interrelated the physiological, hormonal and antibacterial systems evolved by mammals can be. The third group of endocrine hormones are derivatives of the amino acid tyrosine and include the thyroid hormones (T3, T4 etc.) and the catecholamines noradrenaline, adrenaline and dopamine. These latter hormones have neurotransmitter activity.

Endocrine hormones are generally produced by specific glands or glandular tissue and are secreted into the blood. Within the blood they circulate either as the free hormone, or bound to carrier proteins. Albumin is the classic example of a serum protein able to bind a range of circulating hormones. These circulating hormones have no inherent enzymic or other activity. They only exert their actions when they bind to specific cell receptors in target tissues. The polypeptide hormones act at the cell surface by binding to specific plasma membrane receptors. Such binding results in specific intracellular signalling pathways being activated, the general details of which will be discussed in later sections of this chapter. In contrast, the steroid and sterol hormones enter into cells and bind to cytoplasmic receptor proteins which are then able to move into the nucleus, where they act as transcription factors.

Endocrine hormones have a very wide range of physiological actions and only general principles will be discussed here. The control of energy metabolism is a good example of the integration of multiple endocrine hormones. This process involves the production and breakdown of carbohydrate energy stores (the storage carbohydrate glycogen) in the liver and muscle. This is controlled by, among others, the polypeptide hormones insulin and glucagon, and the non-peptidic homones adrenaline and noradrenaline. The devastating effects of the lack of control of this system are shown clearly in diabetes, which, before the introduction of insulin replacement, was a lethal condition. Infection can also alter the hormonal balances in the body and result in energy imbalances. For example, bacterial cytokine inducers such as LPS can promote a state known as cachexia, which is a wasting state where people lose body mass. This is due to the induction of the synthesis of the pro-inflammatory cytokines tumour necrosis factor (TNF)α and interleukin (IL)-1. Cachexia is found in patients with tuberculosis. The common name for this disease was consumption, because patients wasted away, i.e. were consumed by the disease. It is believed that one of the major cytokine-inducing components of *M. tuberculosis* (which does not contain LPS) is cord factor (6,6'-dimycolyltrehalose).

Can bacteria interfere with endocrine signalling?

The large group of intercellular signalling proteins known as cytokines play a major role in the organism's response to both infectious and commensal microorganisms. Certain bacteria and viruses are known to affect neural tissues. For example, *Mycobacterium leprae*, which causes leprosy, and *Treponema pallidum*, the causative organism of syphilis, both have a tropism for nervous tissue and their chronic pathologies mirror this. *M. leprae*, for example, grows within the Schwann cells of peripheral nerves. Have bacteria, or other infectious agents, evolved strategies to target and dysregulate endocrine hormones?

The gastric and intestinal mucosae are tissues which are highly regulated and produce and respond to a wide range of endocrine signals including gastrointestinal hormones such as gastrin, secretin, cholecystokinin and guanylin (a recently discovered gastrointestinal peptide hormone). Many bacteria have the capacity to alter the fluid balance in the intestines and cause diarrhoea. *Escherichia coli* appears to be the most successful bacterium causing diarrhoea and seven different strains have now been defined, which are able to induce pathology by distinct mechanisms (Chapter 1, Table 1.4). Enterotoxigenic strains of *E. coli* produce heat-labile toxin (LT) and heat-stable toxin (ST). The latter toxin binds to the cell surface receptor for the hormone guanylin on intestinal epithelial cells, a membrane-associated guanylyl cyclase, causing an increase in intracellular cyclic guanosine monophosphate (cGMP) which results in uncontrolled fluid loss from these cells (Figure 3.2). It has recently been established that STa is the first bacterial analogue of an endocrine hormone, guanylin, whose function is to activate the intestinal cells' guanylyl cyclase (via the same receptor as STa) and is believed to act in the control of fluid release from these cells, possibly to keep the mucin layer wet (see Chapter 7 for further details). STa mimics the action of guanylin, but in doing so produces diarrhoea. Is STa the first of many bacterial hormone analogues? Another heat-stable guanylin-like toxin called EAST1 has been isolated from enteroaggregative strains of *E. coli* and several other bacteria (*Yersinia enterocolitica, Klebsiella, Citrobacter freundii*) produce structural homologues of STa. Thus the existence of other bacterial endocrine hormone analogues seems likely. One way to search for these would be to look for homologies to mammalian hormones in the bacterial genome sequences that are now available.

Cytokines: early evolved intercellular signals

The term cytokine refers to a large, and steadily enlarging, group of proteins (currently well over 100) which act as cell-to-cell signalling agents and which are key controlling elements in the inflammatory response to infectious agents. Much of our current understanding of infections and of their pathological consequences relies on a knowledge of cytokine biology and so it is important to understand the biology of these proteins and their role in inter- and intra-cellular signalling. There are many cytokines and the terminology used to describe them is confusing. This is a major problem in describing cytokine biology and the authors have tried to minimize the confusion in the hope that the reader will not be put off reading about these key molecules.

The term cytokine is very loosely defined. The following definitions have been taken from current textbooks on cytokines: (i) 'secreted regulatory proteins that control the survival, growth, differentiation and effector function of tissue cells' (ii) 'a diverse group of soluble proteins and peptides which, either under normal or pathological conditions, modulate the

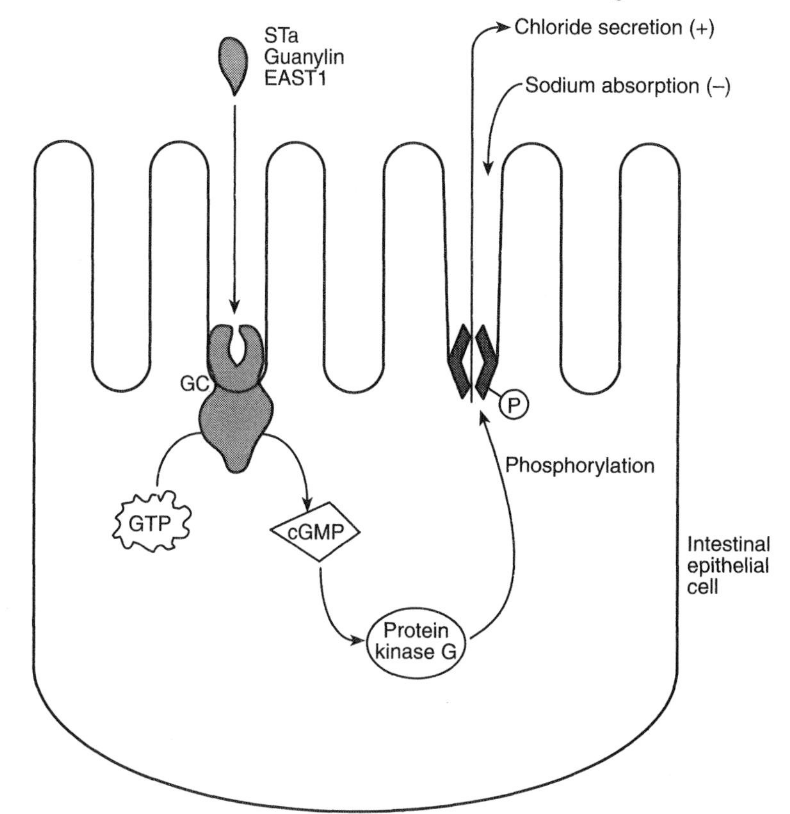

Figure 3.2 Action of the *E. coli* toxin STa. Binding of STa, guanylin or EAST1 to the cell surface guanylate cyclase (GC) results in guanosine trisphosphate (GTP) being converted to cyclic GMP (cGMP). The build-up of cGMP in the cell activates protein kinase G, which phosphorylates a membrane transporter resulting in secretion of chloride and inhibition of sodium absorption

functional activities of individual cells and tissues' (iii) 'polypeptide hormones secreted by a cell that affects growth and metabolism, either of the same cell (autocrine signalling) or of another cell (paracrine signalling) (iv) 'proteins whose overproduction cause disease'. These definitions are not particularly exclusive but they are meant to exclude the various peptide and non-peptide endocrine hormones and the large number of peptide and non-peptide mediators which are associated with inflammation and which are commonly found at sites of infection. Such mediators include the various lipidic mediators: prostaglandins (PGs), leukotrienes, lipoxins, platelet-activating factor (PAF) and the mediators from mast cells, such as histamine and enzymes such as tryptase, and a variety of small peptides known as kinins. While for didactic purposes it is important to try and provide an all-inclusive definition of cytokines, an objective that has not yet been achieved, it should be emphasized that in practice there is significant overlap between cytokines and other mediators. For example, erythropoietin (EPO) is a 30 kDa glycoprotein produced by endothelial and interstitial cells in the kidney that stimulates the production of erythrocytes by acting on erythrocyte progenitor cells. The production of EPO is subject to classic endocrine feedback control, with the kidney controlling the synthesis of this protein by monitoring oxygen tension in the blood. If renal pO_2 is reduced, for example, EPO production is increased. One way of classifying cytokines is on the basis of their cell surface receptors. The EPO receptor is a member of the haemopoietin receptor family which binds a wide range of cytokines (IL-2, IL-3, IL-4 etc.). Thus on this very clear basis, EPO is a cytokine.

A brief history of cytokine research

The discovery of cytokines was made by scientists and clinicians who were studying infectious diseases. The first cytokine to be discovered was interleukin 1 (IL-1) and this was initially discovered, or at least its biological activities were disclosed, in the period from the late 1940s to the early 1960s by investigators, principally in the United States, interested in what caused fever in patients with infections. Experimental studies in the early 1950s revealed that injection of the Gram-negative bacterial cell wall component endotoxin (described in Chapter 2) into animals resulted in the appearance in the blood of a heat-labile (presumed to be proteinaceous) activity which was termed endogenous pyrogen (EP). This protein was partially purified and produced fever when injected into animals. However, EP proved difficult to isolate and it was not certain that its biological actions were due to one protein. In the 1960s immunologists discovered a factor produced by activated macrophages which stimulated the growth of lymphocytes. This activity was called lymphocyte-activating factor (LAF). In the 1970s reports appeared suggesting that EP and LAF shared many properties, although there remained concern that isolated proteins were contaminated with endotoxin. It was not until the late 1970s that EP and LAF, and a number of other acronyms, defining additional biological activities, were brought together and given the name interleukin 1. A few years later it was found that there were two genes encoding IL-1 proteins, the two proteins being termed IL-1α and IL-1β. These two proteins had different physicochemical properties. With hindsight, the fact that the activity of EP or LAF was due to two distinct proteins had contributed to the problems in purifying this biological activity. By the end of the 1980s another member of the IL-1 family had been found. This was a protein which bound to the IL-1 receptor but did not trigger an agonist response. Since this molecule acted as a receptor antagonist it was named interleukin 1 receptor antagonist (IL-1ra). A further proposed member of this family is IL-18, also termed IL-1γ, and the human genome mapping project is revealing further members of this IL-1 group of cytokines.

Another pioneering investigation which contributed to our understanding of cytokines was the discovery in the 1950s of viral 'interference', whereby a viral infection blocks infection by a competing virus. Isaacs and Lindemann in the United Kingdom found that killed virus could inhibit the infection of chorioallantoic membranes by live viruses. The conclusion was that a soluble factor, which they called 'interferon', was produced by cells as a result of virus infection and that this factor could prevent the infection of other cells. This work was initially very controversial and the study of the interferons only really became successful with the introduction of the techniques of molecular biology. We now know that there are three types of interferon with more than 20 different gene products constituting this family of proteins.

The 1960s saw the beginning of cytokine biology which has grown at an unbelievable speed from these early beginnings. In addition to the discovery of interleukin 1 and interferon, the 1960s saw the discovery of growth factors such as nerve growth factor and epidermal growth factor. Colony-stimulating factors were discovered in the late 1960s, as were a range of protein factors that stimulated the growth of lymphoid cell populations.

These were named lymphokines, but this terminology has largely been superseded by the term interleukins. The latest family of cytokines to be discovered are the chemokines, which began to be discovered in the mid-1980s.

Thus it is only during the past 20 years that biologists have become aware of the existence of these enormous numbers of cytokines. More than 100 distinct cytokines have been described and, as most cytokines have their own specific receptor, the number of proteins in the cytokine menagerie is probably well over 200. This story has certainly not stopped, and new cytokines are being discovered every year. The pace of discovery is now being aided by the ability to identify new cytokine genes by homology searching within the expressed sequence tags (ESTs) that have been thrown up as part of the Human Genome Project.

Cytokine nomenclature

The pleiotropy of cytokines, their variegated structures and the finding that known proteins and peptides (e.g. molecular chaperones, antibiotic peptides) may show functional overlap with cytokines makes it important to attempt to classify these proteins in order to bring order to this arena of biology. At the present time cytokines are divided into six 'families' (Table 3.1). These subdivisions are based upon a number of criteria including historical subdivisions, sequence homology, chromosomal localization and biological actions. The term interleukin (*inter*: between; *leukin*: leukocytes) was coined in 1979 to describe protein factors produced by leukocytes which function to modulate the behaviour of other leukocytes. Currently, there are more than 20 interleukins (IL-1 to IL-18) and at least five other proteins have just been found which are likely to be termed interleukins. The cellular sources and functions of these interleukins are provided in Table 3.2. As can be seen, the designation – interleukin – is incorrect for a number of these proteins. For example, members of the IL-1 family, IL-3, IL-6, IL-8, and IL-11, are made by cells other than leukocytes and some of these cytokines can act on cells other than leukocytes. Many interleukins are growth factors for

Table 3.1 Cytokine Nomenclature and Cytokine Families

Family	Examples	Major Biological Activities
Interleukins[a]	IL-1 to IL-18	Mainly lymphoid-lineage growth factors
Cytotoxic cytokines	TNFα, TNFβ, CD40L	Proinflammatory molecules with cytotoxic/apoptotic potential
Interferons	α^{b}, β, γ interferons	Antiviral and immunological actions
Colony-stimulating factors	IL-3, M-CSF, G-CSF, GM-CSF	Myeloid growth and differentiation factors
Growth factors	EGF, TGFα, PDGF	Proliferation of various cell types including epithelial and mesenchymal cells
Chemokines	IL-8, MIP-1α, MCP	Chemotactic proteins for various leukocyte populations

[a] The interleukin family also contains IL-3, which is a colony-stimulating factor, and IL-8, which is a chemokine.
[b] Many subtypes of α interferon exist.

Table 3.2 Cellular source and major functions of the interleukins

Interleukin	Cell source	Major functions
IL-1α (17)[a]	Many cell types	Multiple proinflammatory functions (see Table 3.3)
IL-1β (17)	Many cells types	As IL-1α
IL-1ra (17–22)	MØ, endothelial cell, keratinocytes	Inhibits biological actions of IL-1 (antagonist)
IL-2 (15)	Th_o, Th_1, memory T cells	T cell growth and differentiation
IL-3 (15–17)	T cells, mast cells, epithelial cells	PMN/monocyte colony-stimulating factor
IL-4 (20)	Th_2 lymphocytes	B cell proliferation, MØ inhibition
IL-5 (13)	T lymphocytes	Eosinophil growth/differentiation factor
IL-6 (21)	Many cell types	Multiple functions (overlaps with IL-1)
1L-7 (25)	Bone marrow stromal cells/ thymic cells	B/T cell proliferation factor
IL-8 (8)	MØ, fibroblasts, PMN etc.	T cell/monocyte chemokine/PMN activator
IL-9 (18)	CD4 T cells	Mast cell/T cell growth factor
IL-10 (40 hd[b])	CD4/CD8 T cells, monocytes	Anti-inflammatory, inhibits MØ, T cell activity
IL-11 (23)	Mesenchymal cells	Haematopoietic growth factor, accessory immune factor
IL-12 (70 htd[c])	MØ, B and T lymphocytes, etc.	generates Th_1 lymphocytes
IL-13 (10)	T lymphocytes	Stimulates B cells, inhibits some MØ functions
IL-14 (60)	T and B lymphocytes	B cell growth factor
IL-15	PBMCs	T cell growth factor
IL-16	T lymphocytes	T cell chemoattractant
IL-17	T lymphocytes	Proinflammatory
IL-18 (IL-1γ)	MØ, keratinocytes	Induces Th1 cells, activates NK cells

[a] (), molecular mass of biologically active interleukin (kDa).
[b] hd, homodimer.
[c] htd, heterodimer.

lymphocytes but this family of proteins also contains a colony-stimulating factor (IL-3), the chemokine (IL-8) and a protein with anti-inflammatory properties (IL-10). The reader should have little faith that the name given to a particular cytokine defines its major biological function or functions. This is one of the problems which has arisen with the rapid expansion of cytokine biology.

Tumour necrosis factor (TNF), now recognized as a key protein in infection, was identified in the serum of endotoxin-injected mice as an activity which could kill certain tumour cell lines. This cytokine is probably the active principle in Coley's toxin. Dr Coley, an American physician at the turn of the nineteenth century, reported that injection of bacterial filtrates into cancer patients could cure them. These experiments have been repeated over the past decade with recombinant TNFα and has shown some clinical benefit. The TNF family of cytokines initially included only TNFα and lymphotoxin (also called TNFβ). In addition to their ability to kill certain cell lines, these cytokines are also potent pro-inflammatory molecules and have substantial overlap with IL-1 in terms of biological activity (Table 3.3). Analysis of the TNF receptor gene has revealed that there is a family of TNF receptors which

Table 3.3 Overlapping activities of IL-1 and TNFα

In vivo	Endogenous pyrogens
	Inducers of acute-phase protein synthesis
	Inducers of septic shock-like conditions
	Inhibitors of cardiac function
	Inducers of hyperalgesia
In vitro	Activators of T and B lymphocytes
	Inducers of cyclooxygenase II
	Inducers of metalloproteinase synthesis and release
	Stimulators of fibroblast proliferation
	Inducers of vascular endothelial cell adhesion molecules (ICAM, E-selectin)
	Modulators of vascular endothelial cell coagulation systems
	Stimulators of cartilage breakdown
	Inhibitors of connective tissue matrix macromolecules
	Stimulators of bone resorption
	Induction of cytokine synthesis including autocrine stimulation of IL-1 and TNF

bind a wide range of ligands including nerve growth factor. Interestingly, a number of the ligands of the TNF receptor family are themselves membrane-bound proteins (e.g. CD27, CD30 and CD40). As the key role of the TNFs is cell killing (via induction of apoptosis) or inhibition of cell growth it might be more sensible to redesignate this family as a cell growth inhibitor family and this would bring in a number of other cytokines with similar effects, such as amphiregulin and oncostatin M, which do not bind to the TNF receptor family.

As described, the interferons (IFNs) were some of the first cytokines to be discovered and three families of these cytokines that have a major function in inhibiting the growth and spread of viruses are recognized. The α-IFNs have the most potent antiviral actions. The β- and γ-IFNs have antiviral actions but also modulate immune responsiveness. For example, IFN-γ is one of the most potent activators of macrophages. The mechanism of action of the interferons is shown schematically in Figure 3.3. It should also be noted that the IFNs act against other intracellular parasites including rickettsia, mycobacteria and protozoa.

The colony-stimulating factors (CSFs) are a small group of cytokines which are involved in the control of the growth and differentiation of polymorphonuclear leukocytes (abbreviated to PMN but more often called neutrophils), monocytes and cell populations derived from monocytes (macrophages, dendritic cells and osteoclasts) in the bone marrow. These cytokines also have actions on the mature cells. The role of leukocytes in the defence against infections will be described in detail in Chapters 8 and 9. Briefly, the PMN and the monocyte/macrophage are phagocytic cells which can phagocytose and kill bacteria. The monocyte/macrophage and dendritic cell are aso known as antigen-presenting cells (APCs) and can stimulate antigen-selective T and B lymphocytes.

The term growth factor encompasses a very large number of proteins, including families of proteins such as the TGFβ superfamily (which includes the bone morphogenetic proteins (BMPs) and the activins/inhibins) and the fibroblast growth factor (FGF) family. These cytokines can act on mesenchymal cells (fibroblasts, osteoblasts, chondrocytes) as well as epithelial cells. However, it should be noted that, in spite of the nomenclature, some of the growth factors are able to inhibit the growth of certain primary cells or cell lines. Some growth factors, such as platelet-derived growth factor (PDGF) and FGF, are implicated in the process of carcinogenesis and are classed as oncoproteins (products of oncogenes).

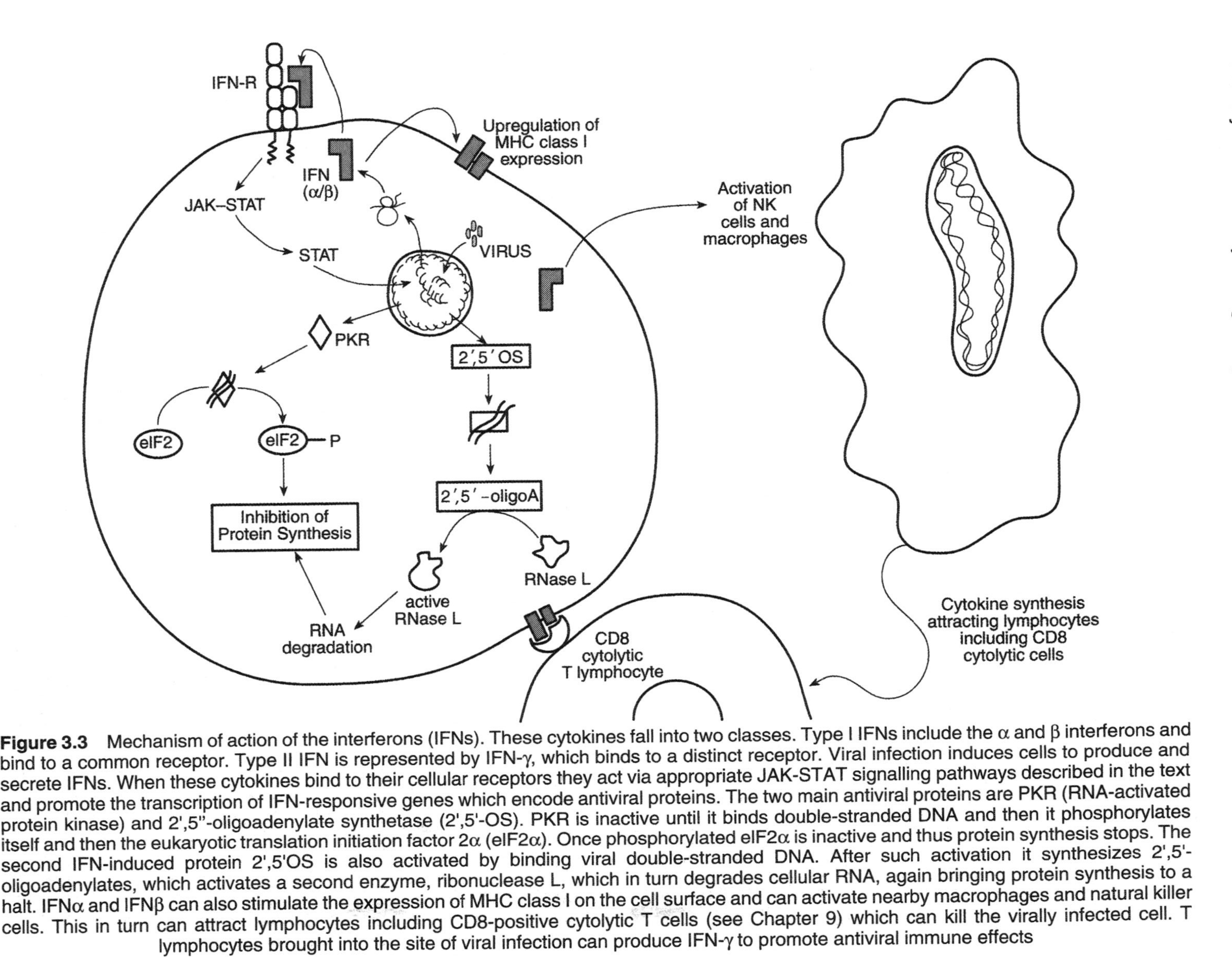

Figure 3.3 Mechanism of action of the interferons (IFNs). These cytokines fall into two classes. Type I IFNs include the α and β interferons and bind to a common receptor. Type II IFN is represented by IFN-γ, which binds to a distinct receptor. Viral infection induces cells to produce and secrete IFNs. When these cytokines bind to their cellular receptors they act via appropriate JAK-STAT signalling pathways described in the text and promote the transcription of IFN-responsive genes which encode antiviral proteins. The two main antiviral proteins are PKR (RNA-activated protein kinase) and 2',5"-oligoadenylate synthetase (2',5'-OS). PKR is inactive until it binds double-stranded DNA and then it phosphorylates itself and then the eukaryotic translation initiation factor 2α (eIF2α). Once phosphorylated eIF2α is inactive and thus protein synthesis stops. The second IFN-induced protein 2',5'OS is also activated by binding viral double-stranded DNA. After such activation it synthesizes 2',5'-oligoadenylates, which activates a second enzyme, ribonuclease L, which in turn degrades cellular RNA, again bringing protein synthesis to a halt. IFNα and IFNβ can also stimulate the expression of MHC class I on the cell surface and can activate nearby macrophages and natural killer cells. This in turn can attract lymphocytes including CD8-positive cytolytic T cells (see Chapter 9) which can kill the virally infected cell. T lymphocytes brought into the site of viral infection can produce IFN-γ to promote antiviral immune effects

The final subdivision of the cytokines is the large group of peptide chemotactic factors known as chemokines. The chemokines have molecular masses in the range 8–10 kDa with 20–50% sequence homology at the protein level and all have conserved cysteine residues involved in intramolecular disulphide bond formation. This large group of proteins can be divided into two major families based both upon the chromosomal location of the genes and the structure of the proteins. The α-chemokine family map to human chromosome position 4q12–21 and the first two cysteine residues are separated by a single amino acid. Because of this, these proteins are called the C-X-C chemokines (C being the single-letter code for cysteine). The β-chemokine family map to human chromosome position 17q11–32. In this group of proteins the first two cysteine residues are adjacent and therefore they are termed the C-C chemokines. The specificity of chemokines for leukocytes is not fully established. In general, C-X-C chemokines tend to attract neutrophils while C-C chemokines attract monocytes. Thus IL-8, which is a C-X-C chemokine, is a powerful neutrophil chemoattractant but will also cause lymphocyte chemotaxis at higher concentrations. Eotaxin is selective for eosinophils. A third family of chemokines first discovered in 1994 has currently one family member called lymphotactin, which is a powerful chemoattractant for T lymphocytes. This protein lacks two of the four characteristic cysteine residues (and is denoted a C chemokine) but in terms of sequence resembles the C-C chemokines. The most recently classified family of chemokines is the CX3C family, whose major member is the membrane-bound chemokine termed fractalkine. The chemokines were and are being discovered at a rapid rate and have had some interesting names applied to them (see Table 3.4 for a list of these proteins, their source and their target cells).

The subdivisions of cytokines just described was derived historically and, given our current knowledge about the biological activity, gene and protein structure and receptor binding of cytokines other ways of classifying these proteins are now possible. For example, the various cytokines can be subdivided into nine groupings which delineate the major biological actions of these proteins (Figure 3.4).

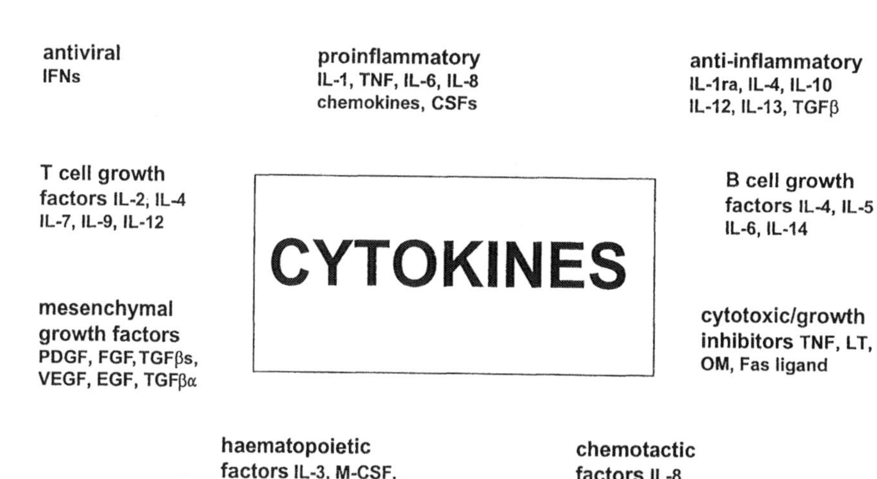

Figure 3.4 Subdivision of cytokines based on their biological actions. This is not a comprehensive list of the known cytokines but a representative sample of cytokines exhibiting each activity. Of course it should be noted that many cytokines have multiple actions (Reproduced from Henderson *et al* (1998) Bacteria-Cytokine Interactions in Health and Disease. Portland Press, London, with permission)

Table 3.4 The chemokines

Name[a]	Family	Major source	Target cell
IL-8	C-X-C	Monocyte/macrophage/lymphocyte	PMN, T cell, keratinocyte
MGSA	C-X-C	monocyte/macrophage/epithelia	PMN, T cell monocyte
PF4	C-X-C	platelet	Fibroblast, platelet, pericyte
NAP-2	C-X-C	Platelet	PMN, fibroblast, endothelial cell
IP10	C-X-C	Monocyte/endothelial cell/fibroblast	Monocyte, endothelial cell, NK cell
MIP-2	C-X-C	Mast cell/alveolar macrophage etc.	PMN, epithelial cell, basophil
ENA-78	C-X-C	Epithelial cell, platelet	PMN
EMF-1	C-X-C	RSV[b]-infected cell	Fibroblast, mononuclear cell
GCP-2	C-X-C	Osteosarcoma cell/kidney tumour cell	Granulocyte
SDF-1	C-X-C	Stromal cells/bone/liver/muscle	Pre-B lymphocyte
MIG	C-X-C	Activated macrophage/monocyte	Monocyte, tumour-infiltrating lymphocyte
MCP-1	C-C	Monocyte/macrophage/fibroblast etc.	Monocyte, T cell, NK cell etc.
MCP-2	C-C	Fibroblast/mononuclear cell etc.	T cell, monocyte, eosinophil etc.
MCP-3	C-C	Fibroblast/platelet/monocyte etc.	Monocyte, T cell, NK cell etc.
MIP-1α	C-C	Monocyte/fibroblast/PMN etc.	T cell, monocyte, PMN etc.
MIP-1β	C-C	Monocyte/fibroblast/lymphocyte etc.	Monocyte, T cell, basophil
RANTES	C-C	Monocyte/T cell/epithelia etc.	Monocyte, T cell, NK cell, dendritic cell
I-309	C-C	T cell/mast cell	Monocyte, macrophage, PMN etc.
Eotaxin	C-C	Macrophage/endothelial cell etc.	Eosinophil
C10	C-C	Macrophage/T cell	T cell, monocyte
HCC-1	C-C	Plasma/spleen/liver etc.	Monocyte, haematopoietic progenitors
Lymphotactin	γ family	Thymocyte/activated T cell	T cell
SCM-1	γ family	T cells	

[a] MGSA, melanocyte stimulatory growth factor (also GRO-α, NAP-3, KC [murine]); PF4, platelet factor 4; NAP-2, neutrophil-activating protein 2; IP10, gamma interferon inducible protein 10; MIP-2, macrophage inflammatory protein 2 (also GRO-β); ENA-78, epithelial-derived neutrophil chemoattractant 78; GCP-2 – granulocyte chemotactic protein 2; SDF-1, stromal cell-derived factor 1; MCP-1, monocyte chemotactic protein 1; MIG, monokine induced by IFNγ; MIP-1α, macrophage inflammatory protein 1α; RANTES, regulation on activation, normal T cell expressed and secreted; SCM-1, single cysteine motif.
[b] RSV, respiratory syncytial virus.

The reader should be aware that modern biology is undergoing a revolution based upon the avalanche of information which is coming from genome sequencing and from the determination of protein structures. This information about gene sequences and about the ways that proteins are folded into their three-dimensional structure is increasingly being used to look for 'homology' between proteins. The reader will probably be familiar with homology/identity in the context of nucleotide or amino acid sequence. With the increasing numbers of protein structures becoming available for comparison (approximately 10 000 at the current time) it is possible to search databases for structural homology. This is the province of structural biology and this disclipline (a combination of protein biochemistry, molecular biology and physical techniques such as X-ray crystallography and nuclear magnetic resonance (NMR) spectroscopy is increasingly being used to search for structural homology in terms of the various folding patterns that linear amino acid sequences can assume. Such structural biological information is now being used to classify cytokines on the basis of (i) their three-dimensional structure and (ii) the structure of the receptors they bind to.

Structurally, cytokines are generally proteins of low molecular mass (the molecular mass range is 5 kDa for TGFα, to 145 kDa for Mullerian-inhibiting substance, MIS) consisting of a single polypeptide chain. Cytokines with structures other than a single polypeptide chain include TGFβ, PDGF, IL-12 and MIS. In addition, TNFα and TNFβ form homotrimers in solution. Over the past decade, sufficient cytokine structures have been elucidated to allow the classification of cytokines into four major structurally defined groups. This raises the hope that the incredible complexity of the cytokine 'zoo' can be mastered and shows the importance of structural biology in modern biomedicine. Members of the first group have structures that comprise four antiparallel α-helical segments; further subdivision is into proteins with long-chain (lc) or short-chain (sc) segments. The second group have long β-sheet structures and can be subdivided further into cytokines with: (i) β-sheet rich structures known as cystine knots; (ii) proteins with the conformation known as β-jellyrolls (called jellyroll (Swiss roll) because the polypeptide chain is wrapped around a barrel core like a Swiss roll), and (iii) proteins with three sets of 4-β strands which have a Y-shaped or three-leaf (trefoil) structure. The third group has short-chain α/β structures and these include members of the EGF family which contain at least two antiparallel β-strands connected to the intervening loops by three disulphide bonds (so-called S-S-rich β-meander structure), insulin-related cytokines which contain a conserved set of three disulphide bonds linking three short α-helices and the C-X-C and C-C chemokines. The fourth group – the mosaic structures – contain a range of domain structures such as immunoglobulin (Ig)-like domains and the so-called 'kringle' domains (found in proteins such as plasminogen).

With a few exceptions, cytokines have no catalytic functions and must bind to a specific cell receptor in order to express biological activity. A large number of cytokine receptors have now been identified and cloned and this has allowed their subdivision into various structural classes. Thus, it is also possible to classify cytokines on the basis of the receptors to which they bind and this is increasingly being used to clarify the biology of these proteins.

Cytokine receptors

Cytokine receptors generally have extremely high affinity for their ligand and only relatively small numbers of individual receptors are found on target cells. Moreover, only a fraction of the available receptors need to be occupied to produce maximal biological effects. These attributes make for an extremely sensitive cell recognition and signalling system. Hormone receptors are normally found only on one target cell population. In contrast, many cell populations have receptors for many different cytokines.

During the past decade or so the sequences and structures of many cytokine receptors have been elucidated and it is now possible to group these proteins into a small number of families based upon sequence homology and structural motifs (Figure 3.5). The cytokines binding to these cytokine receptors are shown in Table 3.5. One of the effects often seen when cytokines bind to their receptors is that the binding event cross-links either two (or more) receptor molecules or else the binding cross-links the receptor and some other protein. For example, when IL-6 binds to its receptor this is not sufficient to transmit the signal into the cell interior as the IL-6 receptor has no intracellular domain and the IL-6 receptor must dimerize with a 130 kDa membrane protein known as gp130. This protein can also dimerize with other cytokine receptors such as that of IL-11.

A growing number of chemokine receptors are being discovered. Currently there are nine receptors for CC chemokines (CCR), five receptors for CXC chemokines and CX3CR1, a receptor for fractalkine.

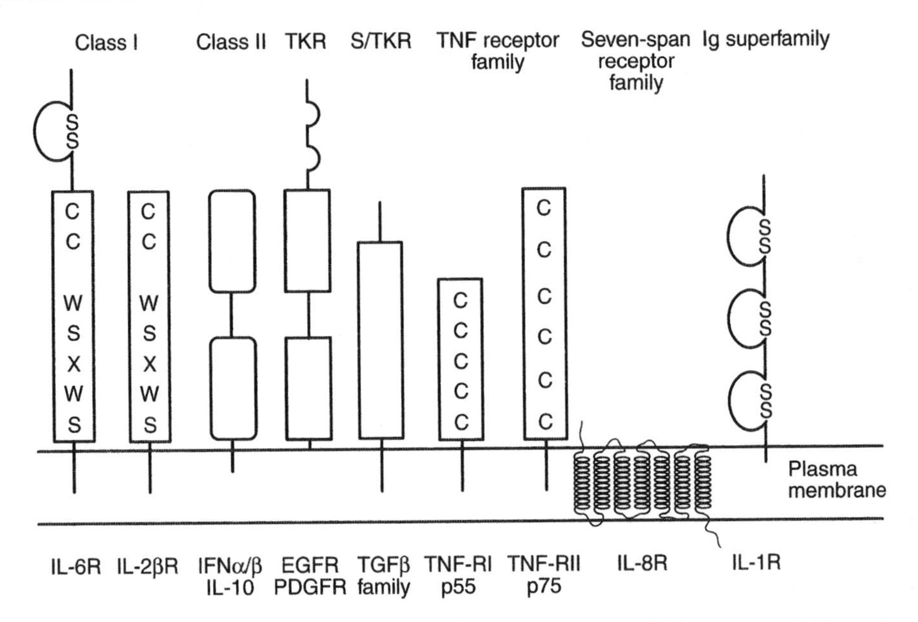

Figure 3.5 Subdivisions of cytokines based on the receptors to which they bind. Class I receptors contain a conserved extracellular sequence of five amino acids (tryptophan-serine-X-tryptophan-serine) or in the single-letter code WSXWS. The 'C's in this and other receptors are conserved cysteines. These receptors bind to cytokines that are folded into globular domains containing four α-helical strands. Class II receptors bind to interferons and IL-10. The TKR receptors are receptors which have an intracellular tyrosine kinase domain. The S/TKR receptors have an intracellular domain which has serine/threonine kinase activity. The large TGFβ family binds to these serine/threonine kinase receptors

One further complication of cytokine receptors is that they can be shed from cells by proteolytic cleavage. Release is thought to be carried out by cell surface metalloproteinases called sheddases. The released soluble receptors are still capable of binding cytokines and can either act as antagonists of soluble cytokines and thus inhibit the activity of cytokines or they can bind to cells which do not normally contain the cytokine receptor and stimulate them. In this way the effect of the cytokine(s) can be extended to bystander cells.

Cytokine signal transduction

The mechanisms of signal transduction will be described in the second half of this chapter but is is worth just briefly describing one particular mechanism of cytokine intracellular

Table 3.5 Cytokines which bind to the various cytokine receptor families

Cytokine receptor class	Cytokines binding to this receptor class
Class I	IL-2 (to the β chain), IL-3, IL-4, IL-5, IL-6, IL-7, IL-8, IL-9, IL-11, IL-12 EPO, G-CSF, GM-CSF, LIF, CNTF, prolactin, growth hormone
Class II	Interferons, IL-10
Receptor tyrosine kinase	Various growth factors (EGF, FGF, IGF, KGF, PDGF, VEGF)
Protein serine/threonine kinase receptors	TGFβ superfamily (30+ proteins)
TNF receptor family (p55/p75)	TNFα, TNFβ, Fas ligand, CD27 ligand, CD30 ligand, CD40 ligand
G protein-coupled receptors	Chemokines (approximately 30 proteins)
IgG superfamily	IL-1α, IL-1β, IL-1ra

There are two distinct TNF receptors of molecular masses 55 and 75 kDa.

signalling. In general, binding of any cytokine to its cellular receptor triggers intracellular kinase activity. This can either be a direct effect of the receptor, which may have its own kinase domain, or it may require the formation of complexes with other intracellular proteins to form a kinase domain. The aim of this is to produce selective signals which can enter into the nucleus of the cell and induce the transcription of particular genes. In the past few years one particular signalling system has been discovered which is beginning to explain how it is possible that cytokines can have the amazing range of actions on cells that they do. This signalling system occurs when cytokines bind to either class I or class II receptors. Having bound to the receptor and induced dimerization of the particular receptor-signalling protein complex, the intracytoplasmic domain of the dimerized receptor complex is phosphorylated on tyrosine residues by a family of bound tyrosine kinases, the Jak (or JAK) tyrosine kinases. This phosphorylation induces binding sites for the SH2 binding domains of particular proteins (see later for explanation of SH domains). In this case the proteins are a group of latent cytosolic transcription factors known as signal transducers and activators of transcription (STATs). There are, at the time of writing, seven STAT proteins (STAT 1–4, STAT 5a, STAT 5b and STAT 6). The STATs bind to the dimerized receptor where they are phosphorylated by the JAKs. This allows them to dissociate and form homo- or hetero-dimers with other STATs, thus forming DNA binding complexes which act as specific transcription activators (Figure 3.6).

By direct study of the STATs involved in cytokine signalling and by use of the knockout of specific STATs to look for functional consequences it appears that some of the STATs may confer specificity on the signalling induced by certain cytokines. Knockout of STAT 4 (using the technique of homologous recombination described in Chapter 4), for example, results in mice which although normal in appearance and fertile have selective deficits in IL-12-controlled functions such as the production of Th_1 lymphocytes, lymphocyte proliferation and IL-12-stimulated production of IFN-γ. As will be explained in Chapter 9, IL-12 plays a major role in the induction of acquired responses to infections and therefore it is important to understand the mechanisms underlying its signalling. Mice made deficient in STAT 6 demonstrate impaired response to both IL-4 and to IL-13. It is not clear whether STAT 4- or STAT 6-deficient mice show any aberrant responses to infection. However, knockout of STAT 1, which is involved in the signalling by the interferons, results in mice which are highly susceptible to infections by viruses and certain intracellular bacteria.

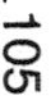

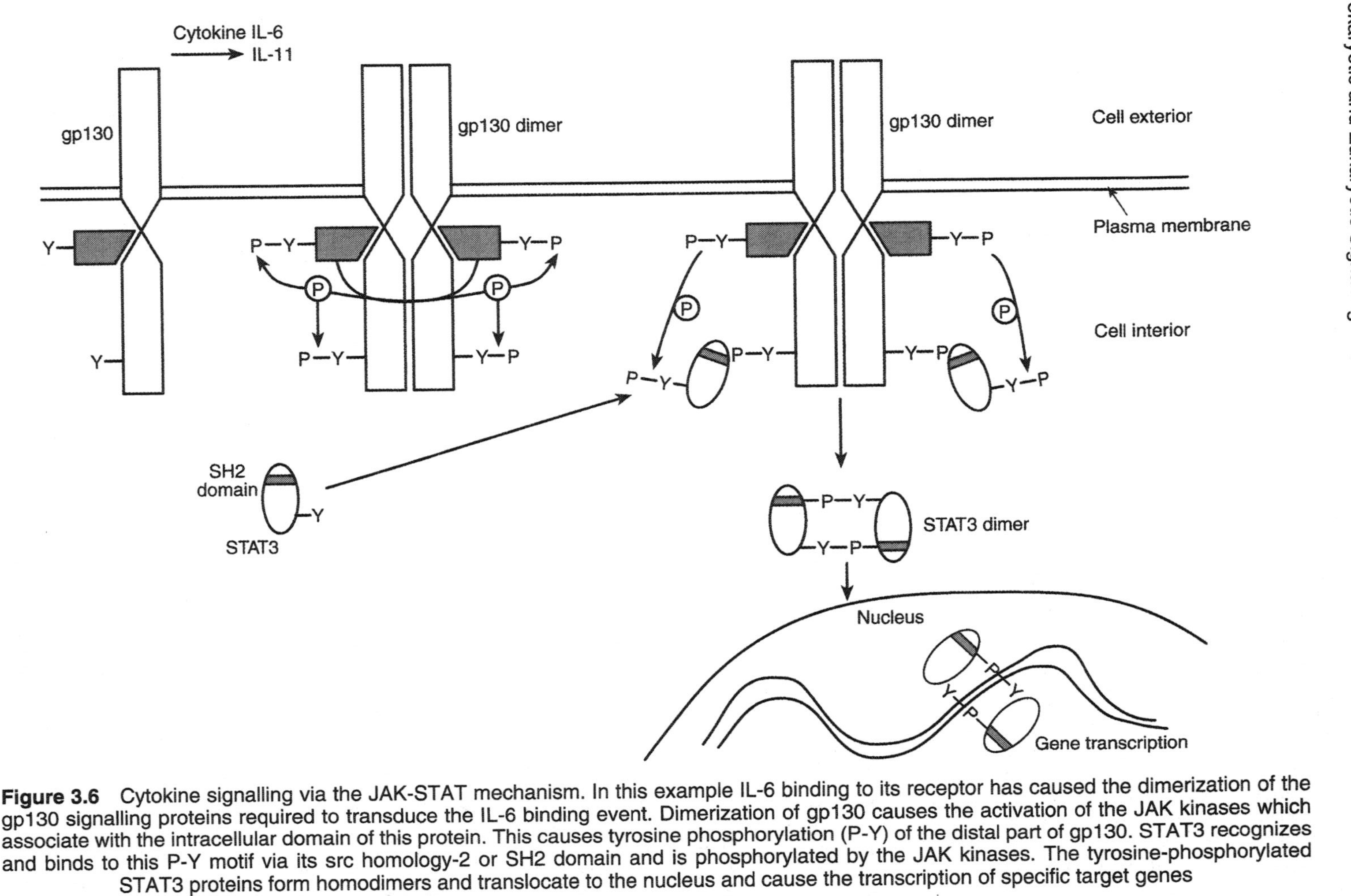

Figure 3.6 Cytokine signalling via the JAK-STAT mechanism. In this example IL-6 binding to its receptor has caused the dimerization of the gp130 signalling proteins required to transduce the IL-6 binding event. Dimerization of gp130 causes the activation of the JAK kinases which associate with the intracellular domain of this protein. This causes tyrosine phosphorylation (P-Y) of the distal part of gp130. STAT3 recognizes and binds to this P-Y motif via its src homology-2 or SH2 domain and is phosphorylated by the JAK kinases. The tyrosine-phosphorylated STAT3 proteins form homodimers and translocate to the nucleus and cause the transcription of specific target genes

It should be noted that the STAT system is only one of the intracellular signalling pathways induced when cytokines bind to their receptors. A more detailed description of intracellular signalling is provided below under 'Cytokine-mediated stimulation'.

Biological actions of cytokines

Although cytokines are often encountered in disease states, these proteins play key roles in normal physiology and physiological development. For example, cytokines are found at all stages of mammalian development. In addition to the cytokines, the cytokine receptors themselves can have biological activities while resident on the cell membrane. There are a growing number of examples of cytokine receptors acting as portals for viral entry into cells. For example, the human immunodeficiency virus enters cells by binding to a number of chemokine receptors and it has also recently been reported that herpes simplex virus 1 enters cells by binding to a novel member of the TNF receptor family (see Chapter 9).

Binding of a cytokine to its specific receptor induces selective intracellular signalling which results in the switching on (or off) of particular genes and the production and release of the products of such genes. These are proteins, such as other cytokines, proteases etc. and/or the products of the activity of these expressed proteins. A good example would be the prostaglandins, produced as a result of the induction of the enzyme cyclooxygenase II (COX II) and nitric oxide (NO) synthesized by inducible nitric oxide synthase (iNOS). The prostaglandins are lipid mediators which have a range of actions on leukocytes and vascular cells. The anti-inflammatory agents known as non-steroidal anti-inflammatory drugs (NSAIDs), of which aspirin and ibuprofen are examples, work by blocking the activity of cyclooxygenases. The major effect of these drugs is pain relief and reduction of fever as the prostaglandins and prostacyclin induce hyperalgesia (lowered threshold in pain nerves) and fever. The binding of a cytokine to its cellular receptor thus results in production of various molecules which could produce pathology (prostaglandin (PG)E_2, NO, tissue plasminogen activator (tPA), plasminogen activator inhibitor (PAI), collagenase). Proteinases such as collagenase and tPA could directly induce tissue damage. NO and PGE_2 are molecules able to modulate the activity of a range of cells including leukocytes, cells associated with the vasculature and cells of bone. The most fascinating action of cytokines is their capacity to induce their own synthesis as well as that of other cytokines (Figure 3.7). These patterns of cytokine production, and the production of secondary mediators induced by cytokines, leads to complicated networks of interactions which will be discussed in more detail in Chapters 8 and 9.

In addition to activating the metabolism of cells, cytokines can modify the behaviour of cells in a wide variety of ways. Growth mediators (interleukins, CSFs, growth factors) can stimulate entry into and completion of the cell cycle. The CSFs and other cytokines (interleukins) are involved in controlling the differentiation of myeloid and lymphoid cell lineages in the bone marrow. A number of cytokines and cell-bound proteins binding to the TNF receptor family are now recognized to act to induce apoptosis, which can be viewed as one aspect of the cell cycle (see Chapter 2). Another major group of cytokines, the chemokines, are involved in the recruitment of the correct populations of leukocytes to sites of injury, inflammation and infection.

Actions of IL-1 on cells

As an example of the range of effects produced by cytokines the biological actions of the prototypic pro-inflammatory cytokine, interleukin 1, will be briefly described. IL-1 has

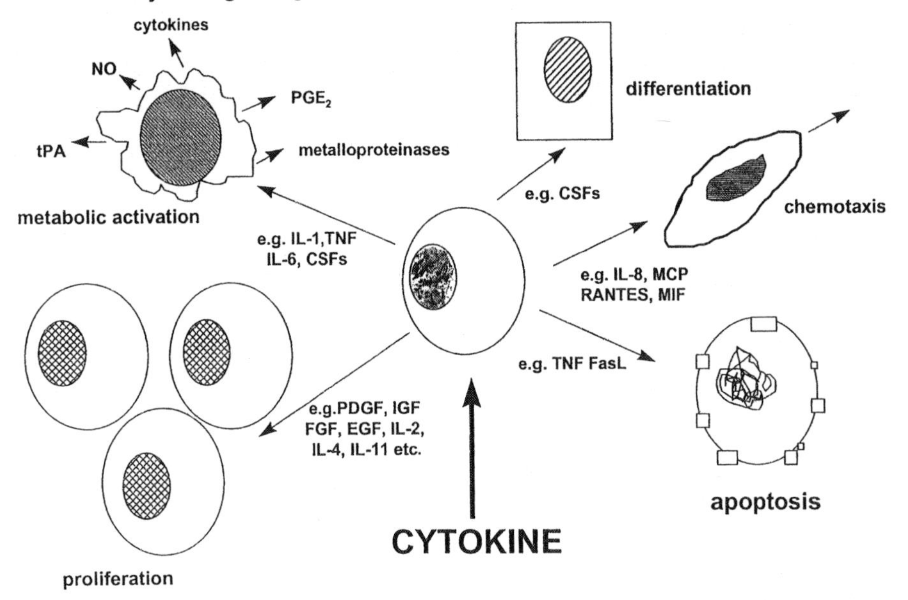

Figure 3.7 The various actions of cytokines on cells (reproduced from Henderson *et al* (1998) Bacteria-Cytokine Inrteractions in Health and Disease, Portland Press, London, with permission)

effects on probably every cell type in the body. The proteins IL-1α and IL-1β are produced as propeptides of molecular mass 31–33 kDa. The IL-1α propeptide precursor is biologically active but the IL-1β propeptide is inactive. Activation of IL-1β is by a specific cysteine proteinase called pro-IL-1β converting enzyme (ICE, also known as caspase 1), which appears also to have a role in the production of secreted IL-1α and of a more recently discovered cytokine IL-18 (or IL-1γ). A number of viruses can selectively inhibit this protease and this ability will be described in Chapters 8 and 9. The proteolytically cleaved products of IL-1 are both biologically active. The third member of the IL-1 gene family is IL-1ra. There are two receptors for the IL-1 molecules. The type I receptor is an 80 kDa signalling receptor but the type II (60 kDa) receptor does not signal and has been termed a decoy receptor. The importance of IL-1 is shown by the evolution of this complex system that involves ICE, IL-1ra and the type II receptor. Indeed, this IL-1 family appears to be unrelated to other cytokines and may be a signalling molecule that evolved early in evolutionary terms. Even *E. coli* has receptors for this cytokine (see Chapter 10). Whether IL-1 started life as a bacterial signalling molecule is an intriguing, although uncertain, possibility. Type I IL-1 receptor occupancy results in the intracellular domain of the receptor making a complex with accessory proteins termed IL-1 receptor-assocated serine/threonine kinases (IRAK)-1 and -2. IRAK becomes phosphorylated and then interacts with TRAF6, a member of the TNF receptor-associated factor (TRAF) family, implicated in the activation of c-Jun N-terminal kinase (JNK) and NF-κB. In an excellent example of the linked nature of biology IRAK-1 and -2 are homologous to the *Drosophila* protein kinase called Pelle, which is important in both dorsal–ventral patterning and in the resistance of *Drosophila* to pathogens. Pelle is essential for the activation of Dorsal, an NF-κB-like protein, which is mediated by the receptor Toll, an IL-1 receptor homologue in *Drosophila*. These findings suggest that the IL-1 system is a very early and central signalling mechanism.

The range of actions of IL-1 is clearly shown by the disparate names that have been coined to describe this molecule (granulocyte pyrogen, endogenous pyrogen, lymphocyte-activating factor, catabolin etc.). The biological activities demonstrated by the mature IL-1 molecules will only be briefly described (see Figure 3.8). Injection of IL-1 into animals or man induces inflammation and tissue destruction. IL-1 is a potent stimulator of macrophage functions and upregulates a variety of genes including those for cytokines, eicosanoids (prostaglandins, leukotrienes) and adhesion receptors.

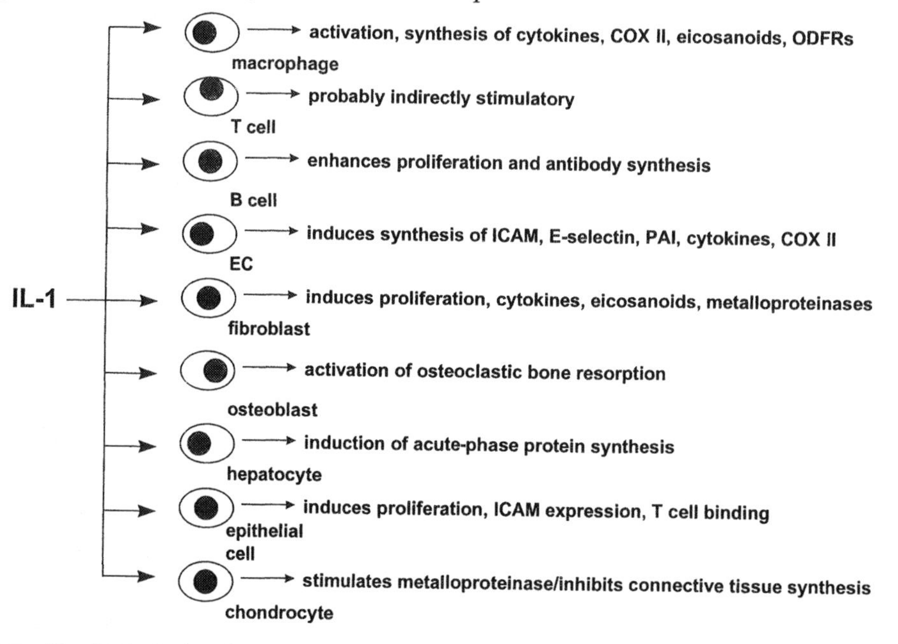

Figure 3.8 The biological actions of IL-1 (reproduced from Henderson *et al* (1998) Bacteria-Cytokine Interactions in Health and Disease, Portland Press, London with permission)

IL-1 is a potent stimulator of prostaglandin synthesis in a range of cells *in vitro*, most notably fibroblasts, monocytes and vascular endothelial cells. Anti-inflammatory drugs such as aspirin and ibuprofen inhibit prostaglandin synthesis and this is believed to be the mechanism of their action.

IL-1 was recognized early as a lymphocyte-activating factor (LAF), and for many years was thought to be a major stimulator of T and B lymphocytes. There is certainly evidence for IL-1 acting to promote B cell function. However, inhibition of IL-1 with receptor-neutralizing antibodies or IL-1ra does not block antigen-driven T lymphocyte activation. Furthermore, IL-1 knock-out mice do not appear to have immunological defects. Consequently, IL-1 is probably best viewed as a vital regulator of innate immune responses.

IL-1 alters the adhesiveness of cultured vascular endothelial cells (VECs) for leukocytes by promoting the synthesis of a number of leukocyte-adhesive molecules including the ICAMs (intercellular adhesion molecules), VCAMs (vascular cell adhesion molecules) and selectins. IL-1 also affects the coagulation system of VECs, favouring a procoagulant state. This involves repression of thrombomodulin, alterations in plasminogen activator, upregulation of tissue factor, increased production of eicosanoids, particularly prostacyclin, and the induction of iNOS, the enzyme responsible for NO synthesis. IL-1 also causes changes in proteoglycan metabolism by cultured VECs. These molecules are important in regulating vascular permeability, in binding chemokines and in interactions with platelets. IL-1 induces

VECs to synthesize a number of cytokines, including IL-1, G-CSF, GM-CSF and various chemokines, and also stimulates vascular endothelial cells to synthesize the vasoactive peptide endothelin (ET)-1, illustrating yet another interaction between peptide hormones and cytokines. IL-1 also increases the expression of the low-density lipoprotein (LDL) receptor on VECs.

IL-1 acts both on the liver and the brain to induce the acute-phase response. This is a stereotyped response to infection (and injury) involving the synthesis of a variety of liver proteins, which act as opsonins, protease inhibitors and free radical scavengers, and the stimulation of cells of the innate and acquired immune response. This will be described in more detail in Chapter 8. IL-1 is also one of the molecules known as endogenous pyrogens and is able to induce fever, which is a classic response to infection. The ability of IL-1 to induce fever is due to its ability to induce the production of prostaglandins in the temperature control centre of the brain.

These various *in vitro* findings are mimicked by injection of IL-1 into animals. Local injection of IL-1 induces leukocytic infiltration, the appearance of low molecular mass mediators of inflammation, the induction of tissue breakdown and the inhibition of connective tissue matrix synthesis. Systemic injection of IL-1 induces the acute-phase response and fever. Proinflammatory cytokines such as IL-1 are extremely potent hyperalgesic agents (molecules which lower the threshold of nerves to painful stimuli) which could contribute to the pain associated with inflammation. This hyperalgesia does not appear to be dependent on the synthesis of prostanoids.

Thus it can be seen from the catalogue of biological effects described that IL-1 is capable of inducing a multitude of effects on cells. The major effect of this cytokine is to prime and activate the defences of the body to enable it to deal with infection. However, a consequence of this system is that IL-1 is often found to be a mediator of tissue pathology and thus a therapeutic target.

The effects of IL-1 and other cytokines on the innate and acquired defences against microorganisms will be described in Chapters 8 and 9.

Cytokines involved in infection and the mechanism of their synthesis

It is believed that the key factors that drive cytokine synthesis are components derived from the infectious agents. For decades it has been known that LPS from Gram-negative bacteria is a potent stimulator of cytokine synthesis. We now know that this molecule binds to cell surface CD14 on monocytes and other cell populations and induces a complex series of signalling pathways resulting in the transcription and release of various pro- and anti-inflammatory cytokines. The major cytokines known to be involved in switching on and swiching off inflammation in infections are shown in Table 3.6. In the case of Gram-negative bacterial infections the release of LPS from the microorganisms can result in the production of proinflammatory cytokines such as IL-1, TNF and cytokines with cytokine network-modulating or anti-inflammatory actions such as IL-6 and IL-10 respectively. It is becoming clear that it is the rate of production of these various pro- and anti-inflammatory cytokines that determines whether patients will develop lethal septic shock or will survive. Studies of the genetics of cytokines have revealed variations (polymorphisms) in the human population in the promoter regions of various cytokines. Such polymorphisms control the ability of individuals to respond to inflammatory stimuli by producing cytokines. The consequence of such polymorphisms is that different individuals will produce different amounts of

Table 3.6 The main cytokines involved in responses to infection

Cytokine	Activity
Proinflammatory	
IL-1	Many and varied – see Figure 3.8
IL-6	Acute-phase response inducer
Chemokines	Leukocyte chemotaxis, chemokinesis, adhesion, activation
IL-12	Macrophage activation, IFN-γ synthesis, Th_1 differentiation
IL-15	Proliferation of NK and T cells
IL-18	IFN-γ inducing factor, Th1 differentiation
TNF	Fever, cell activation, acute-phase response
IFNs	Antiviral activity
IFN-γ	Fever, macrophage activation
Colony-stimulating factors	Production and activation of myeloid cells
Antiinflammatory	
IL-1ra	Antagonizes action of IL-1α and β
IL-10	Deactivates macrophages
IL-13	Inhibits macrophage proinflammatory cytokine synthesis
TGFβ	Inhibits lymphocyte function

cytokines for any given bacterial stimulus. This could result in an underproduction of a key cytokine in one individual, resulting in the failure to mount an immune response sufficient to kill the invading bacterium. In others, the polymorphism could result in overproduction of key cytokines such that tissue pathology is produced. These cytokine polymorphisms are only now being explored but the evidence is accumulating to suggest that they play a role in infectious diseases such as septic shock.

The role of cytokines in infections will be dealt with in much greater detail in Chapters 8–10.

PROKARYOTIC CELL-TO-CELL SIGNALLING: QUORUM SENSING AND BACTERIAL PHEROMONES

The concept that bacteria could talk to one another did not receive much attention until the 1980s when examples were found of cell-to-cell signalling in bacteria. The uses that bacterial cell-to-cell signalling mechanisms are put to are summarized in Table 3.7 and it is very likely that many more examples will exist.

Table 3.7 Cell-to-cell interactions in bacteria

Bacterial mechanism or process	Signalling molecules
Sporulation and fruiting body formation in *Myxococcus xanthus*	Peptide pheromones
Conjugation of *Enterococcus faecalis*	Peptide pheromones
Morphological differentiation in *Streptomyces coelicolor*	Modified peptide
Antibiotic production by *Streptomyces* spp.	?
Autoinducer behaviour in many bacteria	Acyl homoserine lactones (AHLs) Peptide pheromones

Signals controlling conjugation in *Enterococcus faecalis*

One of the three natural mechanisms of transfer of DNA between bacteria is bacterial conjugation, which requires that bacteria come into contact with one another (see Chapter 2). In *Enterococcus faecalis*, a Gram-positive member of the mammalian commensal microflora, the aggregation of bacteria is controlled by the secretion of peptide pheromones. These small peptides induce the production of the adhesins which enable bacteria to form clumps or cellular aggregates within which conjugation takes place. A number of peptide pheromones have been isolated and they are generally hydrophobic octapeptides or heptapeptides. These pheromones are active at extremely low molar concentrations (5×10^{-11} M) and it has been calculated that as few as two molecules of the peptide per cell may be sufficient to produce biological activity. In addition to their potency, these pheromones demonstrate marked selectivity. This agonist–receptor system appears to have significant similarities (in terms of potency and selectivity) to the cytokine signalling systems of eukaryotes.

Signals controlling sporulation in *Myxococcus xanthus*

Sporulation is a fairly common response of bacteria to adverse conditions. Endospores are highly resistant resting forms of bacteria that are produced within the cell rather than on specialized external structures. The endospores of *C. tetani* can be regarded as part of the virulence mechanism of this organism and leads to us being constantly on guard against injuries involving soil which contains such endospores. Now, in contrast to endospores, some bacteria respond to adverse conditions by undergoing complex morphogenetic alterations and forming what are known as fruiting bodies: cyst-like structures composed of an outer thick covering of polysaccharide which makes the cyst resistant to heat and dehydration. The visibility of these fruiting bodies accounts for the fact that it was a myxobacterium – *Polyangium vitellinum* – that was the first prokaryote to be assigned a scientific name in 1809. The best-studied sporulation system is that of the Gram-negative soil bacterium *Myxococcus xanthus*. This organism undergoes a complex life cycle which alternates between the formation of myxospores (fruiting body) and of what are termed vegetative cells. This developmental programme is triggered by starvation and leads to morphological changes within 4 hours as the vegetative cells begin to congregate (reminiscent of the first step in conjugation described above). When a cell density of about 10^5 organisms is reached, a dense mound-shaped structure is formed that is just visible to the unaided naked eye. After 20 hours of starvation cells inside this mound differentiate into myxospores that are heat- and desiccation-resistant dormant cells and form a fruiting body. When favourable conditions are again encountered the fruiting body 'germinates', producing vegetative cells which are tapered and flexible. These cells grow and divide and produce swarms of individual cells whose activity has been likened to that of 'wolf-packs' moving over solid surfaces and devouring bacteria encountered in their paths. When conditions again become unfavourable the formation of the myxospore begins anew. This complex cell differentiation is controlled by extracellular signals. This was discovered when mutants unable to sporulate were incubated with wild-type organisms and regained the ability to form myxospores. A number of signals appear to be involved in this bacterial differentiation system including small peptides and a 17 kDa protein.

Quorum sensing

Quorum sensing is a bacterial cell-to-cell signalling system, with a basic mechanism similar to those described above, which has come to prominence in the last 10 years or so. The dictionary definition of the term *quorum* is 'a fixed number of members of any body, society etc., whose presence is necessary for the proper or valid transaction of business'. Bacterial quorum sensing is a mechanism by which bacteria can take a census of their numbers and having reached a 'quorum' can 'transact business' – only in the case of bacteria the business is the switching on or off of specific genes.

Our current understanding of quorum sensing traces back to studies of the luminescence of the marine bacteria *Vibrio fischeri* and *V. harveyi*. These bacteria form symbiotic relationships with monocentrid fish and with species of squid known as bobtail squids (e.g. *Euprymna scolopes*). Much of our understanding of the relationship between bacteria and marine organisms has come from the study of *Euprymna scolopes*. This is, one might say, an illuminating story and an excellent example of cellular microbiology.

The bobtail squid has a light organ which contains very high concentrations of a monospecific culture of *V. fischeri*. The function of the light organ in the bobtail squid is thought to be part of a camouflaging behaviour called counterillumination, but details of this mechanism are still sparse. The adult light organ is bilobed, each lobe containing three epithelia-lined crypts that house the bacteria. The light organ also has a pair of pores which allows it to make contact with the external environment. Newly hatched squids acquire their symbiotic bacteria from sea water, that is, transmission of the bacteria is horizontal and not vertical (from the parents). Sea water contains a large number of *Vibrio* species. In spite of this *Euprynma scolopes* is only colonized by *V. fischeri* and only certain strains at that. Thus the light organ appears to have a positive selection mechanism choosing only certain *V. fischeri* strains and a negative selection mechanism to exclude colonization by the vast numbers of other bacteria that are present in sea water. How this selection of bacteria is achieved is unclear. One obvious mechanism would be for the epithelium of the light organ to express a specific adhesin for *V. fischeri*. This idea is supported by experiments in which the light organ epithelium is exposed to trypsin, which will remove surface peptides. Such proteolysis prevents colonization of the light organ by *V. fischeri*.

One of the most fascinating findings is that the adult light organ is only formed if the squid is 'infected' with *V. fischeri*. Thus within hours after hatching the nascent light organ is infected or colonized by *V. fischeri*. In response the cells lining the crypts undergo terminal differentiation and the crypt spaces, described above, grow as a result of cell proliferation. However, if animals are treated with antibiotics in the first 8–12 hours following colonization (which kills bacteria in the nascent light organ) the morphogenesis of the light organ does not occur. This suggests that *V. fischeri* produces signals which can directly trigger a specific morphogenetic response in the squid which results in the formation of the complex light organ. This raises the very interesting question: do other commensal bacteria produce morphogenic changes in their hosts and does this occur with human commensals?

Mechanism of quorum sensing

Quorum sensing is a remarkably simple feedback control system. It requires that the bacteria constantly produce small amounts of a signal called an autoinducer. The autoinducers produced by most Gram-negative bacteria are simple organic compounds known as acyl homoserine lactones (AHLs; Figure 3.9). Other bacteria, such as *Staphylococcus aureus*, produce peptide autoinducers. It has recently been reported that *E. coli* and *Salmonella typhimurium*, which were not thought to produce AHLs, also produce a quorum-sensing molecule of molecular mass 1 kDa which is sensitive to base and to heating to 100°C. These autoinducers diffuse into the extracellular environment. The second requirement is that the bacterium has the means of recognizing the presence of the autoinducer. This function is served by a bacterial membrane protein which acts both as a receptor for the autoinducer and as an activator of gene transcription (Figure 3.10). This is related to the His-Asp phosphorelay systems of bacterial sensing and regulation which will be discussed in more detail under 'Intracellular signalling' below.

The best-studied quorum-sensing system is the *V. fischeri* system that produces luminescence. This comprises the *lux* genes which is an operon system involving two main regulatory genes (Figure 3.10) and a number of other genes that synthesize the chemical reagents required to produce photons. The first regulatory gene is *luxI* and encodes a protein which catalyses the synthesis of the AHL. This reaction uses *S*-adenosylmethionine as the donor of the homoserine lactone and a fatty acid moiety linked to an acyl carrier protein. A wide range of AHLs can be produced. For example, the autoinducer for *V. fischeri* is *N*-(3-oxohexanoyl)-L-homoserine lactone. The second regulatory gene is *luxR* and encodes a protein which acts both as a receptor for AHL and as a transducer of the signal activating the other genes present in the *lux* operon. When AHL binds to the LuxR protein the genes *LuxCDABEG* (Figure 3.10) are expressed. The *luxA* and *luxB* genes encode the α and β subunits of bacterial luciferase. The other genes encode polypeptides that are involved in the synthesis of the substrate (tetradecanal) used by the luciferase to generate light.

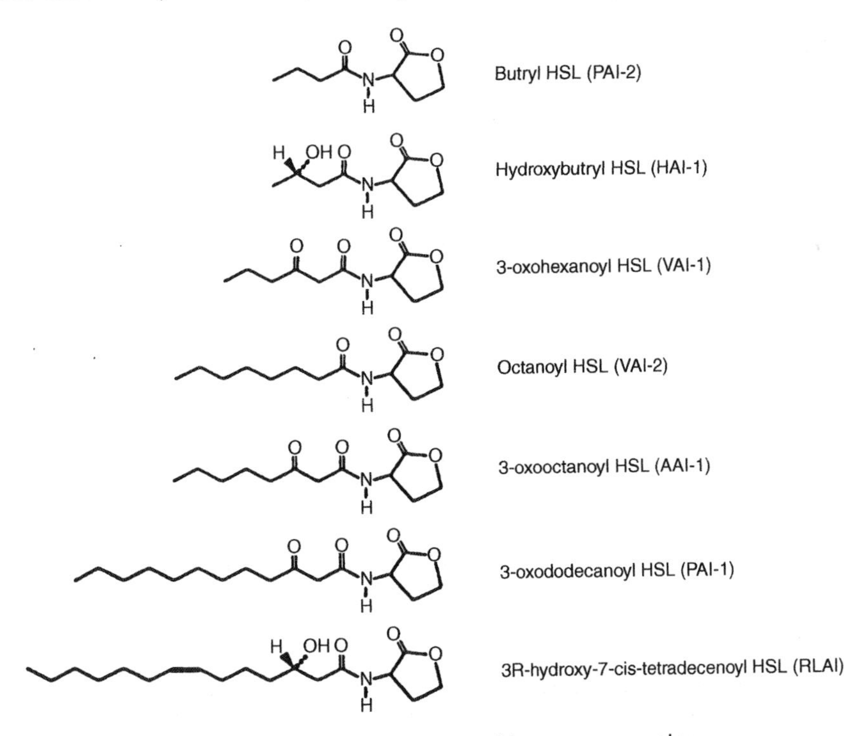

Figure 3.9 Autoinducers used in quorum sensing

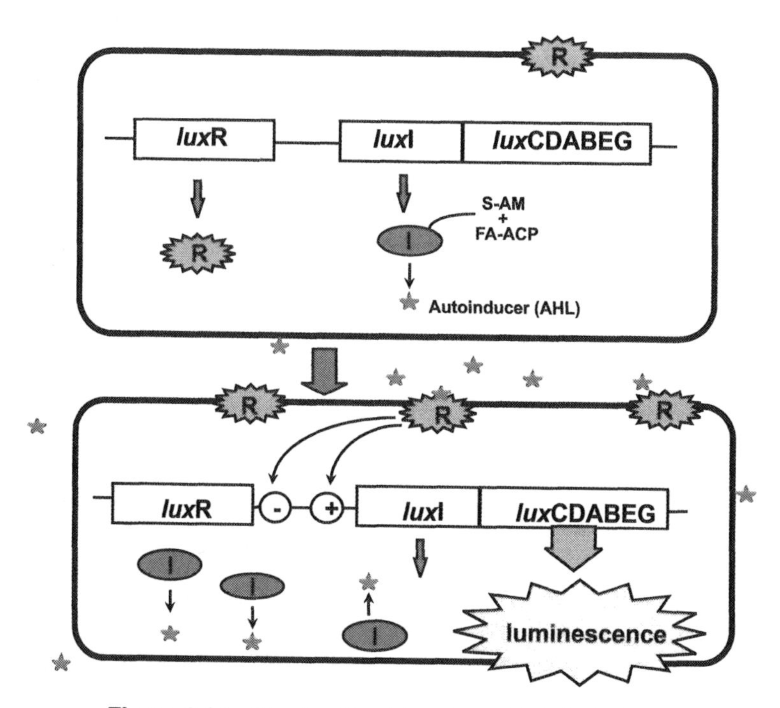

Figure 3.10 The *LuxI/LuxR* system of *Vibrio fischeri*

This mechanism occurs daily in the bobtail squid. Every morning the squid expels 90–95% of the bacteria in its light organ and begins to repopulate these organisms over the next 12 hours, building up the huge numbers, 10^{10} bacteria/ml, required for its nocturnal lifestyle. It is at these astronomical numbers that the quorum-sensing system is maximally stimulated and light is produced.

Quorum sensing as a virulence mechanism

Since the discovery of the *luxI/luxR* system in *V. fischeri* a large number of Gram-negative bacteria have been reported to produce AHLs and presumably to quorum sense. This list includes bacteria of medical importance such as *Pseudomonas aeruginosa, Y. enterocolitica, Proteus mirabilis, Serratia liquefaciens and Cit. freundii.* The luxI/luxR homologues known to be involved in the quorum-sensing systems of these bacteria are shown in Table 3.8. The best-studied pathogenic organisms, in terms of their quorum-sensing mechanisms, are *Ps. aeruginosa* and *Staph. aureus.* The former is an opportunistic pathogen which causes particular problems to children with cystic fibrosis. Unlike *V. fischeri, Ps. aeruginosa* has evolved to utilize at least two quorum-sensing systems which have been termed *las* and *rhl*. The *las* operon encodes the LasR protein, a LuxR homologue which acts as a transcriptional activator in the presence of the autoinducer for this bacterium (*Pseudomonas* autoinducer [PAI]: *N*-(3-oxododecanoyl)-L-homoserine lactone). This AHL is produced by the LuxI homologue, LasI. When concentrations of the *Ps. aeruginosa* autoinducer reach a threshold they switch-on a collection of virulence genes including *lasB, lasA, apr* and *toxA*. The second quorum-sensing system is the *rhl* system, which consists of the transcriptional activator protein RhlR with the

Table 3.8 Quorum-sensing systems in bacteria based on LuxI/LuxR family

Organism	LuxI homologue	LuxR homologue	Functions regulated
Vibrio fischeri	LuxI	LuxR	Luminescence
Agrobacterium tumefasciens	TraI	TraR TrlR	Conjugation Unknown
Chromobacterium violaceum protease	CviL	CviR	Violacein pigment, haemolysin, exported
Erwinia carotovora	ExpI (CarI)	ExpR CarR	Exoenzymes Carbapenem antibiotics
Erwinia stewarti	EsaI	EsaR	Capsular exopolysaccharide
Serratia liquefasciens	SwrI	?	Swarmer cell differentiation
Escherichia coli	?	SdiA	Cell division
Pseudomonas aeruginosa	LasI RhlI	LasR RhlR	Elastase, exotoxin, other virulence factors Virulence factors
Yersinia enterocolitica	YenI	YenR	Unknown

autoinducer being *N*-butyryl-L-homoserine lactone, which is synthesized by RhlI. This quorum-sensing system is involved in the production of additional virulence factors such as the well-known elastase that is capable of cleaving and inhibiting key host defence cytokines such as interleukin 2. Having two quorum-sensing systems opens up the possibility that they may interact, and this is precisely what they do. The *las* system appears to be dominant and PAI can act to inhibit the binding of the second autoinducer (*N*-butyryl-L-homoserine lactone) to the LuxR homologue RhlR (Figure 3.11). Thus there is a hierarchy in the quorum sensing of this bacterium to ensure that the *las* system is activated before the *rhl* system. The reason for such a hierarchical control is not yet clear and awaits a complete explanation.

Several other bacteria quorum sense but do so using signals distinct from the AHLs described above. Several Gram-positive bacteria use oligopeptides as signalling molecules. *Bacillus subtilis* secretes at least two different peptides which are necessary for competence (ability to take up DNA) and sporulation. The opportunistic pathogen *Staph. aureus* has a locus termed *agr* which controls the expression of several virulence factors such as exotoxins, the V8 protease and capsular polysaccharide type 5. The *agr* locus encodes an octapeptide which is believed to be a quorum-sensing autoinducer which activates the *agr* locus (Figure 3.12).

The study of quorum sensing is still in its early phase and much remains to be learned. One recent finding is that the quorum-sensing autoinducer from *Ps. aeruginosa* can interact with host defence systems and can inhibit them, albeit at high concentrations (Figure 3.13). This ability of quorum-sensing autoinducers to both activate bacterial virulence factors and inhibit host defence mechanisms is the equivalent of an evolutionary 'double whammy'.

Formylated peptides in bacterial–eukaryotic signalling

Bacterial protein synthesis is initiated by incorporation of a formyl-substituted methionine residue. This initiating *N*-formylmethionine, with certain other residues (e.g. F-Met-Leu-

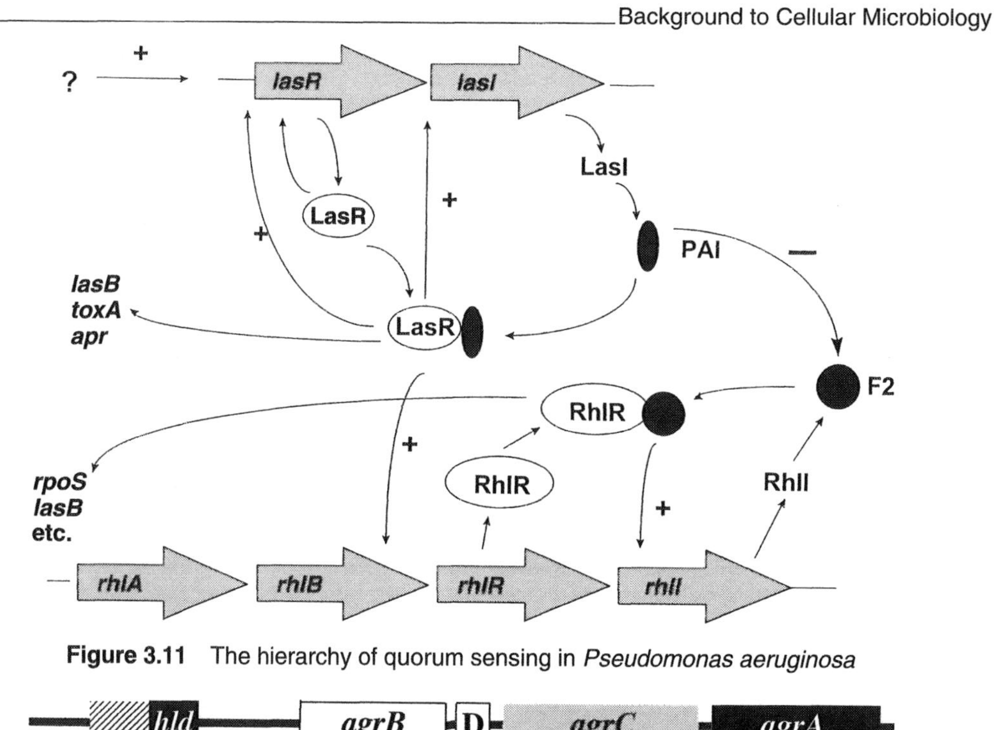

Figure 3.11 The hierarchy of quorum sensing in *Pseudomonas aeruginosa*

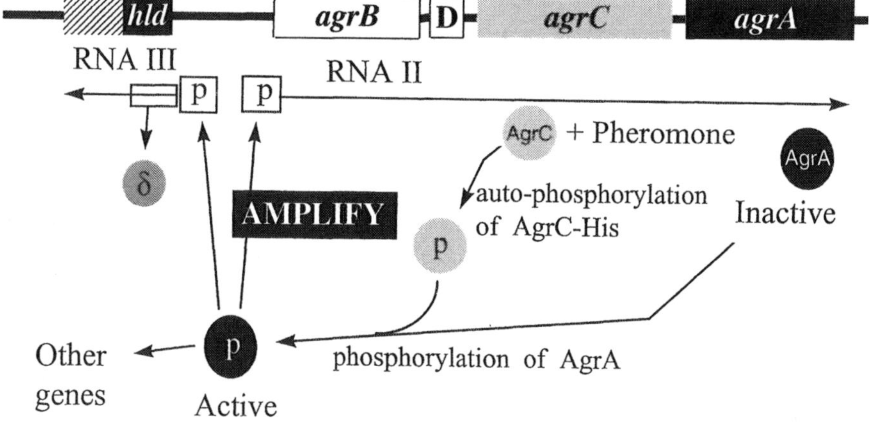

Figure 3.12 The *agr* locus in *Staphylococcus aureus*, which is controlled by an octapeptide 'pheromone'

Phe), is often removed post-translationally and released into the extracellular space. These formylated bacterial peptides are recognized by mammalian leukocytes, and in particular PMNs and monocytes/macrophages, and act as chemotactic signals for these cells which can then move to where the bacteria are. The human receptor recognizing formylated peptides has been cloned and sequenced and is a 38 kDa seven membrane-spanning G protein-coupled receptor and member of the rhodopsin superfamily. It is expressed on neutrophils, monocytes, macrophages and liver parenchymal cells and is 70% homologous to the receptor for the chemokine IL-8. Binding of the bacterial formylated peptides to PMNs, for example, induces chemotaxis, phagocytosis, production of superoxide and release of proteases. The

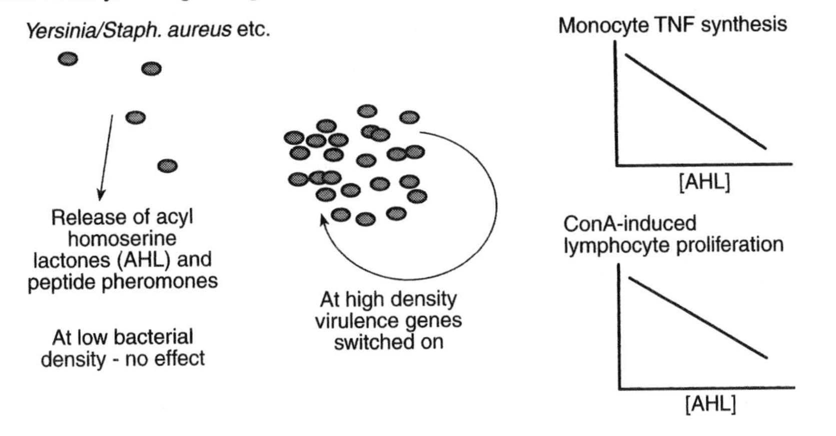

Figure 3.13 Action of AHLs on human immune defence systems. The AHLs can inhibit the synthesis of TNF synthesis by murine monocytes and lymphocyte proliferation induced by concanavalin A (conA)

genes for two additional formyl peptide receptors (FPRL1 and FPRL2) have been mapped to chromosome 19 in humans. FPRL1 is a low-affinity receptor for formyl peptides but the other does not bind such peptides and presumably recognizes other bacterial peptides. Thus it is possible that host cells can recognize a variety of bacterially-produced peptides and that this produces significant cross-talk between bacteria and eukaryotic cells.

INTRACELLULAR SIGNALLING: AN OVERVIEW

Eukaryotic cells and bacteria communicate by the production and release of a wide variety of signals. These signals generally act by binding to protein receptors on the surface of the target cell, although certain signals (steroids and certain autoinducers in bacteria) can enter the cell and interact with signalling systems. This binding event then triggers a series of intracellular reactions which are known collectively as intracellular signalling or signal transduction. It is now realized that to fully understand the operation of any cell, from 'birth' to 'death' will require a detailed knowledge, both spatial and temporal, of its intracellular signalling pathways. It is now 40 years since Sutherland and Rall described the first intracellular signalling molecule, cyclic adenosine monophosphate (cAMP). As more has been learned about the intracellular signalling pathways, their significance has become more obvious. Indeed it could be said that a cell's signalling repertoire defines the cell. Higher eukaryotes have a 'dead man's handle' safety system, whereby their basic life programme is to undergo apoptosis (see Chapter 2 for details) unless they continue to receive signals from other cells. The importance of the intracellular signalling pathways that process such extracellular information was emphasized by the discovery that mutagenic changes in the components of signalling pathways lead to cellular transformation (cancer). Indeed the identification of such mutants in human tumours further highlights the central role of signalling to the control of cellular function. That bacteria have 'noticed' this and chosen to exploit these pathways for their own ends should come as no surprise. Thus we find that bacteria subvert normal eukaryotic cell signalling to enable them to invade cells (Chapter 6) and bacterial toxins target specific signalling molecules to hijack control of host cells (Chapter 7). What has come as a surprise is the discovery of the complexity of bacterial signalling, in particular the 'sophistication' of the signalling which occurs between bacteria.

There are many recent books and reviews on the topic of signalling, both in bacteria and in eukaryotic cells. The aim of this chapter is not to provide a detailed catalogue of the components in these signalling pathways and all that is known about them, but instead to discuss and compare the similarities and differences between prokaryotic and eukaryotic systems and to see how individual signalling modules have been put together to form the complex and versatile systems that enable a cell to respond to subtle changes in its environment. Cell signalling is currently in a state where the scientific community has an enormous amount of information about the process but rather less understanding. Thus this area of biology, with its multiple components and manifold interactions, is complex and its proper understanding requires some knowledge of structural biology. The authors have attempted to provide a general description of the signalling systems in eukaryotes and prokaryotes but it is appreciated that this section of the book may be the most complex and difficult to follow. However, the reader is urged to try and follow the story of cell signalling as it is one of the most fascinating areas of modern biological science. Readers who wish to learn more about protein structural biology are urged to read the excellent general textbook by Sir Max Perutz, one of the pioneers of this field (see 'Further reading').

A complex signalling system is built using a limited number of basic principles

The vast complexity and subtlety of cell signalling mechanisms are analogous to electronic circuits that integrate, modulate and amplify various inputs to generate an output signal, generally the switching on, or off, of specific genes. Despite the intricacies of signalling systems, they are, like electronic circuits, built out of relatively few basic types of module. These are shown in Table 3.9. Indeed each module uses one or more of only four main processes (Figure 3.14). These are: (i) phosphorylation of proteins by enzymes called kinases, (ii) small molecule/protein interactions, often involving phosphates, (iii) protein/protein interactions, that are often mediated by common motifs (specific protein sequences), and which frequently result in membrane recruitment when one component is tethered to a membrane, and (iv) protein/DNA interactions that promote gene expression or its inhibition. By themselves these individual processes are reasonably easy to comprehend. The complexity in intracellular signalling arises when many such signals are interconnected either in series or in parallel with opportunities for crosstalk between pathways, and the ability of certain key checkpoints to act as 'logic gates' that integrate signals from a variety of pathways. It is this interconnectedness which renders the field of intracellular signalling so difficult to comprehend, even to those that work in the field.

In order to show the similarities between prokaryotic and eukaryotic signalling mechanisms, the discussion will start by describing these modules and this will be limited to the general principles whereby these modules are activated and how they process and pass on the signal. How these are woven into pathways in both pro- and eukaryotic organisms will then be described, and at the same time an introduction will be provided to those aspects of signalling that appear to be unique to either prokaryotes or eukaryotes.

Table 3.9 Signalling modules used in intracellular signalling

Modules	Types	Example	Pro- or eu-karyote	How signal is passed on
Receptors	Receptor kinases	Tyrosine kinase	Eukaryotic	Phosphorylation, or contact with other proteins
		Serine kinase	Eukaryotic	Contact with other proteins
		Histidine kinase	Prokaryotic (mainly)	Transfer of phosphate to aspartyl kinase
	Non-kinase receptors	Serpentine	Eukaryotic	Contact with G proteins
		Cytokine	Eukaryotic	Contact with kinases
		His-Asp phosphorelay	Prokaryotic	Transfer of phosphate to aspartyl kinase
Intracellular enzymes	Protein kinases	Src, Jak, cyclin families	Eukaryotic	Phosphorylation, or contact with other proteins
		Asp kinases	Prokaryotic (mainly)	DNA binding, or contact with other proteins
	Lipid-modifying enzymes	PI3K	Eukaryotic	Lipid contact with proteins
		P15K	Eukaryotic	Increased substrate for PLC
		PLC	Eukaryotic	Products activate PKC and calcium release
Other	cyclic nucleotides	cAMP	Both	Binds to and activates proteins
		cGMP	Eukaryotic	Binds to and activates proteins
	metal ions	Calcium	Both	Binds to and activates proteins

The basic building blocks used in signalling

Protein phosphorylation

A feature that pervades practically all signalling modules is phosphorylation of proteins (Figure 3.15). In some cases this is a direct effect, where transfer of the terminal γ-phosphate from adenosine 5'-triphosphate (the high-energy molecule, normally abbreviated ATP) to an acceptor protein, by a protein kinase, modifies the activity of the acceptor, for example the MAP kinases in eukaryotes or the histidyl-aspartyl phosphorelay in bacteria. Addition of a phosphate group can activate, or inhibit, the activity of a particular protein. In others the effect is indirect, for example where a phosphorylated protein is recognized. Good examples are the recognition by an SH2 motif of a phosphorylated tyrosine on a receptor kinase, or where the phosphorylation status of a bound nucleotide is the crucial factor (e.g. in G proteins, where binding of GTP activates their function, whereas GDP binding inactivates). In addition to phosphorylation of proteins, many of the so-called second messengers used in intracellular signalling are small phosphorylated molecules such as the various phosphorylated inositols, or the cyclic nucleotides (cAMP, cGMP). These various areas of signalling will be elucidated.

The study of protein phosphorylation and its close linkage to cellular signalling is a relatively new area of science and most of the seminal discoveries have occurred over the last 10–20 years, although the foundations for the regulatory role of phosphorylation were laid over 40 years ago. The first point to note is the widespread use that all cells make of kinases in order to regulate their function. This is reflected in the large number of different kinases

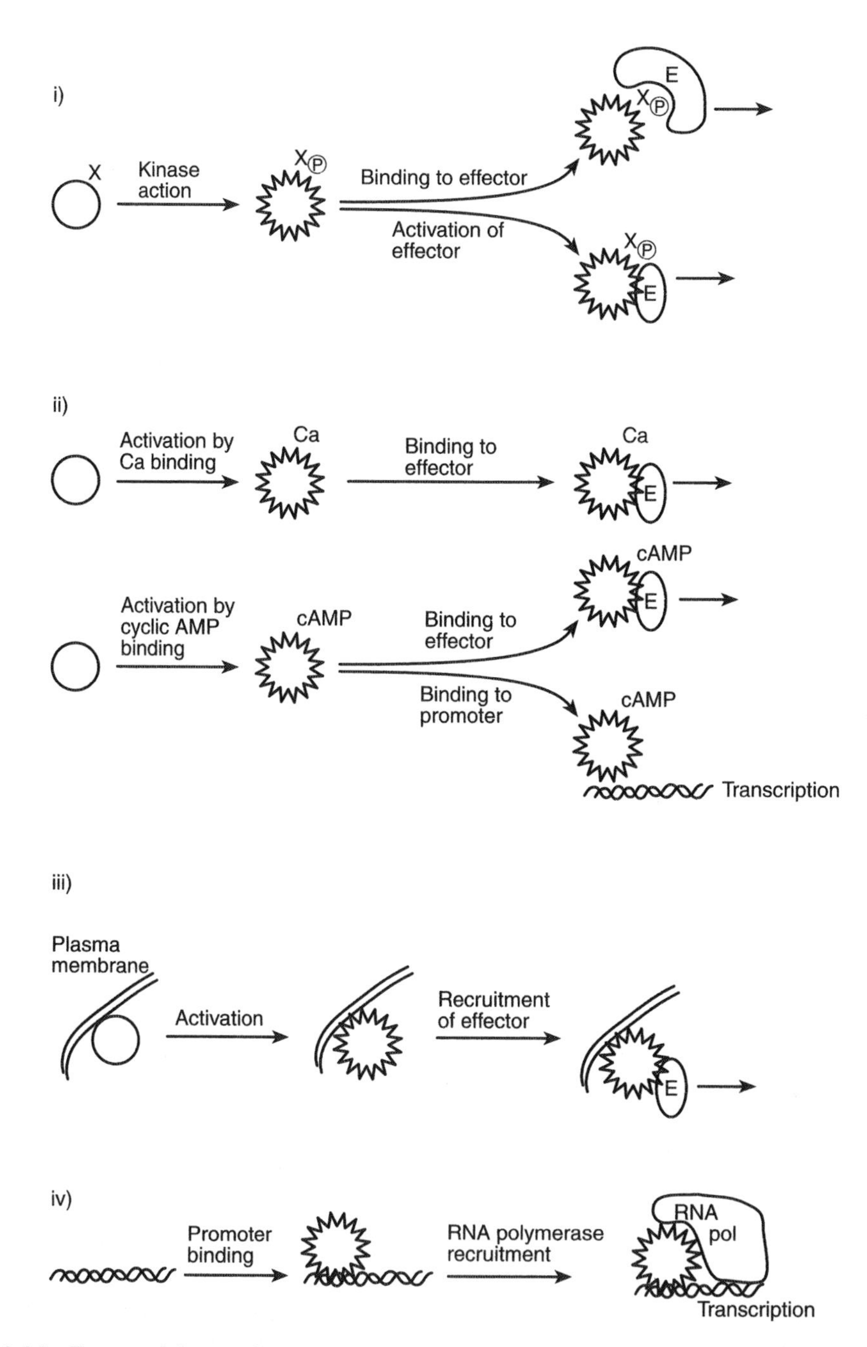

Figure 3.14 Types of interactions used by signalling molecules. E represents the effector. (i) Phosphorylation on one of several amino acids, where X can be Tyr, Ser, Thr, His or Asp; (ii) interactions between small molecules and proteins, for example Ca^{2+} or cAMP; (iii) protein–protein interactions, for example to recruit a component (e.g. transcriptional activator) to a membrane where it is activated, and (iv) interactions between proteins and DNA that regulate transcription

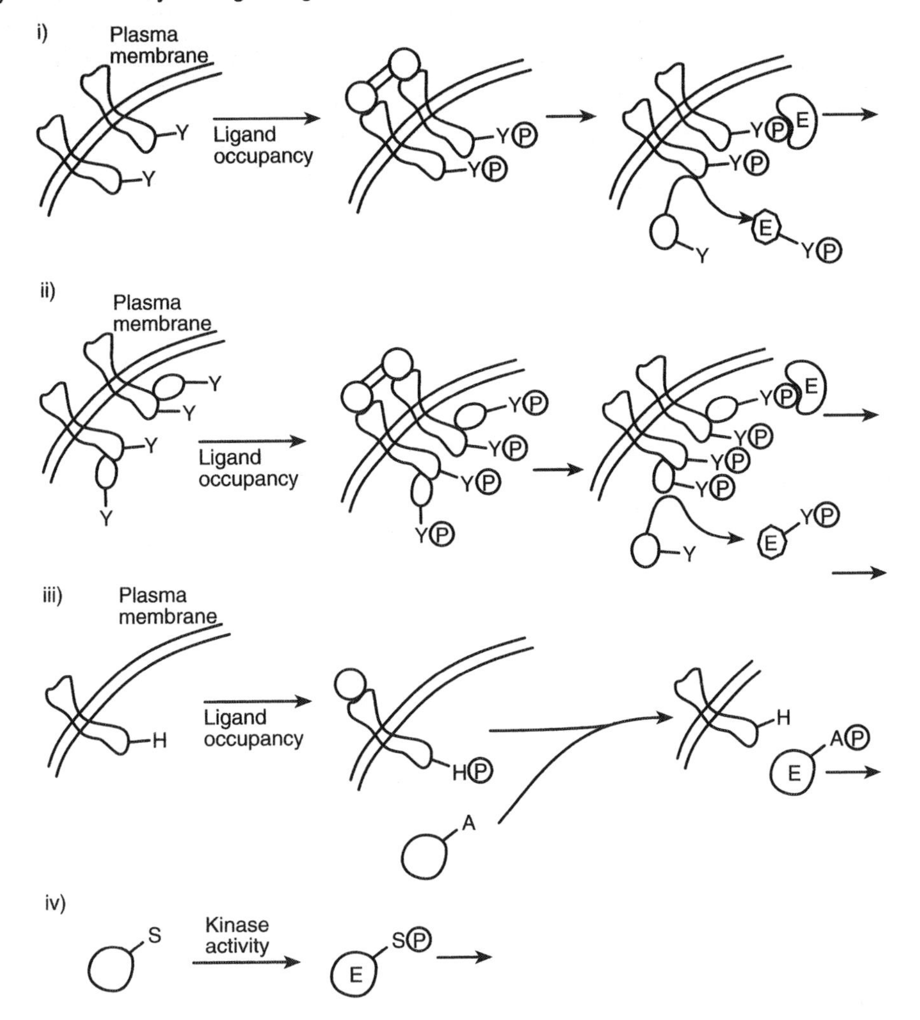

Figure 3.15 A central element of intracellular signalling is activation by phosphorylation which can be triggered in eukaryotic cells by: (i) ligand binding to receptor tyrosine kinases that leads to their dimerization and transphosphorylation. This further activates their kinase activity (for other proteins) and also leads to transmission of the signal by interaction with proteins bearing SH2 domains; (ii) by the association of kinases linked to receptors resulting in events similar to those described in (i); (iii) histidine kinase receptors, found mainly in prokaryotes, bind activating ligands that lead to autophosphorylation and subsequent transfer of the phosphate to a response regulator which is then activated; (iv) soluble kinases can also be activated by phosphorylation. E represents the effector

that have been described. Indeed, an old estimate that the eukaryotic genome encodes 1000 kinases, while appearing to be an excessive claim some years ago, is likely to be an underestimate, and a current estimate is that human cells contain 2000 kinase genes. However, only a limited number of amino acids function as acceptors for phosphorylation by kinases: mainly serine, threonine and tyrosine in eukaryotes, and histidine and aspartic acid in prokaryotes. Nevertheless other acceptor amino acids have been described and the boundaries between prokaryotic and eukaryotic systems is not as clear cut as was once thought.

There is a high degree of amino acid sequence conservation around the catalytic sites of kinases and 12 catalytic subdomains have been identified. Biochemically, kinases have been put into classes depending on the amino acid modified, but they can also be grouped to reflect their organization into pathways, as we shall do later in this chapter. For example, the MAP kinase pathway contains serine/threonine, tyrosine and mixed activity kinases. Phosphorylation is also of importance to bacteria, which are estimated to have in the region of 150 phosphorylated proteins.

Kinases, which are regulatory proteins are, in turn, regulated by one of a number of mechanisms: threonine and/or tyrosine phosphorylation (e.g. the eukaryotic MAP kinases), ligand occupancy that can lead to autophosphorylation, which is then recognized by adapter proteins (e.g. the families of receptor tyrosine kinases in eukaryotes and histidine kinases in prokaryotes) or interaction with small molecules (e.g. cAMP or calcium). The structures of many kinases have been solved and in combination with more conventional molecular analysis this has provided a detailed understanding of how these molecules function and this will now be briefly described. To orient the reader further it is now established that much of intracellular signalling utilizes groups of kinases (and phosphatases) which form pathways by analogy with metabolic pathways. An examples of such a pathway is the MAP kinase pathway.

Serine/threonine kinases

This large family of eukaryotic protein kinases contains some kinases in the MAP kinase pathway and the protein kinase A, B, C and G families, all of which have a major involvement as intermediaries in signal transduction pathways. It also includes the cyclin-dependent kinases that control the regulation of the cell cycle (see Chapter 2 for more details). Relatively recently it has been found that some receptors function as serine/threonine kinases, but the downstream signalling pathways are not as well understood as those of the receptor tyrosine kinase pathways (see below). Some serine and threonine residues are phosphorylated in bacteria and in a few cases the kinases involved have been described.

Tyrosine kinases

Tyrosine kinases in eukaryotes appear at the head of pathways (i.e. as the first component of the pathway) and also in intermediate positions. The receptor tyrosine kinases (RTKs) are regulated by ligand occupancy. Many ligands that bind to surface receptors are dimers, for example PDGF, that induce receptor dimerization leading to autophosphorylation on tyrosine residues. The phosphorylated tyrosine is recognised by proteins bearing the Src homology (SH2) motif (see 'SH2 domain', below). In addition, the phosphorylated RTK is activated to phosphorylate other substrates. Intracellular tyrosine kinases can be regulated by phosphorylation by other kinases (e.g. the MAPK kinases), or by association with receptors (e.g. the Jak kinases). The extensive family of Src tyrosine kinases not only are regulated by various receptors but can also modify receptor function, and thus enable cross-talk to take place between pathways. The discovery, almost 20 years ago, that a known transforming protein, p60-Src from Rous sarcoma virus, was a protein kinase emphasized the importance of protein kinases to cellular function.

Phosphotyrosine has been found in prokaryotes, in particular as part of the stress response.

Histidine kinases

Histidine kinases exist in bacteria as transmembrane proteins that are stimulated to autophosphorylate by ligand occupancy. The phosphate is then transferred to an aspartic acid residue on a response regulator, which is usually a transcription factor. Histidine kinases were once thought to be restricted to the prokaryotic world, but have recently been described in lower eukaryotes and in plants, and there is evidence for their existence in mammals. In eukaryotes, they appear to regulate either differentiation or osmoregulation and this module has been found linked with the activation of a MAP kinase cascade, which is, so far, an exclusively eukaryotic module.

Protein phosphatases

No one would design an electrical or other system with the capacity to be switched on but with no ability to be switched off. Likewise, cells needs to be able to switch off the response to an extracellular signal. As described, protein kinases which add phosphate groups to proteins play a vital role in signal activation. Proteins which remove phosphate groups from proteins are known as protein phosphatases and are increasingly seen as playing key roles in intracellular signalling. Although it was initially thought that phosphatases were passive partners in signal transduction, it is perhaps not surprising that specific phosphatases have now been found that can dephosphorylate phosphotyrosine and phosphoserine/phosphothreonine, and that these have important roles in proliferation, differentiation and cell cycle control. The number of different phosphatases encoded by the eukaryotic genome is predicted to be around 1000. Thus it is clear that the phosphorylation status of phosphoproteins that are used in signalling is moderated by a closely regulated balance of kinases and phosphatases. In bacteria, some kinases are known to function reversibly as phosphatases, although specific phosphatases have also been identified.

Phosphotyrosine phosphatases (PTPs) can either be soluble enzymes or receptors. About 50 soluble PTPs have been described. This list includes some enzymes involved in dephosphorylation of cell cycle control proteins and the Yersinia protein YopH, which is involved in virulence (see Chapters 5 and 6). Receptor PTPs have an extracellular domain that interacts with its ligand, a transmembrane domain and a cytoplasmic region that frequently contains two copies of the enzymatically active domain, though in some phosphatases only one is active. Receptor tyrosine phosphatases are thought to have a role in cell adhesion and the cellular sensing of contact inhibition, since some membrane-bound phosphatases are associated with adhesion molecules and regulate their phosphorylation status. It would be particularly appealing for tyrosine dephosphorylation to be linked to the prevention of growth, since tyrosine phosphorylation events are associated with growth and proliferation. The cytosolic PTPs contain non-catalytic domains that either are directly involved in signalling (e.g. the SH2 domain) or that target the PTP to a particular subcellular location. Both membrane-bound and soluble PTPs are often regulated by Ser/Thr or Tyr phosphorylation.

Four classes of phosphoserine/phosphothreonine phosphatases have been described in mammalian tissues: three are related, of which two appear to have broad specificity (PP1 and PP2A) and the third (PP2B or calcineurin) has a more restricted target range. The fourth group (PP2C) has a broad specificity. All these enzymes are subject to regulation, either via the binding of regulatory proteins, or calcium and calmodulin in the case of PP2B. Some of these proteins are known to have a role in the control of cell cycle progression and signal transduction pathways. A class of potent tumour promoters, of which the seafood contaminant okadaic acid is the best-studied example, inhibit these phosphatases and in consequence potentiate the action of a key protein kinase, protein kinase C (PKC), by blocking dephosphorylation of many of its substrates. Similarly, viral cell transformation proceeds via inhibition of phosphatases, which further demonstrates their importance to the regulation of signalling and suggests that these enzymes can function as tumour suppressors.

Some phosphatases have dual specificity. For example, some phosphatases dephosphorylate members of the MAP kinase family that are phosphorylated on both threonine and tyrosine residues. Phosphohistidine phosphatases have been identified in higher eukaryotes. Some may be specific, and it is also known that the broad specificity phosphoserine/phosphothreonine phosphatases PP1 and PP2A act on phosphohistidine.

Specific phosphohistidine phosphatases in bacteria serve to downregulate the signal prior to transfer to the aspartic acid acceptor. Phosphoaspartate phosphatases have also been identified.

Nucleotide-binding proteins

Three nucleotides have evolved major roles in intracellular signalling. These are guanosine triphosphate (GTP) and the cyclic nucleotides adenosine 3',5'-cyclic monophosphate (cAMP) and guanosine 3',5'-cyclic monophosphate (cGMP) (Figure 3.16).

GTP-binding proteins

GTP binds to a set of eukaryotic proteins (G proteins) that have GTPase activity, removing the terminal phosphate of GTP to produce GDP, which remains bound to the G protein. This reaction is similar to the removal of the terminal phosphate from ATP to produce ADP. The structure of the GTP-bound protein differs from the protein bound to GDP, and in the former state the G protein binds to other, effector, proteins. The cycle (Figure 3.17) is repeated when GDP dissociates from the G protein and GTP binds again. This process can be regulated by factors that stimulate or inhibit the GTPase, or encourage or inhibit exchange of GTP for GDP. There is some, albeit very limited, evidence that GTP-binding proteins operate as regulators in bacteria.

G proteins fall into two categories: the heterotrimeric G proteins that sit at the head of signalling pathways, and the small G proteins or members of the Ras superfamily that are intermediate members of signalling pathways. Some of the Ras superfamily are also involved in transport, but the Ras and Rho subfamily in particular have key roles in signalling processes and, as described in the following chapters, are targets for bacterial virulence mechanisms and bacterial toxins.

Heterotrimeric G proteins. The discovery of G proteins and the realization that they mediate signals from membrane receptors occurred about 20 years ago in studies attempting to elucidate the signalling events linked to carbohydrate metabolism. Subsequent work on the G proteins has been greatly helped by the use of cholera and pertussis toxins which have specific effects on these molecules (described in Chapter 7). The heterotrimeric G proteins consist, as their name suggests, of three different subunits. The α subunit is the largest (39–46 kDa), and the β (35 kDa) and γ (8 kDa) subunits are smaller. Until recently, most attention has

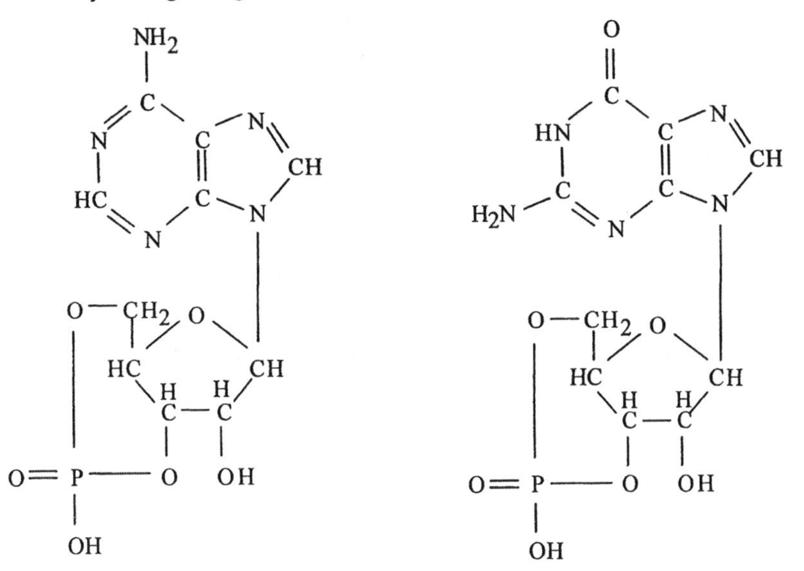

Figure 3.16 The structures of the cyclic nucleotides, cyclic adenosine 3'5'-monophosphate (cyclic AMP) and cyclic guanosine 3'5'-monophosphate (cyclic GMP)

focused on the α subunit, which contains the GTP-binding domain, and which was therefore presumed to be the sole route through which signals reaching the G protein were passed on. The α subunits also show most diversity and the different groups of G proteins are classified according to this subunit. However, despite the prominent role of Gα in signal transduction, it is now recognized that the βγ combined subunit is not an inert partner and also transmits signals by non-covalent interaction with effector molecules. G protein activation follows the exchange of the bound GDP for GTP, and dissociation of the α subunit from the βγ subunits, which operate together as a team (Figure 3.17A). Both the α and βγ moieties are believed to remain membrane associated but can then interact productively with effector proteins.

The heterotrimeric G proteins are membrane-associated and interact with membrane receptors that form a family of structurally related proteins which are also called serpentine receptors as they criss-cross the membrane in a manner reminiscent of a snake (Figure 3.5). The ligands that bind and activate these receptors fall into three categories: small molecules (e.g. prostaglandins, leukotrienes, adrenaline and histamine), glycoproteins (follicle-stimulating hormone) and peptides (neuropeptides, chemokines and opioids). These serpentine receptors have seven membrane-spanning hydrophobic domains, with an extracellular N-terminus that is often glycosylated and an intracellular phosphorylated C-terminus. This structure produces three extracellular loops that can interact with ligands, and three intracellular loops that interact with G proteins. Activation of the G protein by the receptor was thought to be entirely due to a conformational change induced by the interaction, but for one G protein (G_q) at least it is known that tyrosine phosphorylation of the α subunit regulates activity.

The different G proteins are defined by the 21 α subunits so far identified, and have been classified into four subgroups: G_i, G_s, G_q and G_{12}. Several different α, β and γ subunits have

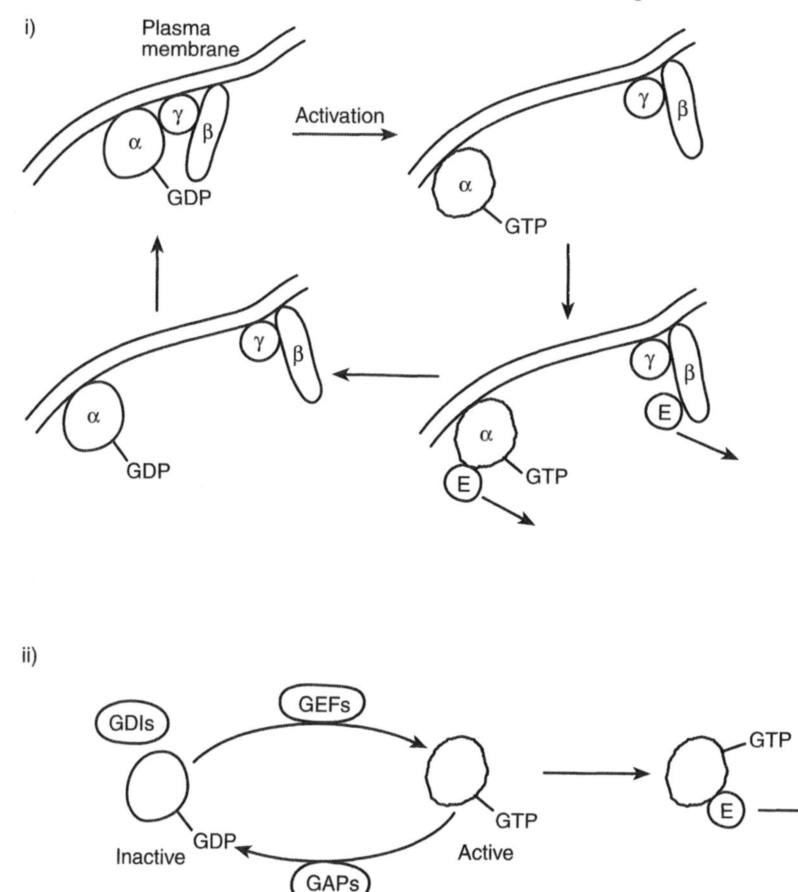

Figure 3.17 The functioning of G proteins. (i) Membrane-bound heterotrimeric G proteins, comprising α, β and γ subunits, are activated by dissociation of the Gα and Gβγ subunits and GDP/GTP exchange on the Gα subunit. The activated Gα and Gβγ subunits can then interact with other signalling molecules. GTPase activity results in the Gα subunit being bound with GDP and subsequent reassociation with Gβγ and downregulation. (ii) Small G proteins interact with guanosine nucleotide dissociation inhibitors (GDIs) that inhibit nucleotide exchange, guanosine nucleotide exchange factors (GEFs) that promote exchange to activate the G protein and GTPase-activating proteins (GAPs) that stimulate intrinsic GTPase activity of the G proteins

been described and though it appears that not all possible permutations of these subunits may occur in nature, this arrangement gives the opportunity for even more variation. Indeed, certain β/γ combinations appear to associate with a greater affinity than others. To date five β and 11 γ subunits have been defined. Four of the β subunits are very similar and differ from each other by a maximum of 17%. The γ subunits display more heterogeneity than the β subunits and the specificity of the Gβγ interactions with effectors may lie in the γ subunit.

The crystal structure of some G proteins has been solved and this helps to explain how the individual subunits and domains within them function. The α subunit has two regions that are affected by the status of the bound nucleotide. The switch II region is a helix that interacts with the terminal phosphate of GTP, but which in the absence of the terminal phosphate (i.e. with GDP) rotates to expose a hydrophobic region that interacts with the β subunit. The switch I region is disordered in the GTP-bound form, but forms a helix that interacts with the β subunit in the GDP-bound form. The β subunit has a seven-bladed propeller structure that comprises several tandem repeats of the WD domain (see 'WD (or WD-40)', below). The β and γ subunits are very closely associated; the γ subunit has few contacts with itself, which explains in molecular terms why they operate as one unit. The α/β interaction causes significant structural changes in the α subunit, which are inferred to be relevant to activation, but the β-propeller is a rigid structure that is not altered upon separation of the functional subunits. Instead, it is believed that the α/β interaction sterically prevents the β subunit from interacting with effector proteins. Both the α and β subunits interact with the membrane receptor which serves to activate the G protein.

What is the advantage to the cell of having such G proteins instead of a receptor that is directly coupled to the next component in the chain? Like all the signalling pathways, extra components give further opportunities for signal amplification and integration and the ability of a given ligand to influence several pathways. Some receptors interact with several G proteins, whereas others are more specific. The G proteins themselves can have different specificities. For example, G_t interacts only with cAMP phosphodiesterase, whereas G_i has a more promiscuous lifestyle. In addition, a wider output is generated by the ability of the closely related family of $G\beta$ subunits to relay signals, which can then be modulated by the more specific signals provoked by $G\alpha$ subunits. Differential tissue distribution of the different α, β and γ subunits also permits further control of the output signal.

Small G proteins. Ras was the first small G-protein that was identified and so the large number of related proteins are also known as the Ras superfamily, or p21 family, because of their molecular weight. Ras was originally defined over 15 years ago in a virus-encoded form from the transforming Harvey and Kirsten sarcoma retroviruses. The cellular homologues were found to be proto-oncogenes that could be permanently activated by mutation at particular residues that are now known to be crucial for the GTPase activity, so that all such mutants remain bound with GTP and in the active state. The discovery that Ras mutation occurred in about 30% of human tumours was crucial evidence for the mutational theory of cancer and showed the importance of the regulation of signalling to that process. The scientific excitement that surrounded these highly significant initial discoveries about Ras cannot be exaggerated.

The Ras superfamily consists of over 50 related proteins that have key roles in several aspects of cellular function. They have been subdivided into a variety of subfamilies (Table 3.10), involved in proliferation and differentiation (Ras subfamily), cytoskeletal organization (Rho), vesicular trafficking (Rab and ARF) and nuclear membrane transport (Ran). Within each family there are further subdivisions but although the individual members of the Ras superfamily have particular roles, there is good evidence that pathways are interconnected. For example, cell transformation by Ras requires the presence of functional Rac.

Several classes of proteins interact with small G proteins to modulate their activity (Figure 3.17B). GDP dissociation inhibitors (GDI) prevent loss of the bound GDP and thereby serve to keep the G protein in an inactive state. GTPase-activating proteins (GAP) stimulate the very low intrinsic level of GTPase activity about 105-fold to attenuate signalling from the activated G proteins. Guanine nucleotide exchange factors (GEF) aid removal of the bound GDP to enable GTP to bind to and activate the protein. These factors are themselves

Table 3.10 Main members of the Ras superfamily

Family	Members of subfamily	Function
Ras	H-Ras	Signalling: proliferation, differentiation
	K-Ras	
	N-Ras	
	Rap1	Antagonizes Ras
	Ral1	?
Rho	RhoA	Actin rearrangement, signalling
	RhoB	Cell cycle, signalling
	RhoC	Actin rearrangement, signalling
	Rac1	Membrane ruffling, signalling
	Rac2	NADPH oxidase
	cdc42H	Filopodia formation, signalling
	Rho G	Cell cycle?
	Rac E	Cytokinesis
	TC10	?
Rab	Many	Vesicular transport
Ran	Ran	Nuclear protein transport

regulated and some have also been shown to be proto-oncogenes. Small G proteins function in the activated GTP-bound state by non-covalent interaction with effector molecules that are generally kinases which sit at the head of kinase cascades. A considerable conformational change between the two forms affects the ability of the G proteins to interact with their effector proteins. This has been shown by the crystal structure of Ras in the GTP- and GDP-bound forms.

The Ras and Rho family proteins are found ubiquitously in eukaryotic cells. Although the basic molecular mechanism of their function has been elucidated, the connection between this and the complex and sometimes opposing biological output remains unclear. For example, Ras can function in both differentiation and proliferation. Possible explanations include the length of time that Ras is activated, cellular differences that might enable only certain effector proteins to be available for interaction with Ras in some cells, and interaction with other pathways that could modulate the signal. For example, this last option is the reason why Ras overexpression leads to proliferation in some cells, and blocks cell cycle progression in others.

Cyclic nucleotide binding proteins

Adenosine 3',5'-cyclic monophosphate, or cAMP, was one of the first intracellular signalling molecules to be identified, in the late 1950s. Its role in mediating hormone action (discovered in the late 1960s) led to the concept of 'second messengers', that is, molecules that transmit the primary signal received at the cell membrane. The roles of this molecule are varied and in one eukaryotic organism, the slime mould *Dictyostelium discoideum*, it even functions as an extracellular signal. Cyclic GMP is not thought to have a role in regulating gene expression in prokaryotes, but it has been suggested that it mediates the response to chemoattractants. The concentration of the cyclic nucleotides is regulated by the adenylate and guanylate cyclases that produce them and specific phosphodiesterases that degrade them to the non-cyclic monophosphates. Much attention is being focused by the pharmaceutical industry on the development of selective inhibitors of phosphodiesterase for the treatment of inflammatory diseases.

In bacteria, cAMP produced by soluble bacterial adenylate cyclases binds to and activates the cAMP receptor protein (CRP), which is a transcription factor. Cyclic AMP is a global signal that affects the expression of a large number of genes, mainly enabling the bacteria to express metabolic enzymes required during growth on a particular substrate. Cyclic AMP concentration also controls the expression of a number of other genes, some of which are involved in pathogenesis.

In eukaryotes, cAMP is produced by membrane-bound adenylate cyclases that are regulated by heterotrimeric G proteins, that in turn are coupled to transmembrane receptors. This complexity reflects the central role that cAMP plays in signalling and perhaps not surprisingly it is interconnected with several other signalling pathways. The main effectors of the cAMP signal are the cAMP-dependent protein kinases (protein kinase A). In the inactive form, without cAMP, these comprise a dimer of regulatory (R) subunits and two catalytic (C) subunits that interact with the R subunit via a pseudosequence that binds in the active site on C. Two molecules of cAMP bind to each R subunit to induce a conformational change that releases activated C subunits which can then phosphorylate a variety of substrates on serine or threonine (Figure 3.18).

Cyclic GMP is produced by soluble and membrane-bound guanylate cyclases. Neither resemble the adenylate cyclases in their organization. The soluble enzymes are heterodimers that are activated by other signalling compounds. The transmembrane guanylate cyclases are homodimeric molecules that respond to various peptides, all of which are involved in salt regulation. Elevation of cGMP concentration operates in a similar fashion to

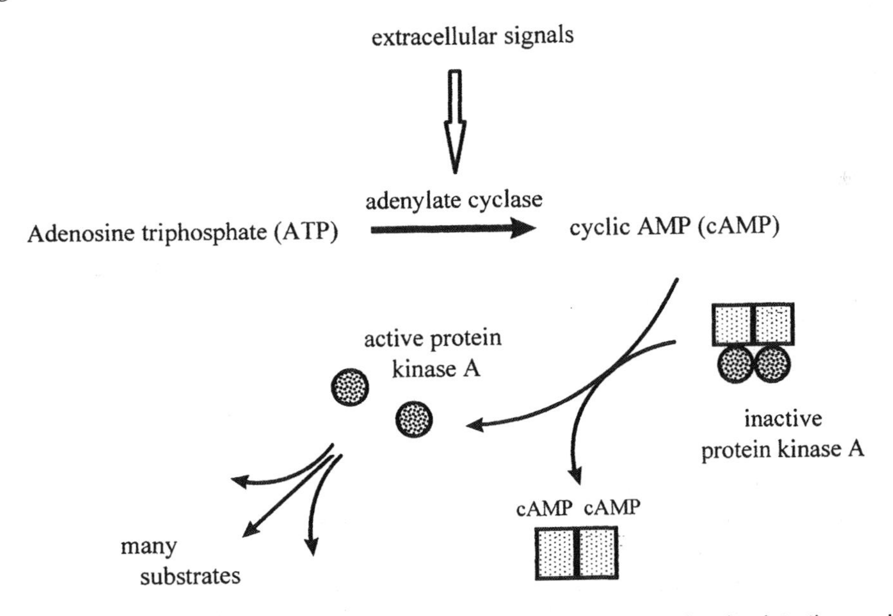

Figure 3.18 Cyclic AMP and protein kinase A (PKA). Extracellular signals stimulate the production of cAMP by adenylate cyclase and this then interacts with PKA to cause dissociation of its subunits and subsequent activation

cAMP to activate a protein kinase, in this case protein kinase G (PKG), which is a serine/threonine kinase. Although PKG is closely related to PKA, it comprises a homodimer where the regulatory and catalytic functions are fused into the same molecule. Full activation requires the binding of four molecules of cGMP.

Both cyclic nucleotides also function in eukaryotes by direct binding to proteins that form cation channels. Binding of the cyclic nucleotide opens the channel. Cyclic GMP also activates a family of phosphodiesterases that degrade both cyclic nucleotides, and which thus function as a feedback mechanism.

Phosphorylated lipids and their role in cell signalling

The organization of lipids in the membranes of higher eukaryotes is not just structural. Lipids are also targets for enzymes that are involved in signalling processes. Distinct cellular phospholipases attack different parts of the membrane lipid moieties to produce a range of signalling molecules. One minor set of lipids, the phosphatidylinositols, play a major role in signalling, and relay signals from in excess of 100 different extracellular ligands. The link between cellular stimulation and phosphatidylinositol lipids was recognized over 40 years ago, but the molecular mechanism has only been clarified during the last 15 years. The polarized head groups of these lipids are the inositols – six-membered carbon rings with a hydroxyl group on each carbon, where a phosphate group can be attached. Several phosphatidylinositols exist, depending on the phosphorylation status of the inositol head group. Their role in signalling is triggered by the activity of at least three enzymes (Figure 3.19). Phospholipase C (PLC) cleaves the phosphatidylinositol 4,5-diphosphate (PIP_2) at the sn3 position to produce soluble phosphoinositols, in particular inositol 1,4,5-triphosphate (IP_3), leaving the remainder of the molecule, diacylglycerol (DAG), in the membrane. Both IP_3 and DAG are highly active in stimulating signalling events. Phosphoinositide 3-kinase (PI3K) is currently attracting a lot of attention. This kinase adds an extra phosphate to phosphatidylinositols to produce a further set of molecules, of which phosphatidylinositol 3,4,5-triphosphate (PIP_3) is the most potent and which sits at the head of a signalling pathway (see 'Phosphoinositide 3-kinase', below). The production of PIP_2 is catalysed by phosphoinositol 5-kinase (PI5K). All these enzymes are regulated by extracellular signals.

Other enzymes are thought to play a more minor role in signalling. Phospholipase A_2 (PLA_2) hydrolyses the bond at the sn1 and sn2 positions of the phospholipid to produce arachidonic acid and lysophospholipid, while phospholipase D hydrolyses lipids at the sn3 position to release the free head group and produce phosphatidic acid, which can be metabolized to DAG.

Phospholipase C and inositol triphosphate

About 10 phospholipase C (PLC) isoforms have been identified and these have been placed into three groups (β, γ and δ), depending on the domain organization and size. PLCβ and PLCγ are regulated by extracellular ligands that either operate via receptor tyrosine kinases in the case of PLCγ, or the heterotrimeric G protein G_q in the case of PLCβ. Both isoforms contain the so-called pleckstrin homology (PH) motifs (see section 'PH domain', below), and PLCγ contains several SH2 domains that interact with phosphotyrosine residues on the activated receptor.

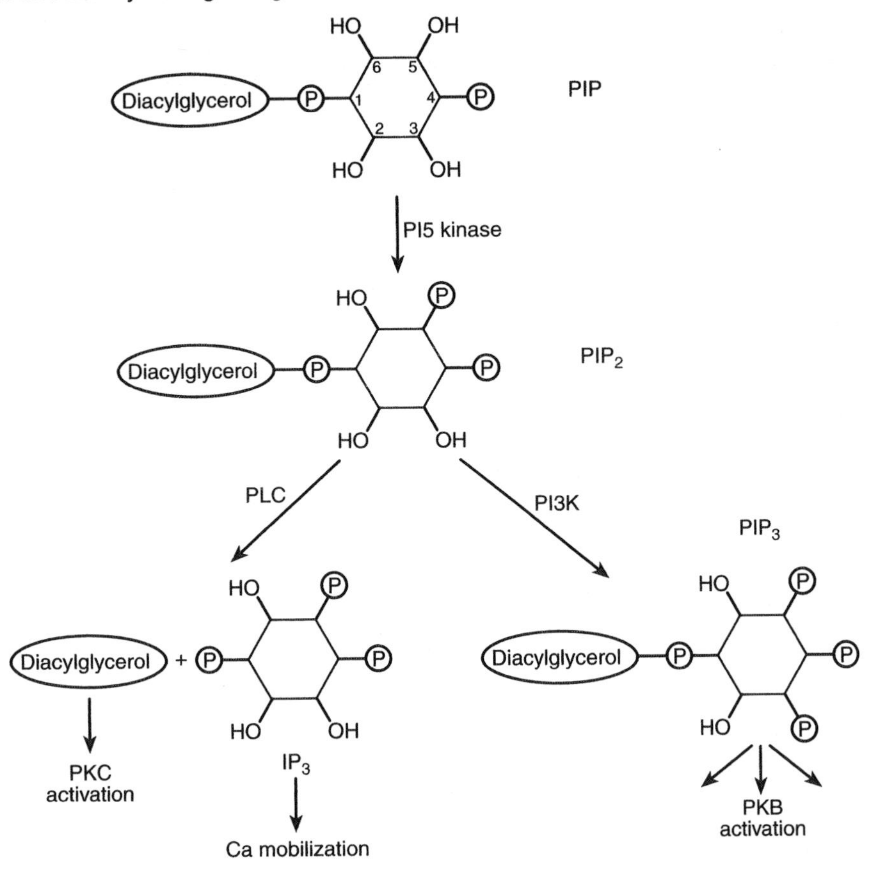

Figure 3.19 Signalling enzymes that modify lipids. The three major enzymes are phosphoinositide 5'-kinase (PI5K) which phosphorylates PIP to produce PIP_2, the substrate for phospholipase C (PLC) and phosphoinositide 3'-kinase (PI3K). PLC action generates IP_3 and diacylglycerol. Both activate further signals

Phospholipase action yields two potent secondary messengers: DAG and inositol triphosphate (IP_3) (Figure 3.19). The intricacies of inositol phosphate metabolism are known, and several species of inositol phosphates have been identified that may have minor roles in signalling. However, IP_3 is the main signalling molecule. It binds to specific receptors on calcium stores. Receptors bound with IP_3 behave as ion channels and release Ca^{2+} ions into the cytoplasm.

Diacylglycerol binds to and activates protein kinase C (PKC), which has wide-ranging effects on a number of substrates (see 'Protein kinase C', below). DAG remains membrane-bound and thus the activation of PKC serves to recruit it to the membrane. Five different types of DAG kinase have been identified that serve to attenuate signals due to activated PKC. Some are calcium-dependent, whereas others interact with different pathways by association with Ras or through PH or ankyrin domains (see 'Non-covalent protein–protein interactions', below).

Phosphoinositide 3-kinase

Phosphorylation of phosphatidylinositol at the 3 position by phosphoinositide 3-kinase (PI3K) has only been recognized as a major signalling event in the last few years, and the molecules involved in activation, the relative importance of the lipid substrate and the effectors that interact with the multiply phosphorylated lipids are not clearly understood. PI3K has a catalytic and a regulatory subunit. The latter has an SH2 domain, an SH3 domain and two proline-rich domains, which can interact with an SH3 domain. It is through the last feature that the cytoplasmic tyrosine kinase Src is believed to activate PI3K. The SH2 domain is believed to receive signals transmitted from receptor tyrosine kinases. It is also clear that the Gβγ subunit binds to the regulatory subunit to activate PI3K, perhaps to facilitate membrane localization that can lead to subsequent activation. PI3K is also believed to be one route whereby stimulation by cell surface integrins is mediated. Several 3'-phosphorylated products are produced by PI3K, depending on the substrate, but it is now generally agreed that the major signalling molecule is generated using phosphatidylinositol 4,5-diphosphate to produce phosphatidylinositol 3,4,5-triphosphate (variously known as PtdIns 3,4,5-triphosphate or PIP_3).

Very recent work has shown that the major effector of PIP_3 is protein kinase B (PKB, also called c-Akt). Its activation by PIP_3 appears to have two elements: activation by non-covalent interaction with PIP_3 via the PH domain on PKB, and phosphorylation by a newly discovered family of kinases that is facilitated by the recruitment of protein kinase B to the membrane. The biological effects attributed to protein kinase B are rapidly increasing.

Phospholipase A_2

PLA_2 is found in a variety of snake and bee venoms, but also in cytosolic form in various cells where it affects signalling. Arachidonic acid produced by the action of released PLA_2 can stimulate cells in an autocrine manner, and gives rise to a very large number of lipid mediators including prostaglandins, leukotrienes and lipoxins. These various lipid mediators were initially believed to be proinflammatory. However, there is increasing evidence that they play key roles in controlling both inflammation and the mucosal response to pathogens. PLA_2 action also provides precursors of various other signalling molecules. Various different groups of PLA_2 have been distinguished.

Phospholipase D

Phospholipase D (PLD) is a membrane-associated enzyme that shows a specificity for phosphatidylcholine to produce phosphatidic acid. The activation of PLD is not fully understood though it appears that activation can either be via PKC or involve the RhoA G protein. In the former case it is not clear whether activation requires phosphorylation to take place.

Intracellular calcium concentration and cell signalling

The cytosolic calcium ion concentration in the eukaryotic cell is tightly regulated and is kept very low (about 10–100 nM), although calcium is sequestered in membrane-bound vesicular stores. Intracellular calcium concentration fluctuates during the cell cycle and can also be raised either by release from the stores or by influx from the much higher calcium concentration outside the cell, although the latter mechanism appears to be restricted to excitable cells. In other cells, transient rises in calcium ion concentration are triggered by activation of receptors that bind either inositol 1,4,5-triphosphate (IP_3R) or ryanodine. Both receptor types are stimulated by calcium itself, which serves to amplify the incoming signal. When bound by ligands, the receptors, which are transmembrane proteins, form calcium-permeable pores. The calcium signal is turned off by energy-requiring pumps that are regulated by the protein kinases A and C and by calcium itself (via calmodulin – see below).

The calcium signal is transmitted mainly by binding to calmodulin (CaM), except in excitable cells where other binding proteins are also involved. Calmodulin is a 16.7 kDa protein that binds calcium with a high specificity through the so-called EF-hand motif. Calcium binding induces a conformational change that enables CaM to interact with a very wide range of effectors, that include among others calmodulin-modulated kinases. The phosphatase calcineurin (PP2B) has been widely studied. It is a heterodimer, with one subunit that contains the iron/zinc-containing catalytic site and a regulatory subunit that binds calcium. The catalytic domain is also regulated by binding to CaM. Calcineurin is involved in many cellular activities, including apoptosis, NO synthesis and activation of T lymphocytes.

Bacteria also have tight control over the intracellular concentration of calcium, which varies with the proliferative state of the bacterial cell. Several proteins with homology to CaM have been identified and it is therefore thought that calcium ion concentration is used as a second messenger as in eukaryotic cells.

Other metal ions can affect gene expression and therefore may be described as signalling molecules. In bacteria there are many iron-regulated promoters, which reflects the essential requirement for iron for bacterial growth and the need for the expression of several proteins to enable the cell to maximize the uptake of iron. Bacteria can also use iron concentration as an indicator of their environment and several virulence determinants (most notably diphtheria toxin) are iron regulated.

Non-covalent protein–protein interactions

We have already introduced, albeit indirectly, the terminology of protein domains or motifs (e.g. SH2, PH domain, ankyrin domain). It has become apparent in recent years that non-covalent interactions between proteins are very important in eukaryotic signal transduction and lead to a conformational change in the effector molecule resulting in activation and/or recruitment to another location (often the cell membrane) for activation there. Several of these interactions use common motifs, but retain specificity by subtle differences in the amino acids that flank them. Other interactions appear to be unique to the pairs of signalling molecules involved, or perhaps common themes have yet to be recognized. These may, for example, involve structural rather than linear sequence similarities.

Several binding domains are commonly found in signalling proteins (Table 3.11). By far the best described are the so-called SH2 and SH3 domains. The PH domain is also widespread. The growing list of such domains also includes WW, WD (or WD-40), ankyrin, SPRY and PTB (or PID) domains. The first two owe their nomenclature to the Src protein (*S*rc *h*omology 2 and 3 domains) PH refers to the protein pleckstrin, a substrate for PKC, from where it was first identified; WW refers to the conserved tryptophan residues (W being the single-letter code for tryptophan); WD domains comprise about 40 residues with a central tryptophan–aspartic acid motif. Ankyrin repeats were first identified in ankyrin, one of the

Table 3.11 Common motifs in signalling proteins

Domain		Interaction	Example
SH2	Src homology region 2	Binds to phosphotyrosine	Grb2
SH3	Src homology region 3	Binds to proline-rich regions	Grb2
PH	Pleckstrin homology	Lipid and protein interaction	Sos
WW	Tryptophan tryptophan	Binds to proline-rich regions	dystrophin
WD(40)	Tryptophan aspartic acid	Binds to PH domains?	Gβγ
Ankyrin	Ankyrin gene	Binds to PH domains?	ankyrin
SPRY	*splA* and *RyR* genes	Binds to proteins?	splA
PTB	Phosphotyrosine binding	Binds to phosphotyrosine	Shc
PID domain	Phosphotyrosine interaction		

many proteins which interact with the actin cytoskeleton (see Chapter 2 for more details). SPRY is made up from the first two letters of each of the two genes (*splA* and *RyR*) in which it has been discovered, PTB stands for *p*hospho*t*yrosine *b*inding and PID is the abbreviation of *p*hosphotyrosine *i*nteraction *d*omain. As these domains are key to signal transduction they will be briefly described.

SH2 domain

These domains recognize and bind to phosphotyrosine and are very widely distributed amongst signalling proteins. However, proteins with the SH2 domain do not attach to every phosphotyrosine. The specificity of the interaction is believed to lie in amino acid residues downstream of the phosphotyrosine. Several SH2 domain structures have been solved and details of the interaction with phosphotyrosine and the flanking regions are known.

SH3 domain

These recognize short (about nine amino acids) proline-rich sequences, and are also widespread in signalling proteins. The generally preferred positions for the proline residues have been worked out – polyproline does not bind with high affinity – but, as with SH2, amino acids C-terminal to the SH3 motif encode specificity.

PH domain

These domains are not clearly defined at the sequence level. They are found in many proteins connected with inositol signalling and it appears that PH domains can mediate interaction with lipids and proteins.

WW domain

This motif appears to be similar to SH3 in that it binds to proline-rich regions.

WD (or WD-40)

These repeats are found in Gβ subunits and in other proteins involved in signal transduction, where it is speculated that they mediate interaction with other proteins, possibly with PH domains. The number of repeats varies between five and eight. They produce a rigid β-propeller structure, which has been particularly analysed in the Gβ subunit.

Ankyrin

The ankyrin repeat has been found in a wide variety of proteins including those involved in cell signalling. Although its function is not understood, it is thought to mediate protein-protein interactions, possibly through a predicted hydrophobic region, and in one case is thought to operate by binding to a PH domain.

SPRY

This motif has been identified by sequence searching and it is assumed that it is involved in protein–protein interactions.

PTB (also called PID)

This motif also interacts with phosphotyrosine. Specificity for particular proteins lies in regions N-terminal to the motif.

Regulation of transcription

The secret to understanding the functioning of prokaryotic and eukaryotic cells lies in how they regulate gene expression in response to the extracellular environment. Cells generally undergo a differentiation programme and only express a subset of all the genes that they encode, unless mutational changes lead to loss of control. This applies not only to multicellular organisms, but also to many free-living eukaryotic cells that have developmental choices, as well as to some bacteria, for example those that sporulate. Moreover, the ability of a cell to respond to any given signal by changing its gene expression is often the end point in a signalling pathway.

Both eukaryotic and prokaryotic cells contain systems that can regulate the expression of individual genes and those that can regulate the expression of large groups of genes. In bacteria, regulators that cause changes in expression of many genes are referred to as global regulators.

The very act of gene expression is also often a means whereby the cell integrates a series of different, sometimes competing, signals and many complex promoters exist in both eukaryotes and prokaryotes that are sensitive to a large number of both positive and negative regulators. In prokaryotes, expression of many of the transcription factors is regulated though post-translational events are also observed, for example with the His-Asp phosphorelay and with cAMP-mediated CRP/DNA interactions. In eukaryotes, much of the control of transcription factors is regulated either by phosphorylation, or protein–protein interactions that serve to recruit other factors to the promoter. There is the added complexity that some factors translocate from the cytoplasm to the nucleus.

The role of the cell membrane

The outer membrane of cells plays an obvious role as the boundary of the cell through which any extracellular signal has to penetrate in one way or another. In a similar manner in bacteria, receptor histidine kinases direct signals from extracellular ligands across the membrane.

As well as the receptors, many other signalling components are found to be membrane-associated. Why is this? One reason may be where the end effect is membrane-associated, such as the organization and reorganization of the cytoskeleton. However, a major advantage of the cell membrane is that complexes of components can be more easily and rapidly organized in a two-dimensional environment. It has become increasingly obvious that these pathways operate in loosely fastened complexes, even though we draw pathways to suggest that each component interacts first individually with its activator and then with its effector. Proteins like the strangely named 14–3-3 proteins in eukaryotes are believed to function by facilitating the association of such complexes and play a significant role in signalling. In many cases, the function of signalling components is to recruit other molecules to the membrane, where they can interact with additional components. For example, Ras activation of the Raf kinase is now believed to be due to the ability of GTP-bound Ras to recruit Raf to the membrane, where it is activated by phosphorylation by membrane-bound kinases.

Prokaryotic signalling mechanisms

Intracellular signalling is analogous to a complex electronic circuit, and thus far we have only provided the reader with a description of the different components (e.g. for transistors substitute kinases, for resistors phosphatases and so on). In the next two sections an attempt will be made to put some of these parts together and the signalling mechanisms, first in bacteria, and then in eukaryotic cells, will be briefly described.

It was recognized many years ago that enzymes responsible for particular catabolic pathways in bacteria were tightly regulated, only being switched on when required. Nevertheless, the extent to which such relatively unsophisticated organisms can recognize and adapt to different environments is astounding. Indeed the ability of a bacterial population to evade the immune response by generating a diverse array of different bacteria, only some of which will survive, shows the complexity of bacterial survival strategies.

Different bacteria have different genome sizes. The larger genomes, of organisms such as *E. coli*, enable the bacteria to be able to choose a wider variety of nutrients and lifestyles, whereas bacteria with smaller genomes have a more limited lifestyle, of which the most extreme examples are the obligate intracellular bacteria *Rickettsia* and *Chlamydia*.

Several generic mechanisms of regulation in bacteria can properly be called signalling systems. These include the histidyl-aspartyl phosphorelay systems, and the cAMP/CRP and FNR DNA binding proteins.

Histidyl-aspartyl phosphorelay systems

The main module that bacteria use to receive and process incoming signals is the histidyl-aspartyl phosphorelay system. This has been found in all bacteria examined to date. The His-Asp phosphorelay was formerly known as the two-component system, but both the number of signalling proteins involved and the diversity of their arrangement make this an inaccurate description. The common features are a histidine kinase that is triggered to autophosphorylate and the subsequent transfer of the phosphate to an aspartic acid residue. The activated phosphoaspartate-modified protein, which is technically an aspartate kinase, can pass on the signal in a variety of ways. Some of these response regulators are DNA-binding proteins, while others operate by protein–protein interactions. Various arrangements of the histidine kinase with respect to the receptor are found. Some transmembrane receptor proteins encode the histidine kinase domain, whereas in others the histidine kinase is a separate intracellular protein that interacts with the receptor. The His-Asp phosphorelay module that was first described comprises a pair of proteins (hence the two-component nomenclature), where the N-terminus of the transmembrane receptor sensor binds to a ligand that activates histidine autophosphorylation in the cytoplasmic C-terminus. This phosphate is transferred to an aspartic acid in the N-terminal receiver domain of the response regulator, and the C-terminus helix–turn–helix motif is facilitated to bind to DNA. The kinase domains show considerable homology, as do the aspartate kinase domains of the response regulators.

Attenuation of the signal takes place either by autocatalysis or by the action of aspartate phosphatases. In the former case the rate of reaction has been shown to vary depending on the nature of the system. Histidine phosphatases that interfere with the signal have also been found.

Each bacterium contains many such systems, and they regulate a wide variety of processes, including chemotaxis, response to osmolarity, oxygen and phosphate and several virulence systems. *E. coli* and *Bacillus subtilis*, for example, are known to contain more than 30 such His-Asp phosphorelays, whereas *M. tuberculosis* contains around 10. We have chosen to describe three systems to illustrate some of the principles involved. These are the chemotaxis signalling pathway of *E. coli*. which has been subject to detailed analysis, the Bvg system involved in the regulation of several virulence genes in *Bordetella* and the EnvZ/OsmR osmoregulatory system.

Chemotaxis

The ability of bacteria, and indeed other cells, to sense gradients of attracting or repelling chemicals and swim accordingly has long fascinated biologists. It has been known for some time that bacteria have two ways of swimming: in a straight line when swimming towards a chemoattractant, or tumbling if the line of travel is not favourable and choosing a new direction at random. The net result is a 'biased walk' that moves the bacterium towards the attractant. It was recognized that the concentration gradient along the short length of the cell was insufficient to be discerned, and that the only possible mechanism was one in which the bacterium took a concentration measurement and compared this with its 'memory' of a previous reading. It was difficult to imagine how this could be achieved by a 'simple' bacterium. The system that enables a bacterium to sense the relative concentration of a variety of chemicals and respond accordingly is fascinating in both its elegance and relative simplicity (Figure 3.20). At least four soluble proteins in the periplasm bind to different chemoattractants. When bound to their cognate ligands these proteins attach to one of five transmembrane receptors that forms part of a signalling complex. This includes the cytoplasmic histidine kinase, CheA, which is activated when the receptor/binding protein is either unbound by ligand or bound with a repellant ligand. Conversely it is downregulated when bound by an attractant ligand. CheA transfers the phosphate to two response regulators, CheY and CheB. CheY leaves the signalling complex and attaches to the flagellar motor to promote tumbling activity. Thus the cell responds to chemorepellants by tumbling, and in the presence of an attractant the His-Asp phosphorelay is inactivated and CheY is not able to leave the signalling complex to bind to the motor to facilitate tumbling and straight line swimming is favoured. The signal is attenuated by CheZ, which acts as a aspartylphosphatase on CheY.

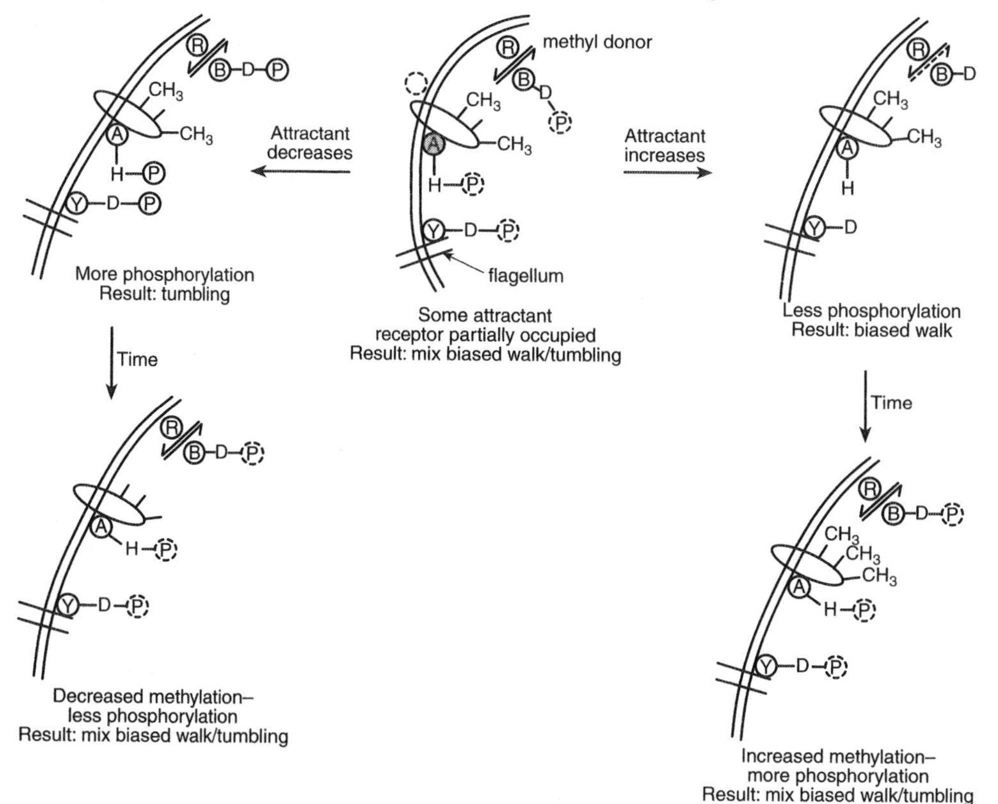

Figure 3.20 His-Asp phosphorelay and chemotaxis in bacteria. This figure is a diagrammatic representation of the processes involved in bacterial sensing and response to chemoattractants. The dashed line round the phosphate group indicates phosphorylation of only some molecules, and therefore only a partial signal. The proteins involved are CheA (A), CheB (B), CheR (R) and CheY (Y). Increasing attractant (right panels) results in less phosphorylation of CheY and thus causes a biased walk. Reduced attractant (left panel) causes increased phosphorylation of CheY and thus increases tumbling. In both cases, changes in the methylation status of the receptor returns the system to baseline

This does not explain how the bacterium can detect chemical gradients, and this is the best part of the story. The sensitivity of the receptor to stimulation by the binding protein is subject to separate control by two enzymes: CheR which methylates multiple glutamate side chains on the cytoplasmic face to increase sensitivity; and CheB, which is a methylesterase. Since CheB is a response regulator that interacts with the CheA histidine kinase, the degree of methylation of the receptor varies with the level of chemoattractant (or repellant). This adaptation part of the pathway enables the cell to measure concentration over time. When the concentration of an attractant is high, CheA histidine kinase is low and consequently the methylesterase CheB is low. The constant activity of CheR methylates the receptor to a high degree and CheA activity is stimulated again, leading to some activated CheB that will moderate the amount of methylation, albeit at a high level, and also some activated CheY that will promote a balance between straight-line swimming and tumbling behaviour. Thus the cell adjusts to the presence of a constant chemical concentration. If the concentration of a chemoattractant decreases, i.e. the bacterium is swimming down a gradient, there will be less ligand-occupied binding protein, increased CheA activity triggered by the highly methylated receptor, and the increase in activated CheY will promote greater tumbling. If the concentration of the chemoattractant increases, i.e. the bacterium is swimming up the gradient, CheA is even less activated, so that there is less activated CheY and thus straight-line swimming is favoured.

The Bordetella *Bvg genes*

Bordetella species produce many virulence determinants, including several adhesins (pertactin, filamentous haemagglutinin and fimbriae) and protein toxins (pertussis toxin, adenylate cyclase and dermonecrotic toxins) (see Chapter 7). Most of these are regulated by a His-Asp phosphorelay system that is triggered *in vitro* by particular salt conditions or temperature. The *in vivo* signal is not known, but may in fact be temperature, since this would allow the bacteria to sense whether they should adapt to life in a warm-blooded animal host or the cooler external environment.

The Bvg system comprises a sensor, BvgS, which is a transmembrane protein, and BvgA, which is the response regulator (Figure 3.21). The histidine kinase activity resides in the C-terminus of BvgS and the aspartate kinase in the N-terminus of BvgA, whose C-terminus has the helix–turn–helix motif typical of a DNA-binding protein. The promoters upstream of the virulence determinants have a classical –10 region, but the –35 region is missing, and thus RNA polymerase does not bind unless BvgA is present. The dinucleotide, TG, at –25 appears to be a common feature in these promoters. The *bvg* locus expresses both BvgS and BvgA

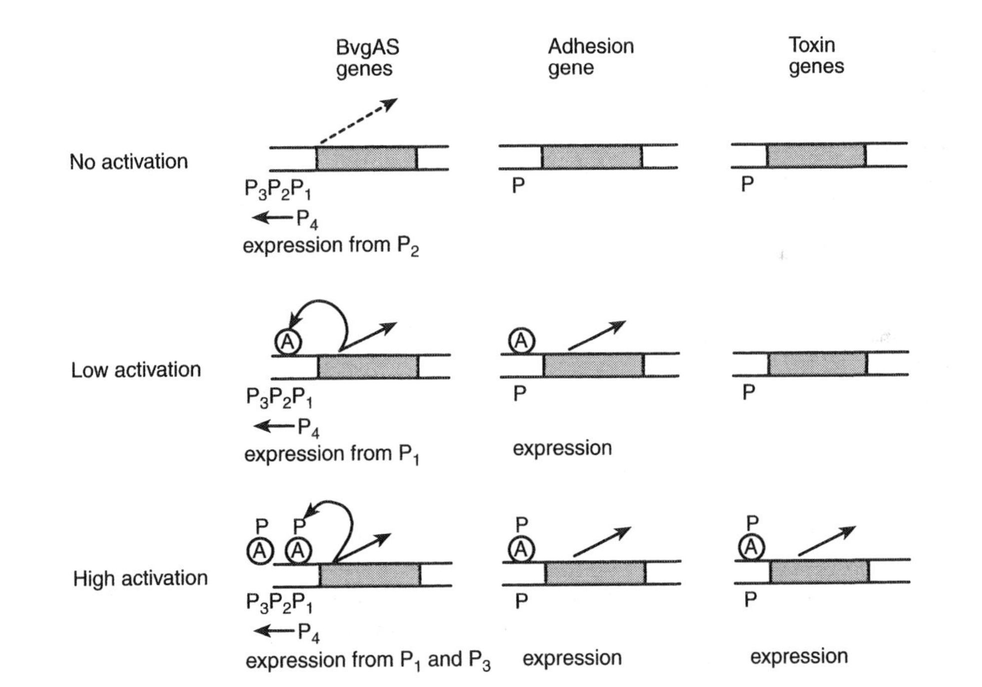

Figure 3.21 Expression of *bvg* genes under the control of BvgAS. Low-level expression of BvgAS occurs when there is no stimulus. Low level of activation drives expression of adhesin genes, but expression from the toxin promoter requires a higher level of both expression and phosphorylation

from a promoter region that comprises four promoters, one of which (P_4) is antisense to the others. P_1 and P_3 are repressed by BvgA, while P_2 and P_4 are activated. The net result is that stimulation of this system causes a slow increase in BvgA over a period of about 24 hours. It was known for some time that induction of the Bvg system leads to expression of the adhesins before the toxins. This made teleological sense, since the bacterium should not waste its energy on making toxins until it has managed to adhere to a suitable surface. It is now known that both the phosphorylation status of BvgA and its concentration are more critical for expression of the toxin genes than the adhesin genes. Furthermore, if the system is turned off, the concentration of unphosphorylated BvgA remains high for a long time, so that a favourable signal can more rapidly turn on toxin gene expression.

Osmoregulation

Bacteria sense osmolarity using the EnvZ/OmpR His-Asp phosphorelay. The membrane-bound EnvZ is a classical histidine kinase which autophosphorylates a histidine on its cytoplasmic domain in the presence of a highly osmolar environment. This phosphate is transferred to an N-terminal aspartate in OmpR which is a DNA-binding protein. At low osmolarity there is a limited amount of phosphorylated OmpR and this binds to a high-affinity site on the *ompF* promoter to facilitate expression of OmpF, which forms a large pore in the bacterial outer membrane. High osmolarity produces greater amounts of phosphorylated OmpR, which is able to bind to a low-affinity site on the *ompF* promoter and repress expression, and also bind to a low-affinity site at the *ompC* promoter to drive expression of OmpF which forms a smaller pore and thus protects the cell from toxic compounds such as bile salts present in the high-osmolarity environment of the host gut.

Crosstalk: opportunities for signal integration

Although each pair involved in a phosphorelay system has a specific partner, the homology between the domains on each protein involved in phosphorelays enables some crosstalk to other systems. Thus stimulation of one system will give some degree of activation of others. This crosstalk has the effect of priming other systems, and thus allows a specific signal to have a wider signalling influence in the cell.

Cyclic AMP and CRP, FNR

Cyclic AMP concentration is a very important signal in bacteria, being estimated to be involved in the regulation of hundreds of genes. Artificial overexpression of adenylate cyclase has been shown to be fatal in some bacteria and there is evidence that intracellular cAMP concentration is tightly controlled, both at the transcriptional and post-translational levels. For example, the *cya* gene in *E. coli* has three promoters, one of which is itself negatively regulated by cAMP. The activity of the expressed adenylate cyclase is influenced by its phosphorylation status, such that the phosphorylated enzyme is active. Phosphorylation is controlled by flux through particular metabolic pathways. The best-known example is that of 'catabolite repression' by growth of *E. coli* on glucose, which leads to a low cAMP concentration and subsequent repression of several genes. Cyclic AMP is degraded by a specific phosphodiesterase.

Cyclic AMP operates by binding to a transcription factor, the cAMP receptor protein (CRP). This was formerly called the catabolite activator protein (CAP) because it was first identified through its role in glucose-induced repression of several catabolic pathways in *E. coli*. It is now known that many genes are affected by CRP, which can operate either to activate or repress gene expression, so that CAP is an inappropriate term. CRP exists as a homodimer and each subunit has two domains. The N-terminal domain of CRP binds one molecule of cAMP and the C-terminus has a helix–turn–helix motif that can bind DNA. Cyclic AMP binding induces a conformational change in CRP, which then binds to short recognition sites on DNA upstream of, or overlapping, promoters. The consensus sequence is TGTGA followed after six bases by the palindromic sequence TCACT, though various promoters that are known to be cAMP/CRP dependent have sequences that diverge considerably from this. In addition, nucleotides flanking these motifs are also important in the interaction. Binding of the CRP–cAMP complex induces DNA bending and can either inhibit or facilitate binding of RNA polymerase and thus gene expression. Both the DNA bending and direct interaction with the RNA polymerase are important for CRP activation. CRP also interacts with other regulators of expression and can serve to integrate signals. Thus the cell is able to interpret incoming signals that modify cAMP by massive changes in gene expression (Figure 3.22).

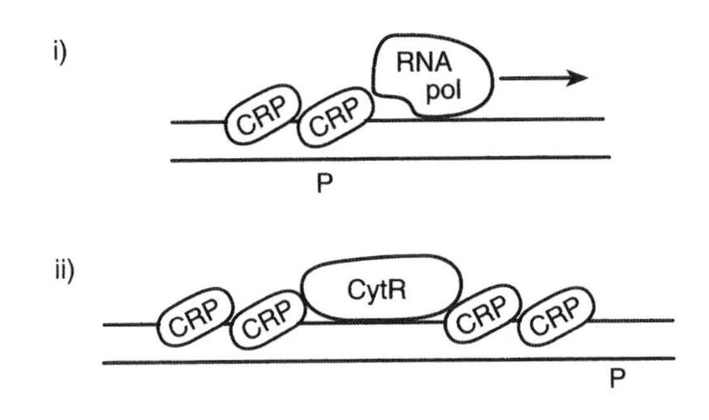

Figure 3.22 Examples of interaction of CRP with DNA. In (i) CRP binds to promoter regions to regulate the binding of RNA polymerase. In (ii) at the *deoP2* promoter, the regulator CytR binds to two appropriately spaced CRP molecules and does not interact directly with the DNA

Where CRP binding is close to the promoter, direct interaction with RNA polymerase occurs and compensates for the absence of a –35 consensus promoter sequence. Such CRP-binding sites are located at –41, –61 or –70, which reflects the importance of CRP positioning on one face of the DNA helix. CRP can also promote expression when the CRP binding site overlaps the promoter region.

When CRP binding is more remote from the promoter, it operates by interaction with other factors. A particularly elegant example is the *malK* promoter. The MalT inducer can bind to three high-affinity sites, that do not facilitate expression, or three adjacent low affinity sites that do. Binding of CRP upstream of these sites interferes with binding to the former sites and forces binding to the low-affinity sites and hence promotes expression. CRP can also facilitate repression from distant binding sites. An example of the complexity possible is provided by the *deoP2* promoter which, unusually, has three interacting elements: the DeoR repressor, the CytR repressor and CRP, which has roles in activation and repression. DeoR and CRP bind to the DNA, but CytR binds to two bound CRP molecules, which explains the previously puzzling observation that there was no consensus binding site for CytR-regulated promoters. Thus CRP exhibits great versatility and has been likened to a eukaryotic regulator.

FNR is closely related to CRP and the consensus binding site, TTGAT, resembles that for CRP but with sufficient differences that crosstalk does not occur. FNR is used by bacteria to regulate genes involved in anaerobic metabolism. Although the structure of FNR is similar to CRP, an extra domain with four cysteine residues binds either two or four iron molecules, depending on oxygen tension. Only FNR complexed with four iron molecules can bind to DNA, where it can activate genes important for anaerobic growth.

Other global regulators of bacterial gene expression

Two other mechanisms exist whereby a bacterium can effect global changes in gene expression. The first mechanism changes the extent to which the bacterial DNA is supercoiled, and thus can affect genes whose promoters are sensitive to subtle changes in DNA topology. The coiled DNA thread is further coiled (supercoiled) to form a coiled coil. The processes of replication and transcription require that the DNA is unravelled for both replication and transcription and this process is controlled by enzymes that actively put in negative supercoils (DNA gyrase, which requires ATP to drive the reaction) and those that relax the supercoiling (DNA topoisomerases). The net result of the action of these two enzymes determines the extent of supercoiling. Because supercoiling affects the juxtaposition of the bases on the double helix, it can alter the ability of some promoters to bind RNA polymerase and transcription factors. The expression of some genes, e.g. *Salmonella* invasion genes, is particularly sensitive to changes in supercoiling. Thus by sensing changes in supercoiling, the bacterium is able to alter the expression of many different genes. It has been shown that temperature affects supercoiling, as does cAMP, and an inhibitor of the DNA gyrase enzyme has also been identified.

The other mechanism that can cause global changes in gene expression is the expression of different sigma factors as described in Chapter 2. These are part of the RNA polymerase and they direct which promoter recognition sequence the RNA polymerase complex will recognize. There are a number of such factors, and the relative ratio can be altered to enable the cell to express a different set of genes. RpoS (KatF or sigma 38) is expressed during starvation and facilitates the expression of a number of genes, including some involved in virulence, for example the *spv* genes in *Salmonella*. Analysis of the recently published genome sequence for *M. tuberculosis* suggests that it encodes 13 possible sigma factors.

Eukaryotic signalling pathways

While the existence of signalling pathways that could enable a cell to interpret its environment has long been obvious, the understanding of how these might operate was not. As small parts of these pathways were identified, starting about 30 years ago, it seemed that the process would be unimaginably complex – indeed too complex to be able to grasp in molecular terms and in its entirety. It is a general truism that when several apparently unconnected parts of a story are known, it is difficult to understand the isolated components and their significance, and it appears that the whole story will be even more difficult to grasp. However, the complete system often displays a logic that is much easier to understand and appreciate. Thus the initial discoveries about Ras activation were not fully comprehensible and were difficult to place in any logical framework, although their importance was evident. In contrast, the description of the whole pathway leading from receptor occupancy through Ras activation to MAP kinase-controlled activation of transcription factors allows all the individual pieces of information about isolated components to fall into place. Though the number of components involved in eukaryotic signalling and the extent of their interconnections are dazzling, it is now possible to have a feel for what is involved and see, even at a superficial level, an overall picture emerging.

The reasons for the explosion of information about signalling are similar to those seen in other aspects of biology: clever new techniques (molecular biology and all its applications, including genome sequencing), a continually increasing number of reagents (isolated components, specific probes, such as antibodies, for individual components and more and more selective inhibitors) and the rapid expansion of the knowledge base on which to hang new discoveries.

Some would say that no pathway has been elucidated in full. It is true that even with the best-characterized pathway – Ras activation and the MAP kinases – many of the details remain to be clarified. Furthermore the interconnections between pathways are such that no single signalling pathway exits without reference to others. Nevertheless several key pathways have been analysed in sufficient detail that the general principles can be seen and marvelled at.

The Ras/MAP kinase pathway

Outline (Figure 3.23)

Ligands binding to the extracellular domain of receptor tyrosine kinases induce dimerization of the receptor and autophosphorylation of tyrosine residues. These phosphotyrosine residues bind to the SH2 domain of Grb2/Sem5, an adapter protein that is bound to the Sos (son-of-sevenless) protein via the SH3 domain of Grb2 and a proline-rich region on Sos. Sos is a GEF (guanine nucleotide exchange factor) for membrane-bound Ras, so that its membrane recruitment leads to interaction with Ras, loss of GDP from Ras and its replacement with GTP. Activated Ras recruits the Raf kinase to the membrane, where it is phosphorylated. The active Raf kinase is a serine/threonine kinase that phosphorylates MEK (or MAP kinase kinase) on two serine residues to activate it. MEK is a dual-specificity threonine/tyrosine kinase that phosphorylates ERK kinase (or MAP kinase) on one threonine and one tyrosine residue. Thus activated, the serine/threonine kinase ERK translocates into the nucleus to phosphorylate transcription factors, of which Elk is the best characterized. Phosphorylated Elk can interact with both DNA and other DNA-binding proteins (in particular the serum response element) to activate expression of *cfos* and other genes. cFos is itself a transcription factor that complexes with cJun to form AP-1, a transcription factor. Another effect is the phosphorylation and activation of PLA_2 leading to wider signalling events.

The reader may be puzzled as to the use of the term son of sevenless (Sos) to describe a guanine nucleotide exchange factor. *Sevenless* is a *Drosophila* gene whose mutation results in one of the eight photoreceptor cells (number 7 as it happens) failing to develop and these mutants are blind to ultraviolet light. Further analysis of these mutants led to the discovery of related genes such as *bride-of-sevenless* (*boss*) and *son-of-sevenless*. This illustrates the ubiquity of these signalling pathways and the fact that information about cell signalling can come from many different scientific disciplines such as, in this case, developmental biology. This does make the literature obscure and there is, unfortunately, little that the authors can do to alleviate this problem.

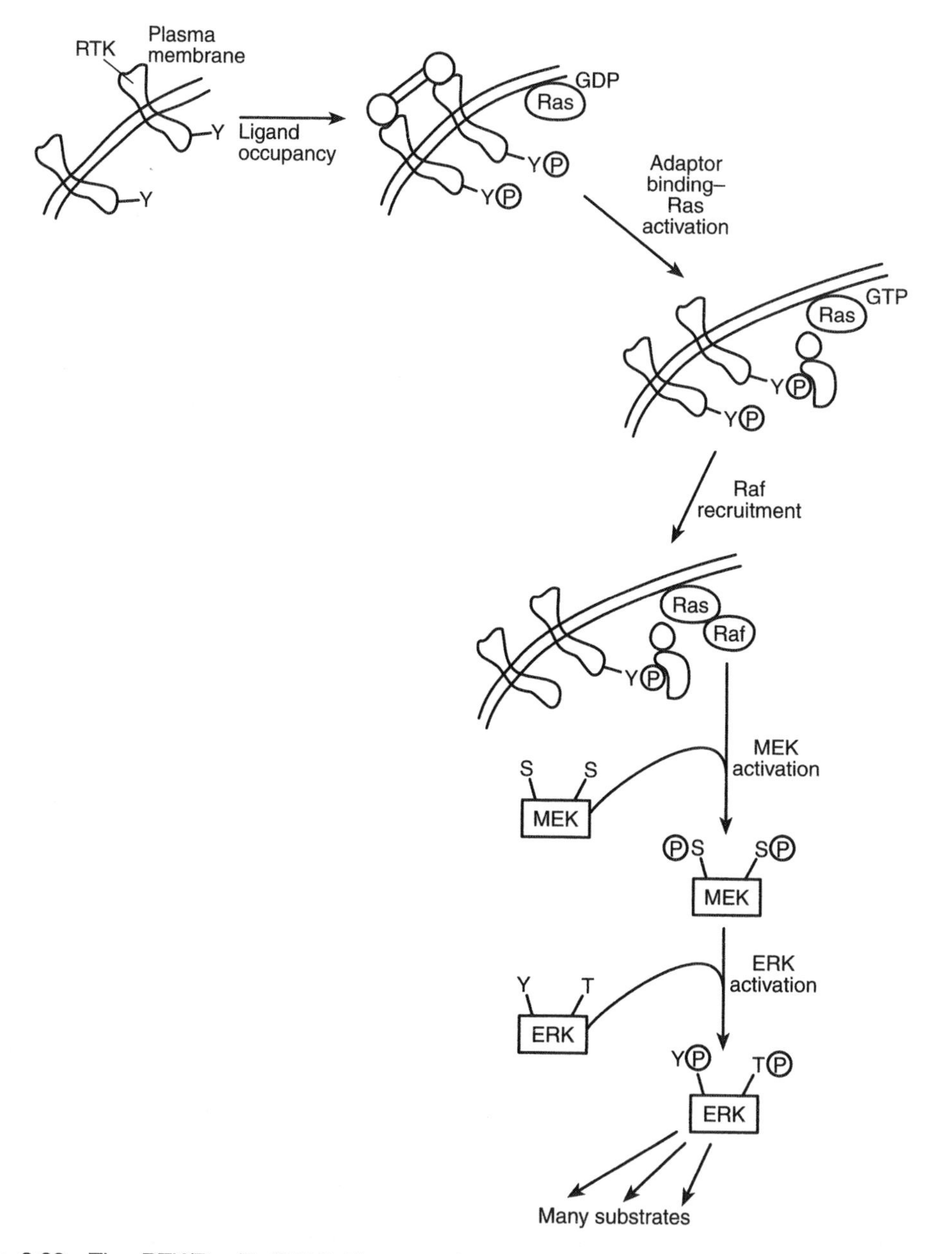

Figure 3.23 The RTK/Ras/Raf/MAP kinase pathway. Ligand occupancy leads to receptor transphosphorylation, membrane recruitment of Grb2/Sos, activation of Ras, membrane recruitment and activation of the Raf kinase, and stimulation by phosphorylation of the MAP kinase pathway, leading to various sequelae in the cell

The various components of the Ras/MAP kinase pathway will now be described.

Receptor tyrosine kinases

Currently, 14 different categories of receptor tyrosine kinases (RTKs) have been described, based on the organization of the extracellular domains. Over 50 receptor ligand combinations have so far been characterized and there are likely to be more. The ligands that activate such RTKs include many growth factors (Table 3.12), several of which are themselves dimers, for example PDGF. Ligand occupancy universally appears to induce dimerization and receptor transphosphorylation. The interaction of the RTK with the SH2 domain on the adapter Grb2 is not the only means by which the RTK promulgates signals. Many RTKs are multiply tyrosine phosphorylated and can signal to other SH2-containing proteins. The PI3K pathway is stimulated via interaction of SH2 domains on the regulatory subunit with phosphotyrosines, as are proteins that interact with Ras (e.g. RasGAP). Various kinases (e.g. members of the Src family) and phosphatases can also be stimulated, and some RTKs stimulate PLCγ, which contains an SH2 domain. In addition, phosphorylation of the RTKs activates their ability to phosphorylate other substrates.

Table 3.12 Receptor tyrosine kinase receptors that bind to growth factors

Receptor tyrosine kinase family		Extracellular features
EGF	Epidermal growth factor	Cysteine-rich
PDGF	Platelet-derived growth factor	Ig-like domain
FGF	Fibroblast growth factor	Ig-like domain and acid box
NGF	Nerve growth factor	Leucine-rich and Ig-like domains
HGF	Hepatocyte growth factor	–
Insulin		Cysteine rich domain, tetramer

The Grb2/Sos interaction

Grb2/Sem5 can associate with other SH2-containing proteins, in particular Shc, another adapter protein that can itself be stimulated by binding to RTKs. Shc provides a link to integrin-mediated signalling. As well as Sos, several other GEFs interact with Ras to activate it (e.g. GRF/cdc25). Some of these are restricted to particular cell types (e.g. cdc25 is reported to be restricted to brain tissue) and interact with only some isoforms of Ras. The phosphorylation status of GRF determines its activity and is thought to be linked to G protein-coupled pathways. It is also reported that GRF is activated by Ca/CaM. Other adapter proteins are known to bind to Sos. Sos also has a DH domain (see 'Activators of Rho', below) that mediates the interaction with Ras. Other adapter combinations that resemble Grb2/Sos have also been found, for example C3G/Crk, which associates with phosphorylated paxillin, and thus, it is thought to link Ras activation to that induced by cellular adhesion. Grb2 interacts with GEFs for other small G proteins (e.g. Vav) to bring about crosstalk between pathways.

Ras and Raf

Ras is a major integrator of incoming signals. As well as the interaction with GEFs described above, Ras also interacts with a variety of GAPs (e.g. p120GAP and neurofibromin (NF-1)) that serve to attenuate Ras function. Many GAPs are large multidomain proteins that display motifs typical of regulated proteins, namely SH2, SH3 and PH domains. P120GAP is stimulated by RTKs and may thereby serve as an important feedback mechanism when only a transient flux through the pathway is required. Indeed during the mid-G_1 phase of the cell cycle, there is a high peak of Ras activation that does not appear to be linked to activation by Grb-2/Sos. In addition Ras activates PI3K.

Ras has a CAAX motif at its C-terminus that is farnesylated on the cysteine (AAX is removed) and other lipid modifications are added. These modifications are required for its membrane location. Such lipid modification of proteins is increasingly seen to be important in cell signalling.

The Raf kinase (sometimes referred to as a MAP kinase kinase kinase) is also a major site for signal integration. The interaction with Ras does not in itself lead to activation. Raf activation is triggered by kinases following its membrane recruitment by Ras. This process also appears to require Raf dimerization and its association with other proteins, notably the 14–3-3 proteins. Not all the kinases involved have been identified, but it is known that members of the protein kinase C family activate Raf, perhaps by phosphorylation or by other means. Similarly there is interaction with G protein-coupled pathways through interaction of $G\beta\gamma$ subunits with Raf, presumably to recruit Raf to the membrane. The cAMP pathway acts in some cells to inhibit Raf by protein kinase A-catalysed phosphorylation of Raf at inhibitory sites, whereas in others the cAMP-linked release of $G\beta\gamma$ subunits from G_i and G_s leads to stimulation.

Raf1 interacts with Ras, but a different isoform (B Raf) is stimulated by a Ras isoform (Rap1) that interacts with Ras in a complex manner. Rap1 and Ras both interact with RGL and ralGDS. The Rap1 interaction antagonizes Ras activation and suppresses Ras-induced transformation, and it is of interest that Rap1 is downregulated during the S phase of the cell cycle. Both Rap1 and RGL are phosphorylated by PKA, which provides yet another link to this pathway. These different isoforms show tissue specificity, which enables different outcomes to be generated by the same initial signalling mechanism.

The MAP kinase pathway

The MAP kinase pathway has attracted significant attention in recent years as it is activated by bacterial factors, particularly LPS, and by a number of cytokines and growth factors, and plays a role in development. The Ras/Raf pathway of activation of the MAP kinases is possibly not the only one, and phosphorylation by other activating kinases may occur, where MAP kinase activation has been reported to be Ras-independent. Furthermore the MAP kinases are not always activated by highly activated Ras, presumably because of the activation of specific phosphatases.

The MAP kinases mediate a number of effects. On one hand there are likely to be several important substrates for the ERK kinase other than the transcription factor Elk, including cytosolic and membrane-bound targets. The MAP kinase pathway also feeds signals into cell cycle control. The timing and duration of signal flux through this pathway will influence the outcome. Inhibition of the MAP kinase pathway blocks cells at the G_1/S boundary. On the other hand, in some cells, MAP kinase induces expression of the cyclin-dependent kinase inhibitor p21 (also called Cip1 or Waf1) that blocks the cell cycle in G_1. This is because normal cells can express functional p53 (see 'Eukaryotic cell cycle progression and signalling', below), which cooperates with Ras expression to induce the production of a cyclin-dependent kinase inhibitor that blocks the cell in G_1. In p53 mutants, the phenotype of many cell lines, the inhibitor is not expressed and proliferation results (see Chapter 2 for a description of the cell cycle).

In cells that are undergoing a developmental programme, MAP kinase activation is involved in passing on signals that specify differentiation.

MEK and ERK kinases also show heterogeneity. MEK 1 and 2 are highly related, as are ERK 1 and 2. There is also heterogeneity in the nomenclature of the members of these pathways. ERK 1 and 2 (*e*xtracellular signal-*r*egulated *k*inase) are also referred to as MAP kinases (*m*itogen-*a*ctivated *p*rotein kinases) or the p44 and p42 proteins (because of their size). MEK is also referred to as MAP kinase kinase. The discovery of several families of such kinase cascades has led to the terms MAP kinase and MAP kinase kinase being used generally, whereas ERK and MEK are used to describe the pathways downstream of Raf.

It is highly likely that multiple examples of the MAP kinase signalling modules exist in cells. Indeed in the relatively simple yeast cell at least five such systems are found. One which has been described in some detail is the MAP kinase pathway induced by stress (see 'Rho effectors', below). This is specifically triggered by other members of the Ras superfamily (the Rho subfamily), though there is also evidence of crosstalk with the p42/p44 MAP kinases.

Regulation of the pathway

Besides the mechanisms already alluded to, there are several other ways that signals can be attenuated. First, receptors are known to be recycled from the membrane through various trafficking routes (i.e. temporarily removed from the cell membrane). There are also a number of phosphatases that can remove phospho-amino acid signals. A large number of factors influence the nucleotide binding status of Ras. It has been shown that activity in this pathway is modulated by the stage in the cell cycle when a signal is applied: whereas Ras activity early in G_1 leads to MAP kinase activation, mid-G_1 activation of Ras appears to occur independently of Grb2/SoS activation and does not stimulate the MAP kinases.

The phospholipase C/inositol triphosphate pathway

Outline (Figure 3.24)

The start of this pathway shows distinct heterogeneity. In one variant, PLCγ is stimulated by its interaction via its SH2 domain with ligand-occupied RTKs (e.g. PDGF receptor) that have dimerized and are tyrosine transphosphorylated. In the other variant, PLCβ associates with Gα_q that is bound with GTP following stimulation by serpentine receptors that are attached to ligands (e.g. neuropeptide binding). Each form of stimulation leads to phospholipase C activity that cleaves PIP_2 to generate inositol 1,4,5-triphosphate (IP_3) and diacylglycerol (DAG). IP_3 binds to and activates calcium channels in membrane-associated calcium stores, and the transient rise in intracellular calcium concentration is sensed by various calcium-binding proteins, of which calmodulin (CaM) is the best-known example. CaM binds to and

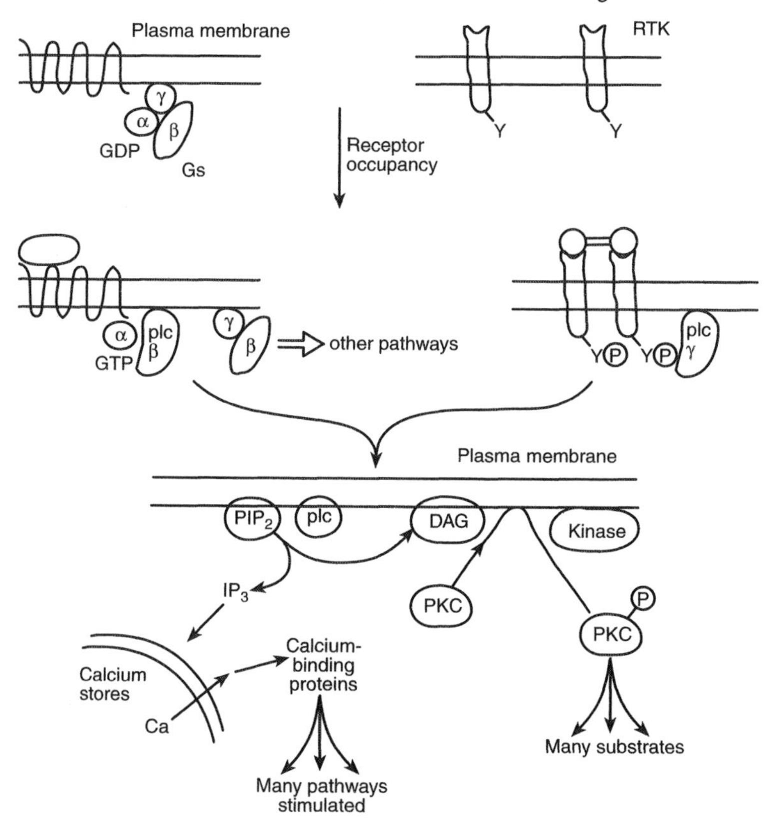

Figure 3.24 The phospholipse C/inositol triphosphate pathway. Phospholipase C *beta* or *gamma* is activated by membrane signalling events and cleaves PIP_2 to produce diaclyglycerol (DAG) and inositol triphosphate (IP_3). These activate calcium ion release and result in activation of protein kinase C (PKC), which phosphorylates many additional protein substrates

activates kinases and other molecules to stimulate a variety of processes. DAG recruits protein kinase C (PKC) to both nuclear and plasma membranes, and the process of binding leads to activation by phosphorylation. Activated PKC has a number of substrates, including the Raf kinase that leads to activation of the MAP kinases, and several cytoskeletal proteins involved in actin organization (e.g. the MARKS protein).

Membrane signalling events

The general principles of RTK heterogeneity have already been discussed. The observation that ligand occupancy of only some RTKs leads to PLC activation further demonstrates the specificity that exists in SH2-mediated interactions.

Ligand occupancy of serpentine receptors complexed with G_q leads to activation of the G protein via GDP/GTP exchange and separation into the two functional subunits, Gα and Gγ. The coupling between the receptor and the G protein is not known in detail, though one mechanism may involve Gα activation by receptor-induced tyrosine phosphorylation. G_q is also known to be tyrosine phosphorylated at other locations by the Src kinase. In addition to activation of PLCβ by

$G_q\alpha$, the $G\beta\gamma$ subunits activate Raf and other effectors. Indeed some isozymes of PLCβ are highly responsive to $G\beta\gamma$ and can thus also be stimulated by other G proteins.

Protein kinase C

Many different PKC isoforms have been described in the 20-year history of research into this important signalling enzyme. These isoforms have been placed into three categories depending on their requirement for calcium and whether they can be stimulated by the phorbol ester tumour promoters (Table 3.13). The isoforms are found in many different species and are therefore believed to provide distinct and useful functions. Some isoforms (e.g. α) are expressed in many tissues, whereas others (e.g. γ) are highly restricted. It is likely that different isoforms mediate different effects (e.g. proliferation, differentiation). In addition, following stimulation, PKC can migrate to plasma membrane, cytoskeleton or nucleus.

Table 3.13 The protein kinase C family

Name	Subgroup	DAG stimulates	Phorbol esters	Calcium
α	conventional (c)	+	+	+
β				
γ				
δ	novel (n)	+	+	–
ε				
η				
θ				
ζ	atypical (a)	–	–	–
λ				
ι				
μ		+	+	–

The Protein Kinase C family

All the PKC isoforms have a similar structure, namely an N-terminal regulatory region and a C-terminal catalytic domain. The N-terminal region contains the DAG binding site (in members of the cPKC and nPKC subfamilies), the so-called pseudosubstrate that blocks the active site until removal by activation, a recognition site for acidic lipids and, in cPKCs, the calcium binding site. The analysis of PKCs has been greatly aided by the observation that the phorbol esters, which are tumour promoters, bind to the DAG binding site. Unlike DAG these are not readily metabolized and remain bound to PKC. It is clear that one function of DAG is to recruit PKC to the membrane. The phorbol esters function by binding to and hiding a hydrophilic region, so that PKC/phorbol ester complex displays a more hydrophobic face that becomes membrane-attached. PIP_3 produced by PI3K action also activates some PKCs. This alternative activation route is likely to be particularly important for PKC isotypes that are DAG-insensitive.

PKCs have a very wide range of targets, and one aspect of their specificity is the site of DAG generation (i.e. at the nuclear or cytoplasmic membrane). PKC feeds into the MAP kinase pathway by its activation of Raf by phosphorylation. A major substrate of PKC is the MARCKS (*m*yristylated *a*lanine-*r*ich *C* *k*inase *s*ubstrate) family of proteins that are associated with focal adhesions and are most readily visualized experimentally. MARCKS proteins bind to CaM/calcium, and to actin which they cross-link. The regulation of actin binding is inhibited by both phosphorylation and CaM binding. Phosphorylation also promotes release from the membrane, subsequent trafficking to lysosomes, and also induces release of CaM. It has been suggested that MARCKS acts as a reservoir for CaM, and thus PKC action serves to release CaM into the cell. In this way it would synergize with another action of PLC, i.e. the stimulation of IP_3 release that leads to a transient increase in calcium which signals in conjunction with CaM. Thus PKC is one of a number of signalling modules linked to membrane and cytoskeletal events. It is becoming clear that such cytoskeletal events are not necessarily end points in signalling but important intermediates.

Regulation of the pathway

Receptor recycling and phosphatase activity will affect the flux of signals through this pathway in a similar manner to others described. In addition PKC is known to decrease the affinity of some receptors for their ligands, in particular the EGF receptor. PKC can also suppress PLCβ activity. Several DAG kinases have been identified and classified into families depending on the factors that regulate them. Some that are calcium-dependent provide an obvious attenuation following PIP_2 hydrolysis that leads to IP_3 formation and calcium ion release. Others have the PH motif or other motifs that mediate protein–protein interactions, and one DAG kinase interacts with Ras.

There are several elements to the regulation of PKC. In the resting state, PKC is phosphorylated on three residues. It is located in the cytoplasm, in a protease-resistant form with the pseudosubstrate blocking its active site. The phosphorylations are required for activity and are regulated by various kinases and phosphatases. Membrane recruitment leads to interaction with phosphatidylserine, pseudosubstrate release and activation. The activated form is protease-sensitive and thus the artificial circumstances of overstimulation by phorbol esters leads to a downregulation (or loss) of PKC. Authentic stimulation by DAG does not lead to such dramatic effects, but should lead to the loss of the stimulated form of PKC.

Adenylate cyclase, cAMP and protein kinase A

Outline (Figure 3.25)

Membrane-bound serpentine receptors (e.g. FMLP or chemokine receptors) upon ligand occupancy stimulate associated G_s G proteins. The activated $G_s{}^*\alpha$, dissociated from the $G\beta\gamma$ subunit, excites membrane-bound adenylate cyclase which catalyses the production of cAMP from ATP. The cAMP produced binds to the regulatory subunits of protein kinase A (PKA), leading to dissociation and thus activation of the catalytic domain. Activated PKA has many nuclear and cytoplasmic substrates and thus mediates a wide variety of effects. In the nucleus it phosphorylates cAMP CREB proteins which are bound to CRE (the cAMP response element) that has the consensus palindromic site TGACGTCA. Phosphorylation induces these proteins to interact with CBP (CREB *b*inding *p*roteins) that act as transcription factors which interact with RNA polymerase.

Membrane events

Attenuation of the pathway is provided not only by phosphatases but also by another G protein, G_i, that is activated by inhibitory signals and functions by inhibiting adenylate

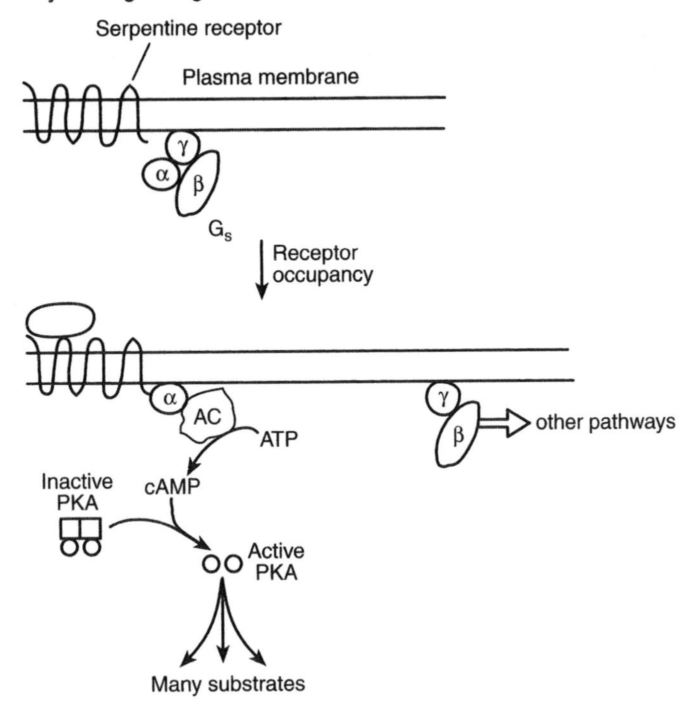

Figure 3.25 The adenylate cyclase, cAMP and protein kinase A pathway. Adenylate cyclase is activated at the membrane by interaction with the activated heterotrimeric G protein, G_s. Cyclic AMP is generated and this binds to and activates protein kinase A (PKA), which phosphorylates many substrates

cyclase. Gβγ subunits can also interact with adenylate cyclase either to inhibit or stimulate its action, and other adenylate cyclases are stimulated by CaM/Ca or protein kinases.

The diversity of protein kinase A effects

The effect of cAMP/PKA on cell proliferation is confusing. It has been reliably reported both to stimulate mitogenesis and to inhibit it. There is very good evidence that raised cAMP (either by inhibition of cAMP phosphodiesterase, cholera toxin action or addition of membrane-permeant modified cAMP) can stimulate mitogenesis. Indeed transgenic expression of cholera toxin led to hyperplastic changes. On the other hand the same range of stimuli leading to raised cAMP concentration can block mitogenic signalling and raised cAMP can even lead to 'reverse transformation'. This situation probably reflects both the complexity and degree of redundancy in signalling pathways that lead to mitogenesis. There is evidence that PKA action on Raf is inhibitory, and thus it is perhaps not surprising that PKA can inhibit signals travelling down the RTK/Ras/Raf/MAP kinase pathway. Conversely, for signals that predominantly signal down other pathways, inhibition of Raf signalling may be less important than the ability of PKA to phosphorylate and activate transcription signals directly. Indeed mitogenic stimulation by cAMP in synergy with insulin does not activate either Raf1 or the downstream MAP kinase pathway. In addition there are several PKA enzymes, and different family members may behave differently in different cells that will have a subtly different pattern of other signalling pathways.

PKA has many substrates. These include other kinases, receptors, phosphodiesterases, cytoskeletal proteins (e.g. vimentin, which is the main protein of intermediate filaments). The important CREB protein family that bind to CRE have been subdivided into several families (CREB, CREM (M for modulator) and ATF1); some function as activators and others as repressors of expression. They have a domain that can be phosphorylated by PKA, PKC and several other kinases, and regions that interact with DNA and the transcriptional machinery. Inactivation appears to be either by the PP1 or PP2A phosphatases.

Integrins, the Rho family and organization of the cytoskeleton

Outline (Figure 3.26)

Soluble ligands arriving at the cell surface are not the only form of signal that a cell receives. From the earliest developmental stage a cell has to know about its position and its neighbours, and this information is relayed by cell surface interactions with other cells and with the extracellular matrix (ECM). Integrins (specifically β_1 integrins) are the transmembrane cell surface molecules responsible for these interactions. Some integrins, such as those that bind fibronectin and vitronectin, promote cell cycle progression, whereas others lead to cell

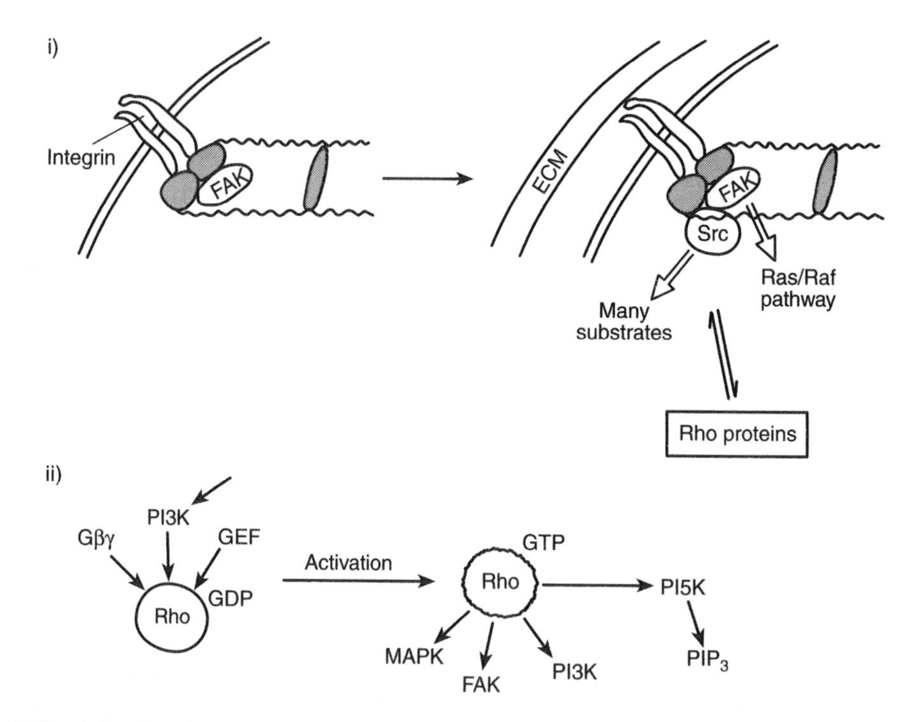

Figure 3.26 Integrins, the cytoskeleton and Rho. In (i) integrins are signalling molecules that interact with the extracellular matrix (ECM) on the outside of the cells and various proteins linked to actin on the cell's interior. The proteins involved include α-actinin, talin, tensin, vinculin and paxillin. On activation a focal adhesion is formed that includes focal adhesion kinase (FAK). The Src kinase is recruited and several proteins in the complex are activated by phosphorylation by Src and FAK. These signals lead to the Ras/Raf, Rho signalling pathways and to cytoskeletal rearrangements. In (ii) some of the possible activators and effectors of Rho proteins are shown

cycle withdrawal. The cytoplasmic regions of integrins link to the cytoskeleton, through which they transduce signals that synergize with other signalling pathways. It is becoming clear that such signals are relayed by the Rho family of small G proteins.

The Rho proteins are members of the Ras superfamily and like Ras they use GTP/GDP exchange as a switch to define their active state. However, both the factors involved in Rho activation and the effectors that they stimulate are still in the process of being identified. Furthermore, one extra complexity is that the Rho proteins can be subdivided into various families that perform subtly different functions. Nevertheless, various signals have been implicated in Rho activation, and as well as signals from integrins. These include the heterotrimeric G proteins, RTKs and PI3K, involved in signalling by growth factors. Activated Rho family members stimulate MAP kinase pathways, in an analogous manner to the Ras stimulation of the Raf/MEK/ERK pathway. The output eventually leads to cytoskeletal rearrangements, but is also linked to cell cycle and growth control like Ras, though these aspects of Rho function are still poorly understood. For example, a survey of the literature places PI3K both upstream and downstream of Rho.

Integrins

Interactions with other cells and the ECM are of crucial importance to the cell and, not surprisingly, integrins have a role in most aspects of both cell function (development, differentiation, growth, migration etc.) and dysfunction (e.g. aberrant growth, metastasis and inflammation). There are at least 20 different integrins, which comprise heterodimers of α and β subunits, and belong to two different protein families: the β_1 integrins and the β_2 integrins. The extracellular domains of the β_1 integrins interact with ECM proteins such as fibronectin and collagen while those of the β_2 family bind to counter-receptors on other cells such as intercellular adhesion molecules (ICAMs) and vascular cell adhesion molecules (VCAMs) (see Chapter 8). Many integrin-binding sites contain the RGD (arginine–glycine–aspartic acid) motif. Interaction with a ligand induces integrin clustering which leads to concomitant clustering of cytoskeletal proteins bound to the short intracellular domains of the integrin. This serves to produce focal adhesions at the cell membrane where actin is localized. As with many other aspects of signalling, the close association of proteins enables phosphorylation events to be triggered. Chapter 2 should be consulted for a description of the cytoskeleton. In addition cell adhesion mediated by integrins is regulated by phosphorylation by the previously described classical signalling pathways.

Rho heterogeneity

The Rho proteins were originally identified by their homology to Ras (Rho stands for *R*as *ho*mology), and have been the subject of recent intense research activity. The Rho subfamilies are believed to control various aspects of the cytoskeleton Thus, the Rho subfamily is believed to control actin stress fibres and focal adhesions, Rac mediates membrane ruffling, while cdc42 mediates filopodia extensions. Currently little is known about the other two members (RhoG and TC10) of the Rho family, although RhoG expression has been shown to be regulated by mitogenic signals and its activity varies through the cell cycle. It is also clear that different tissues express a distinct subset of these proteins. In addition, all the Rho members are involved in the regulation of growth, and it is generally accepted that there is a loose hierarchy of signal transfer from cdc42 to Rac to Rho, with signals from the RTK/Ras and G protein/PKC pathways feeding into Rho family signalling. For example, transformation by Ras requires functional Rac to be present, but Rho has also been implicated in oncogenic transformation. It is of interest that either stimulation or inhibition of Rho activity can promote growth.

The Rho subfamily is further subdivided: RhoA and RhoC appear to be stable proteins whose concentration is not altered during progression through the cell cycle. They are found in both cytosolic and membrane-bound locations and are thought to be recruited to the membrane by interacting proteins. However, RhoB is more highly tissue-restricted and is always found as a membrane-associated protein whose concentration oscillates through the cell cycle, reaching a peak at S phase and falling to its lowest level at mitosis. RhoB is an 'immediate early' protein whose expression is stimulated by various growth factors, for example EGF and PDGF. Its messenger RNA has a short half-life and the RhoB protein is itself unstable. RacE has been specifically associated with cytokinesis.

The crystal structure of several members of the family have been solved. Comparison of the Rac structure with Ras shows that Rac has an extra exposed loop, which is postulated to interact with other proteins.

Many bacterial toxins target the Rho subfamily and the biological actions of these toxins have been very useful in unravelling the cellular biology and microbiology of this group of proteins. The reader should refer to Chapter 7 for a description of the effect of bacterial toxins on Rho.

Activators of Rho

Several guanine nucleotide exchange factors (GEFs) have been identified for Rho. A sequence originally identified in Dbl, the Dbl homology (DH) domain, is found in GEFs that interact with Rho (and Ras) proteins, and is present in Vav, Ect2, Tim and Tiam1, all of which are potentially oncogenic. They also contain a PH motif, and it is believed that DH organizes the interaction with the small G proteins, while PH targets the activated G protein to the cytoskeleton. There has been a lot of work on the Dbl and Vav protein families, which are GEFs for Rho. Dbl associates with Gβγ, and also with the ERM proteins (ezrin, radixin and moesin) that are involved with actin filaments (see below). Radixin may play a regulatory role in Rho activation since it also binds to RhoGDI, which is a GDI for all the Rho proteins. In this regard other GDIs have been identified that have more limited effects, in particular RhoGDI-3 that is specific for RhoB and RhoG and which unusually is membrane-associated. Vav expression is limited to haematopoietic cells, but Vav-2 is more widely expressed, and there are other members of this family. Only phosphorylated Vav functions as a GEF for Rho, and various kinases have been implicated in Vav phosphorylation which link it into several pathways. These include the JAK kinases, stimulated by cytokines (see 'Eukaryotic cell cycle progression and signalling', below), and Tyk-2 that is stimulated by interferon. There is linkage with PI3K, in that association of Vav with PIP_3 but not PIP_2 facilitates Vav phosphorylation, and also some evidence that Vav associates with the p85 regulatory subunit of PI3K. Vav has also been reported to interact with Grb2. These GEFs also appear to operate on cdc42, although a specific regulator (cdc24) for cdc42 has also been found. The Ost oncogene is a GEF that illustrates other possibilities for regulation and crosstalk. Ost activates GDP/GTP exchange on RhoA and cdc42, but also associates with GTP-bound Rac. Thus while some pathways leading to stimulation of the Rho proteins have been identified, it is possible that there will be other routes, for example involving direct interaction with specific Gα subunits of heterotrimeric G proteins.

The proto-oncogene Tiam1, which induces invasiveness, is a GEF for Rac and cdc42, and is mimicked by transfection with V12Rac (a parallel activating mutation to that in Ras that abolishes GTPase activity). So far several GAPs have been found that function on Rho (RhoGAP, p190 and chimerin) to downregulate signalling.

Rho effectors

The Rho family, like Ras, operates by binding to and activating protein kinases. The pathway for Rac and cdc42 was first worked out, and shown to trigger the so-called stress-activated MAP kinase cascade that leads to activation of the *J*un *N*-terminal *k*inases, Jnk 1 and 2, and also the related p38 MAP kinase. Jnk1 and 2 are also referred to as SAPK 1 and 2 (for *s*tress-*a*ctivated *p*rotein *k*inases), or the p46 and p54 MAP kinases. The G proteins bind to PAK p65 (*p*21 *a*ctivated *k*inase) that in turn activates the MAP kinase cascades that activate Jnk and p38. Both Jnk and p38 activate expression of transcription factors.

In addition, other effectors that bind directly to the activated G protein have been identified. For example, NADP oxidase interacts directly with Rac2. Rho is known to stimulate PIP 5'-kinase, leading to elevation of PIP_2 concentration and potentiation of pathways that rely on PIP_2, for example PLC/PKC.

Several kinases have been associated with Rho function. These include ROK, PKN, Rho kinase, Rhophilin and Rhotekin. All the roles of these kinases are certainly not fully understood. However, a possible path through ROK to the cytoskeletal changes observed by Rho activation can be tracked. ROK phosphorylation of myosin phosphatase results in myosin light chain phosphorylation, leading to its interaction with actin and subsequent formation of stress fibrils and focal adhesions. What is clearer is that the assembly of focal adhesions leads to activation of the focal adhesion kinase, $p125^{FAK}$.

The focal adhesion complex contains actin-related proteins, and proteins more overtly involved in signalling. Thus in addition to actin we find the cross-linking protein α-actinin, talin, vinculin, tensin and paxillin. Several kinases can localize to focal adhesions. These include Src, cAbl and the focal adhesion kinase $p125^{FAK}$, one of the principal signalling molecules and an integrator of many different signals.

The focal adhesion kinase is a non-receptor tyrosine kinase. $p125^{FAK}$ has several domains that regulate its activity and location, and which mediate its interaction with a large number of proteins. The N-terminus can bind to integrins, at least *in vitro*, and the C-terminus has a region that is essential for its location to focal adhesions and which binds to some of the proteins within focal adhesions. These include paxillin and $p130^{CAS}$, which are both $p125^{FAK}$ substrates, and also talin and PI3K. Various tyrosine residues on $p125^{FAK}$ have a crucial role in its function. The phosphorylation status of one of the tyrosine residues determines both the association with Src and whether the other tyrosine residues are phosphorylated, so it is likely that Src is responsible for phosphorylation at the other three sites. One of these phosphotyrosines mediates interaction with Grb2, whereas the other two tyrosines, which are adjacent to each other, are thought to be important for the $p125^{FAK}$ kinase activity.

Two proteins phosphorylated by $p125^{FAK}$, paxillin and $p130^{CAS}$, also display several interesting features. Paxillin has a proline-rich SH3-binding domain, a region that interacts with vinculin and other repeat domains. Phosphorylation enables it to interact with Src and Crk, the latter protein linking into the Ras related pathway. $p130^{CAS}$ contains both an SH3 domain and a proline-rich SH3-binding domain, as well as numerous SH2-binding sites, enabling it to interact with Src and Crk.

Integration of integrin signals with those from growth factors

The Rho family proteins are activated upon cell surface engagement of growth factors, but an emerging view is that their primary signalling role is from integrin signals. This highlights the two different but complementary signals (soluble growth factors and cell/cell-mediated interactions) that cells receive, and indeed need to continue to receive, to remain viable. The joint importance of these is understood when we consider what happens when a cell becomes both growth- and contact-independent (the latter is usually called anchorage-independent). Such cells grow out of control and have the ability to metastasize, i.e. they are cancer cells.

Interest in the Rho family is at a peak, but many of the effects observed and the linkages to other signalling pathways are still confusing. One reason that applies also to other pathways is the degree of crosstalk among signalling pathways. It is also true that the subtlety of the role of the different members of the Rho family are not yet fully understood.

Other receptor arrangements

Cytokine-mediated stimulation

Cytokines have been described in detail in earlier sections of this chapter and their various classes of receptors have been detailed. As stated, binding to cell surface receptors often stimulates dimerization (or in some cases oligomerization). Certain cytokine transmembrane receptors have no intrinsic kinase activity, but are closely associated with cytoplasmic protein kinases of the Jak (*ja*nus *k*inases or, as one author somewhat cynically suggested, *j*ust *a*nother *k*inase) family. These autophosphorylate themselves and the receptors upon oligomerization. The kinases also tyrosine phosphorylate members of the STAT family of signal transducers. STAT is named from *s*ignal *t*ransducers and *a*ctivators of *t*ranscription. These molecules possess SH2 domains, so that when phosphorylated form SH2/phosphotyrosine-linked dimers that are able to translocate into the nucleus to act as transcription factors (Figure 3.6).

There is also evidence of crosstalk with other signalling pathways, e.g. the Ras/Raf/MAP kinase pathway, and activation of members of the Src family. There is significant interest in inhibiting the MAP kinases and, for example, scientists at SmithKline Beecham have developed selective inhibitors of p38 which can block the synthesis of TNFα.

Tumour necrosis factor stimulation

The tumour necrosis factor (TNF) family transmit a variety of signals, from proliferation to apoptosis, and are involved in disease processes such as septic and toxic shock and cancer. The two best-known effects of TNF are induction of transcription via NF-κB (*n*uclear *f*actor), and induction of apoptosis via activation of the caspases – enzymes such as the interleukin-converting enzyme (ICE), a cysteine protease, which are pivotal in the control of programmed cell death (apoptosis).

Binding to receptors induces clustering of TNF receptors (TNFRs). These have no known kinase or other enzymatic activity, but possess the 'death domain' (associated with apoptosis) on the cytoplasmic side of the membrane. This associates with TRADD (*T*NF *r*eceptor-*a*ssociated *d*eath *d*omain) and TRAF (*T*NF *r*eceptor-*a*ssociated *f*actor) families. TRADD contains the death domain and associates with TNFR1 through this. The RIP protein also has a death domain and associates with TRADD in a TNF-dependent manner. RIP is a serine threonine kinase, and its activation leads to phosphorylation and destruction of I-κB, an inhibitory protein that is complexed with NF-κB. When no longer bound by I-κB, NF-κB is released to bind to κB elements involved in transcriptional activation. The death domain of RIP is thought to activate ICE and other caspases. NF-κB is important in cell activation by a range of cytokines and by bacterial factors and there is growing evidence that bacterial factors can also inhibit this NF-κB system.

Receptor serine threonine kinases

Members of the TGFβ superfamily bind to serine threonine kinase receptors. The TGFβ superfamily includes TGFβ, activins and bone morphogenetic proteins (BMPs). Ligand occupancy leads to receptor phosphorylation and activation. Much less is known about how these receptors transduce their signals, although some proteins that bind to them have been identified. The end point of activation is linked in to the cell cycle and TGFβ action serves to block cell cycle progression.

PI3 kinase

The components in this signalling pathway have already been described ('Phosphoinositide 3-kinase'), and we have already come across PI3K interaction with other pathways. Activation via the regulatory subunit of PI3K can occur in a variety of ways. PI3K can also be activated by interaction of the catalytic domain with activated Ras. The activated enzyme produces PIP_3 that, like so many signalling molecules, functions by membrane recruitment of PKB, where it is activated. As we have described above, it is likely that PIP_3 has other functions, one of which is to facilitate phosphorylation and thus activation of GEFs that interact with Rho family proteins.

PKB mediates a number of responses that include cell survival and cell stress signals, and lamellipodia formation, as well as respiratory burst in neutrophils. PKB also regulates Rac. PKB has a number of substrates: PKBβ has been shown to translocate to the nucleus upon activation and is presumed to modify transcription factors leading to activation of the serum response factor. The CD95 surface protein that mediates apoptosis is counteracted by the PI3K/PKB pathway.

Other pathways regulate and crosstalk with PKB. These include PKA, which appears to activate phosphorylation of PKB, but not by direct PKA phosphorylation of PKB.

Src kinases

It could be argued that the Src family of kinases do not constitute a pathway. However, it is not in doubt that these kinases play a major role in many processes and impinge on many different signalling pathways. They also have the historical distinction that it was the identification 20 years ago that the transforming protein of the Rous sarcoma virus was a tyrosine kinase that led to the explosion of interest in these kinases.

There are about 10 members of the Src family that display differences with respect to their tissue distribution and also their structural organization. We have already come across important motifs that are named from Src (the SH2 and SH3 domains). In addition most family members also have the SH4 domain, a short region that conveys signals for lipid modification. Activation is mediated by the interactions that the SH2 and SH3 domains make with other proteins. When not engaged with other proteins, these inhibit the active site. In some Src kinases SH2 is bound intramolecularly to a phosphotyrosine.

The range of pathways with which Src impinges is almost boundless, as we have already discussed. They are recruited to many RTKs and their activation can lead to phosphorylation of the receptor and other proteins. They interact with G proteins, the PI3K pathway and integrin signalling. The ubiquity of their action might suggest a lack of specificity, but whether this is provided by the subtlety of their cellular location, or whether they are indeed 'heavy-duty kinases' that can always be called on to carry out their function is not yet clear.

Key components

It is clear from the foregoing description of signalling pathways that certain components crop up in many pathways and thereby provide crosstalk and integration of signals. Some of the more commonly found ones are listed (Table 3.14). At the membrane these include PI3K, Src and FAK kinases and members of the Ras and Rho members of the Ras superfamily.

Table 3.14 Examples of signalling molecules that integrate signals

Component	Input	Output
PI3kinase	SH2, SH3, SH3 binding, Gβγ, Ras	PKB, PKC, Rho family
Src	SH2, SH3	phosphorylates many proteins
focal adhesion kinase	integrins, Src, PI3kinase?	Grb2, phosphorylates proteins
Ras	Sos, cdc25, Crk, p120GAP	Raf kinase, PI3kinase, Ral 1
Raf	Ras, PKA, PKC, Gβγ	MAP kinase, Bcr
Protein kinase C	DAG, PI3kinase	Raf, MARCKS
MAP kinase pathways	Raf, MKP	phosphorylates proteins

Examples of signalling molecules that integrate signals

Outcomes of the activation of signalling pathways

The outcomes of these signalling pathways are relatively obvious. They enable a cell to sense and assess its environment and take appropriate actions. In the case of bacteria this might involve preventing osmotic damage, regulating movement towards favourable substrates, controlling growth and preparation for growth in or on a host (a process that we would call pathogenesis).

In multicellular eukaryotic organisms, signalling performs a similar function, though the signals are more complex both in terms of their number and their type. Some relate to a specialized function, for example immune recognition. Needless to say there is still a huge amount of interest in learning in detail how growth is regulated, with the practical aim of using the information in novel therapies against cancer. It is only relatively recently that the pathways leading from mitogenic signals have been linked into those controlling the cell cycle and we will discuss briefly normal regulation of the cell cycle and what happens when a cell becomes deregulated and transformed.

Eukaryotic cell cycle progression and signalling

The cell cycle has been outlined in Chapter 2. Growth factors affect the cell cycle by modifying the activity of the cyclin-dependent kinases. It is clear that signals from calcium and the Ras/Raf/MEK/ERK pathways, amongst others, affect the activity of these kinases. The phosphatase cdc25, itself regulated by phosphorylation, has been reported to interact directly with Raf.

Cell cycle regulation involves cdk/cyclin interactions and the retinoblastoma protein (Rb). In addition, there are two families of inhibitors that associate with the cyclin/cdk complexes to inhibit their function. p16 is specific for cyclin D, whereas p21/Waf-1/Cip and p27 affect all cyclins. p27 is regulated by mitogenic signals, whereas p21 is regulated by p53. p53 activates a set of genes, of which the best known is p21. p53 also acts to downregulate genes by protein–protein interaction with the TATA-binding protein. p53 contains multiple phosphorylation sites that are thought to be involved in its regulation. For example, it is phosphorylated by the Jnk kinase.

Signalling and cancer

A tumour is caused when a cell that should be relatively quiescent grows inappropriately. A benign tumour is self-limited, whereas a malignant tumour (cancer) is defined by two properties: the ability to invade surrounding tissues and to metastasize (i.e. to spread to remote sites where it can seed fresh tumours). This latter aspect of cancer is most difficult to deal with, since it is impossible to treat effectively with the surgeon's knife. Empirical measures have been largely unsuccessful in the treatment of the common cancers, and it is clear that recent progress and the undoubted optimism for the future lies in acquiring a basic understanding of the processes involved in the initiation and progress of tumour formation. Already the information obtained has thrown up novel approaches to therapy.

About 20 years ago there were several apparently conflicting theories about the causation of cancer, for example chemical, hormonal, genetic, viral. Our present understanding shows that none of these was wrong, but that none was correct. Cancer is now known to arise when a cell accumulates several mutations in key genes so that its growth can no longer be controlled by the rest of the body. Chemical, hormonal, genetic, viral (and indeed bacterial) factors all impinge on this process. They either act as tumour promoters that encourage growth, which provides a greater chance for mutations to occur and remain embedded in the genome, or act as tumour inducers by causing DNA mutation. The signalling pathways are crucial to this process.

The central role of eukaryotic signalling pathways in controlling the cell is obvious, and we have already argued that the signalling pathways of a cell in large measure define its phenotype and function, and the way it interacts with the other cells in a multicellular organism. It is therefore not surprising that perturbation of these pathways by mutational changes (oncogenic mutation) drastically disturbs cell function. Individual mutations that upregulate flux through signalling pathways, or downregulate inhibitory proteins, can transform cells in the artificial conditions of *in vitro* growth. However, accumulation of several such mutations is necessary *in vivo* to lead to a cell that resists apoptosis and grows in an unregulated manner (i.e. is transformed) and has aberrant surface properties that enable it to detach, survive in an anchorage-independent manner and settle and grow in inappropriate sites (metastasize). The differences between the *in vitro* and *in vivo* situation is that many immortalized cells have already accumulated mutations.

The types of mutation that are required for the initiation of cancer fall into several classes: mutations in signalling molecules, mutations in proteins involved in cell cycle control, and mutations that directly affect transcription. The first category includes overexpression of growth factor receptors, aberrant release of locally acting growth factors, and activating mutations in signalling components that are normally regulated by either growth factor or integrin-mediated events. For example, mutations in Ras that inactivate its GTPase activity are found in about 30% of human cancers. Mutations that activate heterotrimeric G proteins and the Src kinase are each found in about 5% of human cancers. Proteins that interact with small G proteins often have oncogenic potential (e.g. members of the Dbl family). Mutations in cell cycle proteins, such as cyclin and cyclin-dependent kinase inhibitors, cause checkpoints to be bypassed and can be oncogenic mutations.

Mutation of factors involved in the regulation of transcription are obvious targets for disastrous mutagenesis. For example, the proto-oncogenes *fos* and *jun* form the AP-1 transcription factor. Two proteins that play a major role in cancer, and thus offer hope for the development of novel therapies, are the p53 and RB proteins, found mutated in 50% and 40% of human cancers respectively. p53 has been described as the 'guardian of the genome' since its role is to monitor DNA damage and instruct the cell either to halt DNA replication until it can be repaired or to undergo apoptosis if the damage is too great. Its discovery and the realization of its therapeutic potential led to an explosion of work on p53 and its naming as 'molecule of the year' in 1993. p53 is responsible for translating the DNA damage induced by radiotherapy or chemotherapy into apoptosis, and it therefore not surprising that tumours with p53 mutations are often resistant to such treatments. p53 is, as one author has put it, 'disconcertingly susceptible to point mutations'. Oncogenic mutations prevent DNA binding and thus the failure to activate either apoptosis or enable DNA repair to take place. Mutation of Rb, or its interaction with viral proteins, releases E2F and can stimulate cell cycle progression in the absence of growth factors. However, Rb has a wider role, in that it appears to regulate at the transcriptional level and is believed to cooperate with p53. Rb appears to have other functions in mediating the choice between growth differentiation and apoptosis.

The relatively new knowledge about components of signalling pathways and the types of genes with which they interact is already being applied in novel strategies to combat cancer. For example, viruses are being engineered to grow only in cells that lack functional p53, and can thus be designed to kill such cells.

CONCLUSION

Communication is paramount in biology and this chapter has highlighted the many varieties of signals that are used to communicate between cells. We have concentrated on one particular set of intercellular signalling molecules, the cytokines, because of their importance in infection and because of the growing realization that bacteria produce a large number of molecules which can modulate the synthesis and activity of cytokines. The effects of these various signals arriving at the cell surface are recognized and integrated by a growing number of proteins which are divided into a small number of functional units: kinases, phosphatases and those proteins with motif recognition properties. There are probably between 2000 and 5000 signal transduction proteins in the average mammalian cell, not all of which have been identified, which accounts for the difficulty, at present, in describing intracellular signalling. In addition to proteins a large number of lipid mediators are also involved in cell-to-cell and intracellular communication. The intracellular signalling proteins integrate signals arriving at the cell's exterior with the activity of the genome and thus form a continuum between the cell's environment and its ability to repond correctly to the environment.

It has come as a surprise to find that bacteria also utilize complex cell-to-cell and intracellular signalling pathways to communicate with each other, and also with eukaryotic cells, and the similarity of the mechanisms used by both prokaryotes and eukaryotes is striking. One of the major discoveries of cellular microbiology is that bacteria have the capacity to utilize eukaryotic cell-signalling pathways during the process of infection. These findings are beginning to show the links between the signalling pathways involved in infections and those responsible for the pathology in idiopathic diseases such as cancer, atherosclerosis and inflammatory conditions.

A major reason for studying intercellular and intracellular communication in cells is that it is likely to provide important clues for the development of novel treatments for disease. This is particularly important in the infectious diseases arena where, with the rapid increase in antibiotic resistance, there is a pressing need to develop new antimicrobial drugs. It is possible that targetting bacterial signal transduction may allow new therapeutic agents to be developed for infections.

FURTHER READING

Books

Aggarwal BB, Puri RK (1995) *Human Cytokines: Their Role in Disease and Therapy*. Blackwell, Cambridge, MA.

Burgoyne PD, Peterson OH (1997) *Landmarks in Intracellular Signalling*. Portland Press, London.

Gerhart J, Kirschner M (1997) *Cells, Embryos and Evolution*. Blackwell, Oxford.

Hancock JT (1997) *Cell Signalling*. Chapman & Hall, London.

Heldin C-H, Purton M (eds) (1996) *Signal Transduction*. Chapman & Hall, London.

Henderson B, Poole S, Wilson M (1998) *Bacteria-Cytokine Interactions in Health and Disease*. Portland Press, London.

Hoch JA, Silhavy TJ (1995) *Two Component Signal Transduction*. ASM, Washington, DC.

Meager, T (1998) *The Molecular Biology of Cytokines*. John Wiley & Sons, Chichester.

Perutz M (1992) *Protein Structure: New Approaches to Disease and Therapy*. Freeman, New York.

Reviews

Berridge MJ (1993) Inositol triphosphate and calcium signalling. *Nature* 361: 315–325.

Botsford JL, Harman JG (1992) Cyclic AMP in prokaryotes. *Microbiol Rev* 56: 100–122.

Bowles DJ (1997) *Cell Signalling. Essays in Biochemistry*, Vol. 32. Portland Press, London.

Clark EA, Brugge JS (1995) Integrins and signal transduction pathways: the road taken. *Science* 268: 233–238.

Fuqua C, Greenberg EP (1998) Self perception in bacteria: quorum sensing with acylated homoserine lactones. *Curr Opin Microbiol* 1: 183–189.

Hartman G, Wise R (1998) Quorum sensing: potential means of treating gram-negative infections? *Lancet* 351: 848–849.

Hunter T (1995) Protein kinases and phosphatases: the Yin and Yang of protein phosphorylation and signaling. *Cell* 80: 225–236.

Lim L, Manser E, Leung T, Hall C (1996) Regulation of phosphorylation pathways by p21 GTPases: the p21 Ras-related Rho subfamily and its role in phosphorylation signalling pathways. *Eur J Biochem* 242: 171–185.
Moore AA, Wolfman A (1994) The 3Rs of life: Ras, Raf and growth regulation. *Trends Gen* 10: 44–48.
Neer EJ (1995) Heterotrimeric G proteins: organizers of transmembrane signals. *Cell* 80: 249–257.
Ruby EG (1996) Lessons from a cooperative bacterial–animal association: the *Vibrio fischeri–Euprymna scolopes* light organ symbiosis. *Annu Rev Microbiol* 50: 591–624.
Schwartz MA (1997) Integrins, oncogenes and anchorage independence. *J Cell Biol* 139: 575–578.
Trends Biochem Sci (1994) Special Issue on Protein Phosphorylation. 19: 439ff.

Papers

Castagliuolo I, Riegler M, Pasha A, Nikulasson S, Lu B, Gerard C, Gerard NP, Pothoulakis C (1998) Neurokinin-1 (NK-1) receptor is required in *Clostridium difficile*-induced enteritis. *J Clin Invest* 101: 1547–1550.
Surette MG, Bassler BL (1998) Quorum sensing in *Escherichia coli* and *Salmonella typhimurium*. *Proc Natl Acad Sci USA* 95: 7046–7050.

CHAPTER 4

Molecular Techniques Defining Bacterial Virulence Mechanisms

INTRODUCTION

As described in Chapter 1, polymerase chain reaction (PCR) amplification and DNA sequencing of 16S rRNA genes are only now revealing the true diversity of bacterial species. However, only a very small proportion of the bacterial species described to date are known to cause disease in humans. Those that do are commonly divided into two groups: primary pathogens and opportunistic pathogens. A primary pathogen has been defined recently by Stanley Falkow as being a microorganism 'whose survival is dependent upon its capacity to replicate and persist on or within another species by actively breaching or destroying a cellular or humoral host barrier that ordinarily restricts or inhibits other microorganisms'. Primary, or overt, pathogens include enteropathogenic shigella and salmonella species, mycobacteria and the pyogenic streptococci. Opportunistic pathogens ordinarily do not cause disease, but do so under extenuating circumstances where host cellular or humoral defences are impaired. Opportunistic pathogens can be members of the normal commensal flora or bacteria whose normal environment is outside the human body. As examples, infective endocarditis is a life-threatening disease frequently caused by colonization of defective heart valves by normal oral bacteria, whereas soil-associated histotoxic clostridia can cause gas gangrene in wound infections.

Considerable detail is known about the host response to infection by both primary and opportunistic pathogens, and this topic is covered in some detail in Chapters 8 and 9. A complete understanding of host–pathogen interactions in disease, however, must necessarily include the identification and characterization of virulence determinants expressed by the bacteria.

Despite continued improvements in health care and the development of antimicrobial agents, pathogens, both primary and opportunistic, still exact a heavy toll in terms of mortality (at least in the Third World – see Table 1.1) and morbidity. It is becoming increasingly evident that we cannot continue to rely on existing antimicrobial agents to keep pathogens at bay. Indeed, due to the continued emergence and spread of drug resistance in pathogens, we may shortly enter a 'post-antibiotic era' where currently treatable infections become far more serious due to the absence of effective antimicrobial drugs. In addition, changes in population demographics and increased numbers of immunosuppressed or immunocompromised individuals has meant that more and more previously benign bacteria and fungi are applying for membership in the pathogen club.

A better understanding of virulence mechanisms and pathogen–host interactions in disease will identify new targets for therapeutic intervention. The challenge to molecular biology is to continue to develop new techniques for dissecting the individual components that, in total, define bacterial virulence.

MOLECULAR TECHNIQUES DEFINING BACTERIAL VIRULENCE

Molecular techniques for defining bacterial virulence mechanisms, while powerful, still require that some basic conditions are met. Firstly, appropriate *in vivo* or *in vitro* experimental models are required that mimic the specific environment of the natural host. In most instances, humans cannot be used as the model system. However, serum taken from patients who have been infected by a particular pathogen can be used to examine the antigens expressed during infection by that pathogen. These antigens may have a role either in pathogenesis or in host immune elimination of infection, either of which will provide important information for developing effective intervention or prevention methods.

Animals such as monkeys, mice and guinea-pigs have long been used in biomedical research as models for human disease. Their use, however, is declining for a number of reasons: they provide only limited information relevant to human disease; there are obvious ethical objections; and there have been substantial improvements in tissue culture technology using cells derived from humans. Tissue culture monolayer studies using primarily epithelial cells and cells of the monocyte–macrophage lineage have added much to current knowledge about the attachment, invasion and intracellular survival and cell division of pathogens. Multiple-layer tissue culture systems using mixed-cell populations (e.g. epithelial and endothelial cells) have the additional advantage of more closely mimicking the challenge faced by a pathogen attempting entry at epithelial surfaces. The system also models more closely the host cell–cell communication that occurs *in vivo*.

A second important factor is the availability of basic genetic manipulation techniques for the pathogen under study. The molecular genetics of certain bacteria such as *Escherichia coli*, *Salmonella typhimurium* and *Streptococcus pneumoniae* are extremely well developed and there are powerful tools and techniques available for these organisms that are simply not available for the majority of bacterial pathogens. What is possible for any given pathogen will define what techniques can be applied.

Prior to detailing the many and varied molecular techniques employed to investigate bacterial virulence mechanisms, an overview of basic molecular protocols is provided.

Basic molecular biology protocols

The legal definition (UK) of gene manipulation is 'the formation of new combinations of heritable material by the insertion of nucleic acid molecules, produced by whatever means outside the cell, into any virus, bacterial plasmid or other vector system so as to allow their incorporation into a host organism in which they do not naturally occur but in which they are capable of continued propagation'. The requirements for gene cloning are encapsulated nicely by this statement, and these include: the ability to specifically cut and join DNA fragments; the availability of cloning vehicles that replicate independently of the bacterial chromosome and which carry markers for the selection of recombinant molecules; and methods that allow introduction of recombinant DNA molecules into bacterial host cells.

A detailed description of gene cloning and basic molecular biology such as restriction, ligation and PCR is beyond the scope of this chapter. Familiarity with basic molecular biology concepts is assumed, and the reader is directed to any number of good molecular biology manuals, some of which are listed at the end of this chapter.

The cloning, then, of bacterial genes presents few difficulties. Gene libraries of pathogen DNA can be constructed readily in *E. coli* or in other host bacteria, and provided suitable and specific probes are available (DNA, oligonucleotide, antibody), the DNA encoding genes of interest can be isolated and sequenced almost as a matter of course. However, difficulties may arise when it comes to genetic manipulation of the pathogenic organism of choice. High on the list of stumbling blocks is not being able to introduce recombinant DNA into the target organism for the purpose of, for example, gene inactivation or complementation analysis (described below), and this problem may comprise several levels. The introduction of recombinant DNA can be achieved using one of several methods. Bacterial competence is the ability of bacteria to take up naked DNA. Many bacteria demonstrate natural competence, including *Strep. pneumoniae*, in which competence was first demonstrated. In this situation, growing bacterial cells are able directly to take up DNA from their environment. Other bacteria, including *E. coli*, require manipulation such as treatment of cells with calcium chloride and brief heat treatment. Many more bacterial species that are neither naturally or artificially competent can be transformed with recombinant DNA by electroporation. In this method, high-voltage electric pulses are used to introduce DNA into pretreated bacterial cells. The pretreatment used will depend on the bacterial species or strain. *E. coli* requires minimal pretreatment, and electroporation is often chosen over the calcium chloride technique since the efficiency with which DNA is introduced is superior. The generation of electrocompetent *Enterococcus faecalis* requires only that cells are grown in the presence of glycine to weaken the cell wall to allow access to DNA during the high-voltage pulse. In contrast, some bacteria will require enzymatic treatment to remove the cell wall to make electrocompetent protoplasts. The generation of protoplasts and their regeneration to osmotically stable wall-bound bacteria can be technically quite exacting. The use of modified bacteriophages as cloning vectors has the advantage that the phage are inherently very good at infecting their bacterial host, and the introduction of DNA by this method is typically very efficient. A final method of introducing recombinant DNA into bacterial hosts that is commonly used is conjugation. This requires specialized plasmids that are transferred from one bacterium to a second bacterium by direct cell–cell contact.

If the target organism contains a restriction-modification system, then incoming recombinant DNA may be degraded rapidly, depending on the DNA sequence and its last host, and no recombinants will be obtained. The genetic markers, such as antibiotic resistance genes, that allow selection of recombinants, must also be appropriate for and properly expressed in the target organism.

Proficiency for homologous recombination to allow the integration of recombinant DNA into the bacterial chromosome also cannot be taken for granted, and transposons may be

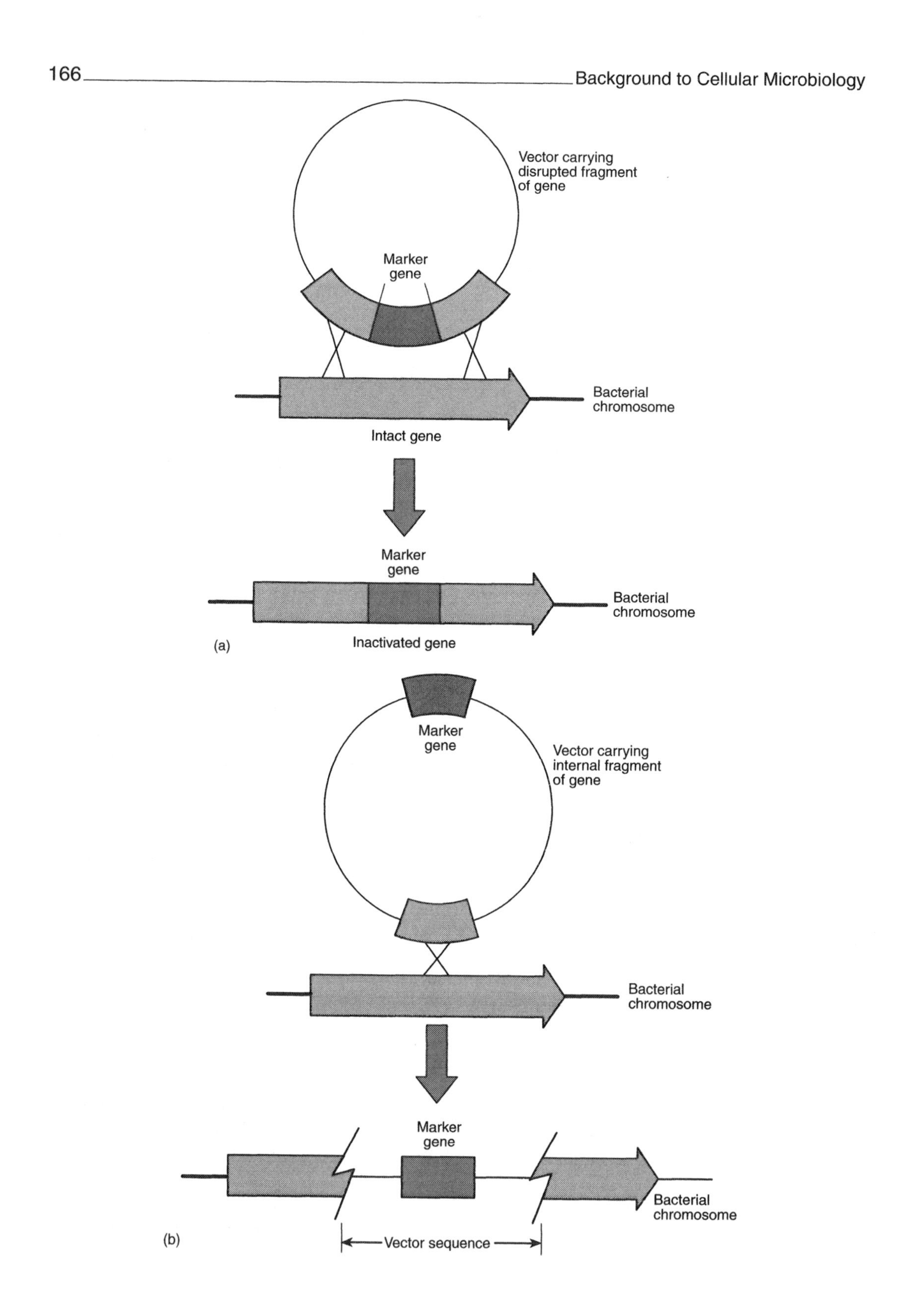
Vector carrying
disrupted fragment
of gene
Marker
gene
Bacterial
chromosome
Intact gene
Marker
gene
Bacterial
chromosome
(a)
Inactivated gene
Marker
gene
Vector carrying
internal fragment
of gene
Bacterial
chromosome
Marker
gene
Bacterial
chromosome
(b)
Vector sequence

required to facilitate DNA integration. Transposons are mobile genetic elements that can insert more or less randomly within the DNA complement of the bacterial genome (plasmid and chromosomal) independently of the host cell recombination system, as described in Chapter 2. Transposons vary in size from those less than 5 kb that encode only genes essential for the transposition event, to those of 18 kb and longer that have multiple open reading frames necessary for inter-bacterial transfer and, often, for multiple antibiotic resistance or toxin production. The virtues of more or less random integration and tagged mutation site have made modified transposons often the choice for generating insertion mutations in target organisms, as will be described shortly.

A number of techniques, such as RNA arbitrarily primed PCR (RAP-PCR), which will be outlined, have been developed recently to counter the barriers to genetic manipulation presented by some bacterial species. These techniques require only that intact RNA can be extracted from bacterial cells. Gene expression profiles can then be created using reverse transcriptase PCR to generate cDNA molecules that can be compared for differentially cultured cells to identify disease-specific gene expression. Ultimately, however, such experiments can only implicate genes in pathogenesis and, in the absence of corroborating evidence of their involvement in the virulence of related pathogens, their contribution to virulence will require gene inactivation experiments. Thus we must continue to develop molecular biological methods for common pathogens.

Mutational analysis

Directed mutation

Directed mutagenesis is probably the most widely used technique for assessing the contribution to virulence of specific bacterial gene products. In a technique often called 'reverse genetics', the gene encoding a presumptive virulence factor that has been identified by conventional biochemical means is disrupted by insertional inactivation or gene replacement. The virulence of the isogenic mutant strain is then compared with that of the parental strain. Clearly this technique requires a means for introducing DNA into the pathogen, as well as suitable selective markers and an inherent capacity for homologous recombination. It is, however, an important method used to confirm the involvement of genes identified by other screening methods in virulence and can equally refute or support the hypothesis.

Figure 4.1 (opposite) Homologous recombination results in the integration of DNA onto the bacterial chromosome. (a) Double crossover event. DNA encoding a portion of the target gene is cloned into a plasmid and disrupted *in vitro* by the ligation of a selectable marker (typically an antibiotic resistance gene) within the gene fragment. The recombinant plasmid is then introduced into the pathogen (electroporation, transformation, conjugation) and, since the plasmid cannot replicate within the pathogen, antibiotic resistance arises following a double crossover recombination event between cloned DNA and the homologous region on the pathogen chromosome. The target gene is inactivated, but no vector DNA is introduced. (b) Insertion duplication mutagenesis. A fragment of the coding portion of the target gene is cloned into a plasmid that does not replicate in the pathogen, as before. If this vector encodes a selectable marker, the construct can be used directly for mutagenesis. Acquisition of marker phenotype by the pathogen occurs following a single homologous recombination (crossover) event that integrates the entire vector. Pathogen DNA cloned in the plasmid is duplicated on the chromosome, and if this includes sequence 5'- or 3'- to the coding portion of the target gene, then a complete gene is recreated, and mutagenesis will be unsuccessful

In directed mutation, an antibiotic resistance marker, or other gene for which there is an easily selected phenotype, is ligated within the coding sequence of a cloned copy of the target gene. This disrupts normal transcription and translation of the gene and additionally provides a marker for selecting recombinant constructs. The construct is then introduced into the target pathogen, where it undergoes recombination with the homologous region on the bacterial chromosome. If the resistance marker is flanked by chromosome-specific sequences, then a double crossover can occur and the marker is introduced in the absence of vector sequence, as shown in Figure 4.1a. In contrast, if a single stretch of DNA is presented, then antibiotic resistance will result from a single recombination (crossover) event, and this is termed insertion-duplication mutagenesis or Campbell-type integration (Figure 4.1B). This method requires that the integrating DNA is present on a (non-replicating) plasmid, and also that internal gene sequence only is incorporated in the plasmid; if the integrating DNA includes non-coding (upstream or downstream) sequences, then an intact copy of the gene on the chromosome is re-formed following plasmid integration. Continued antibiotic selective pressure may also be required to maintain the plasmid in an integrated state. Insertion-duplication mutagenesis has proved useful as a substitution for transposon-based *phoA* fusion and *in vivo* expression technology (IVET) techniques (which will be described), particularly in Gram-positive bacteria.

Since many bacterial genes are organized in transcriptional units and operons, gene disruption can also result in inactivation of those genes in an operon that lies downstream of the insertion site. Such insertions are said to have polar effects on transcription. In these situations the use of markers that lack transcriptional terminators or the introduction of site-specific mutations that do not have an effect on operon transcription may be required to dissect mutational effects more closely.

Random mutation

If no previous knowledge of virulence factors is available, random mutagenesis can be used to identify virulence genes. Traditional methods employed UV or chemical agents to introduce nucleotide changes within the bacterial genome that lead to alterations in gene expression. These are hampered by the need to fulfil two requirements. Firstly, there must be a means for selecting mutants attenuated in virulence properties. Testing of individual mutants is impracticable, and in mass screening it may not be possible to identify individual virulence-defective mutants within pools of mutagenized bacteria. Furthermore, subsequent identification of the mutation site can be technically demanding. Generally, this is achieved by complementation analysis, where a virulence trait is restored with a plasmid-encoded intact copy of the wild-type gene. A plasmid library of wild type genomic DNA is introduced into the mutant strain and transformants that resemble the parental strain in the virulence trait under study are selected in the model system. Difficulties arise if the virulence-defective phenotype is the result of multiple unlinked genetic lesions, where it may not be possible to restore wild-type function with a single gene.

Bacterial transposons have been used extensively for the mutational analysis of virulence in bacteria, particularly in Gram-negative species. Naturally occurring transposons have been modified extensively to improve their facility for mutational analysis; unwanted genes have been removed or replaced with suitable selective markers, useful unique restriction sites have been introduced, and insertion sequences optimized for random integration have been incorporated. Large numbers of single-site mutations can be generated with ease using these modified transposons, and since the mutation site is marked by the transposon DNA flanking the insertion point can be readily recovered for sequence analysis.

If no useful transposon is available for the pathogen under study, then the technique of shuttle mutagenesis can be applied. A genomic library of the pathogen is mutagenized in *E. coli* and pooled mutant plasmids are introduced back into the pathogen. The method is dependent on a means for introducing DNA and on efficient homologous recombination by the pathogen for integration of the interrupted sequences into the bacterial chromosome.

Transposition itself requires only a source of transposon DNA, a target for insertion, and the enzyme(s) that catalyse the transposition event. If these elements are supplied, there is no need for bacteria themselves. This fact has been exploited to develop a commercially available transposon mutagenesis system in kit form that allows cell-free transposon mutagenesis of target DNA sequences for subsequent reintroduction into the host organism. This opens the way for systematic saturation mutation analysis of the entire gene complement of pathogenic bacteria.

Signature-tagged mutagenesis

Signature-tagged mutagenesis (STM) is a further development of transposon mutagenesis that allows the easy identification of virulence-attenuated mutants within pools of mutagenized cells. In STM, individually marked transposons are used for random mutagenesis of the pathogen, and the experimental animal model is challenged with the pool of mutants. Genes essential for growth of the pathogen within the animal model are then identified by differential hybridization analysis of the signature tags amplified from input and recovered pools of bacteria (Figure 4.2).

Random, short (40 bp) sequence tags are introduced into a unique restriction enzyme site within a transposon to create a pool of tagged transposons. The transposon pool is then introduced into the target organism. Individual bacterial clones arising from random transposon integration are selected and arrayed in 96-well microtitre dishes to form a bank of insertion mutants each with a unique 'bar code' signature tag. Duplicate colony blots are prepared from the array of transposon mutants for DNA hybridization. The signature-tagged mutants are pooled and genomic DNA is extracted from the bacteria in this inoculum pool for subsequent analysis. The inoculum pool is then used to challenge the experimental animal, and bacteria are recovered from the spleen or other body site after a specified incubation period by plating on selective medium. Genomic DNA is extracted from this pool of recovered bacteria. The unique signature tags are amplified by PCR using as template the genomic DNA extracted from the inoculum and recovered bacterial pools. The amplified tags are labelled and used to probe, separately, the duplicate colony blots prepared earlier from the array of transposon mutants. Mutants which are unable to grow within the animal model will not be represented, or will only be poorly represented, in the pool of recovered bacteria. Consequently, the signature tag of the transposon that has generated the mutant will be present in the inoculum pool probe but absent from the recovered pool probe, and thus mutants attenuated in virulence will be identified on the colony blots as those bacteria hybridizing with the inoculum probe, but not with, or only very weakly with, the recovered probe (Figure 4.2). STM was used to identified a pathogenicity island of *Sal. typhimurium*, and genes required for cell wall synthesis and nutrient uptake or biosynthesis were shown by STM to be necessary for *Staphylococcus aureus* virulence in mice.

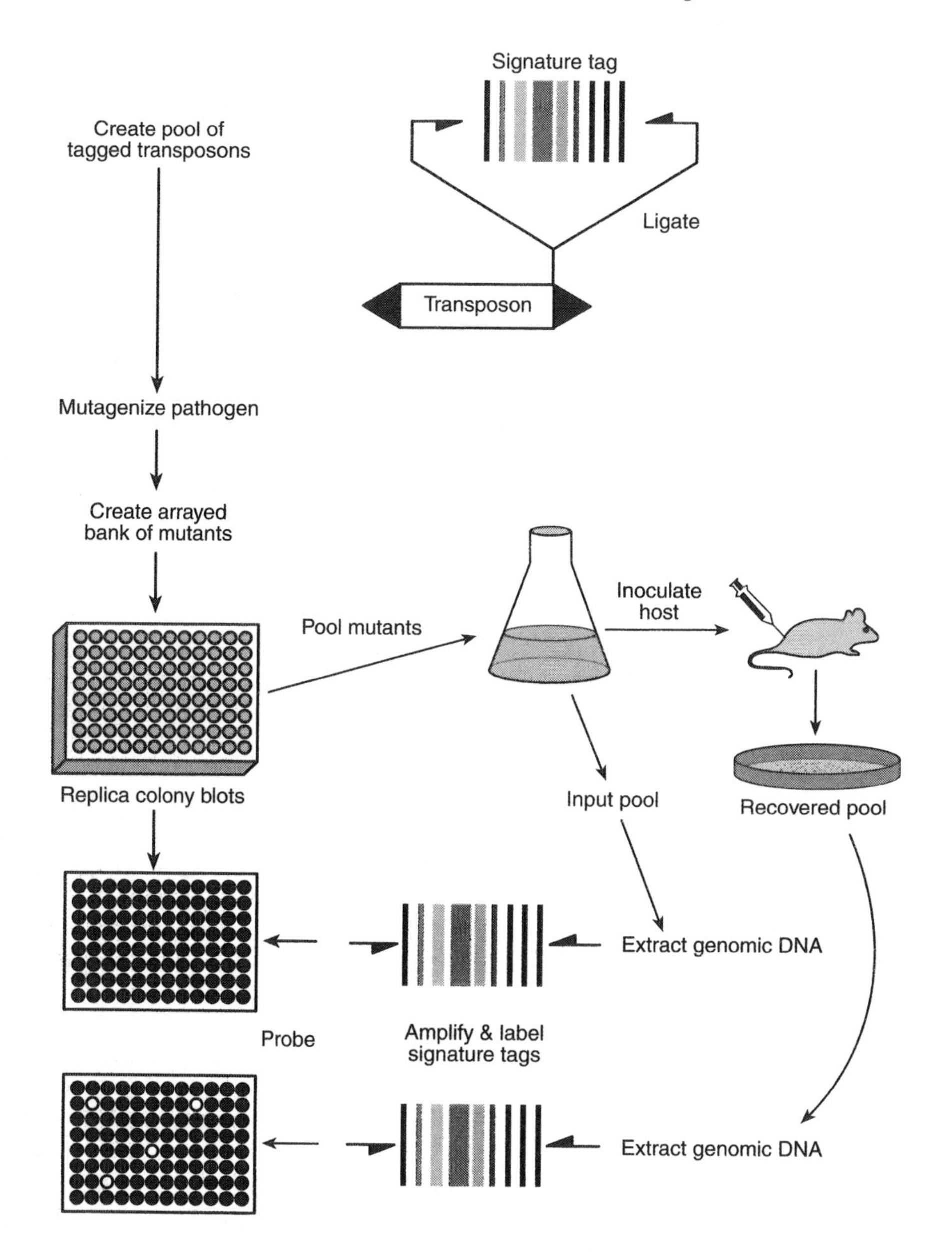

Figure 4.2 Signature-tagged mutagenesis. A pool of tagged transposons is created by insertion of random sequence tags into a unique restriction site, and the pool is used to create an arrayed bank of mutants. Each will have a unique transposon inserted at a different chromosomal location. The mutants are pooled and inoculated into the animal model. Genomic DNA is extracted from inoculum (input) and recovered bacteria and used to amplify the sequence tags. Amplified tags are labelled and used to probe, separately, duplicate colony blots of the arrayed mutant library. Mutants unable to grow within the experimental animal are identified by differential hybridization analysis and can be recovered from the arrayed bank

Gene transfer

Complementation analysis was described earlier in which random mutations are located by gene transfer from wild-type parental organisms. The availability of naturally occurring bacterial strains with reduced or no virulence, or of a closely related species that are not virulent, extends the utility of complementation analysis for the cloning of virulence determinants by gene transfer. Examples include the identification of a single *Yersinia pseudotuberculosis* gene that confers on *E. coli* the ability to adhere to and invade cultured epithelial cells, and the use of a non-invasive *Sal. typhimurium* isolate, or indeed laboratory strains of *E. coli*, as hosts to identify *S. typhimurium* invasin genes. Virulence within an experimental animal system may be a many-component system, and often the genes will be encoded at disparate regions on the bacterial chromosome. Additionally, the transformed genes may not be expressed in the test organism, thus gene transfer has its limitations. In a neat modification of this method, *E. coli* virulence genes have been mapped by loss of function, rather than gain of function, by mating experiments (conjugal gene transfer) between a non-virulent laboratory strain and virulent isolates of *E. coli*.

Protein expression approaches

Bacterial cell surface and secreted components play important roles in virulence, including adherence to host tissues, acquisition of nutrients, immune avoidance or subversion, tissue destruction and adaptive responses to changing environmental conditions. Analysis of cell surface and secreted components of pathogens will identify a subset of virulence factors, and two molecular approaches to identifying virulence factors are based on this premise.

In the first method, sera from patients infected with a particular pathogen are used to screen expression libraries of the pathogen genomic DNA for proteins that are expressed during infection and that are recognized by the host immune system and able to induce an immune response. Putative virulence factors of *Enterococcus faecalis* expressed during enterococcal endocarditis were isolated on the basis of their reactivity with patients' serum. Most of the 38 clones isolated encoded surface or secreted proteins, as might be expected. One large group of clones encoded protein components of ATP-binding cassette (ABC) transporters involved in the transport and/or binding of essential nutrients. It should be noted that ABC transporter function, in addition to serving in the uptake of nutrients, may affect other aspects of bacterial physiology including competence for DNA transformation, sporulation and expression of virulence traits such as adhesion. Also isolated were transcriptional regulators of His-Asp phosphorelay signal transduction systems, and genes encoding cytoplasmic proteins which might indicate release of intracellular proteins by cell lysis, though this group of clones represented only a small proportion of positive clones. Polyclonal antibodies raised in rabbits to extracted surface components of the bacterium grown in broth culture can be utilized to identify factors expressed also by bacteria grown *in vitro*, and thus can select further for those bacterial components expressed only during infection. Alternatively, should an animal model be available, serum can be collected from animals inoculated with live cells or with killed cells and used to differentially screen for surface components expressed during bacterial growth within the animal model.

There are two disadvantages to this procedure. Firstly, the method will generally miss important non-protein antigens such as capsular polysaccharides. Further, the method will not identify non-immunogenic virulence factors.

The second protein expression approach utilizes transposon mutagenesis to identify genes encoding secreted or surface-associated proteins regardless of their immunogenicity, thus avoiding the latter problem described above. Tn*phoA* is a modified transposon that carries a truncated gene (*phoA*) encoding an alkaline phosphatase (AP) enzyme (Figure 4.3a). The AP enzyme is active only if secreted from the bacterial cell but the gene lacks both a promoter and a signal sequence directing secretion of the AP. Tn*phoA* inserts randomly within the genome of the target organism and where insertion occurs in-frame within a gene encoding an exported polypeptide, then an AP-secreting clone is created which can be detected easily by plate assay.

Just because a gene is transcriptionally active during growth of the pathogen within the host or model system it does not necessarily follow that the gene encodes a virulence factor. A further advantage of the Tn*phoA* system is that clones isolated for AP secretion will be disrupted within the gene that provides the promoter and signal sequence for secretion of the AP; such mutants can be tested for virulence directly in the model system. Of course, complications can arise. In one study the protein fused to PhoA was not involved in virulence but this fusion protein became stuck in the bacterial membrane and affected the virulence of the organism.

Transposon mutagenesis has been used less frequently for the analysis of virulence factors of Gram-positive bacteria largely due to the lack of modified Gram-positive bacterial transposons suitable for molecular biology. However, a Tn*phoA*-like system has been developed for identifying secreted proteins of Gram-positive bacteria that is based on insertion-duplication mutagenesis. This method depends instead on the ability to introduce DNA and on the proficiency for homologous recombination, and was used successfully to identify surface and secreted proteins of *Strep. pneumoniae*. Random fragments of pneumococcal DNA were cloned upstream of a truncated *phoA* gene present on a plasmid capable of replication in *E. coli* but not in *Strep. pneumoniae*. Pooled plasmid DNA isolated from independent *E. coli* clones was introduced into *Strep. pneumoniae* with selection for the appropriate antibiotic resistance marker. Since the vector cannot replicate in pneumococci, antibiotic resistance arises from insertion of the plasmid onto the pneumococcal chromosome at a site determined by the insert DNA (Figure 4.3b). Recombinants are screened for AP enzymatic activity using the plate assay, and plasmids from AP-secreting recombinant clones can be rescued and the insert DNA sequenced. This method, as applied to *Strep. pneumoniae*, identified 21 distinct clones amongst which were again components of ABC transporters and two-component sensor regulators (described in Chapter 3). It was noted that despite differences in the promoter and signal sequences utilized by the Gram-negative bacterium *E. coli* and the

Figure 4.3 (opposite) Identification of surface and secreted polypeptides by fusion analysis. (a) Tn*phoA* system. A promoterless *phoA* gene, encoding a truncated alkaline phosphatase that lacks the signal sequence necessary for secretion, is cloned at one end of a transposon. Random integration of the transposon within a gene encoding a surface or secreted protein that creates an in-frame fusion with the truncated *phoA* gene will result in an AP^{+} phenotype that can be identified using a colorimetric plate assay. Additionally, a mutant is created by the insertion. (b) Modified *phoA* system. This method is useful if a transposon is not available for the target pathogen, and relies instead on homologous recombination. A vector is created that carries the truncated *phoA* gene and selectable marker. Random fragments of pathogen chromosomal DNA are ligated into a unique restriction site immediately 5'- to *phoA*, and the pool of recombinant molecules is introduced into the pathogen. Integration of the plasmid occurs by insertion duplication. As before, if the cloned DNA fragment is in-frame with *phoA*, and is part of a surface or secreted protein, AP is secreted. As with insertion duplication mutagenesis, the locus will be recreated if the cloned DNA contains sequence 5'- or 3'- to the coding sequence of the target gene

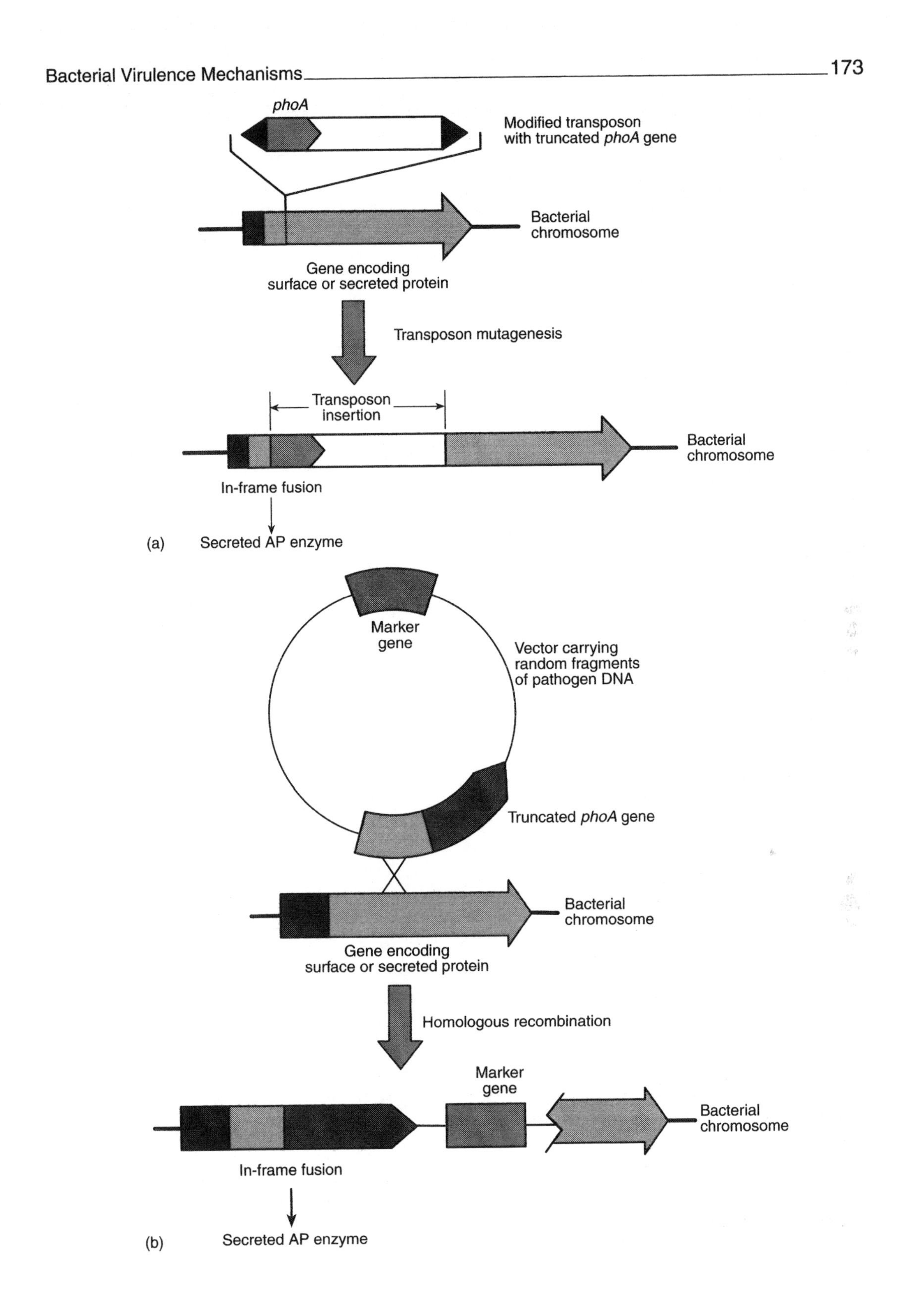
phoA
Modified transposon with truncated phoA gene
Bacterial chromosome
Gene encoding surface or secreted protein
Transposon mutagenesis
Transposon insertion
Bacterial chromosome
In-frame fusion
(a)
Secreted AP enzyme
Marker gene
Vector carrying random fragments of pathogen DNA
Truncated phoA gene
Bacterial chromosome
Gene encoding surface or secreted protein
Homologous recombination
Marker gene
Bacterial chromosome
In-frame fusion
(b)
Secreted AP enzyme

Gram-positive pneumococcus, many pneumococcal promoters and signal sequences functioned also in *E. coli*. In contrast to the Tn*phoA* method, insertion-duplication mutagenesis may not generate knockout mutations depending on the extent of insert DNA.

cDNA approaches

Reverse transcriptase, a retroviral enzyme that synthesizes complementary DNA (cDNA) from RNA templates, has allowed the development of a number of useful screening techniques for identifying genes expressed specifically under *in vivo* conditions.

One set of techniques employs differential hybridization of cDNA molecules prepared from different RNA pools to screen arrayed libraries of pathogen genomic DNA. At its simplest, RNA is isolated from bacteria recovered from infected tissue or from broth culture, and cDNA molecules are generated using reverse transcriptase and short random oligonucleotides to prime the reverse transcriptase (RT)-PCR reaction. cDNA from each pool is labelled and used to probe independently a duplicated colony blot of the arrayed library. Clones encoding genes that are expressed *in vivo* but not *in vitro* (broth culture) are identified by their differential hybridization with the labelled cDNA pools.

In a modification of this method, cDNA molecules common to both growth conditions (*in vivo* and *in vitro*) are first removed from the *in vivo* cDNA pool by subtractive hybridization to create an *in vivo*-specific cDNA probe (Figure 4.4). RNA is isolated from the two populations of bacteria and double-stranded cDNA pools are created as before. The cDNA synthesized from mRNA isolated from broth-grown bacteria is labeled with biotin. This cDNA is made single-stranded by heat denaturation and is hybridized in solution, at a ratio of 10 : 1, with denatured cDNA prepared from RNA extracted from bacteria in infected tissue. Biotin-labelled double-stranded cDNA is then removed using streptavidin-coated magnetic beads. Streptavidin binds avidly to biotin and since the streptavidin is attached to magnetic beads a simple magnet can be used to remove biotin-containing double-stranded DNA. As shown in Figure 4.4, this will result in the removal of cDNA species from the *in vivo*-grown cDNA pool that were present also in the cDNA pool of broth-grown bacteria. To enrich further for *in vivo*-specific cDNAs, the non-biotin-labelled cDNA hybrids are re-amplified and subjected to a further two rounds of subtractive hybridization as above. This final cDNA pool is amplified once more, labelled and used to probe a genomic library of pathogen DNA to identify clones encoding *in vivo*-expressed genes. By applying subtractive hybridization to RNA pools extracted from broth-grown and human macrophage-grown *Mycobacterium tuberculosis*, a single macrophage-specific gene was identified. Several additional clones gave weak but consistent hybridization signals with second- and third-round subtracted probes, and these clones might represent genes that are expressed in broth culture but whose expression is upregulated in macrophage-grown mycobacteria.

The second approach is a modification of the subtractive technique whereby RNA fingerprint patterns from differently grown bacterial populations are compared visually following electrophoretic separation of RT-PCR products on polyacrylamide gels (Figure 4.5). The advantage of this method, known as RNA arbitrarily primed PCR (RAP-PCR), is that less starting material is required and so the procedure can be used to investigate bacterial gene expression in infected tissue or even patient biopsies. Differences in the electrophoretic profiles represent differences in gene expression. DNA in bands of interest is eluted, re-amplified and cloned for sequence determination. As with differential and subtractive

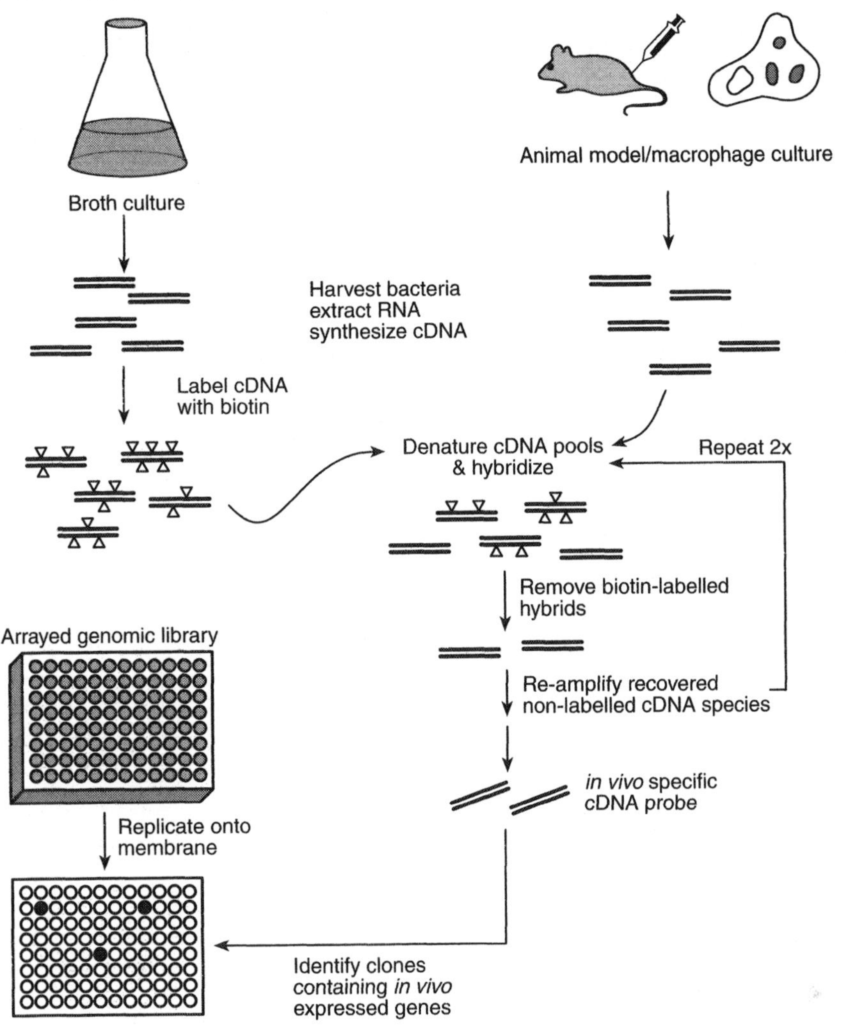

Figure 4.4 Subtractive hybridization. cDNA is synthesized by PCR amplification using RNA extracted from broth culture or *in vivo* system. The cDNA from broth-grown cells is labelled with biotin, denatured by heating and hybridized in solution at a ratio of 10 : 1 with cDNA synthesized from *in vivo*-grown bacteria. Hybrid cDNA molecules are removed using streptavidin-coated magnetic beads, and non-labelled molecules are recovered and re-amplified. Subtractive hybridization is performed a further two times to create an *in vivo*-specific cDNA probe that is used to screen an ordered genomic library of the pathogen

hybridization techniques, this method requires only that RNA can be extracted from the bacterial cells, making these attractive procedures for analysis of virulence in genetically less well-developed pathogens.

In vivo expression technology

In vivo expression technology (IVET) allows sensitive and direct detection of gene promoters that are active during bacterial growth in animal models. The first IVET technique to be

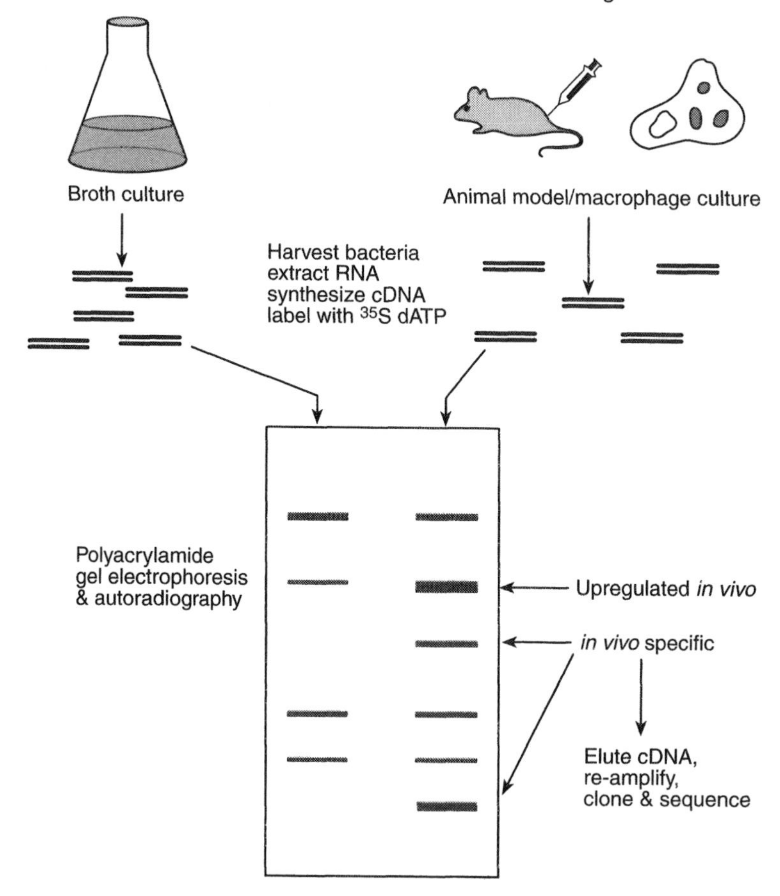

Figure 4.5 RAP-PCR. This method is a simplified cDNA approach for identifying *in vivo*- specific gene expression, that does not require subtractive hybridization. cDNA is synthesized by PCR amplification using RNA extracted from broth culture or *in vivo* system. Labelled cDNA molecules are separated by polyacrylamide gel electrophoresis and bands are visualized by autoradiography. Potentially, this method will identify *in vivo*-specific transcripts as well as upregulated (or downregulated) gene expression. DNA in bands of interest can eluted from the gel, re-amplified, cloned and sequenced

developed required the prior isolation and characterization of an auxotrophic mutant of the pathogen that was unable to grow in the animal model. Auxotrophs are mutants in which the gene(s) encoding a biosynthetic enzyme is inactivated. As a result, the auxotroph cannot synthesize an essential metabolite, but must have it supplied in the growth environment. Commonly used are mutants that are deficient in the *de novo* synthesis of purines but that can be grown *in vitro* by supplementation of the growth medium with purines. Random fragments of pathogen genomic DNA are cloned upstream of a promoterless gene, the product of which complements the auxotrophic mutation. These constructs are introduced onto the pathogen chromosome by homologous recombination to create a pool of random fusion recombinants (Figure 4.6a). The pool of recombinants is introduced into the animal model and bacteria are recovered from the spleen or other body site after a specified incubation period by plating on selective medium. The pool of viable bacteria isolated from animals will be enriched for recombinants which are able to grow within the host, and thus for

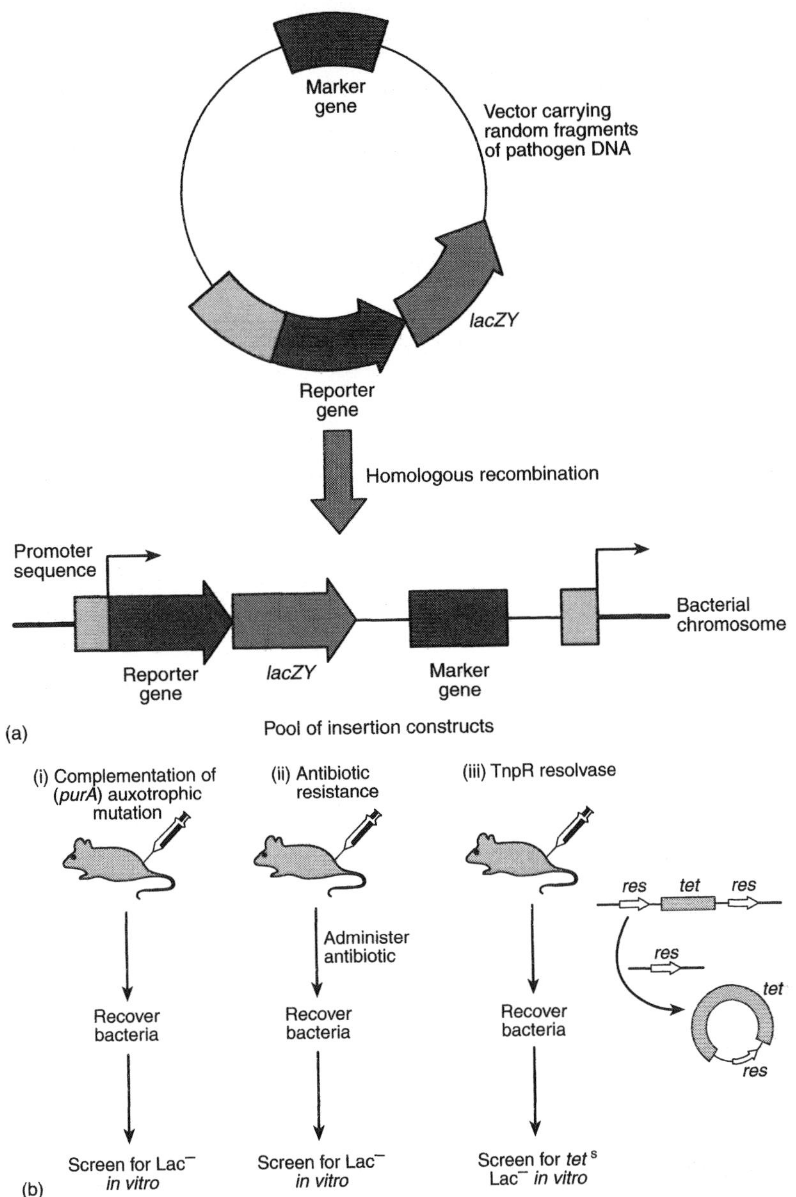

Figure 4.6 *In vivo* expression technology (IVET) methods. (a) Random integration of the IVET vector onto the pathogen chromosome is performed as described for insertion duplication mutagenesis to generate a pool recombinant pathogens. (b) Using the first two IVET methods, complementation of auxotrophic mutation (i) or expression of antibiotic resistance (ii), bacteria isolated from the experimental animal system will by necessity have expressed the selectable marker *in vivo*, thus identifying *in vivo* active-promoters. Isolates can be checked for absence of promoter activity during *in vitro* growth by β-galactosidase activity screening. IVET employing the *tnpR* reporter (iii) will identify *in vivo*-active promoters only following subsequent tetracycline-sensitivity screening of recovered bacteria. Since *in vivo* expression of *tnpR* is not necessary for survival and growth within the animal model, this IVET system provides a more sensitive tool for the identification of promoters expressed transiently or at low levels

recombinants in which the promoterless reporter gene is fused to DNA that encodes an *in vivo*-active promoter. The system is further refined by introducing the *lacZ* gene (encoding β-galactosidase enzyme) downstream of the reporter. By plating recombinants on media containing a chromogenic substrate for β-galactosidase, this construct allows the pre- or post-infection identification of promoters that are active during *in vitro* as well as *in vivo* growth (Figure 4.6a). An obvious limitation of this approach is the need for a characterized auxotrophic mutant. To obviate this, the system has been modified to use a promoterless antibiotic (chloramphenicol) resistance marker for selection of *in vivo*-active promoters in animals whose diet is supplemented with chloramphenicol. The principle here is the same; thus, viable recombinants isolated from the animal will have expressed the chloramphenicol resistance marker during infection, allowing the identification of *in vivo*-expressed genes (Figure 4.6b).

These initial IVET systems select for gene promoters that are expressed throughout infection at high levels. Both methods may miss genes that are transiently expressed, during a specific stage of infection, for example, or genes that are expressed at low levels but that are nevertheless essential for virulence. The use of a promoterless recombinase gene in conjunction with an excisable antibiotic resistance marker has provided an IVET system with exquisite sensitivity for both low-level and transiently expressed *in vivo*-specific promoters. The *tnp*R gene encodes a recombinase (resolvase; Chapter 2) of transposon γδ. This recombinase mediates site-specific excision of any DNA that is flanked by direct repeats of *res*, the targets for the resolvase (Figure 4.6c). For IVET, the target DNA for the recombinase is an antibiotic marker that is included on the IVET plasmid construct, or is integrated on the pathogen genome. Resolvase activity serves to excise this DNA to form a non-replicating mini-circle that is lost from cells during cell division, resulting in antibiotic-sensitive progeny. These are selected by replica plating of recovered bacteria on media with or without the antibiotic. As before, promoters that are active during *in vitro* as well as *in vivo* growth can be identified by plating recombinants on media containing the chromogenic substrate for β-galactosidase. IVET-induced genes fall into a number of categories such as those encoding proteins required for adherence and invasion, for the acquisition or synthesis of essential nutrients as well as those involved in the bacterial response to shock. In many instances directed gene inactivation of IVET-identified genes in conjunction with LD_{50} tests have demonstrated their requirement for virulence in animal models.

IVET in one of its forms has been used to identify *in vivo*-expressed genes of a number of pathogens including *Sal. typhimurium*, *Pseudomonas aeruginosa*, *Vibrio cholerae* and the Gram-positive bacterium *Staph. aureus*.

Gene reporter systems

β-Galactosidase is one of the most commonly used reporters of bacterial gene expression. Promoter – *lacZ* fusions have been widely used to investigate expression of particular genes under defined laboratory conditions thought to mimic the host environment. The development of highly sensitive fluorogenic substrates for β-galactosidase in conjunction with intensified CCD cameras has allowed the analysis of gene expression by bacteria growing within host cells, providing real-time information about both the intracellular environment and the microbial response to it. Other reporters such as luciferase (enzymes that emit photons of light during the oxidation of fatty aldehydes) and green fluorescent protein (gfp) have been developed that further extend the usefulness of this approach.

Strains of *Sal. typhimurium* were engineered to express luciferase from a constitutive promoter and introduced into a mouse model of infection. It was possible to localize bioluminescent bacteria within specific tissues using a CCD camera, and to monitor the progression of infection in live mice over an eight-day period. The *gfp* reporter is amenable, additionally, to fluorescent-based flow cytometry, which can provide a multi-parameter measure of *each* cell within a large sample. This allows the measurement of gene expression by individual bacteria rather than taking the average of a population. Further, by using a fluorescence-activated cell sorter (FACS), a method was devised that uses *gfp* as a selectable marker to identify and separate cells bearing *gfp* gene fusions that are differentially expressed in response to environmental changes.

Genome-based approaches

At the time of writing the genomes of 15 bacterial species have been completely sequenced and the are available in the public domain. The sequencing of a further 48 bacterial genomes is under way. The speed with which bacterial genomes are being sequenced is staggering and Table 4.1 will be out of date when this book is published. The reader should therefore consult the Institute of Genomic Research web page (http://www.tigr.org/) for a current list of sequenced genomes. Amongst those currently sequenced are a number of important human pathogens (Table 4.1). Though complex in themselves, bacterial genomes are remarkably compact (ranging from less than 1 mega-base pairs (Mb) to around 9 Mb) when compared with those of even simple eukaryotes such as brewer's yeast (*Saccharomyces cerevisiae*;

Table 4.1 A list of some bacterial pathogen genomes that have been or are in the process of being sequenced (taken from The Institute for Genomic Research web page http://www.tigr.org/)

Pathogen	Genome size (Mb)
Published genomes	
Borrelia burgdorferi	1.44
Haemophilus influenzae	1.83
Helicobacter pylori	1.66
Mycobacterium tuberculosis	4.40
Mycoplasma pneumoniae	0.81
Mycoplasma genitalium	0.58
Treponema pallidum	1.14
Genomes in progress	
Chlamydia trachomatis	1.05
Listeria monocytogenes	3.00
Neisseria gonorrhoeae	2.20
Neisseria meningitidis	2.30
Salmonella typhimurium	4.50
Streptococcus pneumoniae	2.20
Streptococcus pyogenes	1.98
Vibrio cholerae	2.50

13 Mb) or the single-celled malarial parasite (*Plasmodium falciparum*; 30 Mb), and the time required to complete a bacterial genome sequencing project, technical difficulties notwithstanding, is ever decreasing. Because of the speed of modern DNA sequence data generation, genomic sequencing has outstripped our knowledge of gene function determined by 'wet' experiments. Of the 4286 open reading frames (ORFs) of the *E. coli* K-12 genome, less than half encode proteins of known function. Assigning function to uncharacterized ORFs will necessarily require a better understanding of all aspects of the biology of bacteria themselves.

Broadly speaking, there are three approaches to whole genome sequencing. The first involves prior generation of an ordered library of overlapping clones using cosmid vectors that can carry large (around 40 kb) DNA fragments. Considerable time and expense are involved in the creation of the library. Subsequently, however, since the entire genome is represented, the sequence generated will be without gaps and, further, the degree of sequence redundancy is low. Shotgun cloning, in contrast, is a high-redundancy (5–10-fold) approach where many thousands of random clones are sequenced and the data assembled to generate contiguous sequences. Gaps do occur and these can be closed by direct cloning or PCR amplification of the missing DNA stretches. Although there may be 10-fold redundancy in the sequence data generated, repeated sequencing does lead to very high-quality data, and with high-throughput automated sequencing machines this is the method of choice in most sequencing projects. The third method, termed diagnostic sequencing, is a more 'down and dirty' approach and is consequently much quicker and cheaper than either of the previously described approaches. The strategy is based on sequencing a limited number of plasmid clones carrying inserts randomly distributed over the genome. The data generated is used for the production of longer sequencing templates generated by PCR amplification or as start points for sequencing by primer walking on longer segments of chromosomal DNA carried on cosmid or lambda phage vectors. The redundancy of sequencing using this method may be as little as two-fold although this is tempered by an error rate as high as 1%. Nevertheless, the data can be used for a number of purposes: to generate a detailed physical map of the chromosome in terms of restriction enzyme sites, repeated elements and operon and ORF organization; to derive a comprehensive gene complement of the organism; to provide the framework for high-quality sequencing of the genome with a relatively low redundancy.

DNA sequence retrieval

A small amount of DNA sequence information obtained from clones isolated by IVET or from cDNA fragments of differentially expressed mRNA species identified by RAP-PCR may be sufficient to fish out details of the entire gene and its neighbours in a matter of minutes. However, this may not provide any clues as to the function of the gene. All too often, search results define homology to a putative ORF that is uncharacterized (amusingly called an URF unidentified reading frame by one author), or is itself listed as homologous to an uncharacterized ORF from another bacterium. As an example, IVET was used to identify over 100 *Sal. typhimurium* genes that are specifically expressed during infection in mice. Over 50% had no significant homology with sequences in the database, or encoded putative ORFs with no assigned function. When applied to the genetically less well-characterized bacteria, the proportion of known genes may be even less; of 45 *Staph. aureus* genes induced during infection of mice, six were identical to known *Staph. aureus* genes, a further 11 demonstrated homology to known non-staphylococcal genes and the remainder (62%) had no significant homology to any sequence.

Computer informatics

Sequence homology searches

Genome databases can be searched with nucleotide or amino acid strings that comprise known virulence factors in related pathogens. Powerful search tools are available that permit considerable flexibility with regard to search parameters such as degree of similarity, allowances for gaps in one or other sequence etc. Genes identified *in silico* (i.e. by computer) can be amplified from genomic DNA using PCR primers derived from the database sequence and the cloned DNA used as the basis for gene disruption. The virulence of isogenic mutants thus created can be compared with the wild-type. Since it is not uncommon for proteins of related function to share significant sequence identity at the amino acid level, but not at the DNA level, the ability to search databases with derived amino acid sequences allows the identification of genes that might not be cloned by DNA comparison nor indeed by DNA hybridization experiments. However, amino acid or DNA sequence homology does not necessarily imply conserved function. Nevertheless, the software used to search databases is becoming more and more sophisticated and can, by carefully defining search parameters, indicate biological relationships with a degree of confidence. For example, the presence of particular motifs in a protein sequence can often provide some indication about the function of the proteins. In addition, structural prediction programs can suggest structural arrangements of protein sequences that might indicate the overall function of the protein of interest.

Pathogenicity islands (PAIs, Chapter 2) are large (often >30 kb) genetic elements within the chromosomes of pathogenic bacteria that encode blocks of genes that contribute significantly to bacterial virulence and are thought to be acquired by horizontal gene transfer. These PAIs often demonstrate a different G+C content compared to DNA of the host bacterium and are frequently found in the vicinity of tRNA genes or flanked by insertion sequence elements. These facts offer handles for the identification of possible virulence factors without prior experimentation. PAIs may also be identified by their absence from closely related non-pathogenic strains.

Subtractive hybridization

The genome of *Mycoplasma genitalium* is 0.58 Mb and comprises only 468 identified protein-coding genes that have been dubbed the minimal gene set. However, by comparing the *Myco. genitalium* genome with that of the evolutionarily distant bacterium *Haemophilus influenzae*, a subset of genes conserved in both organisms was generated. The subset comprised 240 genes which was supplemented with some 22 genes for missing steps in critical metabolic pathways that were encoded by dissimilar genes in the two bacteria. Six genes encoding common, functionally redundant, or parasite-specific, genes were removed to generate a minimal set required for modern-type cellular life of only 256 genes. As more and more bacterial genome sequences are completed, this list will be further refined. Subtractive hybridization opens the possibility of defining subsets of genes that specify the phenotypic characters that make the organism unique or that are required for a bacterium to thrive in any given environment, such as within the human body (Figure 4.7).

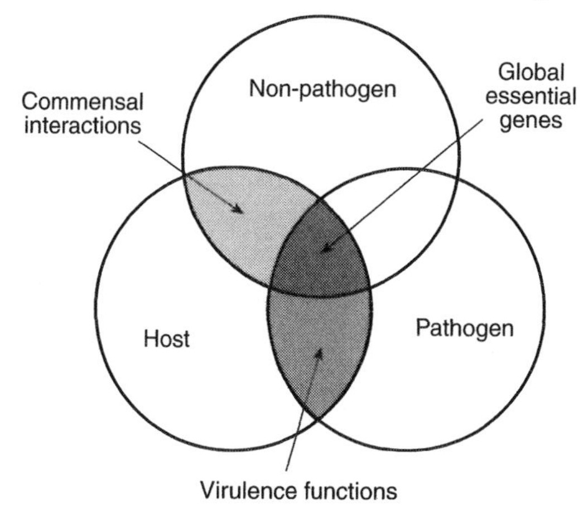

Figure 4.7 Gene expression in health and disease

MOLECULAR BIOLOGICAL TECHNIQUES APPLIED TO EUKARYOTIC CELLS

Various molecular biology techniques devised for eukaryotic cells can also be used to determine the nature of the interaction between bacteria and host cells, or the organism as a whole, and a few of the available techniques will be briefly described. The techniques of subtractive hybridization and RAP-PCR have been described as methods for defining differences in bacterial gene expression under different environmental conditions. This has been used, for example, to determine the genes switched on (or off) when bacteria, such as *M. tuberculosis*, enter macrophages. Similar techniques are available for detecting the pattern of genes switched on by host eukaryotic cells exposed to pathogens or their products. The major difference between bacteria and host cells is the numbers of mRNA molecules they contain. In mammalian cells it is estimated that in 'unstimulated' cells 15 000 mRNA molecules may be present and it is against this background that changes induced by contact with bacteria have to be determined. The current method of choice for identifying genes activated by contact with bacteria or their products is called differential display. This technique is similar to RAP-PCR and uses a set of oligonucleotide primers, one being anchored to the polyadenylated tails of a subset of cellular mRNAs, the other being short and of random sequence such that they anneal at different positions relative to the first primer. Multiple primer sets are used to amplify as many of the cellular mRNA molecules as possible. The mRNAs identified by these primer pairs are amplified after reverse transcription and separated on DNA-sequencing gels, in which state it is possible to compare patterns of DNA fragments from unstimulated and stimulated cells and identify genes transcribed under the given experimental conditions.

Gene reporter systems such as have been described above for bacteria are also increasingly used to determine the transcriptional control of specific genes in cells exposed to bacteria or to bacterial components. A particularly sensitive system is to link the upstream promoter elements of the gene of interest (e.g. a proinflammatory cytokine) to a luciferase and measure the switching on of the gene by measuring the light emitted. It is now possible to use firefly and *Renilla* (sea pansy coelenterate) luciferases linked to different genes within the same cell to examine the independent transcriptional control of two distinct (but possibly mechanistically related) genes.

Yeast two-hybrid or interactive trap cloning in yeast

There are growing numbers of examples of bacteria utilizing type III secretion systems to 'inject' proteins into host cells. The recent finding that the enteropathogenic strain of *E. coli* injects its own receptor (hp90) into host enterocytes is an excellent and exciting example of such bacteria–cell interactions. As described in Chapter 7 bacterial exotoxins can also enter host cells. It is important to determine which intracellular proteins of the host the bacterial proteins interact with. This is difficult using conventional biochemical or immunochemical methods but a recently developed technique called interactive trap cloning or yeast two-hybrid system is a molecular technique able to identify the interacting proteins and which utilizes a reporter gene system. The basis of this system is the modular domain structure of eukaryotic transcription factors which normally have two distinct functional domains: a specific DNA-binding domain (DBD) and a domain that activates transcription (transactivation domain, AD). Both domains are needed for normal gene activation. To identify proteins that interact, two separate hybrid proteins are constructed, normally the two functional domains of the yeast GAL4 transcriptional activator (Figure 4.8). The first hybrid protein is created by cloning a known gene (target gene X) fused with the GAL4 DBD gene. The activation domain vector is used to construct a fusion of the GAL4 activation domain and a target protein Y. Moreover, rather than a single known protein, a library of hybrids with the activation domain can also be constructed to search for new proteins that interact with the cloned target protein. When interaction occurs between the target protein X and candidate protein Y, the two halves of the GAL4 come together resulting in restoration of transcriptional activation of the reporter gene, which can be recognized by the formation of a measurable product.

The yeast two-hybrid technique has been used to investigate the interaction of the listerial protein ActA which induces actin polymerization and to identify the intracellular proteins in epithelial cells to which the opacity-associated proteins of *Neisseria gonorrhoeae* bind. Surprisingly, this latter study showed that the opacity proteins bound to the host metabolic enzyme, pyruvate kinase.

Generation of transgenic animals

The techniques available to inactivate bacterial genes have been described. It is also possible to do this with animal cells and even with whole animals although, as one can imagine, this is an expensive and time-consuming task. Nevertheless, a very large number of so-called transgenic animals have now been produced in which individual (or more than one) genes have been inactivated. These animals are called transgenic knockouts and are extremely useful for defining the purpose of individual genes in the whole animal. It is also possible to produce transgenic animals with supernumerary genes; however, the generation and use of such animals will not be discussed. How does one produce a transgenic animal lacking a functional gene?

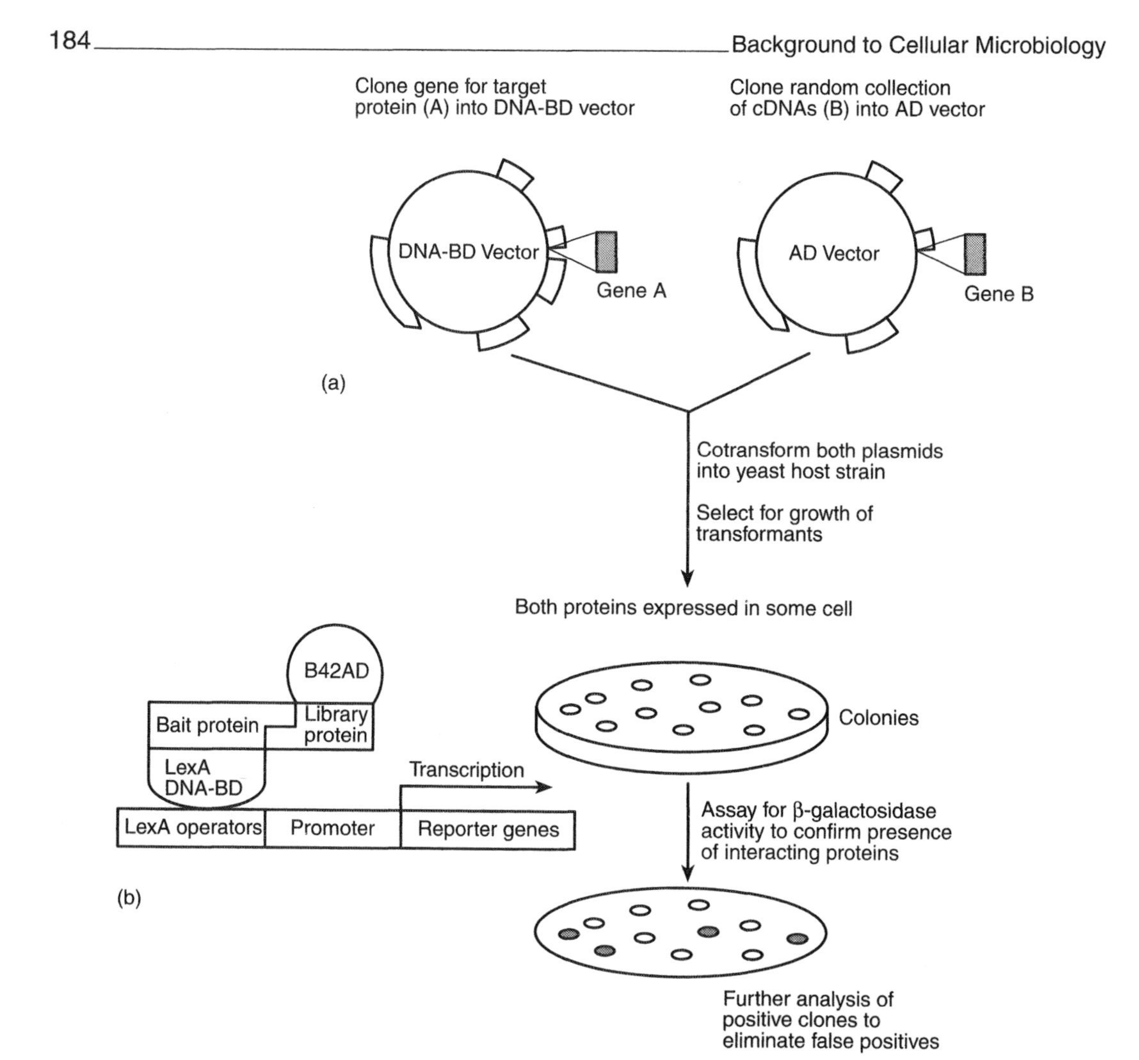

Figure 4.8 Schematic diagram of the yeast two-hybrid methodology. In (a) a flow chart shows the methods of transforming cells and screening to detect protein-protein interactions. In (b) the nature of the molecular interactions producing positive clones is schematically illustrated

Producing transgenic animals requires a combination of skills in embryology, cell culture, molecular biology and animal breeding. In both transgenic knockouts and transgenic supplemented animals (which express exogenously added genes) the DNA of interest is incorporated into fertilized eggs, normally of mouse origin. The secret of knocking out a mammalian gene is to target the gene of interest by a process called homologous recombination. This means that the exogenous DNA is incorporated into the mouse's genome at the same site as the DNA sequence being inactivated. Exogenous genes can be microinjected into recently fertilized embryos. A technique which is more commonly used today is to use totipotent embryonal stem (ES) cell lines, which are undifferentiated embryonal cells capable of differentiating into all types of tissues. These are transfected with the gene of interest and, once it is shown that the gene has been incorporated, the cells can be introduced into a developing embryo by micromanipulation (Figure 4.9). Mice arising from such manipulations are

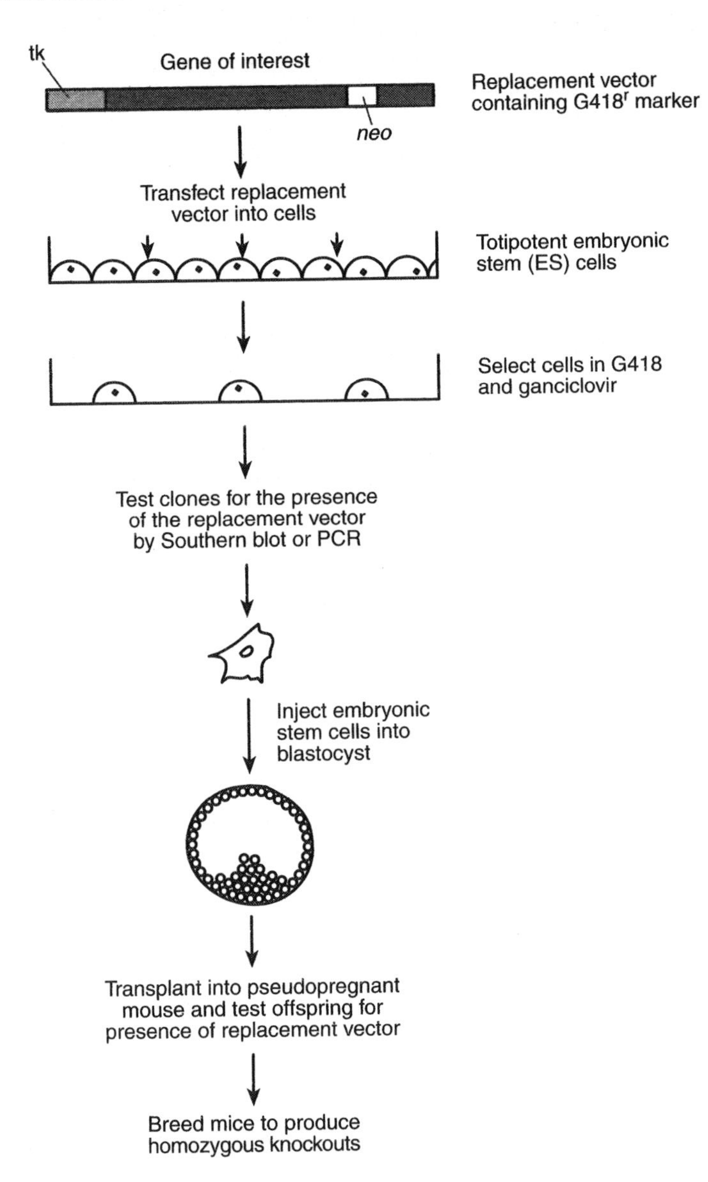

Figure 4.9 The generation of transgenic knockout animals using the process of homologous recombination resulting in targetted gene disruption is illustrated. Totipotent embryonic stem (ES) cells are transfected using a replacement vector containing the enzyme thymidine kinase (tk) and a genomic DNA clone containing the gene of interest in which one of the exons has been replaced with a gene conferring resistance to the drug G418 (*neo*). Expression of thymidine kinase will make cells susceptible to killing with ganciclovir, whilst expression of *neo* will make cells resistant to G418. Consequently, only those cells that have homologously recombined the replacement vector with the gene of interest will grow as these will contain the *neo* gene but not contain the *tk* gene, which is lost during homologous recombination. This eliminates cells that have incorporated the replacement vector randomly. The chimaeric mice produced need to undergo appropriate breeding regimes to produce homozygous mice which can then be used to examine the effect of the removal of this one active gene product

chimaeric, as not all cells will contain the transgene. However, by appropriate breeding, homozygous mice lacking the targetted gene can be produced and, if these animals survive, can be used in appropriate experiments. It is also possible to knock out more than one gene and a number of double knockouts have been produced.

A very large number of transgenic knockout mice have now been produced and are shedding light on the role that individual genes play in our ability to mount effective responses to pathogens (Table 4.2).

Table 4.2 Gene knockouts and their consequences for control of infection

Gene	Phenotype
TCR	Increased susceptibility to *Listeria* and *Mycobacteria*
CD4	Decreased Th_1 activity
RAG-1	Increased susceptibility to *Listeria* and *Mycobacteria*
MHC II	Deficient cell-mediated immunity, inflammatory bowel disease, increased susceptibility to *Listeria* and *Mycobacteria*
β_2-Microglobulin	Increased susceptibility to *Listeria* and *Mycobacteria*
ICAM-1	Impaired inflammatory and immune responses
CoxII	Inhibited response to endotoxin
IL-1β	Resistance to endotoxin shock
ICE (caspace 1)	Resistance to endotoxin
IL-1ra	Decreased susceptibility to *Listeria*
IL-2	Develop ulcerative colitis in response to normal gut flora
IL-4	Deficient Th_2 response
IL-6	Susceptibility to *Listeria*
IL-10	Develop chronic enterocolitis
IL-18	Impaired cytokine and NK response to infection
IFNγ	Increased susceptibility to *Mycobacteria*
IFNγR	Increased susceptibility to *Listeria* and *Mycobacteria*
G-CSF	Increased susceptibility to *Listeria*
TNFR	Increased susceptibility to *Listeria*

CONCLUSIONS

Bacterial virulence is the culmination of multiple and specific gene expression that is dictated, in turn, by bacterial sensing of and responding to the changing host environment. The various molecular techniques outlined in this chapter have contributed significantly to our understanding of bacterial virulence by defining the complement of genes that are expressed during infection or that are necessary for virulence. The methods have also revealed the quantum nature of pathogen evolution through horizontal acquisition of virulence traits.

The bacterial cell surface is the forum for bacterial–host interactions, and surface components that mediate adherence, immune avoidance, invasion, nutrient acquisition and environmental sensing all play an important role in virulence and feature strongly in the list of identified *in vivo*-active genes. Secreted bacterial products that are involved in host signalling and in tissue damage also feature, as do intracellular factors including transcriptional regulators that direct gene expression in response to environmental cues, biosynthetic enzymes, and stress response proteins. What is quite clear, however, from the results of molecular screens for *in vivo*-induced gene expression is that we are still far from fully comprehending the complex interactions that comprise virulence simply because many identified genes have as yet unknown functions. Genomics may offer the way forward but only when matched by laboratory experimentation to characterize the biological functions of these hypothetical ORFs. Systematic sequential inactivation of all genes in a bacterial genome is technically feasible and offers one approach. Perhaps more rewarding would be a closer analysis of the 'accessory' genes over and above the minimal gene set identified by computer-mediated subtractive hybridization. These genes are likely to encode phenotypic traits that make any particular species unique, and will certainly include a complement of genes that permit that species to flourish in its chosen niche.

Interspersed within the chromosome of a number of enteric pathogens are large loci, termed pathogenicity islands, that contribute significantly to the virulence properties of the pathogen. PAIs are often greater than 30 kb in size and may confer virulence traits such as adhesion, invasion, ability to grow in iron-limited conditions and the production of toxins. The different G+C content of PAIs compared to the rest of the chromosome suggests that PAIs were acquired by horizontal gene transfer. Some virulence determinants are plasmid or phage encoded, or are present on conjugative transposons, and again it is easy to envisage the development of a pathogenic phenotype through horizontal acquisition of exogenous DNA. These elements continue to mould pathogen evolution through the interspecies transfer of antibiotic resistance determinants.

This chapter has focused on molecular methods that have been used to define bacterial virulence mechanisms. Is a commensal bacterium defined by the lack of such virulence traits? Or is the commensal lifestyle itself also the result of complex bacteria–host interactions at mucosal surfaces specified by commensal-specific gene expression? Given that the average human body is composed of 10^{13} nucleated cells but contains 10^{14} bacteria, and that this commensal microflora contains a far greater diversity of microorganisms than is present in the limited group that comprise the human pathogens, this might be an even more important question to address using molecular techniques.

FURTHER READING

Books

Asubel F *et al.* (eds) (1995) *Short Protocols in Molecular Biology* Wiley, Chichester.
Bartel PL, Fields S (1997) *The Yeast Two-Hybrid System*. Oxford University Press, Oxford.
Brown TA (1996) *Gene Cloning* (3rd edn). Chapman & Hall, London.
ClarkVL, Bavoil PM (eds) (1997) *Bacterial Pathogenesis*. Academic Press, San Diego.
Cox TM, Sinclair J (1997) *Molecular Biology in Medicine*. Blackwell Science, Oxford.
Dale JW (1998) *Molecular Genetics of Bacteria* (3rd edn). John Wiley & Sons, Chichester.
Dorman CJ (1994) *Genetics of Bacterial Virulence*. Blackwell, Oxford.
Lewin B (1997) *Genes VI*. Oxford University Press, Oxford.
Newton CR, Graham A (1994) *PCR*. Bios Scientific Publishers, Oxford.
Old RW, Primrose SB (1994) *Principles of Gene Manipulation*. Blackwell Science, Oxford.
Sambrook J, Fritsch EF, Maniatis T (1989) *Molecular Cloning: A Laboratory Manual* (2nd edn). Cold Spring Harbor Laboratory Press, Cold Spring Harbor, NY.
Walker MR, Rapley R (1997) *Route Maps in Gene Technology*. Blackwell Science, Oxford.

Reviews

Camilli A. (1996) Noninvasive techniques for studying pathogenic bacteria in the whole animal. *Trends Microbiol* 4: 295–296.

Heithoff DM, Conner CM Mahan MJ (1997) Dissecting the biology of a pathogen during infection. *Trends Microbiol* 5: 509–513.

Hensel M, Holden D (1996) Molecular genetic approaches for the study of virulence in both pathogenic bacteria and fungi. *Microbiology* 142: 1049–1058.

Quinn FD, Newman GW, King CH (1997) In search of virulence factors of human bacterial disease. *Trends Microbiol* 5: 20–26.

Smith H (1998) What happens to bacterial pathogens *in vivo*? *Trends Microbiol* 6: 239–243.

Strauss EJ, Falkow S (1997) Microbial pathogenesis: genomics and beyond. *Science* 276: 707–712.

Valdivia RH Falkow S (1997) Probing bacteria gene expression within host cells. *Trends Microbiol* 5: 360–363.

PART 2

Prokaryotic–Eukaryotic Interactions in Infection

CHAPTER 5

Bacterial Adhesion to Host Cells

INTRODUCTION

In order to remain in or on its host – for example, you, the reader – a microorganism must first adhere to some cell, secretory product or structural component of its intended host or, as is increasingly common because of medical advances, some indwelling prosthetic device. Hence, bacteria are invariably found on the surfaces of epithelial and epidermal cells, either as a result of direct adhesion to host cells or indirectly by binding to either secretory products (e.g. mucin) which often coat such cells or to bacteria which have already adhered. Teeth and connective tissue (the latter exposed as a result of some breach in the protective epithelium or epidermis due to wounds, burns etc.) represent examples of structural components which rapidly become colonized by bacteria while urinary catheters are only one of a number of prosthetic devices which provide surfaces for microbial colonization. Bacteria also adhere to host phagocytic cells and this constitutes one of the host's major defence systems if it results in phagocytosis and destruction of the organism. However, such adhesion is not always to the advantage of the host as many bacteria are able to survive in, and be disseminated by, such cells (constituting a microbial Trojan horse tactic). Once inside the host, a far wider variety of surfaces are available for microbial adherence, including a number of polymers comprising the extracellular matrix (proteoglycans, collagen etc.), bone, endothelial cells and the many specialized cells comprising the various organs of the body. Collectively, bacteria have evolved an array of surface molecules and structures to enable them to adhere to the above-mentioned surfaces. However, a particular organism is usually able to adhere to only a particular surface, or a limited variety of surfaces, and this explains, in part, the remarkable tissue tropism of bacteria, i.e. their ability to colonize (and possibly to induce disease in) only certain host tissues (Table 5.1). Nevertheless, for some organisms, particularly those capable of invading tissues (e.g. *Salmonella* spp. and *Neisseria meningitidis*) a variety of surfaces are encountered during the course of the infectious process and these organisms have evolved strategies to enable them to adhere to these surfaces as and when required. There are, therefore, at least two facets to bacterial adhesion to be considered: the physico-chemical forces involved in the adhesion process and the specificity of the process which dictates which surface an organism will bind.

Table 5.1 Examples of tissue tropism

Organism	Tissue
Neisseria meningitidis	Nasopharyngeal epithelium
Neisseria gonorrhoeae	Urethral epithelium
Vibrio cholerae	Intestinal epithelium
Bordetella pertussis	Respiratory epithelium
Salmonella typhimurium	Intestinal epithelium
Helicobacter pylori	Gastric mucosa
Streptococcus pyogenes	Pharyngeal epithelium
Campylobacter jejuni	Intestinal epithelium
Mycoplasma pneumoniae	Respiratory epithelium

The host is, of course, not a passive bystander during these attempts by bacteria to become established and has evolved a number of strategies to protect itself from microbial colonization. Hence, epidermal surfaces secrete antimicrobial compounds such as lysozyme and antibacterial peptides (see Chapter 8) while evaporation of sweat leaves behind large quantities of salt which many bacteria cannot tolerate. Epithelial surfaces utilize other mechanisms to prevent colonization. The respiratory tract has a ciliated epithelium coated in mucin in which bacteria are trapped, carried to the back of the pharynx by ciliary action and swallowed. The mucin also contains a number of antimicrobial compounds including lysozyme, lactoferrin (a protein which binds iron, so limiting bacterial growth), secretory IgA which prevents attachment of bacteria to epithelial cells, superoxide radicals generated by the enzyme lactoperoxidase and antibacterial peptides secreted by epithelial cells. The urinary tract and oral cavity are regularly flushed with urine and saliva respectively which serve to reduce microbial colonization. Nevertheless, in spite of these defence mechanisms, bacteria do colonize human beings; indeed the average person consists of 10^{14} bacteria and only 10^{13} host cells. Is this because some bacteria consistently manage to outwit their host or does the host allow itself to become colonized by certain bacteria? In support of the latter viewpoint, it has been well established that disruption of the normal microflora by broad-spectrum antibiotics can result in a preponderance of undesirable organisms such as *Candida albicans*, pseudomonads and *Clostridium difficile* in various parts of the body which can induce pathology. Our normal microflora, therefore, would appear to exert a protective effect and, indeed, we usually live in harmony with these bacteria for most of our lives. Has this association arisen by chance or is it the result of co-evolution culminating in a mutualistic relationship to the benefit of both? If colonization by certain bacteria is of benefit to the host, how is such selective adhesion encouraged by the host? Unfortunately, little is known about the mechanisms by which members of the normal microflora adhere to healthy tissues and most of our knowledge of microbial adhesion has been gained by studying the interactions between pathogenic organisms and host cells. Furthermore, although we know something of the means by which pathogens adhere to host cells and the resultant effects on these cells, we know virtually nothing of the effect that host cells have on bacteria as a consequence of these interactions. It must be remembered, of course, that adhesion is often only a prequel to a number of possible outcomes of the bacterium/host interaction. It may be an end in itself, with little effect on the bacterium or the host cell, or it may lead to invasion, killing or the initiation of apoptosis in the host cell. These considerations indicate a number of other facets to bacterial adhesion which are summarized in Figure 5.1. As the remit of this book is limited to cellular microbiology, the adhesion of bacteria to extracellular matrices such as the teeth or biomaterials and which is important in, for example, dentistry and in intensive care medicine will only be briefly mentioned.

The effect of adhesion of bacteria to host cells on the function of both these interacting cells will be discussed in this chapter, although little is known about how adhesion affects bacteria. In many cases, adhesion to a host cell is the first stage in its invasion, so that the eukaryotic cellular response elicited really constitutes part of the invasive process. It would be difficult and somewhat artificial, therefore, to separate discussion of such interactions into two chapters. The topic of this chapter is adhesion and any adhesion process which ultimately results in invasion of the eukaryotic cell, or uptake by it, will be discussed in the next chapter. Bacteria and host cells interact and thereby affect one another's behaviour at a distance by the secretion of a range of products, e.g. toxins, low molecular weight metabolites, hormones, antibacterial peptides, enzymes. Furthermore, a range of cell surface components of bacteria (e.g. lipopolysaccharide (LPS), peptidoglycan, outer membrane proteins) profoundly affect eukaryotic cell behaviour. Such components are likely to exhibit similar activities in the intact bacterium when it adheres to a host cell. However, the conformation of the molecule and synergy with neighbouring molecules in the intact bacterium may be dramatically altered after extraction and so may affect the activity of the component. The interaction of isolated bacterial surface components with eukaryotic cells will not be described in this chapter (but see Chapters 7, 8 and 10), which will only be concerned with changes elicited by adhesion of the whole bacterium to a host cell.

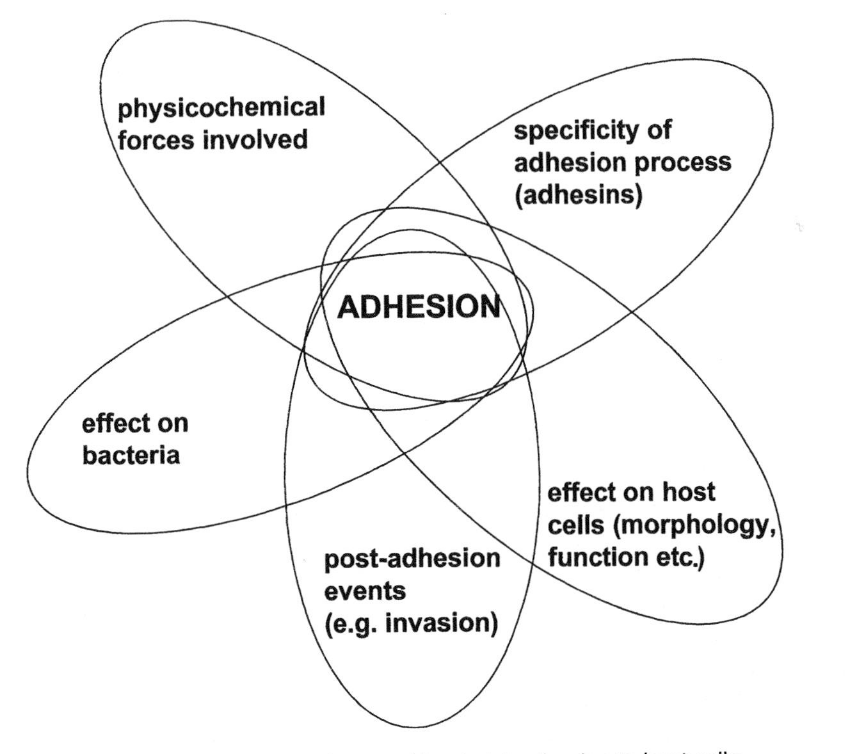

Figure 5.1 The many facets of bacterial adhesion to host cells

BASIC PRINCIPLES OF MICROBIAL ADHESION

Pre-adhesion events

Several stages are involved in the adhesion of a bacterium to a surface. If we imagine a bacterium approaching a surface then at a distance of tens of nanometres the two objects are influenced by two types of forces: van der Waals and electrostatic (Figure 5.2). At a distance of > 50 nm van der Waals interactions occur and as a result of the mutual induction of dipoles in the two objects resulting in their mutual attraction. As the distance between the objects decreases (10–20 nm), electrostatic forces become significant and as most bacteria and surfaces have a negative charge the net effect is repulsion. However, these repulsive forces decrease with increasing ionic strength and in many natural environments the ionic strength is sufficient to reduce or overcome this repulsion. As the bacterium approaches more closely, intervening water molecules will act as a barrier to attachment. However, hydrophobic molecules on the surface of the bacterium, the host cell or both can exclude these water molecules. Hydrophobic interactions between the bacterium and the host cell can then result in adhesion or can enable a close enough approach (<1.0 nm) for other adhesive interactions to occur. The latter include hydrogen bonding, cation bridging and receptor–ligand interactions, i.e. the specific binding of a molecule (ligand) on the bacterial surface to a complementary substrate molecule (receptor) on the host cell surface. Bacterial adhesion to host cells is thought to be mediated primarily by hydrophobic interactions, cation bridging and receptor–ligand binding. While hydrophobic interactions are recognized as being important in the adhesion of bacteria to host cells and to inanimate substrata, more is known about the role of receptor–ligand interactions. The molecules on the bacterial surface responsible for adhesion are known as adhesins.

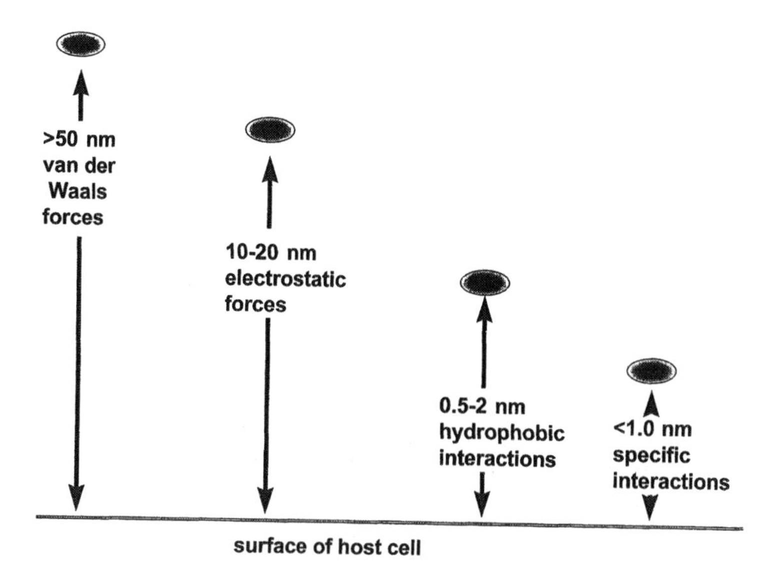

Figure 5.2 Forces affecting adhesion of a bacterium at varying distances from the surface of a host cell

Molecular mechanisms of adhesion

Adhesion between bacteria and host cells is mediated primarily by hydrophobic, ion-bridging and receptor–ligand interactions. These will now been described in more detail. In many cases, however, the nature of the adhesive interaction has not been determined even though it has been established that a particular bacterial adhesin is involved.

Hydrophobic interactions

When non-polar molecules on the bacterial and host cell surfaces approach one another the intervening ordered layers of water are displaced and the consequent increase in entropy results in adhesion being energetically favourable. In contrast to lectin interactions, hydrophobic bonding is often considered to be non-specific as there are no apparent stereospecific interactions between the molecules. Such apparent lack of specificity may simply reflect our lack of knowledge of the identity of the molecules involved and the nature of their physicochemical interactions. The surface components responsible for the interactions are known as hydrophobins and include hydrocarbon groups, aromatic amino acid groups, fatty acids and mycolic acids. It is worth noting that approximately 55% of the accessible surface of an average protein is non-polar. Even in carbohydrates, which are generally considered to be highly polar, hydrophobic regions do exist. For example, galactose sugars contain six adjacent C atoms which form a continuous hydrophobic region and when these sugars interact with proteins these regions are invariably packed against aromatic side chains in the protein.

Cation-bridging

The mutual repulsion between the negatively charged surfaces of bacteria and host cells can be counteracted by divalent metal ions such as calcium ions which thereby act as a bridge between the two. Such interactions are thought to be important in adhesion of oral bacteria to tooth surfaces, in co-aggregation between similar and different bacterial species and, possibly, between bacteria and negatively charged molecules on the surfaces of host cells.

Receptor ligand binding

A molecule (receptor) on the surface of a host cell, whether part of the cell membrane or associated with it, can 'recognize' a molecule (ligand) on the bacterial surface with a complementary structure and form a strong but non-covalent bond. The process resembles the 'lock-and-key' relationship between the catalytic site of an enzyme and its substrate. The process of recognition usually involves only a portion of each of the molecules involved and the molecular structure responsible is known as an epitope. A variety of molecules can function as receptors or ligands including proteins, polysaccharides, glycoproteins and glycolipids. When recognition involves the interaction between a protein and a carbohydrate epitope (which may be on a glycoprotein, glycolipid etc.) the interaction is known as lectin binding. The interaction between a receptor and its complementary ligand is highly specific but it can be inhibited by other molecules with identical epitopes.

Bacterial structures involved in adhesion

Collectively, bacteria elaborate a number of structures which may be involved in adhesion to cell surfaces (Figure 5.3). These include fimbriae (or pili) and proteinaceous fibrils whose primary function appears to be that of adhesion, as well as capsules and flagella which have other functional roles, namely protection and locomotion respectively. All of these structures contain adhesins although the chemical identity of many of them remain to be determined. As well as these structures, the cell walls of many species contain macromolecules which function as adhesins. As will discussed below, a particular species may be able to produce a whole series of adhesives structures (or adhesins) either concurrently or consecutively. In the case of the latter, this may enable the organism to adhere to the different cell types it encounters during the course of the infectious process in which it participates.

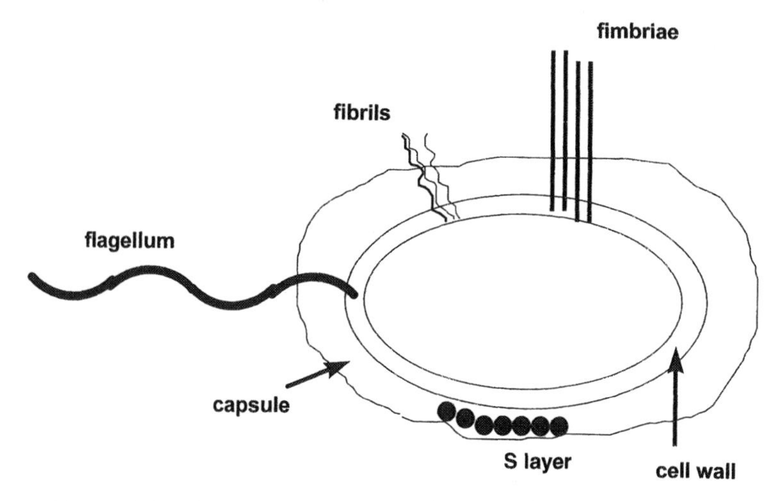

Figure 5.3 Bacterial structures involved in adhesion to host cells

While capsular material is thought to mediate adhesion of a variety of organisms including *Streptococcus* spp., *Staphylococcus* spp., *Klebsiella* spp. and *Bacteroides* spp., there is also evidence demonstrating that capsules can impede adhesion of bacteria such as *N. meningitidis*, group B streptococci and *Haemophilus influenzae* (discussed more fully in Chapters 8 and 9).

S layers consist of a crystalline array of protein, or glycoprotein, self-assembling units external to the cell wall. While this structure has been shown to be important in adhesion of animal pathogens (e.g. *Aeromonas salmonicida, Lactobacillus acidophilus*) to host cells, they do not appear to play a major role in the adhesion of bacteria responsible for diseases of man.

The main function of fimbriae is to enable adhesion of the organism to host cells or to other bacteria, the adhesin being located either at the tip of the fimbria or along the length of the fimbrial shaft. They are widely distributed among Gram-negative genera (*Bordetella* spp., *Salmonella* spp., *Neisseria* spp., *Yersinia* spp., *Pseudomonas* spp., *Porphyromonas* spp.) and have also been detected on streptococci and actinomycetes. They constitute the most frequently used adhesive structure in Gram-negative bacteria. As these structures protrude some distance from the cell surface (beyond capsules and other surface molecules) their ability to interact with host cell receptors may help to overcome electrostatic forces of repulsion between the two cells. A number of attempts have been made to develop a uniform classification system for fimbriae based on their structure and the nature of the adhesin present. None of these have been universally accepted, hence a confusing nomenclature exists. One of the most widely used systems recognizes five major types:

- Type 1: fimbriae are rigid and exhibit mannose-sensitive haemagglutination.
- Type 2: these are similar to type 1 but do not induce haemagglutination.
- Type 3: fimbriae are flexible and mannose-resistant. They are common among the Enterobacteriaceae.
- Type 4: these possess *N*-methylphenylalanine in the amino terminus region of the major subunit.
- Type 5: mannose-sensitive but thinner than Type 1. There are usually few of these per cell.

Examples of the fimbrial types used by various bacteria are given in Table 5.2.

Although flagella are primarily responsible for bacterial motility, in some species (e.g. *Pseudomonas aeruginosa, Helicobacter pylori, Campylobacter jejuni*), they also facilitate adhesion. They consist of a globular protein (flagellin) which aggregates in a helical arrangement to form a long, hollow filament (usually many times the length of the bacterial cell) which terminates in a complex 'basal body' embedded in the cell wall and cytoplasmic membrane.

The cell wall, whose structure was described in Chapter 2, contains a number of molecules which have been shown to have a role in adhesion. Hence LPS, outer membrane proteins, teichoic acids and lipoteichoic acids all function as adhesins. Besides these, hydrophobic molecules in the cell wall (e.g. outer membrane proteins of Gram-negative bacteria) are also thought to be involved in adhesion by interacting with hydrophobic molecules on host cell surfaces or on the surfaces of inanimate substrata. In many cases the receptors on the surface of the host cell which interact with these adhesins have not been identified.

A number of mycoplasmas, which do not have a cell wall, have a specialized adhesive structure for attachment to mucosal surfaces. These organisms display gliding motility and the adhesive structure is formed at the leading edge of the cell. The organelle is an extension of the cell which has a characteristic tapered shape with an electron-dense core that enlarges to form a terminal button, and is formed from cytoskeletal-like proteins. Little is known about the specific functions of the large number of proteins (possibly as many as 25) comprising this organelle. At least two of the proteins, P1 and P30, are known to mediate binding to epithelial cells probably via host sialoglycoconjugates and sulphated glycolipids.

Table 5.2 Examples of fimbriae found on various bacteria

Organism	Fimbrial type	Diameter (nm)	Receptor
Actinomyces naeslundii	2	5	Galactosides
Actinomyces viscosus	1	5	Salivary glycoproteins
E. coli	1	7	Oligomannosides
Klebsiella pneumoniae	3	5	?
Neisseria gonorrhoeae	4	6	?
Ps. aeruginosa	4	5	L-Fucose

Bacterial adhesins

A variety of molecules on the surface of the bacterial cell may function as adhesins and mediate attachment of the organism to a host cell or structure. One of the most extensively studied type of bacterial adhesion mechanism involves the interaction between a bacterial lectin and its corresponding receptor on a substratum. Lectins may be found on the end of pili, in capsules or attached to the bacterial cell wall and have been studied mainly in Gram-negative bacteria. Table 5.3 lists some bacteria in which lectins have been identified together with the identity of the carbohydrate to which the lectin binds. Lectin–carbohydrate interactions have been shown to be involved in the adhesion of bacteria to intestinal epithelial cells, pharyngeal epithelial cells, buccal epithelial cells, erythrocytes, urinary tract epithelial cells, other bacteria and teeth.

Table 5.3 Bacteria which utilize lectins as adhesins

Organism	Receptor
Staph. saprophyticus	*N*-Acetyllactoseamine
Bord. pertussis	Gal-N-acetylglucosamine
Sal. typhimurium	Galβ1–3GalNAc
E. coli	*N*-Acetyl-D-glucosamine, Galβ1,4Glcβ, sialyl–Gal
Fusobacterium nucleatum	Galactose
Ps. aeruginosa	GalNAcβ1,4-Gal, thiogalactosides
N. gonorrhoeae	GalNAcβ1,4Gal
Hel. pylori	NeuNAc2,3Gal
K. pneumoniae	*N*-acetylneuraminic acid and *N*-acetyl-*D*-glucosamine
Haem. influenzae	GalNAcβ1,4Gal
A. naeslundii	Galβ1,3GalNAc
Strep. pneumoniae	GlcNAcβ1,3Gal
V. cholerae	L-Fucose

Lipoteichoic acid (LTA), a cell wall component of Gram-positive bacteria, has been shown to be an important adhesin for a number of species including *Streptococcus pyogenes*, *Staphylococcus aureus*, *Staph. epidermidis* and viridans streptococci. In each of these organisms, the receptor for the LTA has been shown to be fibronectin, a glycoprotein produced by a number of host cells including epithelial cells. Lipoteichoic acids consist of chains of glycerol phosphate, with D-alanine and sugar substituents, attached to a glycolipid (or diglyceride) moiety which is embedded in the cytoplasmic membrane (see Chapter 2). Although the hydrophilic portion of the molecule is usually considered to protrude from the cell surface and contain the adhesive domain, in the case of *Strep. pyogenes* the lipid portion of the molecule is involved in adhesion to host cells and the receptor for this is a fatty acid binding site near the amino terminus of the fibronectin molecule.

Proteins on the bacterial surface may also function as adhesins. For example, *Staph. aureus* produces a 210 kDa surface protein which mediates adhesion to fibronectin. The organism also binds to a number of other host proteins including fibrinogen, mediated by a 59 kDa protein, and laminin, mediated by a 57 kDa protein. *Streptococcus pneumoniae* adheres to glycoproteins of the nasopharyngeal epithelium and this is thought to be mediated by a 37 kDa protein called PsaA. However, antibodies that react with carbohydrates comprising the capsule of the organism can interfere with adhesion. It may be that binding of antibodies to the capsule results in it 'swelling' so that the PsaA no longer protrudes from the capsule, so masking the adhesin. As well as binding to fibronectin via LTA, *Strep. pyogenes* also expresses a number of fibronectin-binding proteins. The major adhesins of *Mycoplasma* spp. are also thought to be proteins and one of their characteristic features is their proline-rich composition. Futhermore, these adhesins have extensive sequence homology to certain mammalian proteins. This molecular mimicry is particularly interesting as it is thought that mycoplasmas provoke an anti-self response, so triggering immune disorders. Patients with *Mycoplasma pneumoniae* respiratory infections demonstrate seroconversion to myosin, keratin and fibrinogen and may develop cardiac abnormalities. Other mycoplasmal adhesins exhibit amino acid sequence homologies with human CD4 and class II major histocompatibility complex lymphocyte proteins, which could generate autoreactive antibodies and trigger cell killing and immunosuppression.

The hydrophobicity of the bacterial cell surface is thought to be an important determinant of adhesion to surfaces, although the identity of the molecules responsible for the hydrophobic character of the surfaces (hydrophobins) have, generally, not been identified. The involvement of hydrophobic interactions in adhesion has been studied most extensively in oral bacteria, in particular the adhesion of streptococci to tooth surfaces. In general, the more hydrophobic an organism is the greater is its ability to adhere to saliva-coated hydroxyapatite. Adhesion of the oral organism, *Porphyromonas gingivalis*, to collagen is also thought to be mediated by hydrophobic interactions. There is also evidence suggesting that hydrophobic interactions are involved in the adhesion of staphylococci to solid surfaces, such as those used in prosthetic devices.

Carbohydrates are known to be common components of the bacterial cell surface and may function as adhesins in certain bacteria. Alginate, an exopolysaccharide of *Ps. aeruginosa*, appears to play a role in adhesion of the organism to tracheal cells and mucins as it binds to both buccal and tracheal cells and to bronchotracheal mucin. Furthermore, antibodies to the alginate inhibit binding of the organism to tracheal cells. *Streptococcus oralis*, a major constituent of dental plaque, adheres to the tooth surface by means of a polysaccharide composed of hexasaccharide repeating units containing glycerol linked through a phosphodiester group to C6 of an α-galactopyranosyl residue and joined end-to-end through galactofuranosyl-β(1→3)-rhamnopyranosyl linkages. Many infections due to *Staph. epidermidis* are associated with prosthetic devices and adhesion to such materials has been shown to be mediated by surface polysaccharides. Biofilm formation on such devices has also been shown to involve the production of an intercellular polysaccharide adhesin.

Several studies have implicated LPS as being important in the adhesion of a variety of Gram-negative bacteria. The LPS of *Campylobacter jejuni*, for example, is thought to mediate attachment of the organism to epithelial cells. Strains which are highly negatively charged and only weakly hydrophobic bind in greater number to a human intestinal cell line than strains that are more hydrophobic. A whole cell lysate of the organism partially inhibited binding to Hep-2 cells and this inhibition was unaffected by heating the lysate to 100°C for 30 minutes or by treating it with proteases. However, treatment of the lysate with periodate (which oxidizes sugars) abolished the inhibitory activity. Adhesion of the organism to intestinal epithelial cells was partially inhibited by fucose and mannose and completely inhibited by LPS from the organism. The binding of *C. jejuni* LPS to the cells was inhibited by fucose or by treating the LPS with periodate. Other organisms in which LPS is thought to function as an adhesin include *Hel. pylori*, *Ps. aeruginosa*, *Salmonella typhi*, *Shigella flexneri* and *E. coli*.

There are an increasing number of reports of enzymes functioning as bacterial adhesins. *Strep. pyogenes* has the glycolytic enzyme, glyceraldehyde 3-phosphate dehydrogenase, on its surface which is able to bind to a number of proteins. The enzyme is though to exist as a 156 kDa tetramer on the surface of the organism and mediates binding to fibronectin, lysozyme, myosin and actin. *Por. gingivalis*, an organism associated with periodontitis, secretes two proteinases, gingipain R (arginine-specific gingipain) and gingipain K (lysine-specific gingipain), each of which has adhesin domains enabling binding to fibrinogen, fibronectin, and laminin. Interestingly, the proteases were able to degrade all three of the bound proteins suggesting that the proteinase-adhesin complexes mediate attachment to target proteins, then degrade them so enabling detachment and then possibly reattachment at a different site. It has also been suggested that the cell surface urease of *Hel. pylori* functions as an adhesin enabling the bacterium to colonize the gastric mucosa. However, recent work has cast doubt on this idea as there appeared to be no difference between the adherence to gastric cells of an isogenic urease-negative mutant and the wild-type organism.

Glucosyltransferase enzymes on the surface of mutans streptococci are thought to be important in the adhesion of these organisms to tooth surfaces. Recently it has also been demonstrated that the enzyme functions in *Streptococcus gordonii* as an adhesin mediating attachment to human endothelial cells and so may enable colonization of the endocardium in infective endocarditis.

An additional protein which has recently been identified on the surface of a number of organisms including *Hel. pylori* and *Haemophilus ducreyi*, the causative organism of the genital ulcer disease chancroid, is chaperonin 60 and it has been established that this oligomeric protein is involved in the binding of these bacteria to epithelial cells. Chaperonin 60 is a heat shock or stress protein whose intracellular levels are increased during periods of stress such as might occur during bacterial colonization and aid in bacterial adhesion.

The above description of adhesins may imply that a particular bacterium displays only one adhesin, or a limited number of such molecules. This is far from being the case. Invasive organisms in particular may need to display a variety of adhesins to enable binding to different cells/tissues/structures during the course of an infectious process. Indeed, some species display a bewildering array of adhesins as exemplified by *E. coli* (Table 5.4). The exact

Table 5.4 Adhesins of enteropathogenic strains of *E. coli*

Adhesin	Location	Molecular mass (kDa)	Genetic location
Type 1	Fimbrial	16	Chromosome
20K	Fimbrial	20	Chromosome
CFA/I	Fimbrial	18	Plasmid
CFA/II-CS1	Fimbrial	15	Plasmid
CFA/II-CS2	Fimbrial	16	Chromosome
CFA/II-CS3	Fimbrial	?	Plasmid
CFA/III	Fimbrial	16	Plasmid
CFA/IV-CS4	Fimbrial	17	Plasmid
CFA/IV-CS5	Fimbrial	21	Plasmid
CFA/IV-CS6	Fimbrial	16	Plasmid
PCF 09	Fimbrial	27	?
2230 antigen	Fimbrial	16	Plasmid
8786 antigen	Afimbrial	16.3	Plasmid
CS7	Fimbrial	22	Plasmid
F1845	Fimbrial	?	Chromosome

function of each of these adhesins remains to be elucidated. Recent studies have revealed that enzymes and heat shock proteins can also function as adhesins.

Nature of the host cell surface

While a considerable amount of information has been gathered about the nature of the surfaces of host cells we must consider ourselves to be 'ignorant' in comparison to the bacteria colonizing such surfaces. We can expect that over many hundreds of million of years of co-evolution bacteria have explored every nook and cranny of these regions in their search for suitable adhesion sites. Broadly speaking, bacteria can adhere to these surfaces in three ways: (1) directly to the lipid bilayer; (2) directly to cell surface receptors whose normal function is to bind host molecules; and (3) indirectly to host molecules already bound to the host cell surface (Figure 5.4).

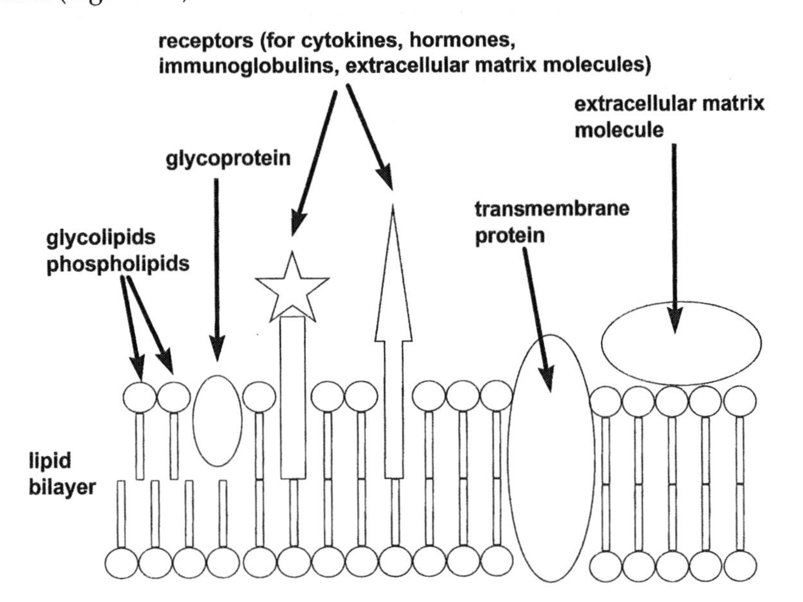

Figure 5.4 Sites on the host cell surface which can function as receptors for bacterial adhesins

Like all biological membranes, the cytoplasmic membrane of host cells has the classic lipid bilayer structure, throughout which proteins are embedded. The major lipids present are phosphatidylcholine, phosphatidylserine, phosphatidylethanolamine, phosphatidylinositol, sphingomyelin, cholesterol and various glycolipids. Several of these contain molecular structures which are recognized by bacterial adhesins. Although outnumbered by lipid molecules, proteins are responsible for most membrane functions including transport of molecules, recognition and binding of hormones, cytokines and extracellular matrix molecules and signal transduction and cell–cell interactions (such as occur in all multicellular tissues). Certain amino acid sequences of proteins and the carbohydrates of glycoproteins also function as important receptors for bacterial adhesins.

Mammalian cells display a wide variety of surface receptors for a range of molecules including hormones, immunoglobulins, cytokines and extracellular matrix molecules (Table 5.5). Bacteria

Table 5.5 Receptors on the surface of mammalian cells

Receptor	Function
Integrins	Interact with extracellular matrix molecules Signal transduction Cytoskeletal arrangement
Cadherins	Adhesion of eukaryotic cells
Selectins	Adhesion of leukocytes to endothelial cells
Intercellular adhesion molecules	Adhesion of eukaryotic cells
Cytokine receptors	Cytokine recognition and signal transduction
Serpentine receptors	Hormone recognition and signal transduction

can adhere to many of these molecules and, as this often induces uptake of the organism by the cell, the phenomenon will be described in more detail in the next chapter.

Host tissues consist of cellular elements embedded in an extracellular matrix (ECM), which comprises a complex mixture of polymers including fibronectin, fibrinogen, collagens, proteoglycans and glycosaminoglycans. The ECM not only has a structural function but, as its components are linked to cell surface receptors (e.g. integrins) it also affects a number of cellular activities including migration, proliferation and differentiation (described in more detail in Chapter 3). Bacteria can adhere to the ECM and hence could, theoretically, also affect the activities of host cells (Table 5.6). However, there appear to be no reports yet of

Table 5.6 Adhesion of bacteria to extracellular matrix molecules (ECM)

ECM	Organism	Adhesin
Fibronectin	*Staphylococcus* spp.	110 kDa protein
	Strep. pyogenes	120 kDA protein Glyceraldehyde 3-phosphate dehydrogenase
	Mycobacterium spp.	32 kDA
	E. coli	
	N. meningitidis	An outer membrane protein (Opc)
Collagen	*Staph. aureus*	133 kDA protein
	Strep. mutans	16 kDa protein
	K. pneumoniae	
Laminin	*Strep. gordonii*	142 kDa protein
	E. coli	Carbohydrate
	Hel. pylori	25 kDa protein
	Staph. saprophyticus	160 kDa haemagglutinin
Vitronectin	*Staph. aureus*	
	N. meningitidis	An outer membrane protein (OpC)
Elastin	*Staph. aureus*	40 kDa protein
Heparan sulphate	*Hel. pylori*	65 kDa protein
Fibrinogen	*Staph. aureus*	92 kDa protein
	Strep. pyogenes	M protein
	Por. gingivalis	150 kDa protein

bacterially induced changes in host cells arising as a consequence of their binding to components of the ECM although, as will be discussed in the next chapter, invasion of host cells by bacteria is often mediated by attachment to host cell receptors (e.g. β_1-integrins) for ECM molecules.

EFFECTS OF ADHESION ON BACTERIA

Although the effect of bacterial adhesion on host cell function has received considerable attention, much less is known about how this interaction affects bacteria and this is summarized in Figure 5.5. Before considering this further it is opportune to pose the question: why should there be an effect on bacteria? There are a number of reasons why we might expect a bacterium to adjust its phenotype on adhering to some substratum on, or within, its host. First of all, prior to contacting its host the bacterium is likely to have resided in a number of environments which may not have been conducive to its growth. These may have included air, water, soil, vegetation, clothing etc. Having reached an epithelial surface (although this may not necessarily be its ultimate destination) the organism will now have to adapt to living, no matter how temporarily, in this new environment and this will involve the upregulation and suppression of a number of gene products. These genomic responses to the environment are discussed in Chapter 4. Of course, this may not necessarily be mediated by the adhesion process itself as bacteria are known to have a number of sensors which would be capable of ensuring adequate adaptation of the organism to its new environment (as described in Chapter 3). Nevertheless, as will be described below, there is evidence to suggest that the adhesion process is used by at least one organism to trigger the expression of gene products which will help it survive in its new environment. Secondly, attachment to a host cell may be only a preliminary to a whole sequence of events involving invasion of the cell, further dissemination of the organism etc. Attachment of the bacterium might, therefore, be expected to be used as a signal for the organism to commence the synthesis of structures

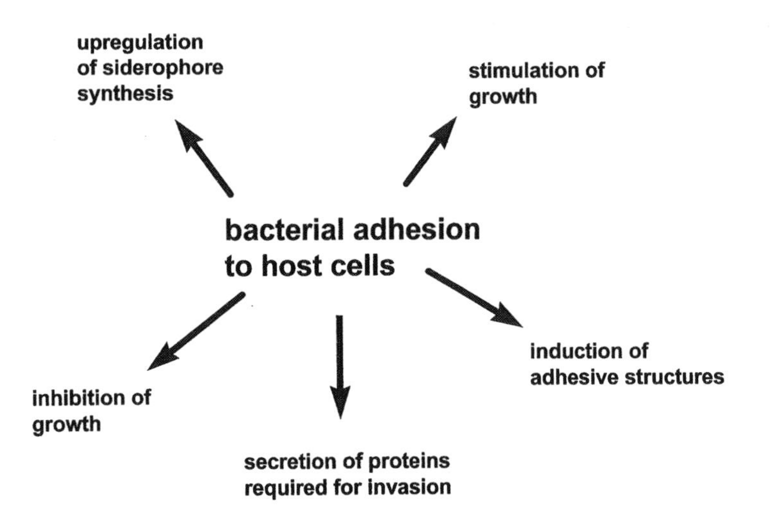

Figure 5.5 Effect on bacterial structure/function of bacterial adhesion to host cells

or signalling molecules necessary to elicit invasion of the host cell. Again, there is evidence to suggest that some bacteria do indeed behave in this manner. However, the number of reports of phenotypic changes in bacteria induced by their adhesion to host cells is very small.

Growth

Adhesion of a uropathogenic strain of *E. coli* to human uroepithelial cells (HUCs) has been shown to inhibit further growth of the organism. When the HUCs were separated from the bacteria by a semi-permeable membrane, however, bacterial growth was unaffected, suggesting that the effect was indeed adhesion-dependent. The bacteriostatic effect did not appear to be dependent on the presence of a specific adhesin–receptor interaction as strains of the organism with pili recognizing different receptors, as well as strains without pili, all elicited a bacteriostatic effect following adhesion to the HUCs. Although the identity of the antibacterial factor(s) responsible has not been determined, there is evidence to suggest that calcium and cAMP are involved in the process. Inhibitors of calcium flux abolished the contact-mediated growth inhibition (see Chapter 3 for a description of calcium-mediated signalling). HUCs obtained from patients with recurrent urinary tract infections were unable to produce an adhesion-mediated antibacterial effect, but their ability to do so was restored by agents which increased intracellular levels of cAMP. As epithelial cells are known to be able to secrete a number of antibacterial peptides (magainins, cecropins etc.) the observed antibacterial effect may have been the result of a contact-induced signalling leading to the upregulation of the synthesis and secretion of such compounds.

In contrast to the above, adhesion to host cells may enhance bacterial growth in some situations. *Neisseria gonorrhoeae*, when attached to HeLa cells (an immortalized epithelial cell line), grows at a rate three times greater than that of the unattached organism. Similarly, *E. coli*, when adherent to epithelial cells derived from the intestine or peritoneal cavity, grows at a faster rate than non-adherent organisms. It has been suggested that such growth stimulation may be due to the accumulation of waste products of the host cell (which can serve as bacterial nutrients) within microscopic invaginations of the host cell membrane. The attached bacteria form a 'lid' preventing the escape of these valuable nutrients and use them for increased growth. Alternatively, it is thought that, due to physicochemical forces, a microzone containing higher concentrations of organic molecules and inorganic ions (i.e. potential bacterial nutrients) exists around all objects submerged in an aqueous environment. This would apply also to epithelial cells lining fluid-filled cavities or mucus-covered surfaces. Attached bacteria could take advantage of this nutrient-rich microzone and so grow at a faster rate.

Structures involved in adhesion or invasion of host cells

It has been shown that adhesion of *Salmonella typhimurium* to an epithelial cell line results in the production of a number of new proteins which are important for continued adherence of the organism. Scanning electron microscopy has also shown that within 15 minutes of contacting epithelial cells *in vitro*, this organism produces surface appendages (termed invasomes) that are distinctly different from known bacterial structures such as flagella or pili, being almost three times thicker than the former and only one tenth of their length. The invasomes are thought to be responsible for mediating internalization of the organism by the epithelial cells and disappear once internalization has commenced. Entry of *Salmonella typhimurium* into non-phagocytic cells is governed by genes of the *inv* locus. Strains of the organism with mutations in the *invG* or *invC* genes are unable to synthesize invasomes and cannot invade the epithelial cells, while mutants with defective *invA* or *invE* genes synthesize the appendages but are unable to retract them – these also are non-invasive. The nature of the signal from the epithelial cells responsible for inducing the formation of invasomes remains to be determined. One of the products of the *Inv* locus, InvJ, is stimulated by contact of the organism with intestinal epithelial cells. The invasion-defective *invG* or *invC* mutants of the organism are unable to secrete InvJ on contact with the cells. Whether InvJ is a component of the invasome, or acts as a signal for its assembly, has not been determined.

Pathogenic bacteria which attach to (infect) mucosal surfaces often form discrete microcolonies and this is particularly evident in the case of enteropathogenic strains of *E. coli*. This phenomenon, known as localized adherence, has been investigated *in vitro* and has been shown to be associated with the induction by the epithelial cells of bundles of filaments (bundle-forming pili – BFP) on the bacteria. The BFP form a network entrapping neighbouring organisms resulting in the formation of a distinctive microcolony. Analysis of the BFP reveal that they consist of 19.5 kDa polypeptide subunits which demonstrate homology with the toxin-co-regulated pili of *Vibrio cholerae* but not with any previously described pilus protein of *E. coli*. Formation of microcolonies could aid the survival of the organism as such structures are known to be less susceptible to host defence mechanisms as well as to antibiotics.

Shigella spp. are the causative agents of bacillary dysentery, an infection characterized by invasion of the superficial intestinal mucosa and diarrhoea. Some of the major virulence determinants of these organisms (those required for invasion and spreading) are encoded on a large plasmid. In the case of *Sh. flexneri*, the plasmid carries the *ipa* operon, which contains genes for three proteins, IpaB, IpaC and IpaD, involved in invasion of epithelial cells. The functions of these genes will be described in more detail in the following chapter but, suffice to say at present, cellular invasion requires the release of these proteins from the bacteria. Recently it has been shown that the trigger for protein release is contact with either epithelial cells (within 5 minutes) or with constituents of the extracellular matrix: collagen, fibronectin or laminin. Although the mechanism involved remains to be determined, it is due neither to enhanced transcription of the *ipa* genes nor to an increase in the quantity of the proteins synthesized. This implies that the proteins are released from a store of pre-synthesized proteins either on the cell wall or within the cytoplasm.

A further example of adhesion-induced changes in a bacterium is provided by the secretion of YopE (a cytotoxin which depolymersizes actin microfilaments) by *Yersinia pseudotuberculosis*. The *yopE* promoter was fused to a *luxAB* operon, the products of which catalyse the emission of photons at a wavelength of 490 nm in the presence of n-decanal. Activation of the *yopE* gene therefore results in light emission. In this visually exciting experiment, *Y. pseudotuberculosis* containing this reporter system was added to HeLa cells and the bacteria were seen to light up on attachment to the epithelial cells. Unattached bacteria failed to emit light. This experiment elegantly demonstrates the attachment-induced production of a key bacterial virulence factor. The role of YopE, and other Yop proteins, in bacterial virulence is discussed in further detail in the following chapter.

C. jejuni provides another example of an organism in which the synthesis of proteins required for invasion of epithelial cells is stimulated by adhesion. In the case of this organism it has been shown that as many as 14 proteins are synthesized within 60 minutes of the bacteria adhering to an epithelial cell monolayer.

Siderophore production

One of the means by which the host restricts bacterial growth *in vivo* is to limit the availability of iron by sequestering it with proteins such as lactoferrin. In order to survive *in vivo*, therefore, bacteria must be able to wrestle iron from host proteins and *E. coli* does this by producing the iron-binding siderophores aerobactin and enterobactin, which are effective at removing iron from host proteins. Recently it has been shown that the production of these siderophores is under the control of a protein encoded by the *bar*A gene and that transcription of this gene is activated by attachment of the PapG adhesin (located at the end of a P-pilus) to its receptor (a Gal∝(1–4) Gal-containing globoside). This clearly illustrates the importance of the adhesion process not only in enabling a bacterium to attach to, and maintain itself within, the host but also in controlling expression of factors essential for its survival in the local environment in which it finds itself.

EFFECTS OF ADHESION ON HOST CELLS

Studies of the effects of adhesion of bacteria on host cells are still at a very early stage but have revealed that dramatic changes can ensue. These can range from alterations in cell morphology to long-term transcriptional changes to programmed cell death (Figure 5.6). As the outcome of the interaction very much depends on the type of host cell with which the bacterium interacts, this section will be divided along these lines.

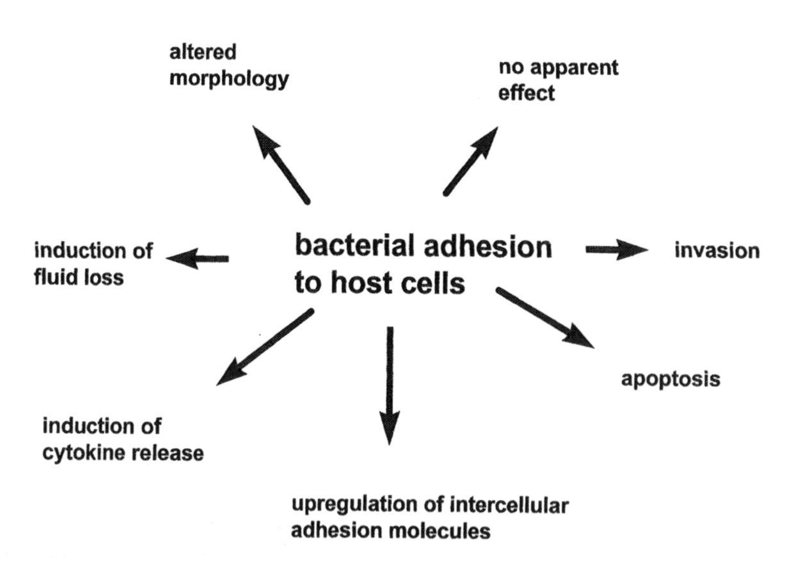

Figure 5.6 Effects on host cell structure/function of bacterial adhesion

Effect on epithelial cells

Apart from the alveoli, bladder, stomach and the urethra, all of the epithelial surfaces of a healthy individual are normally colonized by bacteria. Given that all of these organisms contain molecules which are potentially harmful to mammalian cells (e.g. LPS, peptidoglycan, lipoteichoic acid) it is remarkable that most adherent organisms have, in general, such minimal effects on epithelial cells. Table 5.7 summarizes the reported effects that bacterial adhesion has on epithelial cells and these will be described in more detail below.

Table 5.7 Effects of bacterial adhesion on epithelial cells

Effect on epithelial cell	Example
No apparent effect	Most members of the normal microflora, *Bord. pertussis, V. cholerae*
Altered morphology	Enteropathogenic *E. coli, Tr. denticola, Strep. pyogenes, Hel. pylori*
Induction of cytokine release	Oral streptococci, uropathogenic *E. coli, Hel. pylori*
Expression of intercellular adhesion molecules	Uropathogenic *E. coli*
Invasion	Many examples including: *Salmonella* spp., *Shigella* spp., *Neisseria* spp. *Haemophilus* spp., *Yersinia* spp.

Colonization without any apparent biological effect

Members of the normal microflora

The vast majority of interactions between bacteria and epithelial cells involve members of the normal microflora. Examples include colonization of the mucosal surfaces of the oral cavity by streptococci, of the intestinal epithelium by *Bacteroides* spp., enterococci and *Eubacterium* spp. and of the vagina by lactobacilli. It is generally assumed that adhesion of such organisms to epithelial cells does not induce any dramatic changes in the behaviour of either member of the association. Hence, although adhesion involves mutual recognition (in terms of interaction between adhesins and corresponding receptors) each of the partners then apparently continues its existence in a state of mutual indifference. However, as most studies of adhesion have involved pathogenic bacteria rather than members of the normal microflora, it would be unwise to come to any definitive conclusions about this until further work has been done. In view of the increasing number of reports concerning the production of antibacterial peptides by epithelial cells it is reasonable to speculate that members of the normal microflora colonizing these surfaces may downregulate the production of these antibiotics. Furthermore, members of the normal microflora contain a range of molecules (e.g. LPS, LTA) which are potent inducers of proinflammatory cytokine release from epithelial cells. Why then do these bacteria not induce a state of chronic inflammation? Do these bacteria also contain molecules able to downregulate the release of such cytokines or neutralize the activity of such molecules? Is the synthesis of such molecules controlled by host cells? Clearly, we need to know more about the interactions between host cells and members of the normal microflora (see Chapter 10).

Pathogenic species

A number of bacteria capable of inducing pathology in humans also adhere to epithelial cells without inducing any apparent change in these cells. It must be remembered, however, that although the adhesion process itself may not affect the structure or function of the host cell, the organism itself may produce toxins, enzymes etc. which may ultimately damage these cells.

Bordetella pertussis. Bord. pertussis is the causative agent of whooping cough, an endemic and epidemic respiratory infection of children and adults. Although the bacteria are non-invasive and remain in the respiratory tract, they release toxins (see Chapter 7) responsible for a number of characteristic systemic manifestations of the disease: leukocytosis, abnormal glucose metabolism, weight loss. The organism colonizes the respiratory tract mucosa by adhering to ciliated epithelial cells, which it ultimately kills by releasing tracheal cytotoxin, a peptidoglycan fragment.

A number of surface molecules have been proposed to function as adhesins for epithelial cells including fimbrial proteins, pertactin, pertussis toxin and filamentous haemagglutinin. Although fimbriae are involved in the adhesion of most bacteria to host cells, *Bord. pertussis* appears to differ from this trend in that of the four putative adhesins mentioned above, most evidence favours an involvement of the latter two rather than the fimbriae. Pertussis toxin is an AB_5 toxin (see Chapter 7) consisting of a 26 kDa subunit (S1) responsible for ADP-ribosylation, while the remaining subunits interact with glycoproteins and glycolipids on host cells, thereby enabling adhesion. The S2 subunit binds to lactosamine-containing glycoconjugates and is involved in adhesion to cilia while the S3 subunit binds to 2–6-linked sialic acid and can thereby effect adhesion to macrophages. The filamentous haemagglutinin (FHA) is a 220 kDa protein which contains a number of epitopes able to recognize host cell receptors, one of which is a lectin involved in adhesion to ciliated cells. Both of these adhesins, as well as whole bacteria, can adhere to lactosylceramide and non-sialylated glycolipids isolated from ciliated cells. The structure of the minimal receptor unit for *Bord. pertusssis* on ciliated epithelial cells has been shown to be galactose *N*-acetylglucosamine. The carbohydrate recognition domain of the FHA has been mapped to amino acid residues 1141–1279, while that of the S2 subunit of the PT has been identified as residues 40–54.

Adhesion of *Bord. pertussis* per se does not appear to affect the behaviour of ciliated cells. However, the ultimate outcome of the interaction is cell death due to the activities of pertussis toxin, adenylate cyclase and tracheal cytotoxin.

Vibrio cholerae. V. cholerae is responsible for cholera, one of the most serious gastrointestinal infections of mankind. The disease is contracted by drinking water infected with the organism which then colonizes the intestinal tract, where it secretes an exotoxin which induces massive water loss and electrolyte imbalance resulting, in many cases, in death.

As for many pathogens, a number of structures/molecules have been postulated to function as adhesins for *V. cholerae* including flagella, pili, a haemagglutinin, LPS and various outer membrane proteins. Of these putative adhesins, most attention has been focused on what are termed toxin co-regulated (Tcp) pili. These belong to the type IV family of pilins and are long filamentous structures which form bundles at one location on the cell surface of the organism; their name derives from the fact that the genes encoding them are regulated in a similar fashion to genes encoding the cholera toxin. Mutants which lack Tcp pili are avirulent in human volunteers and mutations in the structural subunit (TcpA) of the pili exhibited decreased colonization in an animal model. As yet, the receptor for Tcp has not been identified. Adhesion of the bacteria to epithelial cells does not appear to have any effect on host cell structure or function.

Enterotoxigenic E. coli. Strains of this organism are responsible for a gastroenteritis with symptoms similar to those of cholera, often referred to as 'traveller's diarrhoea'. As in the case of cholera, the bacteria do not cause any morphological alterations in epithelial cells, nor is there much inflammation in the tissues. Adhesion of the bacteria to intestinal epithelial cells appears to be mediated mainly by fimbriae and diarrhoea is a consequence of the effects of either of two toxins: a toxin known as the heat-labile toxin that is homologous to cholera toxin and a heat-stable toxin (see Chapter 7).

Induction of morphological alterations

Enteropathogenic Escherichia coli

Enteropathogenic *E. coli* (EPEC) strains are the major cause of diarrhoea in children worldwide and infection with these organisms has a high mortality rate in developing countries. It has been estimated that more than one million children die each year due to infection with EPEC. The organism does not produce any enterotoxins or cytotoxins and is not usually invasive but induces a characteristic attaching and effacing (A/E) lesion in epithelial cells in which microvilli are lost and the underlying cell membrane is raised to form a pedestal which can extend outwards for up to 10 μm (Figure 5.7). The bacterial genes involved in the induction of the A/E lesion comprise a 35.5 kb segment of the chromosome known as the locus of enterocyte effacement (LEE). The G+C content of the LEE (38.4%) is very different from that of the *E. coli* chromosome (50–51%), implying that it has been obtained from another source and is a pathogenicity island (described in detail in Chapter 2).

Three stages can be recognized in the interaction between EPEC and epithelial cells: initial adherence, signal transduction and intimate attachment (Figure 5.8). EPEC do not adhere as a monolayer covering the entire surface of an epithelial cell layer but form microcolonies in a patchy arrangement, i.e. adherence is localized. Localized adherence is attributable to the production of bundle-forming pili (BFP) comprised of type IV fimbriae, as mutants unable to produce BFP do not exhibit this behaviour. The production of BFP is induced by contact with the epithelial cells and requires the expression of 14 genes. These are located on a large (approximately 90 kbp) plasmid (known as the EPEC adherence factor plasmid) found in all EPEC strains. Furthermore, the stability of the major subunit of the fimbriae, bundlin – a 19.5 kDa polypeptide – is dependent on the formation of a disulphide bond. The enzyme responsible for this, encoded by the *dsbA* gene, is located in the periplasm and transposon insertion in the *dsbA* locus reduces the level of bundlin and abolishes the ability of the organism to adhere to epithelial cells.

In common with many other intestinal pathogens, EPEC has a type III protein secretion system (described in Chapter 2) which enables transport of proteins directly from the cytoplasm to the cell surface, hence avoiding any periplasmic processing. As secretion is triggered when the bacterium contacts host cells, this secretory pathway is also known as contact-dependent secretion and is used to deliver bacterial proteins directly to their target cell. In the case of EPEC, contact with epithelial cells results in the secretion of a number of proteins via the type III secretion system. Two of these proteins – the 25 kDa EspA and the 37

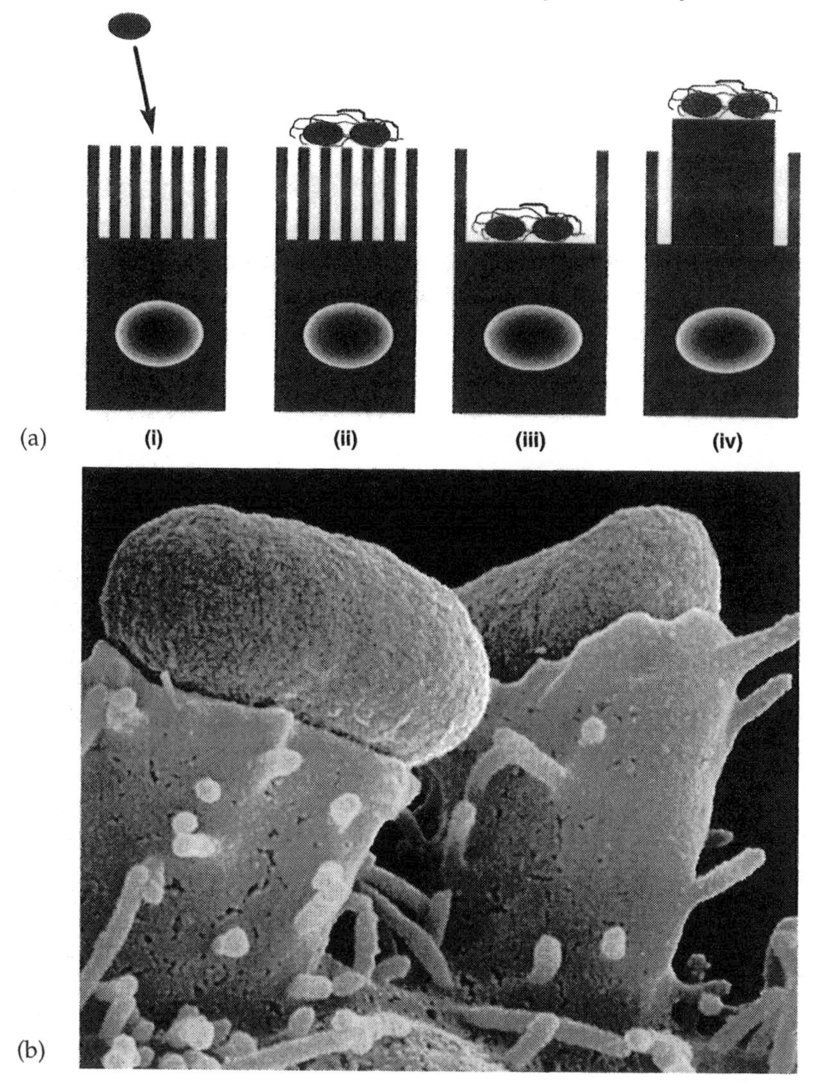

Figure 5.7 (a) Formation of A/E lesions by EPEC. (i) Approach of EPEC towards intestinal epithelial cell; (ii) attachment of EPEC and formation of bundle-forming pili; (iii) loss of microvilli; (iv) production of pedestal. (b) A/E lesion produced by EPEC adhering to enterocytes. Reproduced from Rosenshine *et al.*, EMBO Journal 1996; 15(11): 2613–2624, by permission of Oxford University Press

kDa EspB – trigger a number of responses in the host cell including activation of signal transduction pathways, cell depolarization and the binding of a 94 kDa outer membrane protein, intimin, which is involved in intimate attachment of EPEC, to its receptor. Intimin is encoded by *eae* on the LEE and is an outer membrane protein related to the invasin proteins of *Yersinia* species and is essential for the binding of EPEC. Until very recently, the receptor for intimin was thought to be the tyrosine-phosphorylated form of a surface-exposed host cell membrane protein designated Hp90. However, evidence now suggests that it is actually a 78 kDa bacterial protein (Tir – translocated intimin receptor) secreted by the type III secretion apparatus. So here we have an example of a bacterium carrying around its own

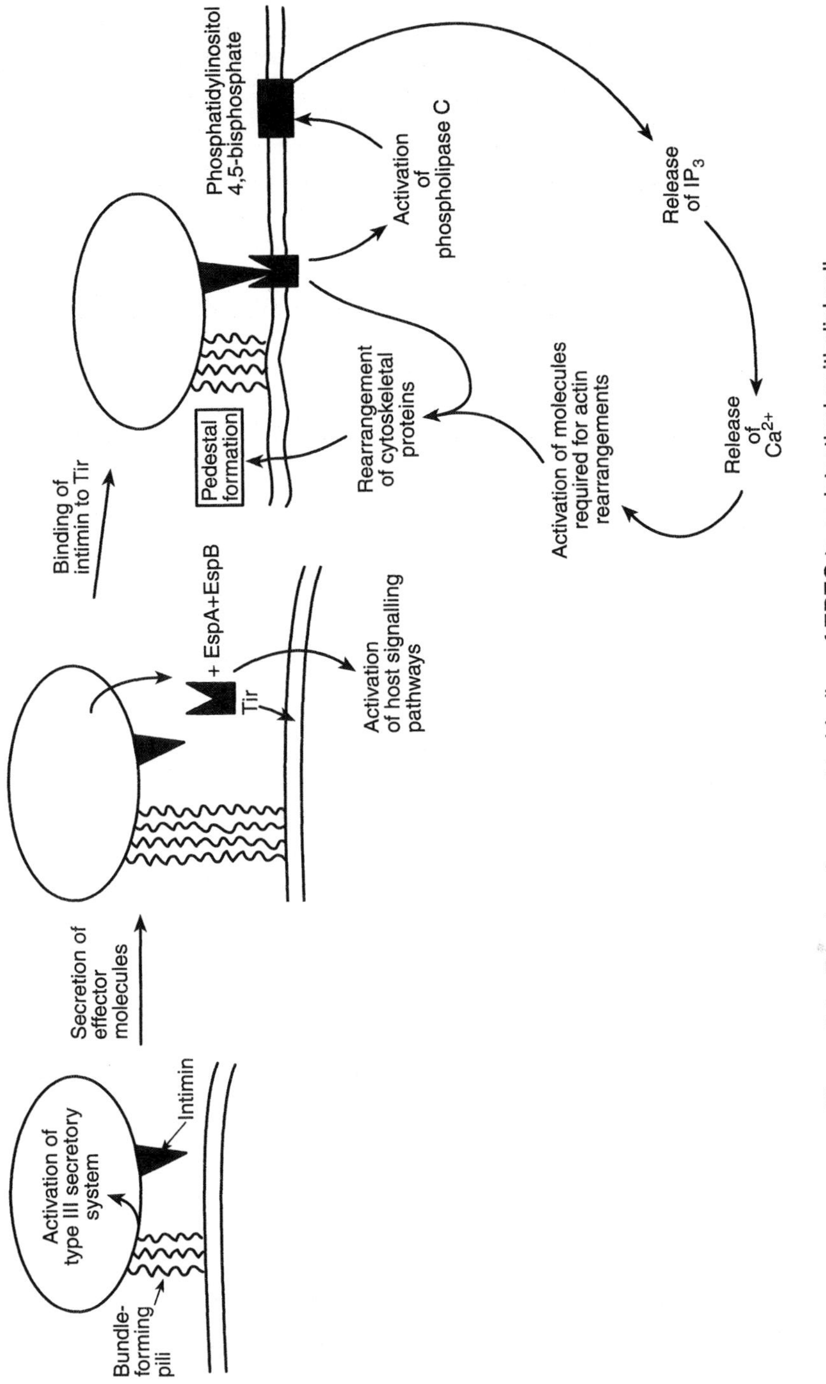

Figure 5.8 Events subsequent to binding of EPEC to an intestinal epithelial cell

host cell receptor! Tir becomes embedded in the membrane of the host cell and then acts as a receptor for intimin. Once intimate attachment has taken place, Tir (probably in conjunction with intimin) directs the re-polymerization of actin monomers and the rearrangement of other cytoskeletal components into the configuration necessary for pedestal formation. Binding of intimin to Tir also results in the phosphorylation of Tir and triggers a number of host cell signalling pathways and the tyrosine phosphorylation of a number of host cell proteins, including phospholipase C. The consequences for the epithelial cell are dramatic. Actin, α-actinin, myosin light chain and molecules that link actin filaments to the membrane (talin and ezrin) accumulate beneath the sites of attachment and this is accompanied by loss of the microvilli. The mechanisms by which these responses are induced have not yet been elucidated. However, it has been shown that the level of intracellular free Ca^{2+} ions is increased and this can activate villin which, is an actin-severing protein. This could account for the breakdown of actinin in the microvilli, leading to their eventual loss. As would be predicted, the release of intracellular Ca^{2+} is mediated by inositol trisphosphate (IP_3), which is generated by the action of phospholipase C. One of several host proteins phosphorylated as a result of EPEC attachment is myosin light chain (MLC). This phosphorylation of MLC induces its dissociation from polymerized actin, which destabilizes actin-containing structures in the microvilli.

As well as formation of an A/E lesion, the other characteristic feature of EPEC infections is, of course, diarrhoea. In the majority of gastrointestinal infections, diarrhoea is a consequence of active ion secretion from host cells and there is evidence that this is also the case for EPEC infections. Hence, infection of epithelial cells with EPEC results in a decrease in resting membrane potential suggestive of either an influx of positive ions or an efflux of negative ions. Other experiments have suggested that efflux of chloride ions, possibly due to activation of protein kinase C, is the more likely explanation. Loss of microvilli would, of course, lead to decreased absorption and this would also contribute to diarrhoea.

The interaction between EPEC and epithelial cells is an excellent example of molecular cross-talk between a bacterium and a host cell. Here we have the epithelial cell signalling the bacterium (by a mechanism to be elucidated) to produce BFP which not only helps it to adhere to host cells but also results in the formation of microcolonies which offer some protection against host defence mechanisms. The bacteria then secrete proteins which perturb host signalling pathways and lead to the cytoskeletal rearrangements which result in loss of microvilli and the creation of a pedestal. Uniquely, one of these secreted proteins, Tir, is taken up by the host cell where it functions as the receptor for the bacterial adhesin, intimin. The interaction also illustrates the fascinating way in which a supposedly 'simple' organism like a bacterium can manipulate host cells for its own benefit. The result of the outcome is that bacteria have established themselves in a suitable habitat for growth and multiplication while the diarrhoea will serve to transmit some of them to new hosts. The function of the pedestal remains obscure.

Other organisms which induce similar A/E lesions include *Hafnia alvei* (another gastrointestinal pathogen of man), *Citrobacter rodentium* (a gastrointestinal pathogen of mice) and enterohaemorrhagic *E. coli*.

Treponema denticola

This organism is an oral spirochaete found in the periodontal pocket (the gap between the gum and tooth) of patients with periodontitis – one of the most prevalent of the chronic inflammatory diseases of man – and is thought to be involved in the pathogenesis of the disease. As ulceration of the gingival epithelium (hence allowing access to underlying connective tissues) is a characteristic feature of periodontitis, considerable interest has been shown in the interactions between *Tr. denticola* and gingival fibroblasts as well as with epithelial cells. While adhesion of the organism to epithelial cells is a well-established phenomenon, the identity of the adhesins responsible has not been definitively established. Treatment of *Tr. denticola* with proteinase K (but not trypsin), heat, glutaraldehyde, formaldehyde and periodate oxidation has been shown to decrease their attachment to rat palatal epithelial cells, suggesting the involvement of protein and carbohydrate moieties. Attachment was also decreased by the presence of D-mannose, N-acetyl-D-galactosamine and sialic acid. These findings suggest that attachment is mediated by trypsin-resistant proteinaceous and carbohydrate adhesins with affinity for D-mannose, N-acetyl-D-galactosamine and sialic acid. Adhesion of the organism to extracellular matrix molecules is also considered to be an important means by which it can attach itself to epithelial cells. Fibronectin, laminin, fibrinogen, collagen I, collagen IV and gelatin can all act as substrata for attachment of the organism and a number of putative adhesins have been suggested as well as flagellin.

The organism appears to adhere to the orally derived KB epithelial cell line mainly via one of its polar regions and this induces a number of changes in these cells, including the formation of membrane blebs and vacuoles, detachment of cells from plastic dishes, a 21% decrease in diameter and a dramatic effect on cytoskeletal components. With regard to the latter, the F-actin and keratin contents of the cells were found to decrease and the proportion of the cells expressing desmoplakin II (a protein involved in maintaining intercellular junctions) also decreased. Such changes, should they occur *in vivo*, would adversly affect the barrier function of the epithelium.

Streptococcus pyogenes

This organism causes a wide range of human infections, including pharyngitis, toxic shock syndrome, erysipelas, impetigo, septicaemia and pneumonia. Skin and throat infection with this organism may be followed by rheumatic fever or acute glomerulonephritis. A large number of adhesins have been identified for *Strep. pyogenes* and in many cases the receptors involved have been identified. The most important of these are shown in Table 5.8. Attachment of *Strep. pyogenes* to human pharyngeal epithelial cells induces membrane ruffling – a feature not observed when the bacteria adhere to buccal epithelial cells. In contrast, adhesion of members of the normal pharyngeal microflora to the epithelial cells has no effect on their morphology. In cases of acute tonsillitis caused by this organism, tonsillar epithelial cells have been shown to exhibit protrusions which appear to 'grip' the bacteria from several directions.

Helicobacter pylori

Interest in this organism has increased dramatically during the last few years as it has been shown to be the major cause of chronic gastritis in humans, is associated with peptic ulcer formation and is a risk factor for the development of gastric carcinoma and possibly also heart disease (see Chapter 10). It has been estimated that in the USA more than 50% of adults over the age of 60 years are infected with the bacterium. Surface molecules such as flagellin,

Table 5.8 Adhesins of *Strep. pyogenes*

Adhesin	Receptor	Target cell/substratum
Lipoteichoic acid	Fibronectin	Buccal epithelial cells, fibroblasts, pharyngeal cells
Fibronectin-binding proteins	Fibronectin	Fibronectin, buccal cells, epidermal Langerhans cells
M protein	Galactose, fucose, fibrinogen	Pharyngeal cells, keratinocytes
Vitronectin-binding proteins	Vitronectin	Pharyngeal cells
Glyceraldehyde 3-phosphate dehydrogenase	Fibronectin, lysozyme, myosin and actin	?
Collagen-binding proteins	Collagen	?
Galactose-binding protein	Galactose	Pharyngeal cells

urease, various outer membrane proteins, the heat shock protein, chaperonin 60 (see above) and an *N*-acetylneuraminyllactose-binding fibrillar haemagglutinin have all been proposed to function as adhesins in *Hel. pylori*. Putative receptor molecules on gastric epithelial cells for these adhesins include phosphatidylethanolamine, lysophosphatidylethanolamine and the Lewis^b blood group antigen. However, the exact mechanism by which *Hel. pylori* adheres to the gastric mucosa has still to be elucidated.

Examination of the gastric mucosa of patients infected with the organism has revealed that it induces effacement of the microvilli of the gastric epithelial cells and adheres to cellular projections known as adherence pedestals. Such ultrastructural changes are similar to those induced in intestinal epithelial cells by EPEC. *In vitro* studies have shown that loss of microvilli and pedestal formation occurs within 3 hours of attachment of *Hel. pylori* to gastric epithelial cells. The pedestals were formed from the bases of the damaged microvilli and consisted mainly of microfilaments. Bacterial adhesion stimulated actin polymerization in the epithelial cells and the actin accumulated immediately below the site of attachment of the bacteria. Furthermore, both α-actinin (a cross-linker of actin filaments – see Chapter 2) and talin (a component of focal adhesions) were found beneath the cytoplasmic membrane at the site of bacterial attachment. As was found with A/E lesions induced by EPEC, attachment of *Hel. pylori* to the epithelial cells induced tyrosine phosphorylation of proteins (of molecular masses 145 and 105 kDa) in the latter and resulted in increased levels of intracellular IP_3.

Induction of functional changes

Cytokine release by oral streptococci

Streptococci constitute one of the numerically dominant genera of bacteria in the oral cavity. A wide variety of species are usually present and most of these rarely cause disease, the exceptions being *Streptoccus mutans* and *Strep. sobrinus* (responsible for caries) as well as *Strep. sanguis*, which may cause endocarditis when it gains entry to the bloodstream of patients with pre-existing structural or anatomical cardiac abnormalities. Because of their involvement in caries, there has been tremendous interest in the means by which these bacteria adhere to teeth and a great deal is known about the adhesins responsible. However, less is known about the adhesins involved in the colonization of mucosal surfaces by these organisms. Among the many adhesins expressed by streptococci are a group of rhamnose/glucose polymers and a group of antigenic proteins (antigens I/II) with molecular masses of 180–200 kDa. These appear to be present on the surfaces of many oral streptococci and have been shown to mediate their binding to human epithelial cell lines.

Adhesion of any of 11 strains of oral streptococci to epithelial cells was found to stimulate release of the CXC chemokine, interleukin 8 (IL-8 – see Chapter 3, Table 3.4), but not IL-6. IL-8 was also released from the cells following binding of purified adhesins (RGP and protein I/IIf) from *Strep. mutans*. In the case of protein I/IIf, this release was abrogated by fucose and *N*-acetylneuraminic acid, implying that molecules on the surface of the epithelial cells containing these epitopes functioned as receptors for protein I/IIf. The release of IL-8 by epithelial cells following adhesion of oral streptococci would be expected to induce an inflammatory response accompanied by an influx of PMNs. If this were to happen the epithelia of our mouths would be continuously inflamed. As this is not the case, other factors must be operating *in vivo* to counteract this response. Although the epithelial cells clearly released IL-8 following adhesion of streptococci, it must be remembered that it is extremely difficult to predict the outcome of such an event *in vivo*. Cytokines have many and varied effects on the different cells comprising a tissue and the release of one cytokine by one type of cell can often induce the release of different cytokines from other nearby cells. This will include molecules with anti-inflammatory as well as proinflammatory activities and results in the establishment of what is known as a 'cytokine network', the net effect of which may be a pro- or anti-inflammatory response in the tissue or indeed a state of 'no change'. In the experiment described, the investigators did not determine which, if any, other cytokines were released following bacterial adhesion.

Urinary tract infections due to E. coli

Adhesion-mediated cytokine release. One of the main defence mechanisms of the urinary tract is the flushing action of urine and this constitutes a major deterrent to colonization by infecting organisms. The ability to adhere to urinary epithelial cells is, therefore, a prime requirement for a uropathogenic organism. The most frequent cause of urinary tract infections (UTIs) is *E. coli* and adhesion of this organism to urinary epithelial cells has been studied extensively. *E. coli* exhibits a variety of adhesins which can be broadly classified as fimbrial, fibrillar and non-fimbrial. Those considered to be important in adherence to urinary epithelium are the fimbriated type, mainly P (receptor: Galα1–4Gal), S (receptor: NeuAcα2–3Gal), Dr (receptor: decay accelerating factor, a complement regulatory protein) and type 1 (receptor: D-mannose). Binding of *E. coli* to urinary epithelial cells via P, S or type 1 fimbriae has been shown to elicit the release of the proinflammatory cytokines IL-6, IL-8, IL-1α and IL-1β. As urinary IL-6 levels increase in patients with UTI and in volunteers infected with *E. coli*, and as P-fimbriated *E. coli* are more potent at inducing IL-6 than S- or type 1-fimbriated strains, studies of adhesion-induced cytokine release by *E. coli* have concentrated mainly on IL-6 induction by P-fimbriated strains of the organism. IL-6 is generally considered to be a proinflammatory cytokine which can stimulate the release of acute-phase proteins, activate B and T cells, stimulate immunoglobulin synthesis and stimulate the proliferation of fibroblasts, although there is also evidence that this cytokine has anti-inflammatory effects (see Chapters 8–10).

P fimbriae bind to Galα1–4Galβ-containing epitopes present on the oligosaccharide portion of the globoseries glycosphingolipids which are bound to ceramide in the outer leaflet of the epithelial cell membrane. Binding of *E. coli* to these receptors has been shown to be essential for IL-6 release. Inhibitors (e.g. D-threo-1-phenyl-2-decanoylamino-3–1-propanol) of the synthesis of glycosphingolipids not only reduce their levels in epithelial cells but also result in reduced adherence of *E. coli* and reduced secretion of IL-6. Recently, ceramide has been shown to function as a second messenger in cell signalling. Agonists such as TNFα and IL-1β interact with their receptors and activate sphingomyelinases to release ceramide from sphingomyelin. The released ceramide can activate NF-κB or protein kinases belonging to the Ser/Thr family. These events ultimately result in cell activation, cytokine release or apoptosis (see Chapter 3). P-fimbriated *E. coli*, but not an isogenic non-fimbriated control, has been shown to stimulate ceramide release, and its conversion to ceramide 1-phosphate, in a human kidney epithelial cell line. Staurosporine (an inhibitor of Ser/Thr kinases, mainly protein kinase C), but not genistein (an inhibitor of tyrosine kinase), markedly decreased the amount of IL-6 synthesized by the cells in response to the bacteria, implying the involvement of Ser/Thr kinases. Interestingly, IL-6 release induced by *E. coli* with type 1 fimbriae was not affected by staurosporine, suggesting the involvement of a different signalling pathway. Type 1 fimbriated *E. coli*, unlike P-fimbriated strains, is known to bind to terminal mannose residues and these are found on glycoproteins rather than glycolipids. The transmembrane signalling pathway involved in IL-6 release induced by type 1-fimbriated strains remains to be elucidated.

Adhesion-induced expression of intercellular adhesion molecules. A characteristic feature of any bacterial infection is the migration of PMNs (also commonly referred to as neutrophils) towards the site colonized by the infecting organism and, in the case of UTIs, this leads to the presence of PMNs in urine. In response to a chemotactic gradient from the site of the infection, PMNs adhere to and then traverse the vascular endothelium (described in more detail in Chapter 8). While a great deal is known about the mechanisms involved in these processes, very little is known about how PMNs then cross the epithelial lining into the mucosal space, an event occurring in infections of the urinary, intestinal and respiratory tracts. Adhesion of *E. coli* to epithelial cells can trigger the release of IL-8, which is a potent chemoattractant for PMNs and will therefore stimulate the migration of PMNs to the urinary epithelium. Recent work has given us some idea of the mechanism involved in PMN traversal of the epithelial layer. An *E. coli* strain with both P and type 1 fimbriae was shown to adhere to kidney and bladder epithelial cells and induce transepithelial migration of PMNs in a time- and dose-dependent manner. This was accompanied by the expression of ICAM-1 on the surfaces of the epithelial cells. Migration was reduced by antibodies to ICAM-1, CD18, CD11b and IL-8. These findings suggest the following model to explain the transepithelial migration of PMNs (Figure 5.9). Adhesion of *E. coli* to epithelial cells induces the release of the PMN chemoattractant IL-8, which induces the migration of PMNs from the bloodstream. At the same time, the adherent *E. coli* upregulates the expression of ICAM on the surface of the cells which binds to the CD11b/CD18 integrin on the surface of the arriving PMNs. Details of the mechanism by which the PMNs then traverse the epithelium are, however, still sketchy.

Helicobacter pylori

Exposure of a number of different epithelial cell lines to *Hel. pylori* results in the secretion of IL-6 and IL-8 but not TNFα. Interest has centred on the stimulation of IL-8 secretion as this

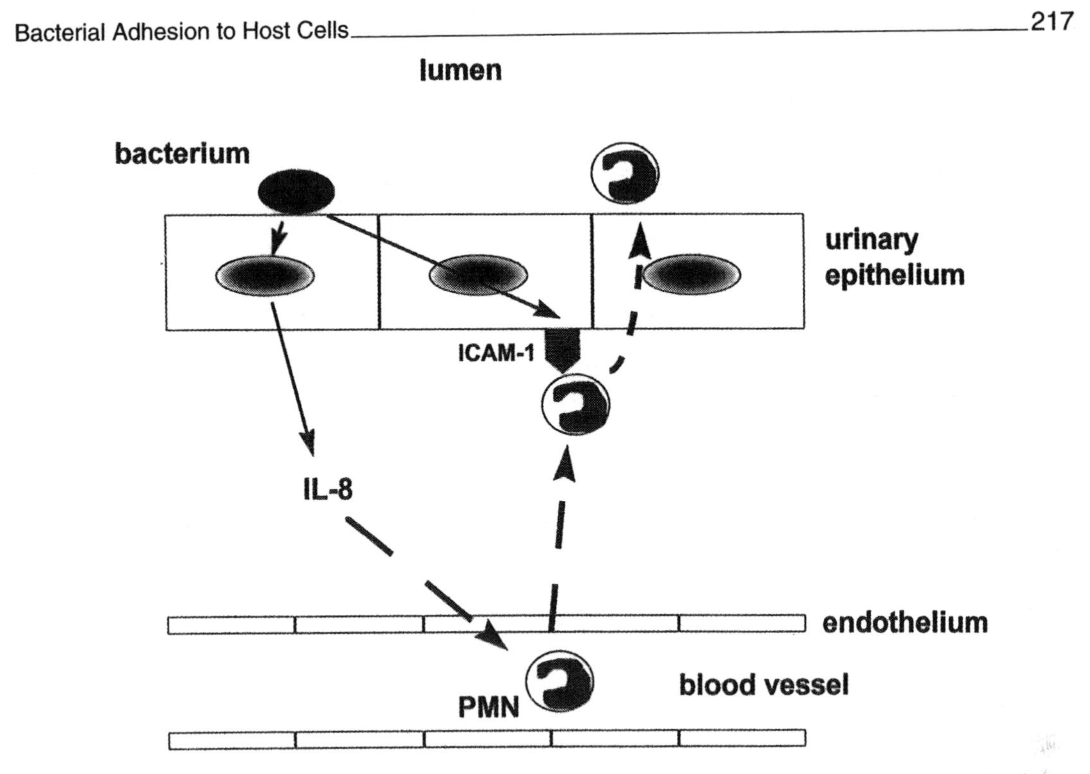

Figure 5.9 Events following adhesion of uropathogenic strains of *E. coli* to the urinary epithelium

cytokine is a potent inflammatory mediator by virtue of its ability to attract and activate PMNs, which would then contribute to mucosal damage. Immunohistochemical studies have also shown that expression of IL-8 in the gastric epithelium of patients infected with *Hel. pylori* is much greater than in those patients in which the organism could not be detected. A number of components and secretory products of *Hel. pylori* have been proposed as being the inducers of IL-8 secretion including LPS, an extracellular polysaccharide, the *cagA* gene product (an immunodominant antigen), the surface-associated urease, a lipoprotein and the vacuolating cytotoxin. Using a series of isogenic mutants, as well as isolated components, it has been shown that none of these is responsible for IL-8 induction. However, an aqueous extract of the bacterium containing surface-associated proteins was a potent inducer of IL-8 synthesis. It may well be that the IL-8-inducing molecule(s) are secreted from the organism and induce IL-8 synthesis by epithelial cells. One possible inducer of IL-8 synthesis is chaperonin 60, a protein-folding oligomer which is normally found intracellularly, but which, in *Hel. pylori*, is found on the cell surface where it may act as an adhesin. The authors have shown that bacterial chaperonin 60 proteins are potent inducers of cytokine synthesis.

Cellular invasion

Adhesion of pathogenic bacteria to epithelial cells often is a preliminary to invasion. As adhesion itself often triggers in the host cell a series of changes leading to uptake of the organism, it would be artificial to discuss only the adhesion aspect in isolation from subsequent events. Consequently, for those organisms which invade epithelial cells the adhesion mechanisms involved will be discussed in the following chapter.

Adhesion to fibroblasts

Fibroblasts have long been known to be key cells in the production of extracellular matrix material and the maintenance of connective tissue integrity. More recently it has been demonstrated that they can also secrete cytokines and other inflammatory mediators and so are important in host defence and the maintenance of a cytokine network compatible with health. Interference with their function, therefore, can have profound immunobiological and structural consequences. Bacteria can gain access to, and interact with, fibroblasts either by invading mucosal surfaces or via breaches in the integrity of the epithelium due to wounds, abrasions or burns. It is surprising, therefore, that the interaction between bacteria and fibroblasts has received so little attention.

Treponema denticola

A 64 kDa protein from the outer sheath of *Tr. denticola* may function as an adhesin mediating attachment of the organism to human gingival fibroblasts (HGFs). Interestingly, cyanogen bromide fragmentation of the protein produced a 42 kDa fragment containing the sequence TLDLALD which was identical with a sequence in the β_2 integrin subunit of human leukocyte adhesion glycoprotein, p150,95. A separate 58 kDa protein from the outer sheath of the organism may be involved in the attachment of the organism to extracellular matrix molecules. The protein was cloned and expressed in *E. coli* and the recombinant protein adhered to laminin and fibronectin. Pretreatment of these immobilized matrix components with the protein resulted in increased attachment of the organism.

Attachment of *Tr. denticola* to HGFs induces a number of alterations in these cells including: (i) retraction of pseudopods followed by rounding of the cells and the formation of membranous blebs; (ii) rearrangement of filamentous actin networks into a perinuclear array; (iii) detachment of the cells from their substratum, possibly as a result of the degradation of fibronectin by surface-associated proteases; and (iv) cell death. Death of the cells would, obviously, have adverse consequences for tissue integrity if it occurred *in vivo*. However, the cytoskeletal changes induced by the organism would also be detrimental as they would inevitably interfere with the ability of fibroblasts to phagocytose collagen fibres – an important aspect of tissue remodelling.

Haemophilus ducreyi

Haem. ducreyi is the causative agent of the sexually transmitted disease chancroid. The organism gains access to the dermis via microbreaks in the epidermis and there it induces ulcer formation. *In vitro* studies using human foreskin fibroblast (HFF) monolayers have shown that the organism forms microcolonies on the surface of these cells and often becomes embedded in deep invaginations of the cell membrane. The bacteria can also penetrate between cells but do not invade them. Attachment to the HFFs results in a distinctive cytopathic effect – disruption of cell monolayers, changes in cell morphology and cell death. *Haem. ducreyi* produces a cell surface-associated haemolysin and, as haemolysin-defective mutants have no adverse effects on HFFs, this suggests that the haemolysin may be responsible for the adherence-induced cytopathic effects. Interestingly, adhesion of the bacteria to HFFs induces changes in the outer membrane protein profile of the organism, although nothing else is known about this.

Porphyromonas gingivalis

Adhesion of this organism to HGFs is mediated by the major fimbrial protein, FimA. Inactivation of the *fimA* gene results in mutants which are unable to produce fimbriae and exhibit decreased adherence to HGFs. Within 60 minutes of adhering to HGFs, *Por. gingivalis* induces dramatic morphological changes in the cells: long microvilli are formed which surround clumps of bacteria. No such changes are induced by the afimbriated mutant.

Adhesion to vascular endothelial cells

Vascular endothelial cells form a continuous monolayer lining the blood vessel walls and function not only as a barrier between the blood and the vessel walls but also in the regulation of blood vessel tone and permeability, coagulation of blood, leukocyte and platelet reactivity, angiogenesis and the source of vascular mediators (e.g. nitric oxide, prostacyclin, endothelin). In order to colonize tissues, blood-borne bacteria have to adhere to and traverse the endothelium. *Staph. aureus, N. meningitidis, Haem. influenzae, Strep. pyogenes, Borrelia burgdorferi, Listeria monocytogenes, E. coli* and *Ps. aeruginosa*, for example, have all been shown to adhere to, and invade, endothelial cells and these will be discussed in the following chapter.

Adhesion to phagocytic cells

An understanding of the mechanisms involved in adhesion of bacteria to phagocytic cells is vital as this is the mechanism by which bacteria can be effectively disposed of. However, some bacterial species can avoid this important host defence system by actively colonizing phagocytic cells. Adhesion appears to occur either directly (i.e. involving the interaction between adhesins and receptors on the phagocyte) or indirectly by interacting with host components (e.g. immunoglobulins or complement components) which then bind to phagocyte receptors. Whichever mechanism is involved, adhesion induces endocytotic uptake of the bacterium into a phagosome which then fuses with a lysosome resulting, usually, in the death of the organism (see Chapter 8 for more details). Some bacteria have evolved means of surviving within phagocytes and examples include *Mycobacterium tuberculosis, Bord. pertussis, Yersinia enterocolitica* and *Lis. monocytogenes*. Other, more audacious species, cock a snoop at phagocytes and actually invade them e.g. *Legionella pneumophila, Salmonella* spp. and *Shigella* spp. Both invasion and phagocytosis involve profound adhesion-induced changes in the phagocyte and are more appropriately discussed in the next chapter. It has been shown, nevertheless, that adhesion of some invasive organism (including *Legionella pneumophila* and *Sal. typhimurium*) induces cytokine release by the host cell prior to internalization. Attachment of *Leg. pneumophila* or *Sal. typhimurium* to murine macrophages in the presence of cytochalasin (an inhibitor of actin polymerization which prevents uptake of bacteria) induces increased expression of the cytokines IL-1β, IL-6 and GM-CSF and the chemokines MIP-1β, MIP-2 and KC. Interestingly, the presence of α-methyl-D-mannoside (MM) inhibited the induction of mRNA for the cytokines but not chemokines. These data imply that adhesion-induced cytokine synthesis is mediated by a mannose-sensitive receptor–ligand interaction but that some other interaction is responsible for chemokine gene transcription. In the case of *Sal. typhimurium,* it is likely that the mannose- sensitive adhesin is located on the flagellum as mutants without flagella failed to induce increased levels of cytokine mRNA and, furthermore, isolated flagella were able to stimulate cytokine (but not chemokine) mRNA synthesis in the macrophages. This suggests that different receptor ligand systems are responsible for adhesion-mediated upregulation of cytokine and chemokine synthesis in macrophages. There is also evidence that a second receptor-ligand system may be involved in the induction of increased cytokine (but not chemokine) synthesis by *Sal. typhimurium.* Hence, treatment of the organism with serum was found to induce higher levels of GM-CSF message in macrophages than when untreated bacteria were used. Not all of this enhanced synthesis of message was abolished by the presence of MM, suggesting the involvement of complement and/or Fc receptors.

Another interesting consequence of bacterial adhesion to phagocytic cells which does not involve bacterial internalization is apoptosis. This has been reported to occur when the enteric pathogen, *Y. enterocolitica,* binds to macrophages. Once this organism has invaded the intestinal mucosa it resists phagocytosis and maintains an extracellular lifestyle.However, it does adhere to macrophages and this induces death of these phagocytes which display all of the features characteristic of apoptosis: shrinkage of the cytoplasm, condensation of nuclear chromatin and fragmentation of DNA (see Chapter 2). Apoptosis was dependent on a functional type III secretion system (see Chapter 2), suggesting that one or more of the secreted Yop proteins was responsible for the induction of cell death. The ability of *Y. enterocolitica* to induce the suicide of one of the key effector cells of the host defence system is likely to constitute an important survival strategy for this extracellular pathogen.

CONCLUSIONS

The first stage in any infectious process is adhesion of the causative organism to a host cell or to polymers secreted by host cells. Bacteria synthesize adhesins which enable them to bind to specific receptors on the target cell or secreted polymers and this accounts (together with local environmental factors) for the tissue tropism exhibited by most organisms. Increasing knowledge of the mechanisms underlying bacterial adhesion could lead to the development of novel means of preventing bacterial infections based on interfering with this adhesion process. Once bound to their target cell/molecule, bacteria then engage in an exchange of signals with their host (molecular crosstalk) which can have one of several consequences: (i) mutual indifference and the continued existence of both members of the association in an, apparently, unaltered state; (ii) profound alterations to the target cell (and to the infecting bacterium) resulting in either localized or systemic (by release of toxins etc.) pathology; (iii) invasion of the host cell and induction of pathology elsewhere. The first possibility is characteristic of the interaction of members of the normal microflora and epithelial cells. However, we hypothesize that the apparent inactivity exhibited by the two members of the association may cloak an intense exchange of signals between them. The scant information available on such interactions is the result of a lack of interest in how multicellular organisms interact with their normal microflora – the study of pathogens is thought to be far more exciting and relevant to disease. We consider this to be misfounded, as a knowledge of how epithelial cells can live in harmony with our normal microflora could lead to the development of new approaches to the treatment of infectious diseases. The interaction inevitably involves host cells controlling the growth of these bacteria as well as the expression of potentially harmful bacterial products. The second possibility has been the subject of more extensive research as it involves interactions between disease-inducing organisms and their host. After all, who could fail to be interested in an adhesive process which induces the levitation of a bacterial microcolony on a pedestal? However, although the effects of adhesion on host cell structure and function are receiving considerable attention, little is known about the concurrent effects on the bacteria. The third possibility, involving invasion of the host cell, is the subject of the chapter which follows.

FURTHER READING

Books

Ofek I, Doyle RJ (1994) *Bacterial Adhesion to Cells and Tissues*. Chapman & Hall, London.

Reviews

Baldwin TJ (1998) Pathogenicity of enteropathogenic *Escherichia coli*. *J Med Microbiol* 47: 283–293.

Donnenberg MS, Kaper JB, Finlay BB (1997) Interactions between enteropathogenic *Escherichia coli* and host epithelial cells. *Trends Microbiol* 5: 109–114.

Gilsdorf JR, McCrea KW, Marrs CF (1997) Role of pili in *Haemophilus influenzae* adherence and colonization. *Infect Immun*, 65: 2997–3002.

Jacques M (1996) Role of lipo-oligosaccharides and lipopolysaccharides in bacterial adherence. *Trends Microbiol* 4: 408–409.

Jenkinson HF, Lamont RJ (1997) Streptococcal adhesion and colonization. *Crit Rev Oral Biol Med* 8: 175–200.

Ljungh A, Moran AP, Wadstrom T (1996) Interactions of bacterial adhesins with extracellular matrix and plasma proteins: pathogenic implications and therapeutic possibilities. *FEMS Immunol Med Microbiol* 16: 117–126.

Mouricout M (1997) Interactions between the enteric pathogen and the host: an assortment of bacterial lectins and a set of glycoconjugate receptors. *Adv Exp Med Biol* 412: 109–123.

Sharon N (1996) Carbohydrate–lectin interactions in infectious disease. *Adv Exp Med Biol* 408: 1–8.

Sharon N, Lis H (1995) Lectins–proteins with a sweet tooth: functions in cell recognition. *Essays Biochem* 30: 59–75.

Smyth CJ, Marron MB Twohig JM, Smith SG (1996) Fimbrial adhesins: similarities and variations in structure and biogenesis. *FEMS Immunol Med Microbiol* 16: 127–139.

St Geme JW III (1997) Bacterial adhesins: determinants of microbial colonization and pathogenicity. Adv Pediatr 44: 43–72.

Svanborg C, Hedlund M, Connell H, Agace W, Duan RD, Nilsson A, Wullt B (1996) Bacterial adherence and mucosal cytokine responses: receptors and transmembrane. *Ann NY Acad Sci 797: 177–190.*

Virji M (1996) Microbial utilization of human signalling molecules. *Microbiology* 142: 3319–3336.

Whittaker CJ, Klier CM, Kolenbrander PE (1996) Mechanisms of adhesion by oral bacteria. *Ann Rev Microbiol* 50: 513–552.

Wilson M, Seymour R, Henderson B (1998) Bacterial perturbation of cytokine networks. *Infect Immun* 66: 2401–2409.

Chapter 6

Bacterial Invasion of Host Cells

INTRODUCTION

What do we mean by bacterial invasion? The dictionary definition of the word *invade* is 'to enter for hostile purposes'. Is the presence of a bacterium inside a host cell evidence that an invasive process has occurred? For cells not normally considered to be phagocytic, such as epithelial cells, the answer to this question is likely to be yes. However, the situation is not so clear-cut when it comes to phagocytic cells. Hence, the presence of bacteria within a macrophage may indicate that the host cell is merely performing its normal function, i.e. disposing of bacteria, or it may indicate that the macrophage is being invaded, as for some bacterial species this is their preferred habitat. Invasion of host tissues by bacteria is a characteristic feature of a number of infectious diseases, including those listed in Table 6.1. This prompts the obvious questions 'Why?' and 'How?' We know quite a lot about the how but far less about the why. With regard to the latter, it has been suggested that invasion enables bacteria to evade host defences, seek out new supplies of nutrients or find new tissues to colonize. However, this is an anthropomorphic perspective and, while invasive bacteria are likely to benefit in the ways listed above, there is little evidence available to support these hypotheses. With the question how, we are on firmer ground and during the last decade great strides have been made in unravelling the mechanisms used by bacteria to invade host cells. While, collectively, bacteria have evolved a variety of invasive mechanisms, some common elements in these processes are evident as most involve the manipulation of normal host cell cytoskeletal components such as actin and tubulin, resulting in the invagination of the host cell membrane to enclose the bacterium within a vacuole (see Chapter 2). This often occurs by interference with the intracellular signalling pathways (detailed in Chapter 3) either by stimulation of inhibition of signal transduction, or both. However, a limited number of pathogens (e.g. *Rickettsia prowazekii*) use the less subtle tactic of simply digesting away part of the cytoplasmic membrane. More diversity is evident among invasive pathogens when the events subsequent to invasion are considered. Hence the organism may remain inside the vacuole (*Mycobacterium tuberculosis*), escape from the vacuole and colonize the cytoplasm (*Listeria monocytogenes*) or escape from the vacuole and the cell and then spread systemically (*Yersinia enterocolitica*).

The first stage in an invasive process is the adhesion of the invading organism to a host cell. This process, and the mechanisms involved, have been described in Chapter 5. In this

Table 6.1 Examples of invasive pathogens and the diseases caused by them

Organism	Disease
Shigella spp.	Dysentery
Salmonella spp.	Typhoid fever and gastroenteritis
Escherichia coli (enteroinvasive strains)	Similar to dysentery
Streptococcus pneumoniae	Meningitis, pneumonia
Haemophilus influenzae	Meningitis, pneumonia
Listeria monocytogenes	Listeriosis
Mycobacterium tuberculosis	Tuberculosis
Brucella spp.	Brucellosis
Yersinia spp.	Plague, gastroenteritis
Clostridium perfringens	Gangrene
Borrelia burgdorferi	Lyme disease
Treponema pallidum	Syphilis

section we will consider only those adhesive processes which ultimately lead to invasion of the host cell.

INVASION MECHANISMS

The mechanisms involved in invasion of host cells by bacteria are many and varied but can be broadly classified into three main groups on the basis of whether they involve the microfilaments or microtubules of the host cell or entail forced entry into the cell. Certain features of the invasion process are also dependent on the type of host cell being invaded. In this section we will describe the ways in which bacteria invade epithelial cells, endothelial cells and macrophages.

Invasion of epithelial cells

The epithelial surfaces constitute the body's outer integument and so represent the interface between the controlled, stable *milieu intérieur* and the constantly changing external environment. As is well known, this interface is colonized by an enormous population of microbes that constitute the normal microflora. The epithelial interface will periodically be visited by microbes from the external environment, which may be highly pathogenic. The epithelium, therefore, constitutes the first physical barrier, preventing access of microbes, pathogenic or commensal, to deeper tissues of the body. Invasion of the cells comprising this physically rather fragile barrier is, therefore, the first step in the initiation of a systemic infection.

Cell invasion involving actin rearrangements

The majority of invasive bacteria gain entry into epithelial cells by inducing the rearrangement of microfilaments of the cytoskeleton (see Chapters 2 and 3). A number of bacteria, particularly *Yersinia* spp., *Salmonella* spp., *Shigella* spp. and *Lis. monocytogenes*, which utilise this means of entry, have been the subject of intensive study during recent years and the main features of their invasive interaction with epithelial cells are summarized in Table 6.2.

Table 6.2 Comparison of invasion mechanisms used by important intestinal pathogens

Feature	*Yersinia* spp.	*Salmonella* spp.	*Shigella* spp.	*Lis. monocytogenes*
Attachment to host cell	Multiple contact Zippering	Localized contact	Localized contact	Multiple contact Zippering
Morphological alterations in host cell	Pseudopod formation Transient effect	Pseudopod formation Membrane ruffling	Pseudopod formation Membrane ruffling	Pseudopod formation Transient effect
Fate of bacteria	Reside in vacuole Little/no replication	Reside in vacuole Replicate in vacuole	Escape from vacuole Multiply in cytoplasm Move through cytoplasm	Escape from vacuole Multiply in cytoplasm Move through cytoplasm
Intercellular invasion	No	No	Yes	Yes

Yersinia spp.

Three species of the genus *Yersinia* can cause disease in man: *Y. pestis* (plague), *Y. enterocolitica* (gastroenteritis) and *Y. pseudotuberculosis* (gastroenteritis). Although all three species are invasive pathogens, the two enteropathogenic species usually invade only the submucosal tissues, in contrast to *Y. pestis* which invariably spreads systemically. Another important difference between the enteropathogenic species and *Y. pestis* is their mode of transmission. *Y. enterocolitica* and *Y. pseudotuberculosis* are transmitted via contaminated food or water whereas *Y. pestis* can be contracted from a flea bite or by inhalation of an aerosol generated by an infected individual.

As this section is concerned with invasion of epithelial cells, only the enteropathogenic strains will be considered here. When *Y. enterocolitica* adheres to an epithelial cell, close contact is made at many points on the bacterial surface – a process described as 'zippering'. This induces the uptake of the organism into an endocytic vacuole and the bacteria appear to sink into the membrane of the epithelial cell. Within a few minutes of entry, the host cells exhibit a normal appearance.

Adhesion of these bacteria to host cells is mediated by a number of adhesins: invasin (encoded by the chromosomal *inv* gene), the Ail protein (encoded by the chromosomal *ail* – attachment-invasion locus – gene) and YadA (encoded by the *yadA* – yersinia adherence – gene located on a 75 kb virulence plasmid). Invasin is an outer membrane protein which binds to $\alpha_5\beta_1$ integrins – a particular subset of the β_1-integrin family of cell surface molecules whose main function is to anchor host cells to extracellular matrix molecules such as fibronectin and collagen (see Chapter 3). Binding to integrins is used by a broad range of microbes to enable colonization, and sometimes invasion, of host cells. For example, *Bordetella pertussis, Leishmania mexicana, Histoplasma capsulatum, Borrelia burgdorferi*, echovirus 1 and adenovirus all use integrins as receptors. Invasin consists of 986 amino acids, of which the 192 C-terminal residues are involved in adhesion and invasion. Interestingly, the N-terminus of this protein shows homology to the host cell-binding proteins (intimins) of enteropathogenic and enterohaemorrhagic strains of *E. coli*. Near the C-terminus they have two cysteine residues separated by 74–77 amino acids which, in the case of invasin, form a disulphide bond essential to the proper functioning of the protein. Invasin binds to its integrin receptor with an affinity approximately 100-fold greater than that of the natural ligands for the receptor. The YadA adhesin (which forms a fibrillar matrix on the surface of the bacterium) also binds to $\alpha_5\beta_1$-integrins while the receptor for Ail, an outer membrane protein, has not been identified. Binding of invasin to its integrin receptor at a number of points along the bacterium/host cell interface ('zippering') induces uptake of the bacterium (Figure 6.1). The zippering necessary for inducing invasion is a consequence of the high receptor density at the bacterium/host cell interface due to the recruitment of additional β_1-integrin molecules to this site. This clustering of integrins induces protein tyrosine kinase activity, which is essential for invasion as shown by the fact that inhibitors of host tyrosine kinases prevent invasion. However, the specific kinases involved in the process have not yet been identified. Endocytosis is also inhibited by cytochalasin, implying the involvement of actin microfilaments. The early stages of internalization are characterized by the formation of clathrin lattices immediately beneath the bacteria. The resulting vacuole is surrounded by polymerized actin and other proteins including filamin and talin (see Chapter 2). Internalized bacteria are thought to survive, but not reproduce, inside the vacuoles. Although *Y. enterocolitica* readily invades intestinal epithelial cells *in vitro* there is doubt concerning whether this occurs *in vivo* in the manner described above. This is because integrins are not found on the apical surfaces (i.e. facing the intestinal lumen) of epithelial cells but only on the opposite (basolateral) surfaces. This rather neatly serves to illustrate the danger of extrapolating too far from inadequate laboratory models. Whenever possible, *in vitro* investigations need to be backed up by *in vivo* studies, preferably in man. When this is not possible, an animal model may be useful, especially if the disease process in the animal is similar to that occuring in man. Studies of invasive disease due to *Yersinia* spp. in animals have revealed that the organism attaches to and invades the microfold cells (M cells) of Peyer's patches. M cells,

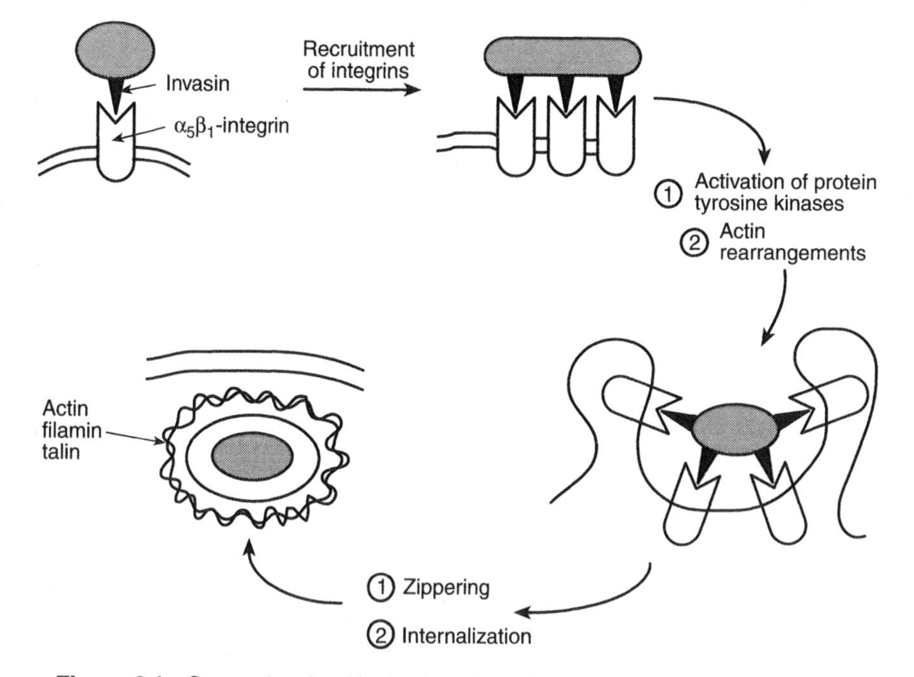

Figure 6.1 Stages involved in the invasion of epithelial cells by *Yersinia* spp.

together with associated macrophages, T cells and B cells, form what are known as follicles and collections of follicles are known as Peyer's patches. M cells are naturally phagocytic and their main function is to take up bacteria and other antigenic material from the intestinal lumen and present these to the associated cells of the immune system. Invasin, or perhaps YadA or Ail, may mediate binding of bacteria to M cells so facilitating their uptake and traversal of the epithelial layer. *Y. enterocolitica* is also an enteropathogen of rodents, and yersiniosis in mice closely resembles human *Yersinia* infections. Using an experimental mouse infection model, it has been shown that *Y. enterocolitica* selectively invades the Peyer's patches via M cells – bacteria were observed inside M cells within 1 hour of infection. As *Yersinia* can evade phagocytosis (see 'Extracellular lifestyle', below), the bacteria are able to multiply in the underlying tissue and this is accompanied by a massive recruitment of phagocytes and abscess formation. Destruction of the follicle-associated epithelium was observed several days after infection. Whether this was directly due to the bacteria or to the discharge of the lysosomal contents of the phagocytes remains to be determined.

It would appear, therefore, that invasion of the intestinal epithelium by *Y. enterocolitica in vivo* occurs via the M cells of Peyer's patches. Once the organism has traversed the epithelium, it may, of course, invade epithelial cells from the basolateral surface and this could take place in a manner similar to that described for invasion of epithelial cells *in vitro* (Figure 6.2). In this section we have described how *Yersinia* spp. invade host cells. However, an essential feature of the pathogenesis of diseases due to these organisms is their ability to *resist* uptake by phagocytic cells and this will be the subject of a later section (Extracellular lifestyle').

Salmonella spp.

Some species of the genus *Salmonella* are invasive pathogens of man and animals. Examples include *Sal. typhi*, the causative agent of typhoid fever, a systemic infection contracted from contaminated food or water, and enteric pathogens typified by *Sal. typhimurium* which rarely penetrate beyond the submucosal tissues in man and cause gastroenteritis following ingestion in food, milk or water.

Invasion of epithelial cells by *Salmonella* spp. occurs following adhesion to microvilli. Unlike *Yersina* spp., however, *Salmonella* spp. can invade the apical surfaces of the intestinal epithelium. Within 1 minute of bacterial contact, pseudopods are formed which extend from the host cell surface to engulf the bacteria, which are then internalized within a vacuole. The host cell membrane initially displays a ruffled appearance but this disappears once the bacteria have been internalized. The vacuoles often coalesce and this may be followed by bacterial multiplication and, eventually, either destruction of the cell or movement of the vacuole to the basolateral pole of the cell and release of the bacteria into the lamina propria.

Adhesion of *Salmonella* spp. to intestinal epithelial cells appears to be due mainly to mannose-specific type 1 fimbriae. It is thought that this rather loose association acts as a stimulus for the synthesis of as yet unidentified adhesins which mediate the close contact necessary to induce invasion of the host cell. Fimbria-mediated contact with the host cell has been shown to induce the appearance of bacterial surface appendages (termed invasomes) comprised of unidentified proteins which disappear immediately prior to invasion (Figure 6.3). Mutants unable to produce invasomes are able to adhere to, but not invade, epithelial cells. The receptor(s) for the adhesin(s) responsible for inducing bacterial uptake has not yet been identified. The epidermal growth factor receptor on epithelial cells may function as the invasin receptor but evidence for such a role is conflicting.

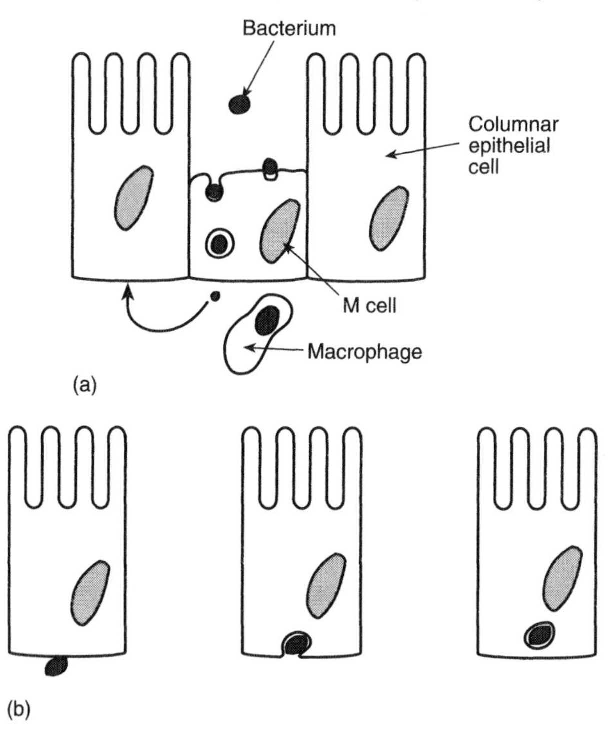

Figure 6.2 Possible mechanisms of intestinal invasion by *Yersinia* spp. (a) Uptake of bacterium by M cell and traversal through the epithelium. (b) Invasion of columnar epithelial cell by zippering following binding of invasin to β_1-integrin receptors on basolateral face of host cell

The exact sequence of events involved in invasion, and their control, remains to be elucidated. However, some information is available concerning the initial host cell signal transduction events following bacterial adhesion and these are outlined in Figure 6.4. One of the initial events is an increase in the level of inositol trisphosphate, possibly due to bacterial stimulation of host cell phospholipase C, which induces an increase in the intracellular level of Ca^{2+} ions; chelators of intracellular Ca^{2+} ions are known to prevent bacterial invasion (see Chapter 3 for a detailed discussion of intracellular signalling mechanisms). Ca^{2+} ions are involved in actin polymerization, and cytochalasins, which inhibit microfilament formation by blocking actin polymerization, prevent bacterial invasion. The actin rearrangements underlying pseudopod formation and phagocytosis are also regulated by Rho family GTPases. Cytoskeletal proteins other than actin, e.g. α-actinin, talin, tubulin, tropomyosin and ezrin, also appear to be involved in invasion as they accumulate in the host cell at the site of bacterial adhesion. Several host cell membrane proteins, including MHC class I heavy chain, β_2-microglobulin, the fibronectin receptor and CD44 (a receptor for the glycosaminoglycan (GAG) hyaluronan) form aggregates in the vicinity of bacterial attachment. The net result is uptake of the bacterium by macropinocytosis – a process which also involves the intake of large quantities of extracellular fluid. The organism, therefore, comes to reside in a large, fluid-filled vacuole known as a spacious phagosome, which is surrounded by polymerized actin and other proteins such as talin and α-actinin (Figure 6.5). During the formation of the phagosome considerable sorting of macrophage membrane proteins takes place. Hence,

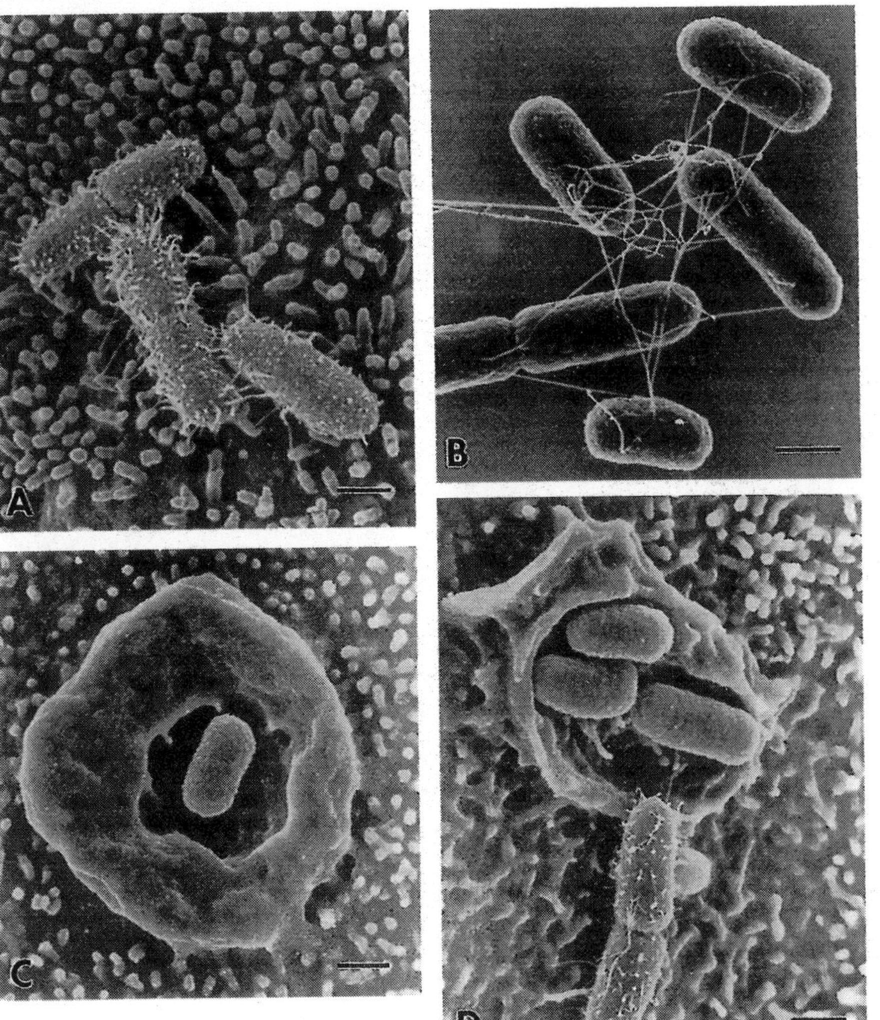

Figure 6.3 Invasion of epithelial cells by *Salmonella typhimurium*. Appendages can be seen on bacteria associated with epithelial cells (A) but not on bacteria which have not made contact with these cells (B). 30 minutes after infection, membrane ruffles are apparent on the epithelial cells (C, D). Bacteria associated with the membrane ruffles have lost their surface appendages (C, D) in contrast to those bacteria which have not yet induced ruffling (D). Reproduced with permission from 'Contact with epithelial cells induces the formation of surface appendages in *Salmonella typhimurium*' Ginocchio CC, Olmsted SB, Wells CL, Galan JE (1994) *Cell* 76: 717–724, copyright Cell Press

MHC class I heavy chain molecules are included in the resulting phagosome while transferrin receptor, β_2-microglobulin, Thy-1, CD44 and the fibronectin receptor are excluded. As the phagosome matures, the MHC class I heavy chain molecules are lost from the phagosome while lysosomal phosphatase and lysosomal glycoproteins are incorporated into it. Incorporation of the lysosomal proteins is the result of fusion of the phagosomes with vesicles originating from the trans-Golgi network. However, the phagosomes do not fuse with endosomes displaying mannose 6-phosphate receptors. The vacuoles do not become acidifed and

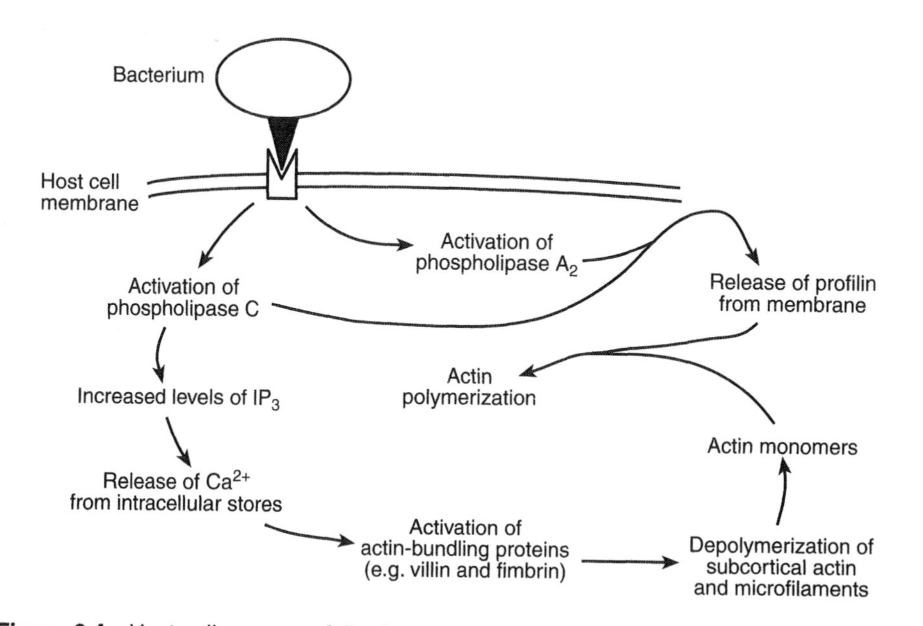

Figure 6.4 Host cell response following attachment of *Salmonella* spp. to epithelial cells

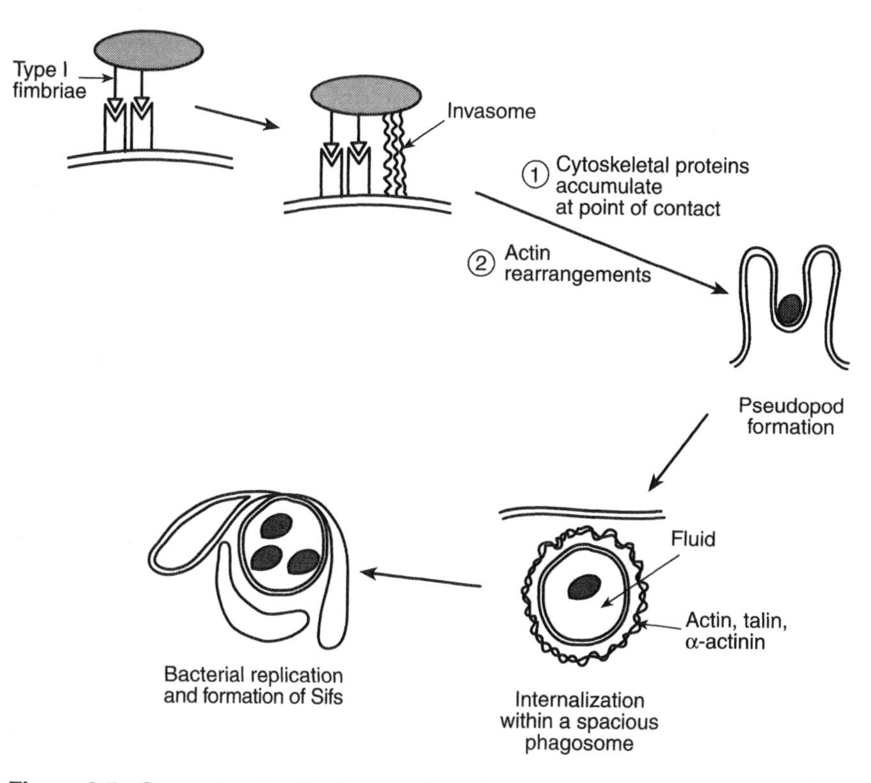

Figure 6.5 Stages involved in the invasion of epithelial cells by *Salmonella* spp.

do not contain lysosomal hydrolytic enzymes 4–6 hours after invasion bacterial proliferation takes place and this is accompanied by the formation of lysosomal glycoprotein-containing fibrillar structures (salmonella-induced filaments – Sifs) attached to the phagosome. The formation of Sifs, which are thought to be involved in the delivery of nutrients to the bacteria, is a microtubule-dependent process.

A large number of gene loci have been shown to be essential for invasion and more than 25 of these are located at 59 min on the bacterial chromosome; this constitutes a 40 kb pathogenicity island known as SPI-1 (*Salmonella* pathogenicity island 1 – see Chapter 2). One of the loci of SPI-1 (*inv/spa*) is responsible for the formation of the surface structures produced when the bacterium adheres to the host cell. Other genes of SPI-1 encode the proteins necessary for a type III secretion system as well as the proteins actually secreted by this system which induce the invasion process. As outlined in Chapter 2 ('Pathogenicity islands'), a type III secretion system consists of effector molecules (which induce some change in host cell function), cytoplasmic chaperones for these proteins, cytoplasmic and outer membrane proteins (involved in transport of the effector molecules across the cytoplasmic and outer membranes respectively), an ATPase (to energize the system) and proteins which regulate the operation of the system. The key outer membrane components of the system include InvG, which is thought to form a channel in the outer membrane through which the effector molecules are exported, as well as PrgH and PrgK, whose functions remain to be determined (Figure 6.6). Another outer membrane protein, InvH, probably plays a role in adhesion of the organism to the host cell. The cytoplasmic membrane protein InvA, like InvG, probably forms a channel through which the effector molecules are transported, a process which is carried out by a translocase system comprising the membrane proteins SpaP, SpaQ, SpaR and SpaS. Energy for the process is provided by InvC, which is an ATPase. The effector molecules of the system are protected from degradation, and maintained in the correct conformation for export, by the molecular chaperones SicA and InvI. Molecular chaperones have the function of catalytically folding and refolding proteins such that they take up their correct three-dimensional conformation. SicA is a chaperone for SipA, SipB and SipC while InvI is a chaperone for InvJ and SpaO. The secretory process is regulated in some way by the outer membrane protein InvE, which has homology to YopN an important regulator of the type III secretory system of *Yersinia* spp.

Proteins known to be secreted by this system include those which are involved in the operation of the secretion system itself and those which affect host cell function in some way, i.e. they are effector molecules. The former include InvJ, SpaO and SipD. SipD shares homology with the IpaD protein of *Shigella* spp., which regulates the secretory system in these organisms by forming a plug in the pore. It may be that SipD has a similar function in *Salmonella* spp. SipA, SipB, SipC and SptP are thought to constitute the main effector molecules exported by the secretory system although little is known of their functions. SipB shares significant sequence homology with IpaB of *Shigella* spp. and YopB of *Yersinia* spp. and, as the latter is known to form a pore in the target host cell membrane, SipB may play a similar role. SptP has been shown to have tyrosine phosphatase activity and therefore be involved in modulating host cell signal transduction pathways. SipC is homologous to the IpaC protein of *Shigella* spp. which induces membrane ruffling and endocytosis, suggesting that SipC may have a similar function in *Salmonella* spp.

Since the early 1990s, it has been known that regulation of invasion is profoundly affected by a number of environmental factors including oxygen levels, osmolarity and bacterial growth phase. Hence invasion is enhanced under anaerobic conditions, when the bacteria

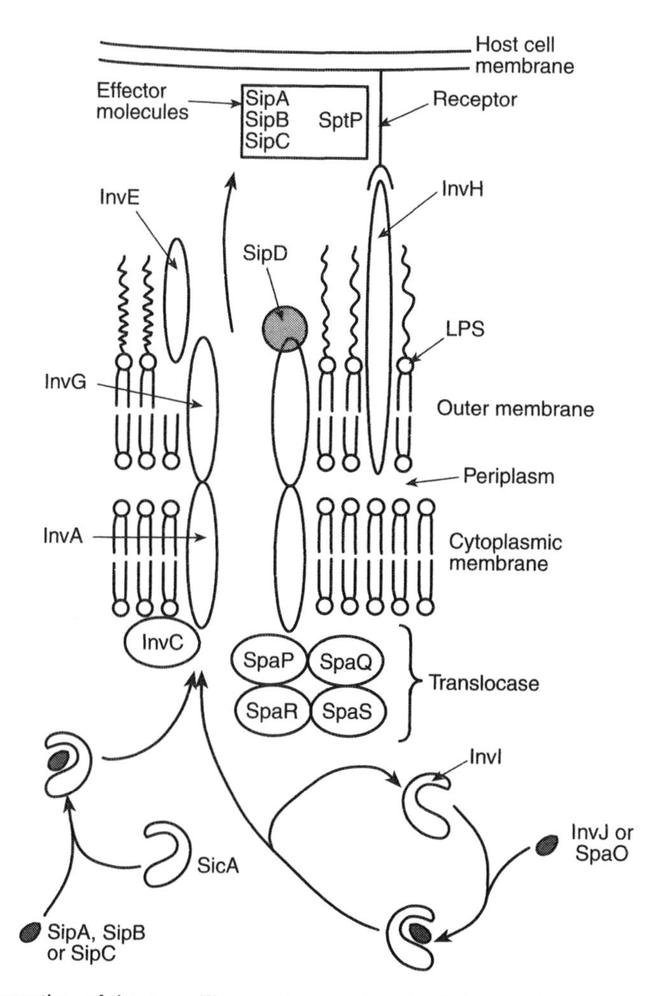

Figure 6.6 Operation of the type III secretory system in *Salmonella* spp. (see text for details)

are in the stationary growth phase and when the osmolarity is high. It is now apparent that regulation of invasion gene expression is under the control of at least three systems the two-component regulatory system, PhoP/Q, and the transcriptional factors HilA and InvF. The PhoP/Q system represses the transcription of genes encoding the type III secretion system as well as some of the SsP proteins in response to environmental signals which activate the PhoQ sensor kinase. The genes repressed by PhoP/Q are designated *prg* (PhoP repressed genes) genes. InvF promotes transcription of the *sspBCDA* operon while HilA activates the expression of PhoP-repressed invasion genes such as *prgHIJK, sspA, sspC, invF* and *orgA*. Transcription of *hilA* is under the control of SirA, a protein containing 234 amino acids which is similar to response regulators of His-Asp phosphorelay regulatory systems. Interestingly, the *sirA* locus is at centisome 42.4, i.e. it is not on SPI-1. SirA is activated by phosphorylation by an unidentified sensor protein in response to some as yet unidentified environmental signal(s).

Shigella spp.

This genus consists of four species *Sh. dysenteriae, Sh. flexneri, Sh. boydii* and *Sh. sonnei* – all of which can cause gastrointestinal infections in man. The severity of the infection depends on which species is involved but all induce a diarrhoea which may contain mucus and/or blood. The disease is contracted by ingestion of contaminated water or food and is generally self-limiting in adults. In children, however, it can be fatal and it is a major killer of infants in some developing countries.

Following attachment of bacteria to the epithelium of the colon, pseudopod formation can be seen and these structures can extend from the cell surface for tens of microns. The pseudopods extend around the bacterium, so enclosing it in a large vacuole. After internalization of the vacuole the appearance of the cell surface returns to normal. The ingested bacteria escape from the vesicle, multiply and then move through the cytoplasm to induce a protrusion into an adjacent cell. The membranes of the protrusion are then lysed, so enabling bacteria to gain access to the adjacent cell (Figure 6.7). The bacteria do not normally spread systemically but are confined to subepithelial tissues.

Little is known of the adhesive structure or the adhesin(s) responsible for mediating attachment of *Shigella* spp. to colonic epithelial cells although LPS has been implicated as an adhesin and β_1 integrins have been suggested as possible sites of attachment. In the case of *Sh. flexneri* it has been suggested that a Congo red-binding protein may function as an adhesin although its receptor on epithelial cells has not been identified. Although the adhesin of *Sh. dysenteriae* has not been identified, the receptor is thought to be an *N*-acetylneuraminic acid-containing molecule.

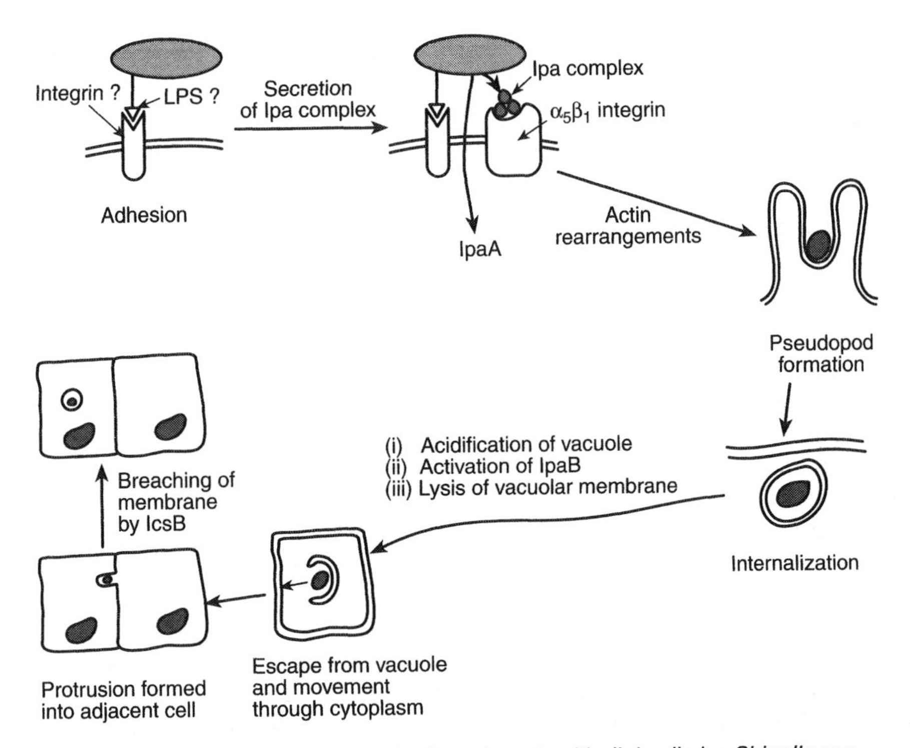

Figure 6.7 Stages involved in the invasion of epithelial cells by *Shigella* spp.

Invasion of epithelial cells by *Shigella* spp. requires the presence of a 220 kb plasmid which contains 32 invasion-associated genes. These genes are arranged in two divergently transcribed regions. One of these contains the *ipa* (invasion plasmid antigen) genes which are responsible for inducing bacterial entry into the host cell, while the other consists of the *mxi-spa* (membrane expression of Ipas–surface presentation of antigens) locus which is involved in the formation of an entry-associated secretion and translocation system and is an example of a type III secretion system. Little is known of the organization or control of this secretory system in *Shigella* spp. However, as the components of such systems are highly conserved at the protein sequence level, comparison of *Shigella* proteins with those of known function from other organisms gives us some idea of their likely function (Table 6.3).

Table 6.3 Components of the type III secretion system displaying homology among intestinal pathogens

Salmonella spp.	*Shigella* spp	*Yersinia* spp.	Possible location	Possible function in one or more species
InvA	MxiA	LcrD	Cytoplasmic membrane	Forms channel in cytoplasmic membrane
SicA	IpgC	SycD	Cytoplasm	Chaperone
InvC	Spa47	YscN	Cytoplasm/membrane	ATPase – energizer of secretion system?
InvG	MxiD	YscC	Outer membrane	Forms channel in outer membrane
InvE	MxiC	YopN	Outer membrane?	Regulation of secretion
PrgK	MxiJ	YscJ	Outer membrane	?
SpaP	Spa24	YscR	Cytoplasmic membrane	Translocase?
SipB	IpaB	YopB	Outer membrane or secreted	Pore formation in host cell membrane?

The Ipa proteins constitute the secreted effector proteins. IpgC functions as a molecular chaperone, while IpaB and IpaD probably function in the regulation of the secretion of effector molecules by forming a plug for the secretory channel. IpgC is a 17 kDa molecular chaperone of the 62 kDa IpaB protein and the 43 kDa IpaC protein and prevents aggregation and degradation of these key invasion-associated proteins. Within 5 minutes of contact with a host cell, Ipa proteins are transported to the bacterial cell surface and into the environment via a channel formed by products of the *mxi-spa* locus (Figure 6.8). Secretion of these proteins is dependent on the bacterial growth cycle, with more proteins secreted when the bacteria have reached the division stage. Initial secretion occurs in the region of the septation furrow of the cell. Once in the extracellular environment the IpaB and IpaC proteins (and possibly IpaA but not IpaD) form a soluble complex known as the Ipa complex. This complex induces pseudopod formation in epithelial cells following binding to the $\alpha_5\beta_1$-integrin. The formation of these extensions is dependent on actin reorganization within the host cell and inhibitors of F-actin, such as cytochalasin, prevent invasion by *Shigella* spp. F-actin polymerization occurs beneath the point of contact with the bacterium and in the extensions which form around it. Actin filaments are cross-linked in the projections due to the involvement of the actin-bundling protein plastin, indicating that the mechanism of pseudopod formation is similar to that of microvillus formation. Other cytoskeletal proteins which accumulate at the site of attachment of the bacterium include α-actinin, vinculin and talin. It is possible that one of the Ipa proteins, IpaA (a 78 kDa protein), enters the host cell and binds directly to vinculin, which induces the accumulation of α-actinin and talin to the membrane of what is destined to become the bacterium-enclosing vacuole. The protein Src tyrosine kinase, $pp60^{c-src}$, which is involved in actin cytoskeletal rearrangements, also accumulates at the site of bacterial entry and is thought to be responsible for tyrosine phosphorylation of cortactin. This protein co-localizes with newly formed actin filaments and is probably involved in the cytoskeletal remodelling essential for bacterial uptake. Another enzyme involved in cytoskeletal rearrangements, the small GTPase Rho, has also been implicated in *Shigella* invasion. An inhibitor of this enzyme prevents the formation of projections and reduces uptake of *Shigella* spp. by epithelial cells. This contrasts with invasion by *Salmonella* spp., which is unaffected by inhibitors of Rho. Invasion of *Shigella* spp. also differs from that seen in *Salmonella* spp. in that cells are more invasive when in the early log phase of growth rather than the stationary growth phase.

After internalization, the bacteria escape from the vesicle into the cytoplasm and this is probably mediated by the IpaB protein as *IpaB* mutants are unable to exit from the vesicle. IpaB has lytic activity and this is maximal at low pH. It is likely, therefore, that the lowering of the pH which usually occurs after vesicle formation could act as a trigger for lysis of the vesicular membrane and release of the bacteria. The bacteria then multiply rapidly and begin to spread throughout the cell. Movement of the bacteria is achieved by a mechanism

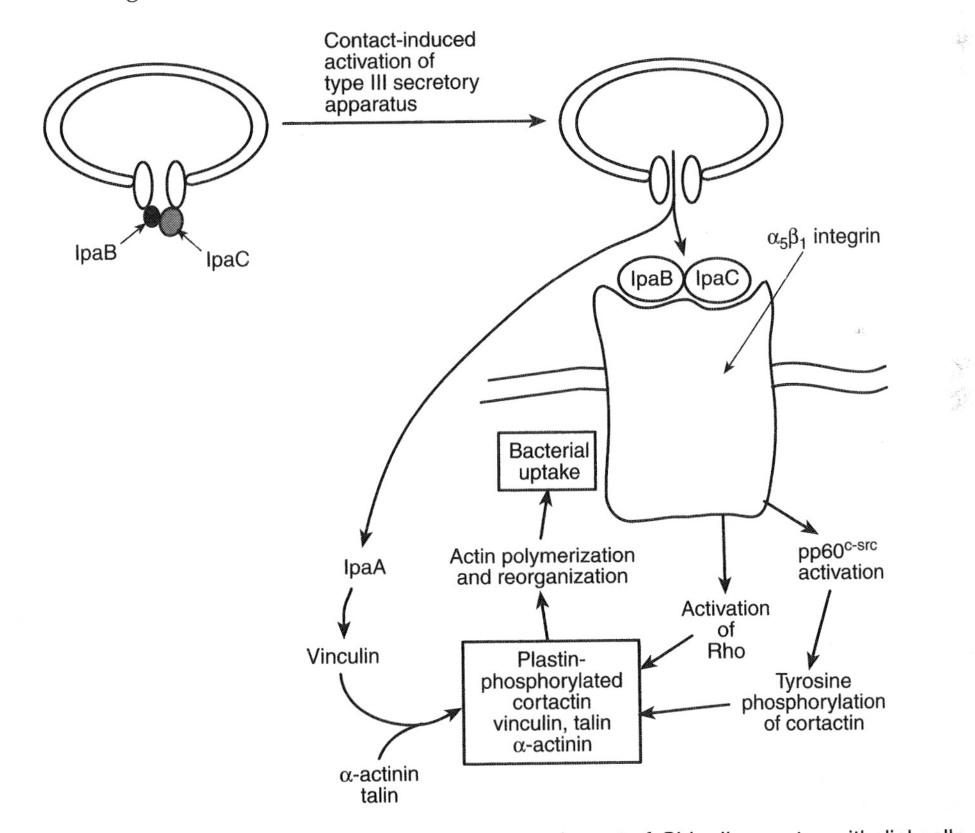

Figure 6.8 Host cell signal transduction following attachment of *Shigella* spp. to epithelial cells

involving the recruitment of actin to form a 'tail' at one end of the bacterium which propels it through the cytoplasm and into an adjacent cell. The formation of the actin tail is dependent on the presence of a 120 kDa outer membrane protein IcsA (VirG) which accumulates in the bacterial outer membrane at the point of tail formation. The N-terminal 706 amino acids (the α-domain) of the molecule protrudes into the cytoplasm, while the remaining 344 amino acids (the β-domain) are embedded in the outer membrane. An N-terminal 95 kDa fragment (formed by cleavage of an Arg–Arg bond at position 758–759, close to the junction of the α- and β-domains) is slowly released from the molecule due to the action of IcsP, a *Shigella* protease. The 95 kDa fragment has ATPase activity and is thought to induce actin condensation, hence propelling the bacterium through the cytoplasm ahead of the actin tails (see Chapter 2 for a description of actin polymerization).

In addition to *Shigella* spp., a number of other bacterial species are also able to utilize polymerization of actin to propel themselves within a host cell. These include *Lis. monocytogenes* (see next section), *Lis. ivanovii* and some *Rickettsia* spp. Certain viruses also utilize this system. It has long been known that actin polymerization was involved in movement and shape change in normal, uninfected mammalian cells. However, in mammalian cells, the polymerization of actin is initiated at the plasma membrane after a signal has been received, whereas actin polymerization and movement by intracellular bacteria do not involve signal transduction at a membrane. On contacting the cytoplasmic membrane, the bacteria are pushed into the adjacent cell, forming a protrusion. The membranes separating the two cells are then breached, possibly by the *Shigella* protein IcsB, and the bacteria invade the adjacent cell.

As was the case for *Yersinia* spp., animal studies, this time in monkeys, suggest that *Shigella* spp. invade the intestinal mucosa via M cells. Invasion of epithelial cells may then take place via their basolateral surfaces where integrins, the putative receptors for shigella adhesins, are located. This would enable the organisms to avoid phagocytes and other arms of the host defence system.

Listeria monocytogenes

This Gram-positive organism is the cause of an uncommon, but serious, form of food poisoning with a mortality rate as high as 70%. It is one of the few bacteria capable of crossing the placenta and this can result in pre-term labour, stillbirth or infected infants.

Uptake of the organism by epithelial cells takes place in a manner similar to *Yersinia* spp., as it involves the formation of pseudopodia which associate closely with the bacterium, resulting in its engulfment within a vacuole. There is no membrane ruffling and, although cytoskeletal rearrangements are involved, these are highly localized. The bacteria then escape from the vacuole, multiply and move within the cytoplasm by a mechanism involving the formation of actin tails.

Very little is known about the means by which the organism binds to host cells, although it has been suggested that the adhesin may contain α-D-galactose residues which bind to complementary mammalian cell receptors. More recently, evidence has accumulated implicating heparan sulphate proteoglycans (e.g. heparin and heparan sulphate) as key host cell receptors involved in the invasion process. Pre-treatment of bacteria with either of these molecules inhibited invasion of epithelial cells, and heparinase treatment of host cells resulted in a substantial reduction in bacterial adhesion and invasion. The bacterial surface protein ActA, which is involved in intercellular movement of the bacterium (see below), may be the complementary adhesin of the heparan receptor as it contains several clusters of positively charged amino acids which could function as binding domains.

Invasion of intestinal cells is mediated by an 80 kDa surface-associated/secreted protein known as internalin (InlA). Mutations in the *inlA* gene render the organism incapable of invading intestinal cells. InlA contains multiple tandem copies of a 22 amino acid leucine-rich motif known as a leucine-rich repeat and its C-terminus is similar to that of the M protein of *Streptococcus pyogenes*. Like the Ipa proteins of *Shigella* spp., InlA mediates the invasion of polarized epithelial cells via their basolateral surface. The mammalian cell receptor for this molecule is E-cadherin, a transmembrane glycoprotein involved in the formation of intercellular junctions. The molecule consists of a cytoplasmic domain which interacts with cytoskeletal components, a transmembrane region and an extracellular domain which mediates adhesion. The binding of the natural ligands to E-cadherin does not lead to their uptake. *Lis. monocytogenes* has a number of internalin homologues, one of which, InlB, is involved in the invasion of hepatocytes and endothelial cells. The host cell receptor for InlB has not yet been identified. Invasion of intestinal cells is blocked by tyrosine kinase inhibitors, suggesting the involvement of these signalling molecules, although the identity of the specific kinases has not been established. Invasion also requires phosphoinositide 3–kinase (PI3K), which is involved in the control of cytoskeletal rearrangements through the Rac GTPase (Figure 6.9 – refer also to Chapter 3 for discussion of PI3K).

Escape from the vacuole is mediated by the actions of a haemolysin, listeriolysin O, together with a phosphatidylinositol-specific phospholipase C and a broad-range

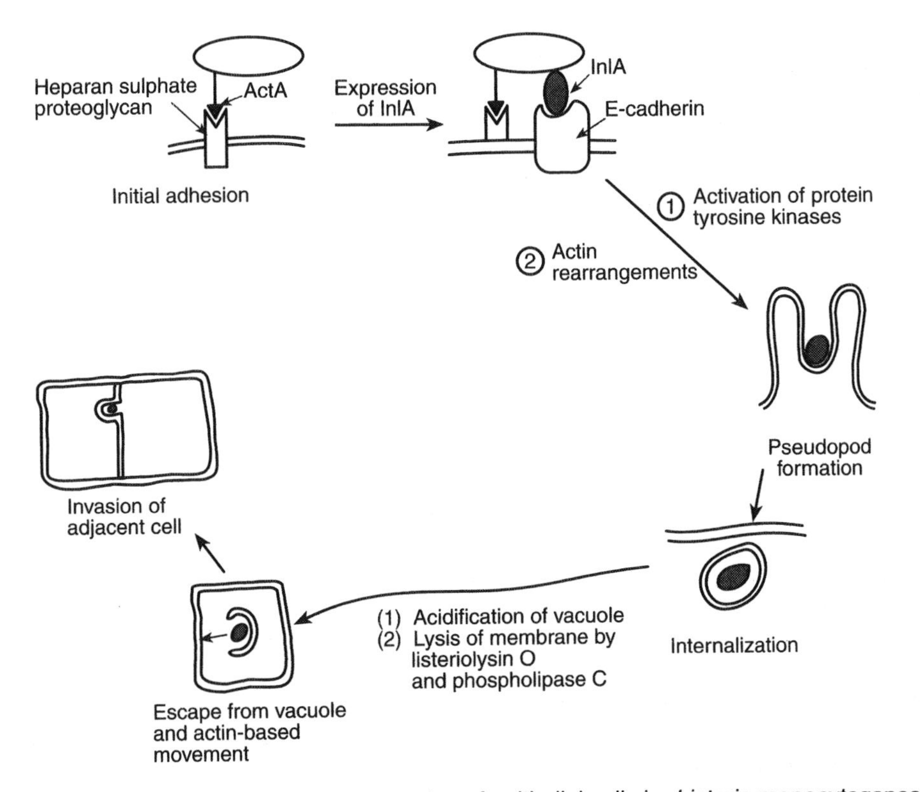

Figure 6.9 Stages involved in the invasion of epithelial cells by *Listeria monocytogenes*

phospholipase C. In contrast to other members of the haemolysin family, listeriolysin O displays optimum activity at an acidic, rather than neutral, pH – an adaptation suited to its role in lysing the membrane of the acidic vacuole. Once the internalized bacteria have been released from vacuoles they replicate (with a mean generation time of 50 minutes) in the cytoplasm and become coated in actin filaments. The actin then forms a long tail which trails behind the bacterium and can move at a rate of between 6 and 60 μm/min (Figure 6.10). The greater the speed of the bacterium the longer the actin tail. Movement is the result of continuous assembly of microfilaments followed by their release and cross-linking behind the bacterium (see Chapter 2). It has been estimated that the turnover rate for the actin filaments which form the tail is 33 seconds. The bacterial protein responsible for actin polymerization, and hence bacterial movement is ActA, which accumulates at the poles of the organism, with the highest concentration at the pole adjacent to the actin tail. The N-terminal and a proline-rich region of ActA have high homology with regions of human vinculin – a cytoskeletal protein. The actin tails have been shown to contain a variety of proteins including α-actinin, tropomyosin, fimbrin, profilin, vinculin, ezrin, talin, gelsolin, VASP and the Arp2/Arp3 complex. However, the role of many of these proteins in enabling bacterial movement has not yet been established.

On contacting the cell membrane, the bacterium continues to move forward, resulting in a protrusion into the adjacent cell. The bacterium then penetrates these two membranes, possibly as a result of lipid hydrolysis due to bacterial phosphatidylcholine-specific phospholipase C, and so invades the adjacent cell.

Mycobacterium tuberculosis

Tuberculosis kills approximately 3 million people annually and it has been estimated that approximately one third of the world's population is infected with the causative organism, *M. tuberculosis*. The disease is primarily a lung infection and is transmitted via aerosols generated by infected individuals. The bacteria may then invade the epithelium of the alveoli and/or the alveolar macrophages and are subsequently transported throughout the body within monocytes and macrophages. Although invasion of macrophages by *M. tuberculosis* has received most attention (see 'Survival in remodelled vacuoles', below), studies have been directed at establishing the means by which the organism invades epithelial cells. The type II pneumocyte is one of the major cells of the alveolar epithelium and this has been shown to be invaded by *M. tuberculosis in vitro*. The bacteria are found both inside vacuoles as well as inside the cytoplasm of the pneumocyte.

Burkholderia cepacia

This bacterium causes potentially fatal lung infections in patients suffering from cystic fibrosis. It is virtually impossible to eradicate once an individual becomes colonized. It can also invade the respiratory epithelium, causing a necrotic infection that is associated with septicaemia. *In vitro* studies have shown that these bacteria may invade human alveolar epithelial cells in a manner similar to the invasion of gut epithelial cells by *Salmonella* spp. Invasion of epithelial cells by *Burk. cepacia* involves adhesion of the bacterium to the cell's microvilli followed by endocytotic engulfment. Invasion appears to involve microfilaments but not microtubules, as inhibitors of the former (cytochalasin) but not the latter (colchicine) prevented bacterial invasion. The bacteria do not appear to escape from the vacuoles but can grow and reproduce within them.

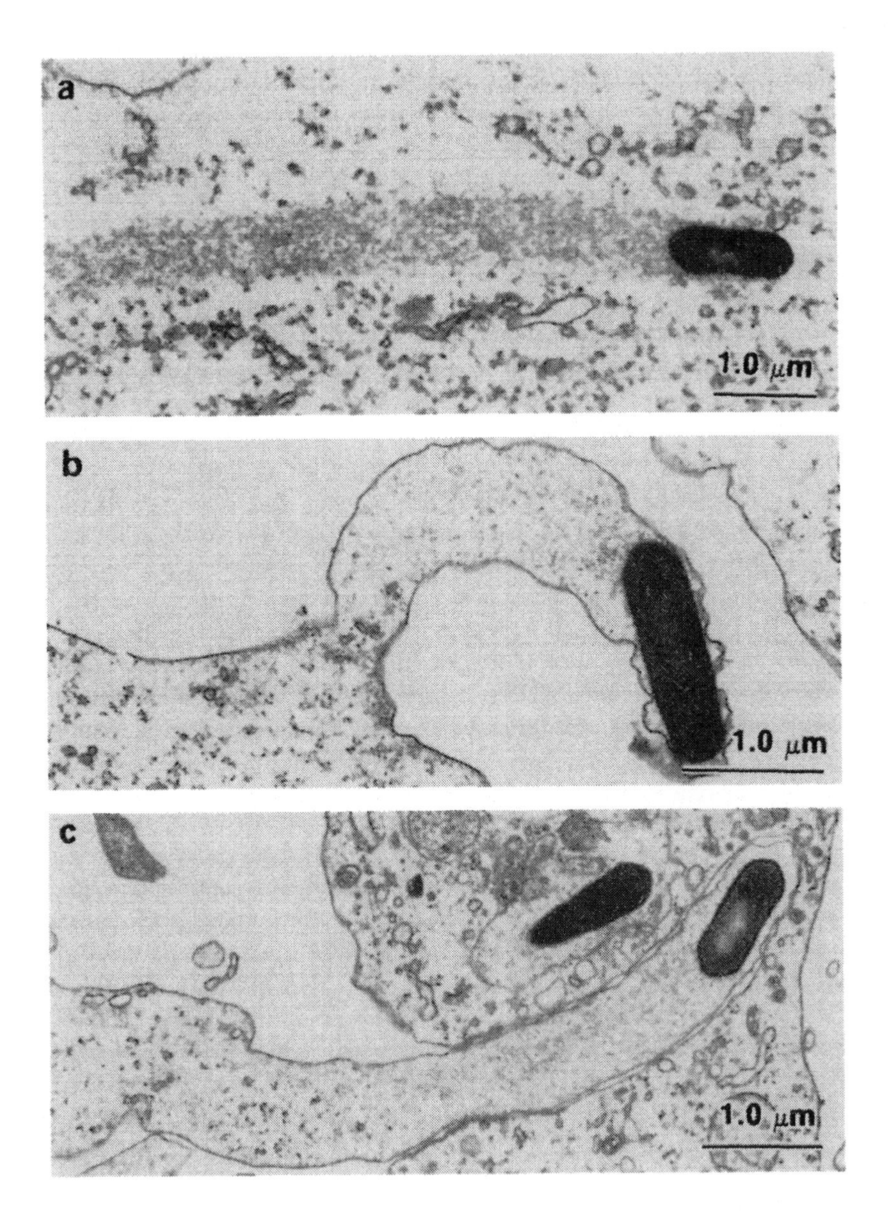

Figure 6.10 Thin sections through the murine macrophage cell line J774 which is infected with *Listeria monocytogenes.* In (a) one can see a moving bacterium which clearly shows the F-actin comet tail. In (b) a bacterium has encountered the plasma membrane of the macrophage and has formed a characteristic protuberance. In (c) a bacterium is moving from one cell to another by the engulfment of the protuberance containing the bacterium. As can be seen the engulfed bacteria is transported from one cell to another but never leaves the cytoplasm of the macrophage. Reproduced from 'Host–pathogen interactions during entry and actin-based movement of *Listeria monocytogenes*' by Ireton K and Cossart P. With permission, from the *Annual Review of Genetics*, Volume 31, © 1997, by Annual Reviews

Cell invasion involving microtubules

Although microfilaments appear to be involved in the invasion of host cells by many bacterial pathogens, in some cases invasion also involves rearrangement of cytoskeletal microtubules (see Chapter 2 for a description).

Campylobacter jejuni

This organism is one of the main causes of human diarrhoeal disease and is usually contracted by ingestion of contaminated poultry or by consumption of contaminated drinking water. The clinical symptoms can vary from a watery diarrhoea to a febrile state with blood-containing diarrhoea. Invasion is likely to be a feature of some cases as bacteraemia can occur and the organism has been detected within intestinal biopsies from patients. Invasion of a number of intestinal epithelial cell lines by various strains of *C. jejuni* has been demonstrated and, interestingly, invasion by one of these bacterial strains used was unaffected by the microfilament depolymerizer, cytochalasin D. In contrast, the 'microtubule depolymerizer' colchicine caused a marked reduction in invasion by this strain. The synthesis of new bacterial proteins (possibly adhesins to mediate intimate cell contact), but not new eukaryotic cell proteins, was found to be essential for invasion, which was accompanied by the formation of clathrin-coated pits (see Chapter 2). It is important to note that the invasion mechanism, particularly with regard to the involvement of microfilaments and microtubules, was very much dependent on the particular bacterial strain and epithelial cell line used – with some strain/cell line combinations, microfilament formation was crucial to the invasive process.

Porphyromonas gingivalis

Por. gingivalis is one of several oral bacteria implicated in periodontitis, the most common chronic infectious disease of humans, which involves the destruction of tooth-supporting tissues eventually resulting in tooth loss. Many studies have demonstrated the ability of this organism to invade oral epithelial cells *in vitro*. Invasion proceeds more efficiently in primary cell cultures than in transformed cells, suggesting that transformation-induced changes in cell surface receptors or in intracellular signalling pathways may be detrimental to the invasion process. Invasion requires bacterial, but not host cell, protein synthesis and is more effective with log- and stationary-phase bacteria than lag-phase cells. The bacteria adhere to epithelial cells via fimbriae and this is followed by the appearance, beneath the point of attachment, of an electron-dense region in the epithelial cell, suggestive of the formation of coated pits. Uptake of the bacteria would therefore appear to involve receptor-mediated endocytosis and this is supported by the finding that monodansylcadaverine, an inhibitor of this process, prevents bacterial invasion. The invasion process involves the tyrosine phosphorylation of a 43 kDa protein, which may be a mitogen-activated protein (MAP) kinase, and invasion is inhibited by colchicine but not by cytochalasin D, suggesting the involvement of microtubules rather than microfilaments. As in the case of *C. jejuni*, however, the invasion mechanism appears to be strain-specific as uptake of some strains also requires the involvement of microfilaments. Invasion also induces a transient increase in epithelial cell cytosolic [Ca^{2+}], a characteristic shared with *Sal. typhimurium* and enteropathogenic *E. coli*. This increase in intracellular [Ca^{2+}] is not the result of the influx of extracellular Ca^{2+} but appears to be due to release of Ca^{2+} via activation of the smooth endoplasmic reticuluum

Ca^{2+} ATPase (SERCA) pumps. The bacteria rapidly become located in the cytoplasm and appear to be able to replicate there but intercellular spread has not been observed. Periodontitis is a chronic, localized infection and the bacteria do not, apparently, invade deeper tissues. Invasion of epithelial cells may provide the organism with a protected, nutritionally rich environment enabling it to persist within the oral cavity for long periods of time and to occasionally induce acute episodes of tissue destruction.

Klebsiella pneumoniae

This organism is an opportunistic pathogen associated mainly with nosocomial pneumonia and bacteraemia. However, it is also known to be a frequent cause of urinary tract infections in women. Invasion of both bladder and intestinal epithelial cells by this organism has been demonstrated *in vitro* and was inhibited by G-strophanthin, suggesting that uptake was by receptor-mediated endocytosis. Although cytochalasin inhibited invasion, suggesting the involvement of actin microfilaments, it was also found that taxol (which stabilizes microtubules) and colchicine (which depolymerizes microtubules) reduced invasion, implying that microtubules were also involved. Once internalized, the bacteria remained in vacuoles and multiplication was evident. The adhesin mediating attachment of the organism to epithelial cells has not yet been identified but is thought to bind to the *N*-acetylglucosamine residues of a glycoprotein receptor.

Forced entry

Rickettsia prowazekii is a Gram-negative obligate intracellular parasite which is the causative agent of typhus – a disease contracted from the faeces of infected lice either by inhalation or transmission through broken skin. The organism binds to a cholesterol-containing receptor in the host cell membrane and then penetrates into the cell – a process which requires actively metabolizing bacteria and host cells. Invasion is also dependent on phospholipase A as an inhibitor of this enzyme, phentermine, prevented invasion.

Paracytosis

This is a process in which bacteria pass through layers of cells without actually penetrating individual cells of these layers. It is therefore an example of tissue invasion rather than cellular invasion. A good example of an organism displaying this behaviour is *Haemophilus influenzae*, which resides in the upper respiratory tract of man. This organism is responsible for a wide range of diseases including localized respiratory tract infections and systemic infections such as meningitis, arthritis and cellulitis. Evidence that paracytosis may occur during the disease process *in vivo* comes from examination of biopsies of the respiratory mucosae of patients with bronchitis due to this organism. These have revealed the presence of bacteria in the subepithelial layers although the epithelium itself was undamaged. An *in vitro* study using a human lung epithelial cell line has shown that the organism can pass through layers of these cells (up to three cells thick) without affecting their viability or the integrity of the layer (Figure 6.11). Bacteria could be seen between the cells. Penetration is thought to involve disclosure of the intercellular junctions and is prevented by inhibitors of bacterial protein synthesis.

Other examples of paracytosis include the passage of *Treponema pallidum* and *Borr. burgdorferi* across endothelial cell monolayers.

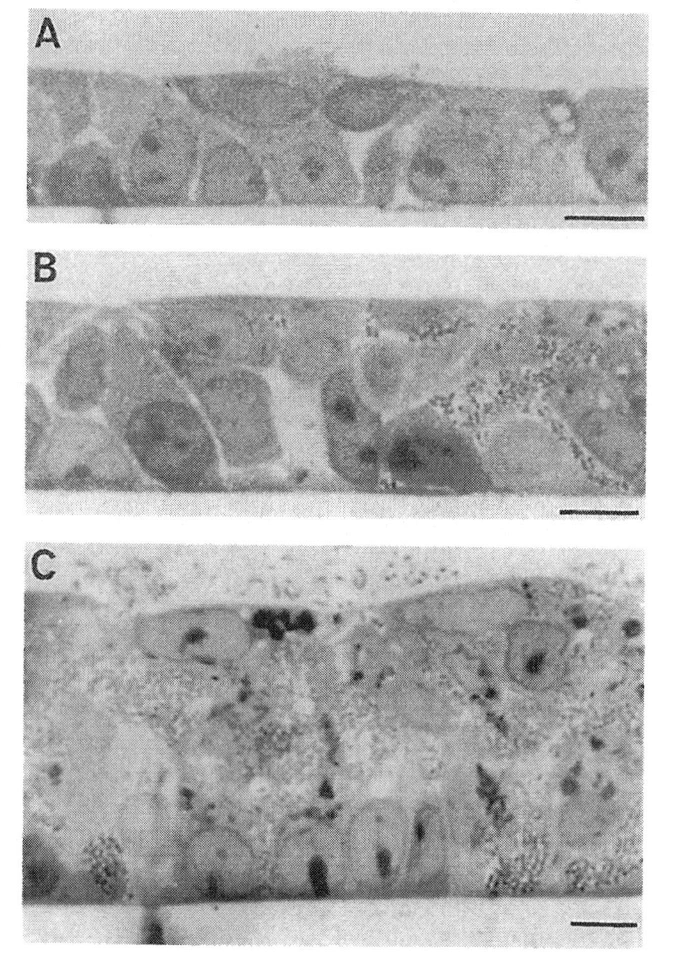

Figure 6.11 Paracytosis of a human lung epithelial cell line by *Haemophilus influenzae*. In (A) the cell layer is uninfected while in (C) the cells have been infected with an adherent strain of *Haem. influenzae* for 24 hours. Bacteria are clearly visible between, and on the surfaces of, the epithelial cells. In (B) cells have been cultured in the presence of a non-adherent strain of the bacterium and only occasional clusters of bacteria can be seen between the cells. Reproduced with permission from 'Paracytosis of *Haemophilus influenzae* through cell layers of NC1-H292 lung epithelial cells' by van Schilfgaarde M, van Alphen L, Eijk P, Everts V, Dankert J. (1995) *Infect Immun* 63: 4729–4737

Invasion of endothelial cells

Many organisms, having traversed an epithelial surface, spread throughout the body causing pathology at sites distant from the site of entry, e.g. *Neisseria meningitidis, Streptococcus pneumoniae, Haem. influenzae, Sh. dysenteriae, Sal. typhi* etc. In order to do so they must enter the bloodstream, which entails crossing another cellular barrier – the endothelium. Although invasion of endothelial cells by bacteria is not as well documented as invasion of epithelial cells, invasion appears to follow one of four main courses: (i) invasion followed by intracellular persistence without multiplication (e.g. *Staphylococcus aureus, Pseudomonas aeruginosa,* group B streptococci); (ii) invasion followed by intracellular replication (*Rickettsia rickettsii*, *Bartonella henselae*); (iii) traversal without cell disruption (e.g. spirochaetes); and (iv) invasion within phagocytes (e.g. *Lis. monocytogenes*).

Neisseria meningitidis

Meningitis is a life-threatening infection of the central nervous system characterized by inflammation of the meninges (the membranes enclosing the brain and spinal cord) and invasion of the subarachnoid space. The commonest cause of the infection in the UK is *N. meningitidis*, an organism which is found in the nasopharynx of as many as 15% of the healthy population. The organism possesses a number of adhesins which have been postulated to play a role in the invasion of host cells (Table 6.4). The expression of the various adhesins is affected by the host environment as different adhesins/invasins will be needed at different stages of the disease process. Access to the meninges by this organism requires that bacteria invade the endothelium and exit the blood vessels. Which of the many putative adhesins/invasins mediates this process is a controversial area but there is now considerable evidence that the major adhesin involved is the 28 kDa outer membrane protein Opc. Interestingly, Opc shares sequence homology with the invasin of *Yersinia* spp. Opc, however, does not interact directly with the endothelial cell surface but binds to serum proteins containing Arg-Gly-Asp (RGD) sequences (probably vitronectin) which then interact with the $\alpha_v\beta_3$-vitronectin receptor on the surface of the endothelial cell. This triggers the uptake process, which involves the formation of membrane protrusions that enclose the attached bacteria resulting in their internalization within vacuoles (Figure 6.12).

The ability of Opc to mediate adhesion (and hence invasion), however, is affected by other bacterial surface molecules and this has been demonstrated using mutants or variants defective in the expression of one or more putative adhesins. Hence, sialylation of LPS was found to significantly inhibit the ability of Opc to mediate adhesion to endothelial cells, possibly as a result of steric hindrance of receptor-binding epitopes or because of alterations in surface charge. The presence of a capsule also significantly reduced adhesion to, and invasion of, endothelial cells. In contrast, pili significantly increased invasion. The most invasive

Table 6.4 Adhesins/invasins of *Neisseria meningitidis*

Adhesin/invasin	Target cell	Function	Receptor
LPS or LOS	Epithelial cell	Adhesion, invasion	Asialoglycoprotein receptor
Opa family	Epithelial cell, PMN endothelial cell	Adhesion, invasion	CD66 family, Glyosaminoglycans
Opc	Epithelial cell, PMN Endothelial cell	Adhesion, invasion	Vitronectin
PilC	Epithelial cell Endothelial cell	Adhesion	CD46 (C3b/4b receptor)
Pilin	Epithelial cell, PMN Endothelial cell	Adhesion, invasion?	?
Por (PI)	Epithelial cell, PMN Endothelial cell	Invasion	Direct insertion

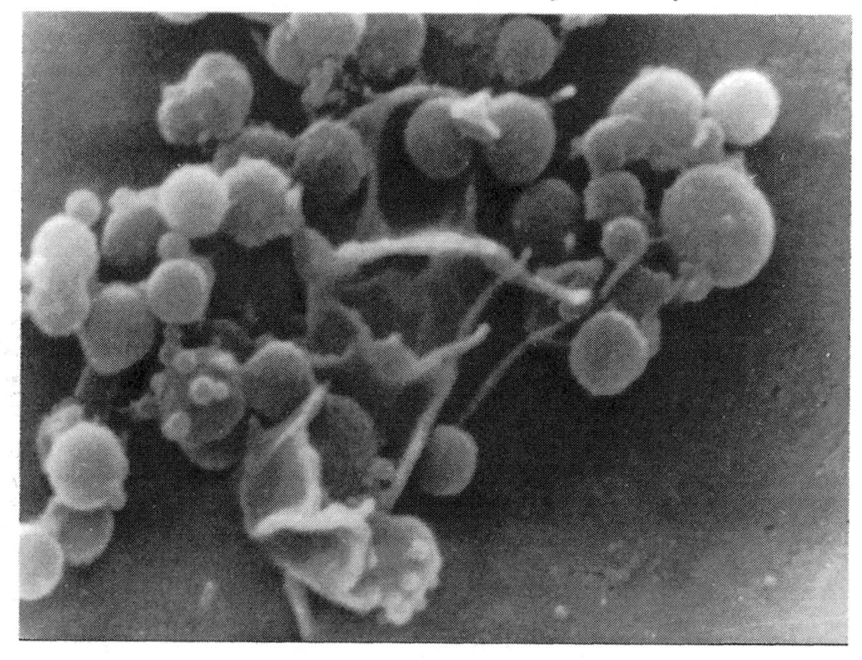

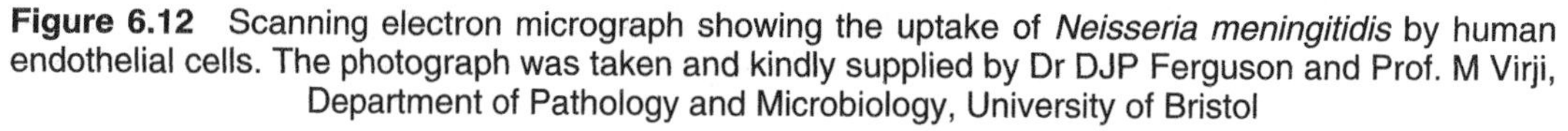

Figure 6.12 Scanning electron micrograph showing the uptake of *Neisseria meningitidis* by human endothelial cells. The photograph was taken and kindly supplied by Dr DJP Ferguson and Prof. M Virji, Department of Pathology and Microbiology, University of Bristol

phenotype, therefore, was found to be the Opc-expressing, piliated, non-capsulated strain which did not have sialylated LPS. Variation in the expression of surface structures, including capsule, pili and sialylation of LPS, occurs during the course of infection with *N. meningitidis*. Invasion is inhibited by cytochalasin D, implying the involvement of actin microfilaments, and is also inhibited by antibodies against Opc. Subsequent to invasion, the bacteria have been shown to emerge from the cells and can successfully traverse human umbilical vein endothelial cell (HUVEC) monolayers.

Bartonella henselae

This organism is the causative agent of cat-scratch disease, which is the most common cause of chronic lymphadenopathy in children and adolescents. The disease, which is contracted from cats, is characterized by a pronounced lymphadenopathy that is often accompanied by fever, fatigue and headaches. In immunocompromised patients it can induce skin tumours due to proliferation of endothelial cells in response to colonization by the organism. Following adhesion to a human umbilical vein endothelial cell *in vitro*, the bacteria are transported by the leading lamella of the cell to a site just ahead of the nucleus where they form an aggregate (Figure 6.13). The bacteria are then engulfed by fusion of membrane protrusions which surround and then extend over the aggregate. The resulting globular structure (5–15 μm diameter) is known as an invasome. The formation of the protrusions involves the rearrangement of the actin cytoskeleton but is independent of microtubules. Both actin and phosphotyrosine accumulate in the membrane protrusions and intercellular adhesion molecule (ICAM)-1 accumulates at their tips, while actin stress fibres are twisted around the basal region of the structure. The overall morphology of the cells is also affected and they become spindle-shaped. The whole process is relatively slow, requiring approximately 24 hours for its completion. Little is known about the angiogenesis (i.e. the formation of new blood vessels) induced by this organism and it is not clear whether it is due to the secretion of an angiogenic factor by the organism, the induction of the release of an endogenous angiogenic factor (e.g. fibroblast growth factor) or interference with the inhibitory control of angiogenesis.

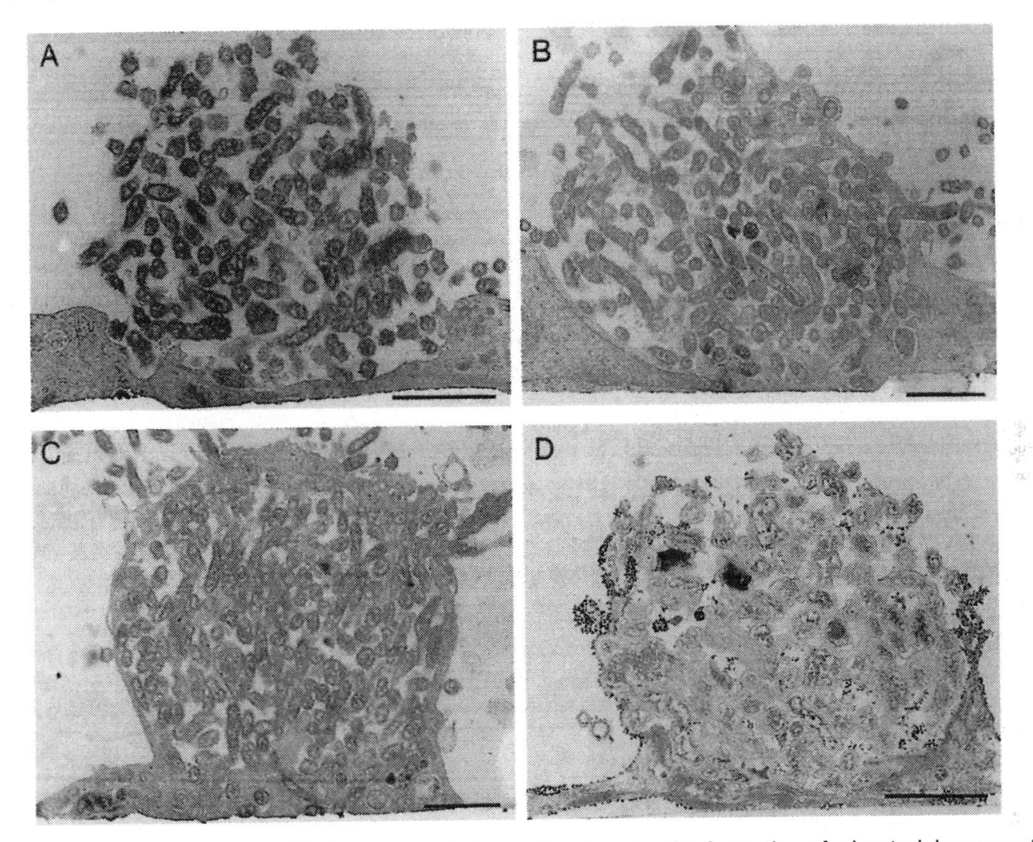

Figure 6.13 Invasion of HUVECs by *Bartonella henselae* showing the formation of a bacterial aggregate (A) followed by engulfment (B, D) and internalization (C) by means of an invasome. Reproduced with permission from 'Interaction of *Bartonella henselae* with endothelial cells results in bacterial aggregation on the cell surface and the subsequent engulfment and internalisation of the bacterial aggregate by a unique structure, the invasome' by Dehio C, Meyer M, Berger J, Schwarz H, Lanz C. *J Cell Sci* 1997; 110: 2141–2154, Company of Biologists Ltd

Escherichia coli

This organism is well known for its ability to cause diarrhoeal diseases and urinary tract infections but it is also one of the most frequent causes of neonatal meningitis. The disease is preceded by a bacteraemia but little is known of the way in which it crosses the blood–brain barrier. Experiments using brain microvascular endothelial cells (BMEC) have shown that a 35 kDa outer membrane protein A (OmpA) of the organism plays a central role in the invasion of these cells. Invasion of BMEC was blocked by an antibody to OmpA and an OmpA mutant showed considerably reduced invasion frequency. The OmpA protein is thought to mediate binding of the organism to BMEC receptors containing the *N*-acetylglucosamine β1 4 *N*-acetyl–glucosamine epitope. Interestingly, polymers of β1–4-linked *N*-acetylglucosamine were able to prevent entry of *E. coli* into the cerebrospinal fluid of neonatal rats. The use of receptor analogues could therefore form the basis of a novel therapeutic approach for meningitis.

Listeria monocytogenes

Invasion of endothelial cells by *Lis. monocytogenes* takes place in a similar manner to that described for its invasion of epithelial cells. Following binding to the endothelial cell via the invasion protein, internalin, the bacteria are endocytosed. They then escape from the vacuole, replicate within the cytoplasm and exhibit actin-based motility which they use to spread intercellularly. It has also been demonstrated that monocytes containing the organism can enter endothelial cells, after which the bacteria escape from the monocytes into the cytoplasm of the endothelial cell.

β-*Haemolytic streptococci*

These streptococci characteristically produce zones of haemolysis on blood agar and are subdivided into a number of groups on the basis of the antigenic nature of their cell walls. Two of the groups, Group A and Group B, are major pathogens of man.

Group B streptococci are a frequent cause of septicaemia and pneumonia in neonates. They are able to invade human BMEC *in vitro* by an endocytotic mechanism which involves both microfilaments and microtubules. The bacteria remain within vacuoles and do not appear to multiply, but do survive for up to 20 hours, and cause damage to the endothelial cells, possibly due to the action of the haemolysin they produce. Interestingly, when different strains of the organism were compared, it was found that the serotype III strains (which account for 90% of strains isolated from meningitis cases) were more invasive than strains belonging to serotypes Ia, Ib, II and V.

Group A streptococci are responsible for a variety of topical infections of mucosal and epidermal surfaces but can also cause disseminated infections such as puerperal fever and neonatal sepsis. They have been shown to be capable of invading HUVECs in a manner similar to that described for Group B streptococci.

Borrelia burgdorferi

The spirochaete *Borr. burgdorferi* is responsible for Lyme disease, a chronic infection characterized by a skin rash, fatigue, headaches, dizzy spells, facial paralysis and arthritis. The disease is a zoonosis (a disease acquired from an animal), and is transmitted from an infected animal (e.g. a deer) to man via a bite from the deer tick *Ixodes dammini*. After the tick bite, the organism may spread through the skin or may be disseminated in the bloodstream, from which it can gain access to a range of tissues by crossing the endothelium.

Penetration of an endothelial cell monolayer by the spirochaete can be enhanced by the presence of plasminogen acquired from plasma. The plasminogen molecule binds to the organism via a lysine residue, and each spirochaete has almost 3000 binding sites distributed over its surface. Once bound, the plasminogen can be converted to the enzymically active form, plasmin, by a host-derived plasminogen activator. The resulting plasmin is a potent serine protease which cannot only lyse fibrin clots (its main function) but can also hydrolyse other proteins such as laminin, which is a major component of the basement membrane. Studies *in vitro* have shown that plasmin-coated *Borr. burgdorferi* have a greater ability to penetrate endothelial cell layers than plasmin-free spirochaetes. This is an excellent example of the way in which a bacterium can manipulate normal host functions for its own benefit – in this case use is made of an extracellular protease system.

Whether or not passage of the organism across a monolayer occurs via a paracytotic or transcytotic pathway is the subject of controversy – possibly both pathways are involved. Hence, the organism has been reported to penetrate monolayers of HUVECs via the intercellular junctions. However, in another study of the passage of the organism across a HUVEC monolayer, only 7% of 500 bacteria observed were found in intercellular junctions; the remaining 93% were seen to be intracellular. In some cases it was evident that the bacteria were enclosed within host cell membranes.

Streptococcus pneumoniae

This organism is a frequent cause of three life-threatening invasive diseases meningitis, pneumonia and septicaemia. The way in which it adheres to, and invades, endothelial cells is particularly interesting in the context of host–bacteria interactions, as the organism makes use of the consequences of cytokine–host cell interactions to increase both its adhesive and invasive potential. *Strep. pneumoniae* can bind to either of two receptors on HUVEC which have epitopes consisting of GalNAcβ1–3Gal and GalNAcβ1–4Gal. Treatment of HUVEC with either TNFα or IL-1α was found to increase adhesion of the organism to these cells by almost 200% and 100% respectively. This increased adhesion was accompanied by the appearance of a new HUVEC receptor for the bacteria – one that could be blocked by *N*-acetylglucosamine. Evidence suggests that this newly displayed receptor is actually the receptor for the proinflammatory lipid mediator platelet-activating factor (PAF). PAF is produced when phospholipases release aracidonic acid from cellular phospholipids and the resulting phospholipid is appropriately acylated. PAF receptor antagonists could inhibit the enhanced adhesion of the bacteria to the cytokine-activated HUVEC and PAF is one of a growing number of examples in which phospholipid-derived mediators are being shown to have a controlling action on bacteria–host interactions.

The bacterial adhesin for this new receptor is the phosphorylcholine residue on its teichoic acid.Invasion of HUVECs which had not been treated with cytokines was minimal (<0.1% of adherent bacteria) but significant invasion (3.7% of adherent bacteria) was observed after they had been exposed to TNFα – again invasion was prevented by a PAF receptor antagonist.

This is yet another example of the way in which an organism can utilize host defence systems – cytokine production and PAF activity – to its own advantage.

Invasion of macrophages

One of the principal functions of the macrophage is to engulf antibody- or complement-coated (opsonized) bacteria, internalize them in a vacuole and then kill them. Adhesion of an antibody-coated bacterium to the macrophage is mediated by an interaction between the Fc domains of the antibodies and the Fcγ receptors on the macrophage surface. This binding triggers internalization of the bacterium within a vacuole which then fuses with a lysosome to form a phagolysosome. The bacterium is killed inside the phagolysosome by the action of enzymes, antimicrobial peptides, reactive oxygen species and low pH (see Chapters 2 and 8). However, a number of bacterial species have the ability to invade macrophages, so avoiding this sequence of events and thereby surviving within them. Other species are taken up by the normal phagocytic route but are able to survive within the macrophage – whether this should be classified as an invasive process is debatable.

Bordetella pertussis

Although *Bord. pertussis* does not invade the respiratory epithelium, the organism does invade, and survive within, macrophages and this is thought to account for the prolonged nature of the infection (i.e. whooping cough) in some patients. As described in Chapter 5, the main adhesins of the organism are two secreted proteins: filamentous haemagglutinin (FHA) and pertussis toxin (PT). The S2 and S3 subunits of PT are both involved in adherence to macrophages and recognize different macrophage receptors. Binding of S2 and S3 is reduced by galactose and sialic acid, respectively. These two adhesins also show homology with the N-terminal region of the lectin domains of a group of cell surface molecules known as selectins. The selectin on endothelial cells (E-selectin) interact with carbohydrates on the surface of neutrophils to induce rolling of the latter along the endothelium. This is accompanied by activation of the integrin CD11b/CD18 on the neutrophil surface which then induces migration of the neutrophil across the endothelium into the tissue space (see Chapter 8). PT mimics the action of selectins and induces the activation of CD11b/CD18 on the macrophage, which acts as a recptor for FHA. The FHA of *Bord. pertussis* has sequences which mimic regions of the natural ligands for CD11b/CD18: those containing the amino acid triplet Arg-Gly-Asp (RGD) and those containing factor X-like sequences. The lectin-binding domain of the FHA is also involved in adhesion of the organism to the macrophage and accounts for approximately 50% of the binding of the organism to macrophages. The receptor is an unidentified lactosylceramide-containing molecule. Once the organism has adhered to the macrophage, internalization takes place by a mechanism which still remains to be described.

Legionella pneumophila

This organism is found in a variety of natural aquatic habitats and in man-made water systems such as air conditioners. Inhalation of the bacteria in aerosols can result in a potentially lethal pneumonia known as Legionnaires' disease, which has a mortality rate of greater than 25%. Once in the lungs the organism invades alveolar macrophages and can spread systemically inside blood monocytes.

The main adhesin mediating attachment to macrophages appears to be a 29 kDa major outer membrane protein (MOMP), which is a porin attached to the underlying peptidoglycan. The MOMP binds to the complement components C3b and C3bi (products of complement activation by the alternative pathway), which then mediate attachment to complement receptors CR1 and CR3, respectively, on the macrophage membrane. Use of these receptors for invasion enables the parasite to avoid activating the oxidative burst of the phagocyte.

Attachment of the organism to these receptors induces the formation of long pseudopods by the macrophage which coil around the bacterium and draw it into the cell (Figure 6.14). Once internalized, the external membranes of the coil disintegrate, leaving the bacterium within a phagosome of conventional appearance. This 'coiling phagocytosis' has also been observed to mediate the uptake of a number of other organisms including *Borr. burgdorferi* and *Chlamydia psittaci*. A considerable degree of membrane sorting occurs during the formation and internalization of the coil, probably under the influence of as yet unidentified bacterial products. The major histocompatibility complex (MHC) class I and class II molecules and alkaline phosphatase are excluded from the inner face of the coil (which is destined to be the phagosomal membrane), while 5'-nucleotidase accumulates.

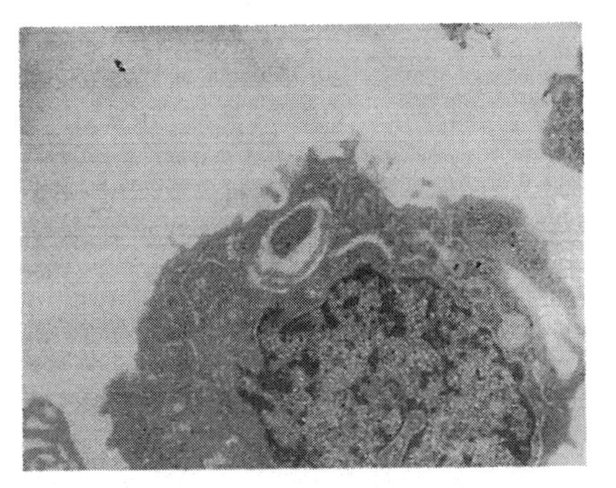

Figure 6.14 Coiled phagocytosis during invasion of a human monocyte cell line by *Legionella pneumophila*. Reproduced with permission from 'Growth of *Legionella pneumophila* in *Acanthamoeba castellanii* enhances invasion' by Cirillo JD, Falkow S, Tompkins LS. *Infect Immun* 1994; 62: 3254–3261

A key factor contributing to the invasiveness of the organism is the presence of a 24 kDa surface protein known as Mip (macrophage infectivity potentiator). Invasion experiments have demonstrated that isogenic mutants without an intact *mip* gene have greatly reduced ability to infect macrophages. Mip is a peptidyl-prolyl *cis/trans* isomerase which catalyses the *cis/trans* interconversion of prolyl peptide bonds in oligopeptides. Its role in invasion has not been clarified but it has been suggested that, because of its strong polycationic properties (polycations are known to induce phagocytosis of inert particles) it may promote phagocytosis.

Following internalization, the organism resides and multiplies within the resulting phagosome, as will be decribed later ('Survival in remodelled vacuoles').

Salmonella typhimurium

Following adhesion of the organism to the macrophage surface, dramatic ruffling of the host cell surface occurs and this is followed by macropinocytosis. In contrast to the localized membrane ruffling characteristic of internalization of *Sal. typhimurium* by epithelial cells, however, ruffling occurs over a much larger area and is not confined to the immediate vicinity of the individual organism. The internalized bacteria come to reside in vacuoles known as macropinosomes, which then fuse to form spacious phagosomes containing more than one bacterium. Bacteria are also found in smaller vacuoles, the membranes of which are tightly apposed to the bacterial surface. Eventually all of the internalized organisms are observed to reside in vacuoles with tightly apposed membranes.

Uropathogenic Escherichia coli

Uropathogenic strains of *E. coli* with type I fimbriae can, in the absence of opsonic antibodies, bypass the normal phagocytic pathway of macrophages which involves binding to Fc or C3b receptors, internalization and subsequent killing. Using the fimbrial protein FimH as an adhesin, the organism binds to CD48, a glycosylphosphatidylinositol (GPI)-linked membrane glycoprotein, and is internalized within a tight-fitting phagosome which is unlike the spacious phagosome formed when antibody-coated bacteria are phagocytosed. The resulting phagososme is not processed in the usual way and is not acidified, so enabling the bacteria to survive. Such a process could only occur in opsonin-deficient environments such as the urinary tract.

Listeria monocytogenes

Deposition of the complement components C1q or C3b on the surface of *Lis. monocytogenes* enables binding of the organism to corresponding macrophage receptors and results in internalization and subsequent killing of the organism. However, adhesion of the organism to the macrophage surface by other adhesins enables its internalization and intracellular survival. At least two bacterial proteins are thought to be involved in this process – internalin and a 60 kDa major extracellular protein, p60, encoded by the *iap* gene. A role for p60, which is a murein hydrolase required for cell division, in the invasion process has been deduced from experiments using p60-defective mutants which demonstrated that the mutant exhibited decreased adherence to, and invasion of, macrophages. Its precise function in the invasive process remains to be established but it has been suggested that it aids adhesion of the bacterium to the macrophage by reducing the surface charge of the host cell membrane. Once the organism has adhered to the macrophage surface, internalization is thought to occur by a mechanism described previously (see under '*Legionella pneumophila*'). Subsequent escape from the vacuole, actin-based motility and escape from the phagocyte take place as described previously for epithelial cells.

CONSEQUENCES OF INVASION

After an organism has invaded a host cell what happens next? Obviously, there are consequences for the host cell and for the bacterium and these will be discussed separately.

Effect on host cells

Invasion of a host cell by a bacterium can affect the former in a variety of ways, ranging from the barely perceptible to the ultimate effect – death. In many cases, however, the effects are intermediate between these extremes and may be transient or long-lived depending mainly on whether the bacterium takes up permanent residence within the cell. To catalogue all of the changes in host cells resulting from bacterial invasion would be an enormous task; therefore this section is limited to a discussion of those changes likely to have an effect on disease progression.

Cytokine release

A variety of cells respond to bacterial invasion by secreting cytokines (see Chapters 3, 5, 8 and 9) and the release of these inflammatory mediators may serve to activate host defence systems or, because of their overproduction, may have adverse consequences for the host. Human colon epithelial cells, for example, have been shown to secrete four proinflammatory cytokines – IL-8, monocyte chemoattractant protein (MCP)-1, GM-CSF and TNFα – in response to invasion by *Salmonella dublin, Sh. dysenteriae, Y. enterocolitica, Lis. monocytogenes* or enteroinvasive *E. coli*. This appears to be a specific effect as the synthesis of a wide range of other cytokines (IL-1, IL-2, IL-4, IL-5, IL-6, IL-10, IL-12p40 or IFNγ) was not affected by bacterial invasion. As IL-8, MCP-1, GM-CSF and TNFα are all involved, directly or indirectly, in chemotaxis and activation of inflammatory cells, the upregulation of these cytokines in epithelial cells would activate the mucosal inflammatory response to deal with the invading organisms. In the case of *Y. enterocolitica*, it has been shown that virulent strains stimulate the release of significantly lower levels of IL-8 than non-virulent strains. This would be of obvious benefit to the invading organism, especially during the early stages of infection. The suppression of IL-8 release was found to be dependent on the presence of both YopB and YopD, although the mechanism involved has not been determined.

Endothelial cells can also respond to bacterial invasion by increased cytokine gene transcription. Internalization of *Staph. aureus* by HUVECs, for example, results in the synthesis of mRNA for both IL-6 and IL-1β. Treatment of the cells with cytochalasin D inhibited cytokine gene expression, demonstrating that cytokine induction was a consequence of bacterial invasion rather than adhesion.

Growth of the obligate intracellular pathogen, *Chlamydia pneumoniae*, inside human monocytes induces the production of TNFα, IL-1β, MCP-1, IL-8 and IL-6. Heat-inactivated bacteria failed to stimulate production of any of these cytokines.

One of the key cytokines involved in the control of infections is TNFα. This stimulates macrophage function, enhancing their bactericidal activity, and is involved in the induction of a specific immune response to intracellular pathogens. Invasion of macrophages by a number of *Brucella* spp. has been shown to inhibit the production of TNFα: the molecule responsible is thought to be a 45–50 kDa protein secreted by the organism. This protein was also able to inhibit TNFα synthesis when macrophages were invaded by *E. coli* or when they were exposed to LPS. Interestingly, another invasive pathogen, *Y. enterocolitica*, has been shown to suppress TNFα secretion by macrophages; this was attributable to pore formation in the macrophage membrane induced by the YopB protein secreted by the organism.

The chemokine IL-8 is a potent chemoattractant and activator of polymorphonuclear leukocytes (PMN) and the stimulation of its secretion from epithelial cells by invading bacteria serves as an early warning system for the immune system. One of the oral bacterial species responsible for periodontitis (see above), *Por. gingivalis*, fails to induce the secretion of this cytokine when it invades oral epithelial cells. Furthermore, epithelial cells infected with the organism were unable to secrete IL-8 in response to other bacteria which normally induced such a response. This phenomenon, which can be described as local chemokine paralysis, would severely impair mucosal defence, so enabling colonization of host tissue.

A truly fascinating example of the way in which bacteria can turn host defence systems to their own advantage was the revelation that cytokines, the mediators of the host's response to infectious agents, could actually be used by bacteria as growth factors! Hence TGFα1 and IL-6 have been shown to stimulate the growth in macrophages of *M. tuberculosis* and *M. avium*, respectively. In the case of *M. avium*, growth of the organism in culture medium was also stimulated by IL-6. This organism appears to have a receptor for IL-6 with a dissociation constant (K_d) of 50 nM and there are approximately 15 000 receptor sites per bacterium. Growth of *Lis. monocytogenes* in macrophages is also stimulated by CSF-1 and IL-3 and TNFα has been shown to stimulate the growth of *M. tuberculosis* in human monocytes. The role of cytokines as bacterial growth factors is discussed in more detail in Chapter 10.

Prostaglandin release

Diarrhoea is one of the interesting, but unpleasant, consequences of bacterial invasion of the intestinal mucosa. It can be viewed either as a host protective mechanism in that it results in the expulsion of the infecting organism from the gut, or as the means by which bacteria convey themselves to other hosts. The diarrhoeal response to infection is mediated, in part, by prostaglandins (including prostaglandin E_2 – PGE_2), which are important regulators of gastrointestinal fluid secretion by virtue of their ability to induce chloride secretion from mucosal cells. The prostaglandins are derived from prostaglandin H, which is produced by the action of the enzyme cyclooxygenase (COX) on arachidonic acid, which is itself derived from membrane phospholipids. The cyclooxygenase activity is the product of two distinct genes: the constitutively expressed COX I and the inducible COX-II. Infection of intestinal epithelial cells with the invasive organisms *Sal. dublin, Y. enterocolitica, Sh. dysenteriae* or enteroinvasive *E. coli* both *in vitro* and *in vivo* has been shown to rapidly upregulate expression of COX II and the secretion of PGE_2 and $PGF_2\alpha$. Non-invasive organisms had little effect on COX II expression. The supernatants from bacteria-infected intestinal cells were able to increase chloride secretion from epithelial cell monolayers – a process linked to increased fluid secretion by these cells.

Effect on the expression of adhesion molecules and neutrophil adhesion

The epithelium of the gastrointestinal tract, as well as constituting a physical barrier to bacteria, also functions as part of the host's immune system. For example, it produces components of the complement system and can process and present antigens. Furthermore, as mentioned previously, these cells secrete proinflammatory cytokines in response to bacterial invasion, so signalling the presence of bacteria and initiating the recruitment of neutrophils and mononuclear phagocytes to the site of bacterial invasion. It has recently been shown that invasion of human colon epithelial cells by *Sal. dublin, Y. enterocolitica, Lis. monocytogenes* or enteroinvasive *E. coli* upregulates the expression of ICAM-1. This molecule is the receptor for the β_2-integrins LFA-1 and Mac-1 on neutrophils, and increased neutrophil adhesion to the epithelial cells was apparent following bacterial invasion. Upregulation of ICAM-1 expression due to bacterial invasion could, therefore, serve to keep neutrophils at the epithelial surface, so helping to combat invading bacteria.

In contrast, invasion of oral epithelial cells by *Por. gingivalis* has been shown to diminish the expression of ICAM-1. PMNs added to the system failed to transmigrate across the epithelial cell monolayer and transmigration induced by stimuli such as *E. coli* was also inhibited following invasion of the cells by *Por. gingivalis*. This ability to block neutrophil migration across the epithelial barrier may critically impair the ability of the host to deal with bacteria accumulating on the tooth surface and so may contribute to the initiation of periodontal diseases.

Invasion of HUVECs by *Chl. pneumoniae* induces the upregulation of a range of adhesion molecules including E-selectin, ICAM-1 and vascular cell adhesion molecule 1 (VCAM-1). Maximal expression of E-selectin was found approximately 6 hours after invasion whereas ICAM-1 and VCAM-1 were expressed approximately 14 hours later. The ability of the organism to stimulate the expression of adhesion molecules essential for the recruitment of leukocytes to sites of inflammation may be important in the pathogenesis of diseases associated with this organism, e.g. atherosclerosis (see Chapter 10). Invasion of HUVEC by *Lis. monocytogenes* also induces increased expression of E-selectin and ICAM-1 and it is thought that listeriolysin O contributes to this upregulation in some, as yet unidentified, manner.

Cell death

Intracellular growth and replication within host cells often lead to the death of the colonized cell as a result of nutrient depletion, degradation of important intercellular constituents and the build-up of toxic end products of bacterial metabolism – this is often termed necrotic cell death. However, bacterial invasion of host cells can sometimes induce a second form of cell death known as apoptosis, a process described in Chapter 2. Apoptosis, otherwise known as programmed cell death, is a natural process by which the host controls cell numbers and it is also important in embryogenesis and the response to genetic damage. In contrast to necrotic cell death, apoptosis is a highly regulated process which involves the active participation of the cell and is controlled by complex signal transduction pathways. Some bacteria have evolved means of inducing this process in host cells – again illustrating how they can pirate the normal functions of host cells and turn them to their advantage. For example, apoptosis of macrophages, a key effector cell in host defence, could be considered to be of obvious benefit to the invading organism – but would it? Macrophage apoptosis could also be regarded as a host defence strategy: by undergoing auto-destruction, the host is depriving the invader of the opportunity to reproduce in its preferred intracellular environment. This is one mechanism used by virally infected cells. Some bacteria trigger apoptosis 'at a distance' by secreting apoptotic toxins, e.g. *Bord. pertussis, Corynebacterium diphtheriae, Pseudomonas aeruginosa* and *Actinobacillus actinomycetemcomitans*. In this case, it is easy to see the potential advantage to the bacterium: a decrease in the number of bacteria-destroying host cells. Other bacteria, including *Sal. typhimurium* and *Sh. flexneri*, induce apoptosis only after they have invaded the macrophage – is this to the ultimate advantage of the organism (by killing host defence cells) or the host (by sacrificing infected cells)?

Invasion of macrophages by *Sal. typhimurium*, for example, has been shown to induce apoptosis (Figure 6.15). Within approximately 15 minutes of infection, the macrophages lost their phagocytic ability and after a further 30 minutes as many as 50% had died. The cells exhibited the characteristic appearance of apoptosis: membrane blebbing, typical chromatin condensation and fragmentation and the presence of apoptotic bodies. There is some controversy with regard to the effect of cytochalasin D, an inhibitor of invasion, on apoptosis – one group reporting that it prevented apoptosis while another found that it did not. Apoptosis is dependent on the bacteria having a functional type III protein secretion system and a

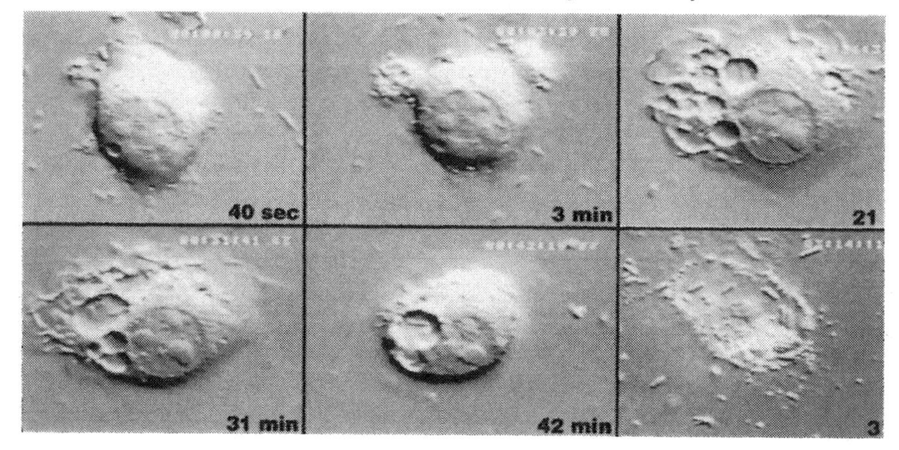

Figure 6.15 *Salmonella typhimurium* infection of the macrophage cell line, RAW264.7. The images are frames taken from a video obtained on a microscope equipped with differential interference contrast (Nomarski) optics. Dramatic membrane ruffles are seen on the surface of the macrophages at 3 minutes. Bacteria are seen swimming inside vacuoles within the cell at 21 and 31 minutes. At 42 minutes the macrophages begins to shrink and by 3 hours the macrophage is dead. (Images were kindly provided by Denise Monack, Barbel Raupach and Stanley Falkow of the Department of Microbiology and Immunology, Stanford School of Medicine, Stanford, CA)

mutation in the *sipD* gene, which encodes one of the proteins secreted by this system, renders the organism unable to induce apoptosis. It is also affected by the activation state of the macrophage: activation (e.g. by IFNγ and LPS) dramatically increases their susceptibility to apoptosis.

Macrophages are also induced to undergo apoptosis by *Sh. flexneri*. In this case, apoptosis is dependent on the internalized bacteria escaping from a vacuole into the cytoplasm. Once in the cytoplasm, the bacteria secrete the 62 kDa protein IpaB which is disseminated throughout the cytoplasm, although its concentration is greatest around the bacterium, and binds to macrophage-derived IL-1β-converting enzyme (ICE), now known as caspase-1. The *ipaB* gene is one of several invasin genes located on a plasmid which also encodes a type III secretion apparatus essential for the export of IpaB. ICE is a cysteine protease and is translated as a zymogen which is capable of auto-cleavage and so its activation must be tightly controlled in the cell. One well-characterized function of ICE is to cleave the biologically inactive pro form of IL-1β to generate IL-1β. However, its activation by IpaB is also crucial for the induction of apoptosis – inhibitors of ICE prevent apoptosis.

Macrophages are not the only host cells to undergo apoptosis due to bacterial invasion. Following invasion by *Staph. aureus*, epithelial cells show all of the characteristic features of apoptosis once the bacteria have escaped from the endosome. Invasion is essential for the induction of apoptosis, as cytochalasin inhibited both processes.

The paradigm of Lis. monocytogenes

The above sections give some idea of the range of effects that bacterial invasion can have on host cells. However, bacterial invasion is a multi-stage process which induces a series of host cell responses that differ at each stage. The invasion of macrophages by *Lis. monocytogenes* will now be used as a paradigm of this sequential series of responses (Figure 6.16).

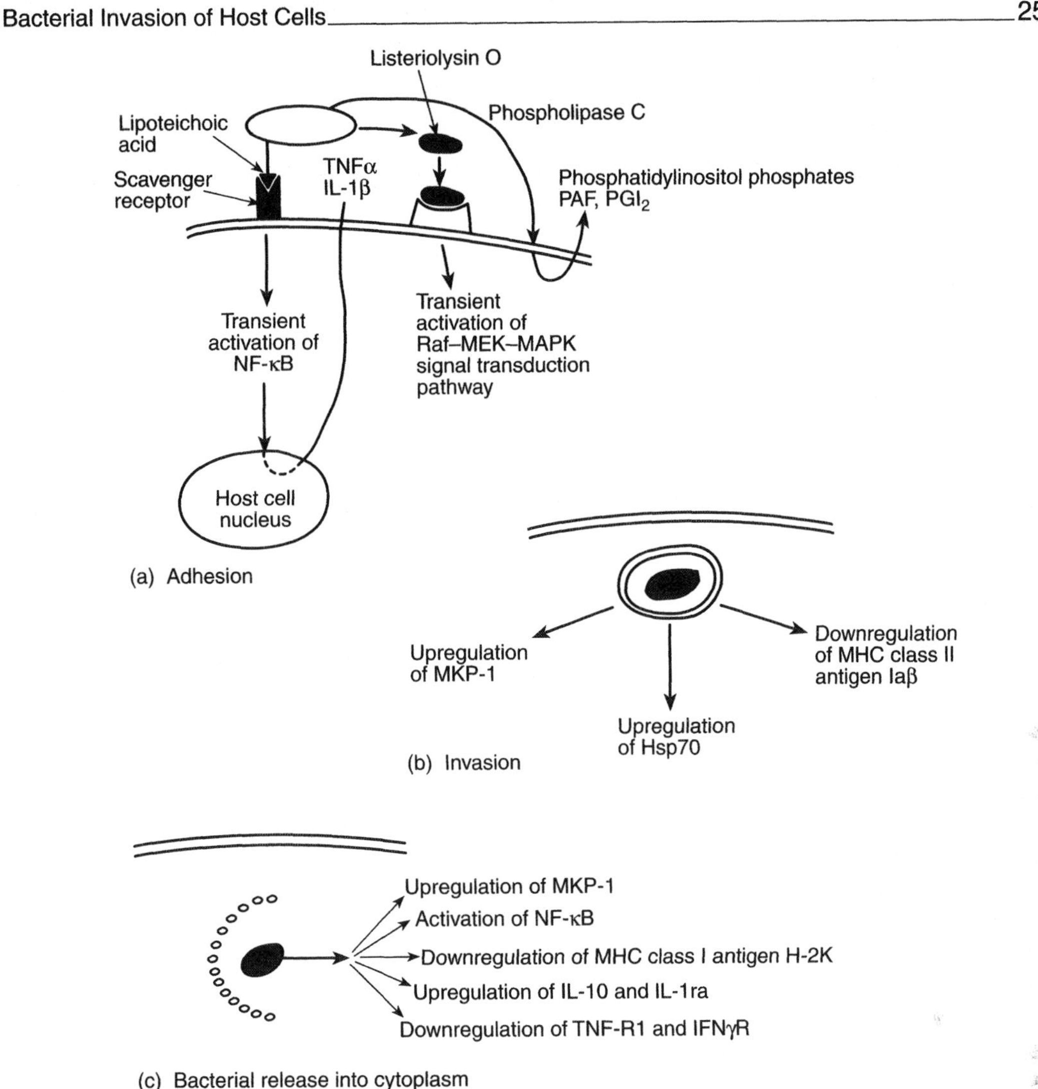

Figure 6.16 Host cell responses to adhesion and invasion by *Listeria monocytogenes*

Adhesion of the organism, the first stage in the invasion process, induces the transient activation of the major host cell transcription factor NF-κB and the transcription and translation of genes encoding the proinflammatory cytokines TNFα and IL-1β. Transcription of the genes encoding these cytokines is dependent on NF-κB. These events are likely to be initiated by binding of the lipoteichoic acid of the organism to the scavenger receptor of the macrophage as it has been shown that lipoteichoic acid isolated from the organism can itself induce both the activation of NF-κB and cytokine synthesis. Listeriolysin O produced by the organism activates the kinase cascade of the Raf-MEK (mitogen-activated protein kinase kinase)–MAPK (mitogen-activated protein kinase) signal transduction pathway (see Chapter 3 for details). Furthermore, listeriolysin, together with phospholipase C, induces the release of platelet aggregating factor, PGI_2 and phosphatidylinositol phosphates.

Phagocytosis of the bacterium induces transcription of the macrophage Hsp70 gene as well as downregulating the gene expressing the MHC class II antigen Iaβ. The latter would impede recognition of infected macrophages by CD4 T lymphocytes and so help the bacteria to undermine this arm of the host defence system. Phagocytosis also upregulates (weakly) the gene encoding MKP-1 which dephosphorylates activated MAPK, thereby blocking the activation of the Raf-MAPK pathway.

Following the release of bacteria from the phagosome, expression of MKP-1 is strongly upregulated. This could be as a result of activation of stress-activated protein kinase (SAPK), which is involved in the control of the gene encoding MKP-1. Listeriolysin and the products of the genes (*mpl, actA* and *plcB*) comprising the lecithinase operon are needed for the long-lasting activation of NF-κB which has been demonstrated in infected macrophages. It is possible that host cell gene products essential for multiplication of bacteria within the cytoplasm are under the control of this transcription factor. To the further benefit of the invading organism, the host cell then upregulates the expression of two anti-inflammatory cytokines (IL-10 and IL-1ra) and downregulates the expression of TNF receptor type I and the IFNγ receptor. These would, in effect, help to counteract the effects of the proinflammatory cytokines induced on adhesion of the organism to the macrophage. This brief description of the series of responses made by a host cell to an invading bacterium will give some idea of the complexity of the interactions which occur at the various stages involved in the invasion process.

Effect on bacteria

Once a bacterium has invaded a host cell, it has a limited number of options regarding where it subsequently survives and grows. It may continue to live within the vacuole enclosing it, it may exit from the vacuole and live in the cytoplasm or it may exit the cell and maintain an extracellular existence. Examples of bacteria opting for these various habitats are given in Table 6.5. However, it must be remembered that some species described as intracellular may also spend some of their time extracellularly as they make their way, or are transported, to other sites within the host.

Table 6.5 The various habitats of invasive pathogens and examples of organism adopting these habitats

Habitat	Organism	Disease
Vacuole	*Mycobacterium tuberculosis*	Tuberculosis
	Salmonella typhimurium	Gastroenteritis
	Coxiella burnetii	Q fever
	Legionella pneumophila	Legionnaires' disease
	Chlamydia trachomatis	Trachoma
	Brucella spp.	Undulant fever
	Burkholderia cepacia	Pulmonary infections
	Francisella tularensis	Tularaemia
Cytoplasm	*Listeria monocytogenes*	Listeriosis
	Shigella spp.	Dysentery
	Rickettsia spp.	Typhus, Rocky Mountain spotted fever
	Haemophilus ducreyi	Genital ulcers
Extracellular	*Yersinia* spp.	Plague, gastroenteritis

Whichever of the above options is chosen, the invading organism now has to adapt to the new environmental conditions in which it finds itself and this will involve the expression of gene products needed to survive in its new habitat and the downregulation of those that are no longer required. This is achieved by means of regulatory systems described previously (Chapter 3). These systems can respond to changes in a variety of environmental factors that the pathogen may encounter including temperature, pH, osmolarity and the concentration of oxygen, carbon dioxide, various macro- and micro-nutrients and antibacterial molecules such as reactive oxygen species and antibacterial peptides. Any given invading organism may encounter several different sets of environmental conditions during its transit from a mucosal surface, into a cell and, possibly, through different intracellular compartments. Furthermore, different gene products will need to be expressed to enable it to adhere to, and invade, a host cell.

A considerable amount of information is now available on the regulation of genes encoding products needed by *Salmonella* spp. to invade and survive within a number of host environments and this will be used as an example. *Salmonella* spp. are ingested in contaminated water or food and their first habitat within the host is the small bowel, from which they invade the mucosa and so give rise to systemic disease. Invasion of epithelial cells by *Salmonella* spp. is dependent on the production of secreted invasion proteins (Sips) which are exported by a type III secretion system activated by contact with the host cell membrane. All of these proteins are encoded by genes located on the *Salmonella* pathogenicity island 1 (SPI-2) and their expression is under the control of the HilA protein (see Chapter 2). Expression of *hilA* (which is also located on SPI-1) is influenced by a number of environmental factors including oxygen concentration, osmolarity and pH. High levels are expressed in response to low oxygen, high osmolarity and an alkaline pH, i.e. conditions characteristic of the lumen of the small bowel. Little is known of the way in which these environmental factors regulate *hilA* expression but the latter is controlled, in part, by *sirA*, which encodes a two-component response regulator that responds to environmental signals. Once it has been internalized, the organism remains within a vacuole (where it can replicate) until it exits from the basolateral surface of the epithelial cell. Whilst inside the vacuole, expression of both the iron-regulated gene, *iroA*, and the magnesium- regulated gene, *mgtB*, are increased, implying that the vacuole contains low levels of free Fe^{2+} and Mg^{2+}, respectively. Expression of the *cadA* gene (which encodes for lysine decarboxylase and is induced by low pH) is also upregulated, suggesting that both oxygen and lysine are present in the vacuole and that its pH is approximately 6.0. The possible roles for the products of these genes in the survival and/or virulence of the organism remain to be established.

The next key stage in the life history of the organism is its internalization by, and survival within, macrophages. The expression of a large number of genes is needed to ensure survival and growth within this cell. After uptake by the macrophage, the organism finds itself in an environment in which nutrients are limited, the osmolarity is high and the pH is low. This results in a lag phase lasting several hours during which time very little bacterial growth occurs. The organism responds to this by producing an alternative sigma factor for the bacterial RNA polymerase, σ^s (encoded by the *rpoS* gene) which replaces the vegetative σ^{70} in the RNA polymerase (see Chapter 2). The increased levels of σ^s result in global changes in gene expression rendering the organism more resistant to the adverse conditions in which it finds itself. σ^s also induces the expression of the plasmid-encoded *spv* genes (found in certain *Salmonella* spp. associated with systemic infections) which enables the organism to reproduce within the macrophage. The *spv* locus consists of five genes: *spvRABCD*. SpvR is a transcriptional activator which not only positively regulates its own synthesis but also the expression of the *spvABCD* operon. SpvA is a repressor. Efficient transcription of both operons is also dependent on high levels of σ^s. The means by which the Spv proteins induce proliferation has not been determined. As the bacteria-containing phagosome is processed within the macrophage the concentration of Mg^{2+} ions falls and this induces a response by the two-component signal transducer PhoP/PhoQ. PhoQ phosphorylates the response regulator PhoP and this leads to the induction of *pags* (PhoP-activated genes), which are essential for long-term survival of the organism in the macrophage. Two of these genes, *pmrA* and *pmrB*, encode another two-component regulatory system, PmrA/PmrB, which is involved in conferring resistance to phagosomal cationic antimicrobial peptides. PhoP also represses *hilA*, hence downregulating the expression of invasion genes which, of course, are not at this stage required by the organism.

Investigations of the ways in which bacteria respond to life within their host have been helped enormously in recent years by the development of a number of molecular and genetic techiques (described in Chapter 4). Bacterial genes expressed when the organism is inside a host cell can be identified by isolation of bacterial mRNA from which cDNA can be synthesized and probed for the gene(s) of interest. Using this technique, genes expressed by *Leg. pneumophila* and *M. tuberculosis* within macrophages have been identified. Other techniques (described in Chapter 4) include signature-tagged mutagenesis, differential fluorescence induction and *in vivo* expression technology. The latter has been used to identify genes expressed by *Sal. typhimurium* in mice and in macrophages, some of which are listed in Table 6.6.

Table 6.6 Examples of *Salmonella typhimurium* genes expressed *in vivo*

Gene	Function	Role in pathogenesis
phoP	Virulence regulator	Invasion/macrophage survival
cadC	Cadaverine synthesis	Acid tolerance
cfa	Membrane modification	Stationary phase survival
otsA	trehalose synthesis	Osmoprotectant
recD	Recombination/repair	Macrophage survival
hemA	Catalase cofactor	Peroxide resistance
entF	Enterobactin synthesis	Iron acquisition
fhuA	Iron transport	Iron uptake
spvRABCD	Unknown	Intracellular survival

Less is known concerning the effects of the host cell environment on gene expression in other organisms. Transfer of *Brucella abortus* from a tissue culture medium to bovine macrophages has been reported to have a dramatic effect on protein expression. Twenty-four new proteins were detectable while 50 proteins were repressed. Furthermore, the level of expression of 19 proteins was increased while that of 54 other proteins was decreased. One of the proteins whose synthesis was upregulated inside macrophages was a homologue of the cell stress protein, chaperonin 60 (GroEL). The organism also increased the synthesis of another stress protein, DnaK, inside the macrophage. Indeed, a *dnaK* deletion mutant of *Brucella suis* was unable to survive inside macrophages, showing the importance of this stress protein for intracellular survival.

BACTERIAL SURVIVAL AND GROWTH SUBSEQUENT TO INVASION

Broadly speaking, once an organism has invaded an epithelial cell it may then follow one of two pathways: either it proliferates within the cell and then makes its exit into the external environment, or it crosses the epithelial layer, invades deeper tissues and, possibly, spreads to other sites within the host. The former represents, in many ways, the 'safer' option for the bacterium as the latter route will mean that it will be faced with, and must combat, a variety of additional host defence strategies. In a normal, healthy adult, infections by many organisms (e.g. *Staph. aureus, E. coli, N. gonorrhoeae, Shigella* spp., *Campylobacter* spp., *Helicobacter pylori, Mycoplasma* spp., many *Salmonella* serotypes) are confined to the mucosal surfaces and it is only when the host defence systems are compromised as a result of some underlying systemic disorder, poor nutrition, advanced age or following the administration of immunosuppressive drugs that these bacteria invade deeper tissues. Such organisms will often replicate within the epithelial cell and then exit the cell (or are released following cell lysis) to the mucosal surface, on which they spread and then invade neighbouring cells.

However, there are, of course, some bacteria (e.g. *M. tuberculosis, Y. pestis, Tr. pallidum, Brucella* spp., *Sal. typhi*) which invariably cause systemic diseases. Once these organisms have crossed the epithelial layer they are confronted by products of the inflammatory response (phagocytes and serum factors) designed to dispose of them. If the organism remains extracellular it will be continually exposed to antibacterial factors present in blood or tissue fluid such as complement components (resulting from activation by the alternative pathway), antibodies, antimicrobial molecules discharged by phagocytes and, of course, phagocytic cells. Bacteria have evolved a variety of strategies for dealing with phagocytes, ranging from avoidance of phagocytosis to the production of toxins which kill the phagocyte. Some of these are discussed below and also in Chapters 7 and 8. An alternative approach adopted by many invasive bacteria is to opt for an intracellular lifestyle and to colonize the very cells designed to eliminate infectious organisms. As PMNs have only a short lifespan (2–3 days), the longer-lived macrophages are a more attractive target for these bacteria. The macrophages may be utilized either as a 'permanent' habitat or as a temporary refuge and means of transport to other tissues.

Intracellular lifestyle

As can be appreciated from Table 6.5, a wide range of bacteria are able to survive inside vacuoles within the host cell and this constitutes the most common habitat of intracellular pathogens. The mechanisms used by these most audacious of organisms to enable their survival in a host structure designed to dispose of them may be classified into two broad groups. Some species (e.g. *Coxiella burnetii, Brucella* spp.) simply adapt themselves to this hostile environment and withstand the bactericidal components of the phagolysosome. Other organisms inhabit a different type of vacuole which either does not acidify (e.g. *M. avium, Nocardia asteroides, Leg. pneumophila*) or does not fuse with lysosomes (e.g. *Mycobacterium* spp., *Leg. pneumophila, Sal. typhimurium, Chlamydia* spp.).

Survival in phagolysosomes

Cox. burnetii, the only species of the genus *Coxiella*, is a small, Gram-negative bacterium which is incapable of replicating outside host cells. It has been traditionally regarded as belonging to a large group of organisms with similar characteristics known as rickettsias, although recent 16S rRNA gene sequence analysis has shown it to be more closely related to *Leg. pneumophila*. It is transmitted in aerosols or by contact with infected cattle or sheep and causes Q fever – a disease characterized by pneumonia, fever, headaches and muscular pain. It is also responsible for approximately 10% of cases of infective endocarditis. Although the organism can invade fibroblast cells *in vitro*, the main habitat of the organism *in vivo* is the macrophage, in which it can grow and multiply (Figure 6.17).

Phase variation of the LPS results in two antigenically distinct forms of the organism. Phase I bacteria have a smooth form of LPS, are highly virulent and are isolated from infected humans and animals. Phase II bacteria have a rough LPS, have reduced virulence and are obtained following passage of the organism in cell culture (see also Chapter 8). The phase I form of the organism binds to a complex consisting of the leukocyte response integrin ($\alpha_v\beta_3$) and integrin-associated protein, whereas the Phase II variant also binds simultaneously to β_2, integrins. Following adhesion, the organism is endocytosed by a cytochalasin-sensitive process and the resulting phagosome is processed in the normal manner, leading to the formation of an apparently normal phagolysosome, i.e. it contains membrane proton ATPase, LAMP-1, LAMP-2 and lysosomal enzymes (Figure 6.18). The parasitized cell is capable of division, giving rise to one daughter cell containing the vacuole and one parasite-free cell. The vacuole-containing cell may then rupture, liberating the bacterium, while the other cell can be infected, hence repeating the cycle. The organism is able to reproduce in the acidic phagolysosome (pH = 5.2) but little is known of how it manages to do so in this harsh environment. It has been shown that the bacterium actually requires an acidic environment to enable nutrient uptake: raising the pH of the vacuole using amines has a bacteriostatic effect. Internalization of the organism is accompanied by a reduction in both the oxidative metabolism of the phagocyte and its production of superoxide. This may be due to the bacterium's acid phosphatase (a 91 kDa periplasmic enzyme) which has been shown to inhibit superoxide production by PMNs stimulated with formyl-Met-Leu-

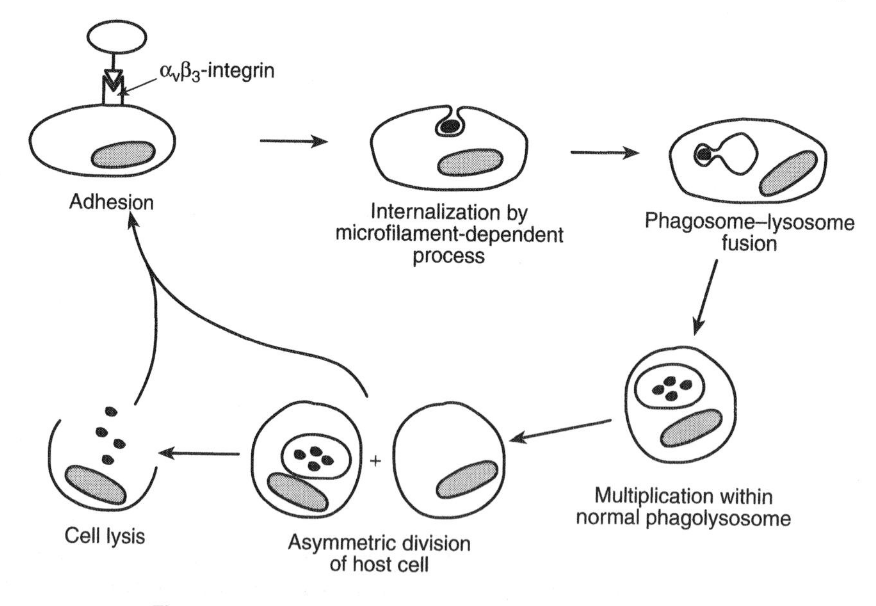

Figure 6.17 Stages in the life cycle of *Coxiella burnetii*

Phe – possibly by disrupting tyrosine phosphorylation/dephosphorylation reactions associated with the signal transduction pathway responsible for induction of the metabolic burst. The organism also has a superoxide dismutase and catalase which would help protect it from residual reactive oxygen species produced by the phagocyte. More recently it has been shown that the organism has a gene encoding a protein similar to the macrophage infectivity potentiator of *Leg. pneumophila*.

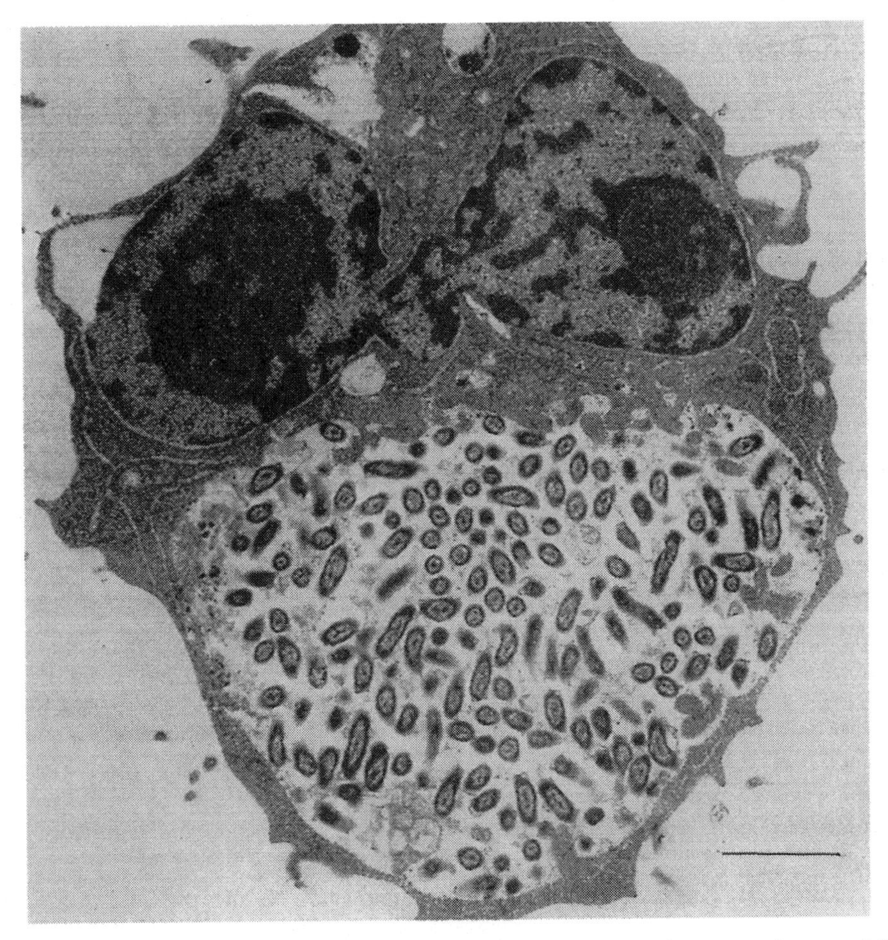

Figure 6.18 *Coxiella burnetii* proliferating within the phagolysosome of a macrophage cell line. Reprinted from *Trends Microbiol*, vol. 2, Baca OG, Li YP, Kumar H, Survival of the Q fever agent *Coxiella burnetii* in the phagolysosome, pp. 476–480, Copyright 1994, with permission from Elsevier Science

Survival in remodelled vacuoles

Once a bacterium has been taken up by a phagocyte, the resulting phagosome usually then enters the endocytic pathway, which involves a series of changes (described in detail in Chapters 2 and 8) leading ultimately to the formation of a phagolysosome within which the bacterium is killed. Intracellular bacteria have, of course, evolved ways of circumventing or disabling this potent antimicrobial mechanism.

The means by which *Sal. typhimurium* accomplishes this has been demonstrated in a series of experiments involving uptake by macrophages of live and dead bacteria and latex beads. Phagosomes containing dead bacteria (or latex beads) followed the normal endocytic pathway of the macrophage whereas those containing live bacteria did not. Specifically, although the vacuoles contained lysosomal glycoproteins and acid phosphatase, they were devoid of the mannose 6-phosphate receptor and cathepsin L which are indicative of the maturation of the phagosome into a late endosome/prelysosome. Studies by other investigators have also shown that the *Sal. typhimurium*-containing phagosomes do not fuse with lysosomes. The bacteria therefore reside in a remodelled vacuole within which they are able to reproduce. Another interesting finding was that phagocytosis of a non-invasive mutant of *Sal. typhimurium* by macrophages resulted in the formation of bacteria-containing vacuoles with characteristics identical to those formed by the invasive wild-type organism. Vacuole remodelling, therefore, does not depend on the method of uptake but is a consequence of the activities of the internalized bacteria. The vacuoles in which the bacteria reside are only slowly acidifed and it takes up to 5 hours for the pH to drop to below 5.0, whereas vacuoles containing dead bacteria acidify rapidly – the pH falling to less than 4.5 within 1 hour. The PhoP/PhoQ regulatory system responds to low pH by inducing the formation of at least five bacterial products which are thought to aid the survival of the organism. It has also been shown that more than 30 bacterial proteins are induced when bacteria are exposed to macrophages; many of these may be involved in protecting the organism from the bactericidal activities of these phagocytes.

As described previously, *Leg. pneumophila* are internalized by macrophages by coiled phagocytosis. Due to selective incorporation and exclusion of macrophage membrane proteins, the resulting vacuolar membrane has a distinctive composition (Figure 6.19). Approximately 15 minutes after internalization, the phagosome becomes surrounded by smooth vesicles and then by mitochondria (after approximately 1 hour). The phagosome does not become acidified (remaining at a pH of approximately 6.1) and does not fuse with lysosomes, remaining devoid of endosomal (transferrin receptor) and late endosomal/lysosomal markers (CD63, LAMP-1 and LAMP-2). It has been suggested that the polycationic protein, Mip, of the organism may be involved in the prevention of phagosomal acidification and the inhibition of phagosome-lysosome fusion. Four to eight hours after internalization, the phagosome is surrounded by ribosomes and ribosome-containing vesicles and bacterial multiplication begins with a doubling time of approximately 2 hours. Recruitment of host cell organelles to the phagosome appears essential for bacterial multiplication as mutants unable to induce this recruitment are unable to grow inside macrophages. Although the interaction between *Leg. pneumophila* and macrophages has been well characterized at the morphological level, we know little of the means by which the organism manages to survive and grow intracellularly. It has been reported that internalization of the organism within macrophages results in the synthesis of at least 35 new bacterial proteins and the repression of at least 32 proteins. The identity and functions of most of these proteins remain to be determined.

Francisella tularensis is responsible for tularaemia, a disease which can be contracted from wild or domesticated mammals either directly or indirectly from the ticks of these animals. It has been estimated that inhalation of as few as 10 bacteria can initiate an infection. Although often described as being facultatively intracellular, the organism is rarely detected extracellularly *in vivo* but appears to spend most of its time inside macrophages. Little is known about the means by which it invades macrophages but, once internalized, it is able to prevent

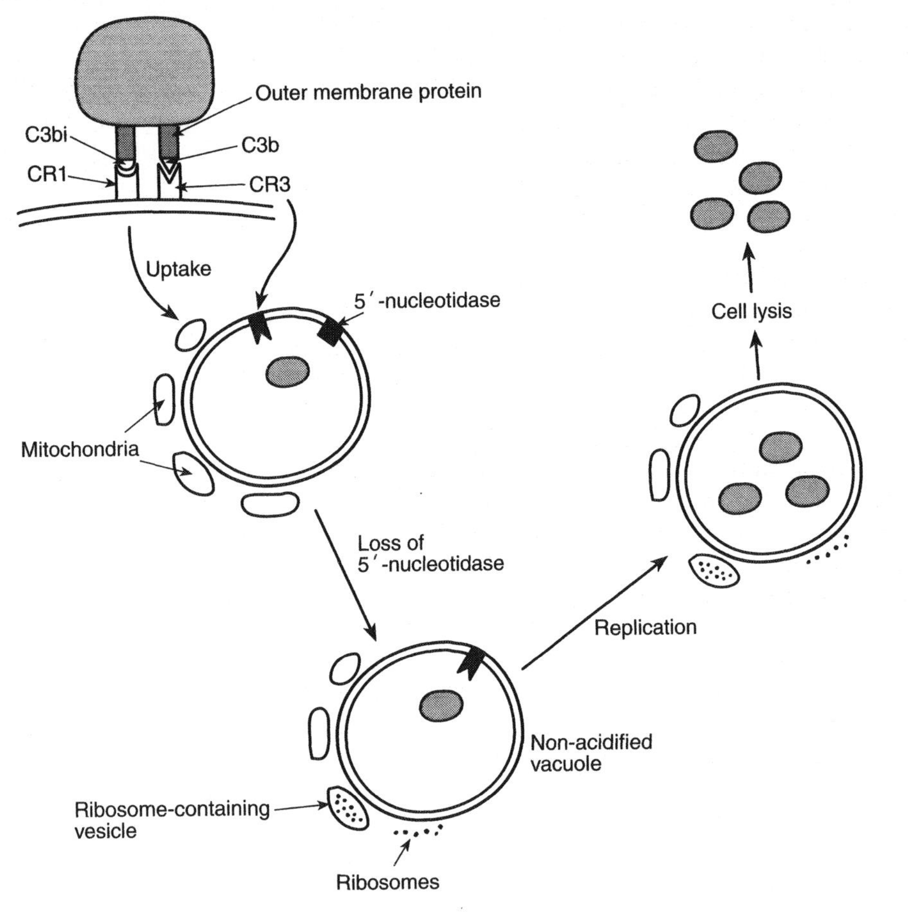

Figure 6.19 Invasion of macrophages by *Legionella pneumophila*

phagososme–lysosome fusion. It inhabits acidifed endosomes and is able to replicate within these vacuoles. Indeed, acidification of the endosome was shown to be essential for the growth and reproduction of the organism as inhibitors of endosome acidification also inhibited growth of intracellular bacteria. The acidified vacuole of a macrophage also comprises part of the iron transport system of the cell. Iron transferrin is internalized via a receptor into an endocytic vacuole and ferric ions are released only after acidification of the vacuole. One of the means by which the host limits growth of bacteria which have penetrated epithelial surfaces is to limit the availability of extracellular iron – an element essential for bacterial survival and growth. It may well be that one of the reasons why the organism uses acidified endosomes as its preferred habitat is because of the ready availability of iron. Addition of iron chelators to the macrophages was found to inhibit growth of the intracellular bacteria.

Chlamydia spp. are responsible for a range of infections in man and other animals. Trachoma, an ocular infection due to *Chl. trachomatis,* is the major cause of preventable blindness. The organism is also an important cause of genital tract infections. *Chl. trachomatis* is an obligate intracellular parasite which exists in two forms – an elementary body and a reticulate body – both of which are bounded by an outer membrane similar to that found in other Gram negative bacteria, except that it does not contain peptidoglycan. The elementary body (EB), which consists of a rigid structure approximately 0.3 μm in diameter, constitutes the infectious stage of the organism. The reticulate body (RB) is the intracellular, non-infectious, form of the organism and resembles a typical Gram-negative coccus with a diameter of approximately 1.0 μm. The adhesin(s) involved in the attachment of the organism to mucosal cells has not been determined although the 40 kDa major outer membrane protein (MOMP) is a likely candidate (Figure 6.20). The MOMP is a porin, forming a trimer enclosing a pore in the outer membrane, and can bind to heparan sulphate proteoglycans on epithelial cells.

Following attachment, the EB is endocytosed by the host cell. Little is known about this process other than it involves rearrangement of microfilaments and is accompanied by tyrosine kinase activation. There is evidence suggesting that *Chlamydia* spp. may have a type III secretion system: chlamydial genes encoding proteins homologous to those associated with this type of system in other organisms have been identified. Internalization is associated with an increase in respiration and glucose catabolism by the host cell. Chlamydia appear to be devoid of any energy-generating system and so are energy parasites being totally dependent on ATP synthesized by the host cell. The ATP translocase of these organisms takes in ATP and excretes ADP – the opposite of the vast majority of other bacteria.

Approximately 6 hours after being enclosed within an endosome, an EB will have differentiated into an RB. This process is probably triggered by the reducing environment within the endosome which affects the MOMP, causing an increase in the pore diameter, thereby allowing ingress of nutrients required for differentiation. There is considerable evidence that the vacuole in which the organism resides is not processed along the normal phagosome route. Hence, the mannose 6-phosphate receptor (a late endosomal–prelysosomal marker) cannot be detected in the vacuole, acid phosphatase and cathepsin D (lysosomal enzymes) are absent and the lysosomal glycoproteins LAMP-1 and LAMP-2 are not found. The endosome is not acidified and does not fuse with a lysosome but is transported to the Golgi apparatus, where it fuses with sphingomyelin-containing exocytic vesicles. The sphingomyelin is incorporated into the outer membrane of the bacteria. The subsequent fate of the bacteria-containing endosomes is dependent on the particular *Chlamydia* species. In the case of *Chl. trachomatis*, the endosomes fuse to form a large inclusion, whereas for *Chl. psittaci* the endosomes remain as single units. Depending on the species and, indeed, the particular strain of a species, vacuole fusion may or may not involve cytoskeletal components such as microfilaments. An RB may undergo binary fission within its endosome and these daughter cells then give rise to EBs which are released on cell lysis.

Comparatively little is known about chlamydia–host cell interactions as the organism is very difficult to work with: it cannot be cultivated in cell-free media, there are no well-characterized mutants available and genetic manipulation techniques have not been devised for the organism.

Members of the genus *Brucella* are Gram-negative, facultative, intracellular bacteria that induce chronic infections in humans and other animals. After invading the reticuloendothelial system, the bacteria develop intracellularly within mononuclear phagocytes generally in specific locations within the body (spleen, brain, heart and bones). In humans, the main pathogenic species is *Br. melitensis*, which is contracted from sheep and goats and their products. The disease, known as undulant fever, is characterized by an intermittent

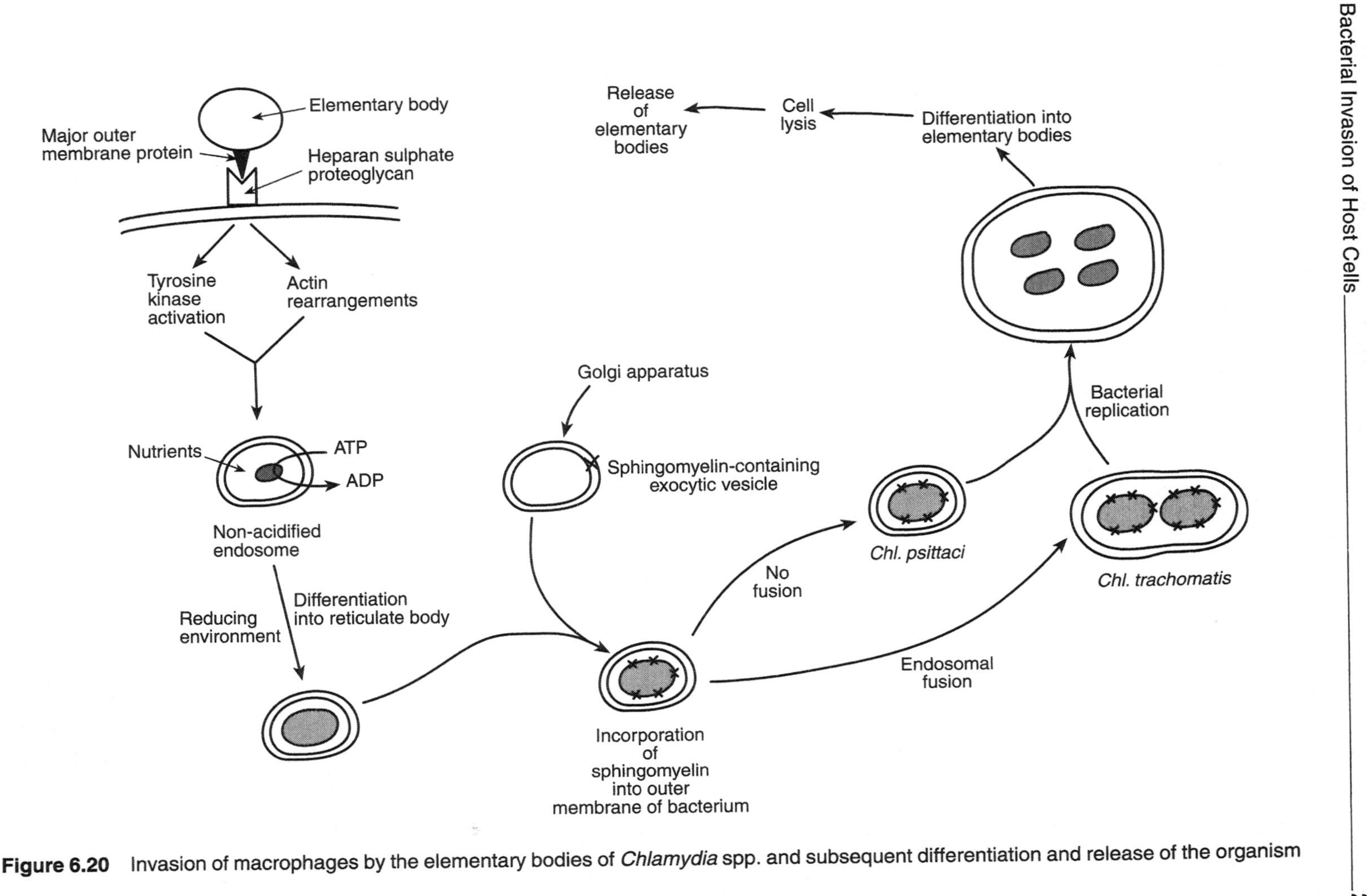

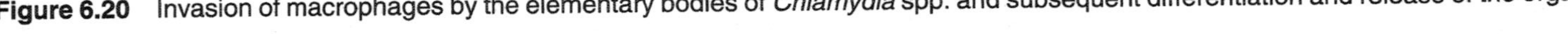

Figure 6.20 Invasion of macrophages by the elementary bodies of *Chlamydia* spp. and subsequent differentiation and release of the organism

fever but displays a wide variety of symptoms affecting many different organ systems, so that diagnosis relies on the isolation of the organism from the bloodstream. Little is known about the mechanisms involved in the uptake of *Brucella* spp. by macrophages although recently it has been reported that cytochalasin inhibits uptake of *Br. suis*, suggesting the involvement of actin microfilaments. Following internalization, the bacteria appear to reside in phagosomes which do not fuse with lysosomes; this is thought to be achieved by bacterial products such as sulphatides or adenosine and guanosine monophosphate. Survival of the organism within macrophages is associated with the synthesis of a number of proteins but their specific functions have not yet been determined. A contributing factor, however, must be the ability of *Brucella* spp. to inhibit the production of TNFα, a cytokine shown to be important in the macrophage-mediated killing of these bacteria (see 'Cytokine release', above).

The genus *Mycobacterium* consists of Gram-positive bacilli with a distinctive cell wall characterized by the presence of unusual glycolipids. A number of mycobacteria are pathogenic for man but the most important is undoubtedly *M. tuberculosis*, the causative agent of tuberculosis. Tuberculosis is one of the most prevalent infectious diseases of man and it has been estimated that approximately 2 billion people are infected by the organism and this results in the death of approximately 3 million people each year. In developing countries, almost 20% of adult deaths are due to tuberculosis. Another important human pathogen is *M. leprae*, which causes leprosy, a chronic disease which affects approximately 10 million people, mainly in developing countries. Tuberculosis is a systemic disease contracted by inhalation of *M. tuberculosis*-containing aerosols produced by infected individuals. It progressively destroys the lungs and can spread systemically to infect any other part of the body. Investigations of the virulence factors of the organism and details of the mechanisms underlying its interactions with host cells are hampered by its slow growth rate and the difficulty in producing mutants defective in a single gene. One of its most important virulence factors is undoubtedly its ability to survive and replicate within phagocytes, particularly macrophages (Figure 6.21). Activation of the alternative pathway of complement results in the deposition of the complement components C3b or iC3b on the surface of the organism. These mediate binding to the macrophage via the complement receptors CR1, CR3 or CR4 and this triggers internalization of the organism into a phagosome. There is also evidence suggesting that adhesion to, and invasion of, macrophages could involve another adhesin – the terminal mannosyl units of lipoarabinomannan, one of the components of the cell wall of the organism. Analysis of the resulting phagosome has revealed that MHC class I and class II molecules, as well as the transferrin receptor, are present. This contrasts with normal primary phagosomes, which usually lose these molecules. The phagosomes then fail to fuse with lysosomes, as demonstrated by their low content of the lysosomal markers LAMP-1, LAMP-2, the mannose 6-phosphate receptor and cathepsin D. Experiments with phagosomes containing dead bacteria or inert particles have shown that fewer fusion events take place when compared with phagosomes containing live *M. tuberculosis*. However, some lysosomal glycoproteins have been detected in phagosomes containing live *M. tuberculosis*. These are thought to arise as a result of the phagosome fusing with vesicles formed from the trans-Golgi network. *Mycobacterium* spp. are also able to inhibit vacuolar acidification which usually results in a pH of 4.5. *M. avium*, for example, appears to be able to exclude the proton-ATPase which is responsible for this acidification and the vacuolar pH remains above 6.3. In the case of *M. tuberculosis*, release of glutamine synthetase into the vacuole may help to inhibit acidification. Little is known about the subsequent growth and reproduction of the organism within the modified phagosome which it inhabits.

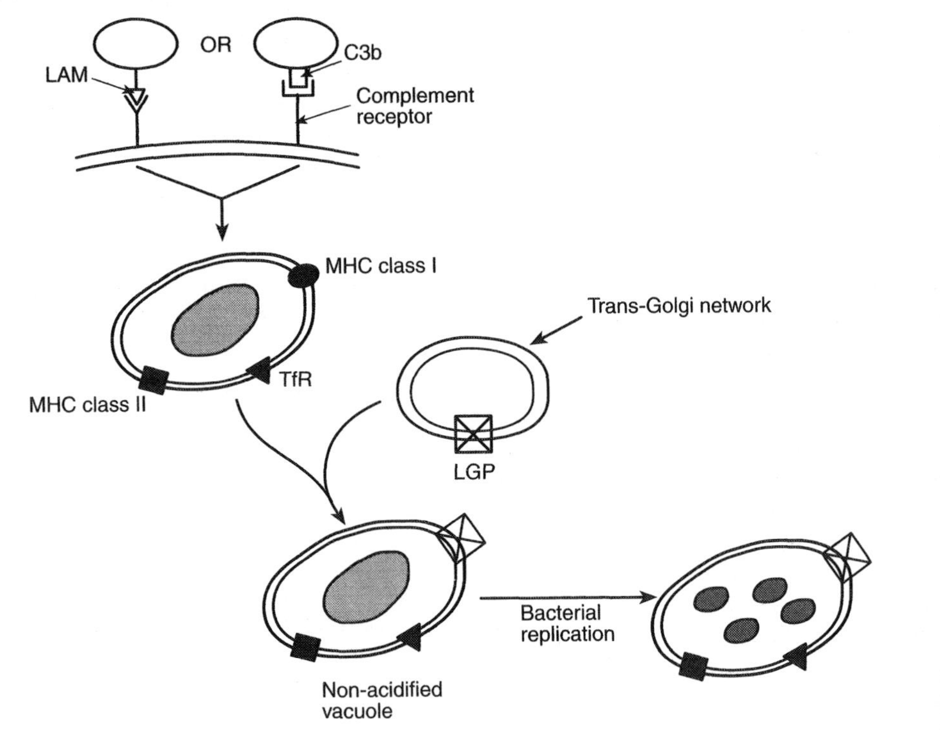

Figure 6.21 Fate of *Mycobacterium tuberculosis* following macrophage invasion. LAM-lipoarabinomannan, LGP-lysosomal membrane glycoprotein, TFR-transferrin receptor

Survival in the cytoplasm of the host cell

Following invasion of a host cell, a number of bacteria escape from the resulting vacuole and colonize the cytoplasm. Examples of organisms exhibiting this behaviour pattern (*Lis. monocytogenes* and *Shigella* spp.) have already been described above. Additional examples include *Staph. aureus*, group B streptococci, *Rickettsia* spp. and *Haemophilus ducreyi*. Of these, only *Rickettsia* spp. are obligate intracellular parasites. *Rick prowazekii* colonizes endothelial cells and macrophages *in vivo* although it can also invade a number of epithelial cell types *in vitro*. The organism reproduces within the cytoplasm of the host cell until huge numbers are produced and the cell lyses. *Rick. rickettsii*, the causative agent of the tick-borne Rocky Mountain spotted fever, also grows in the cytoplasm of host cells and can move from cell to cell propelled by F-actin filament formation in a manner similar to *Lis. monocytogenes* and *Shigella* spp.

Extracellular lifestyle

As can be seen from the above sections, many bacteria have developed the ability to live inside phagocytes. An alternative strategy, exemplified by *Yersinia* spp., is to avoid phagocytosis in the first place – once they are phagocytosed they are generally killed. *Yersinia* spp. have evolved a highly sophisticated system for preventing their uptake by phagocytic cells. This comprises at least 12 secreted proteins known as Yops (*Yersinia* outer proteins) encoded by genes located on a 70 kb virulence plasmid (pYV). Transcription of *yop* genes is inhibited by high intracellular levels of the protein LcrQ. The pYV plasmid also encodes a type III secretion system (Ysc) consisting of at least 22 proteins involved in the transport of the Yops out of the bacterium and into the host cell (Figure 6.22). Not all of the Ysc proteins have been identified but some are plasma membrane proteins (e.g. YscR, YscU) and others (e.g. YscC) are components of the outer membrane. YscC is thought to self-assemble to form a large channel in the outer membrane through which the Yops are secreted. This channel is normally blocked by YopN until contact is made with a phagocyte. YopN then probably interacts with a receptor on the phagocyte, thereby inducing opening of the channel so releasing LcrQ, hence lowering its intracellular concentration and allowing expression of Yops. One of the released proteins, YopB, forms a pore in the membrane of the phagocyte through which a number of Yops (known as effector proteins) gain entry to the host cell cytoplasm. Secretion of the effector proteins is dependent on the involvement of chaperones which prevent intracellular aggregation of the proteins and maintain them in the correct conformation for export. SycE (specific Yop chaperone E) and SycH are the chaperones for YopE and YopH, respectively, while SycD functions as a chaperone for both YopB and YopD. Of the four effector proteins (YopE, YopH, YopM, YopO), YopE and YopH are responsible for the anti-phagocytic activity of the organism. Phagocytic mechanisms involving integrins, Fc receptors and complement receptors are all affected. The 23 kDa YopE disrupts the actin microfilament structure of the cell, possibly by modifying small GTP-binding proteins which are involved in regulating the actin network. YopO is an 81 kDa serine/threonine kinase and probably interferes with a signal transduction pathway of the phagocyte. The 51 kDa YopH is a protein tyrosine phosphatase and is thought to be the main anti-phagocytic effector molecule – its absence enables bacterial phagocytosis. This enzyme dephosphorylates the tyrosine phosphorylated forms of the focal adhesion proteins $p130^{cas}$ and FAK (focal adhesion kinase) which are involved in the regulation of cytoskeletal rearrangements that mediate phagocytosis. The β_1-integrins which are receptors for the bacterial adhesin, invasin, are linked to the cytoskeleton via multimolecular complexes known as focal adhesions (see Chapter 2). Binding of invasin to the β_1-integrins induces tyrosine phosphorylation of $p130^{cas}$ and FAK and their recruitment to the focal adhesions and this triggers the cytoskeletal rearrangements responsible for bacterial uptake. By dephosphorylating the two focal adhesion proteins YopH inhibits these rearrangements, so enabling the bacterium to maintain an extracellular existence. As well as preventing phagocytosis, there is also evidence to show that YopH inhibits the oxidative burst of PMNs. Another Yop, YopE, and YadA are also able to inhibit the PMN oxidative burst. Control of protein secretion relies not only on the feedback inhibition system involving LcrQ but also on temperature. A temperature of 37°C induces the transcription of the chromosomal gene *ymoA* which is thought to encode a histone-like protein. The product of this gene regulates the expression of the plasmid-encoded gene *virF* and VirF activates the expression of most of the *yop* and *ysc* genes.

A less sophisticated, but certainly effective, means of avoiding phagocytosis is to produce a capsule. The presence of this structure renders the organism less susceptible to phagocytosis and this tactic is used by many organisms known to cause systemic infections. The capsule is more often than not a polysaccharide (e.g. *Staph. aureus, K. pneumoniae, Strep. pyogenes, E. coli* and *Staph. epidermidis*) but proteinaceous capsules have also been shown to confer decreased susceptibility to phagocytosis (e.g. *Bacillus anthracis, Campylobacter fetus*). More information on the anti-phagocytic mechanisms of bacteria is given in Chapter 8.

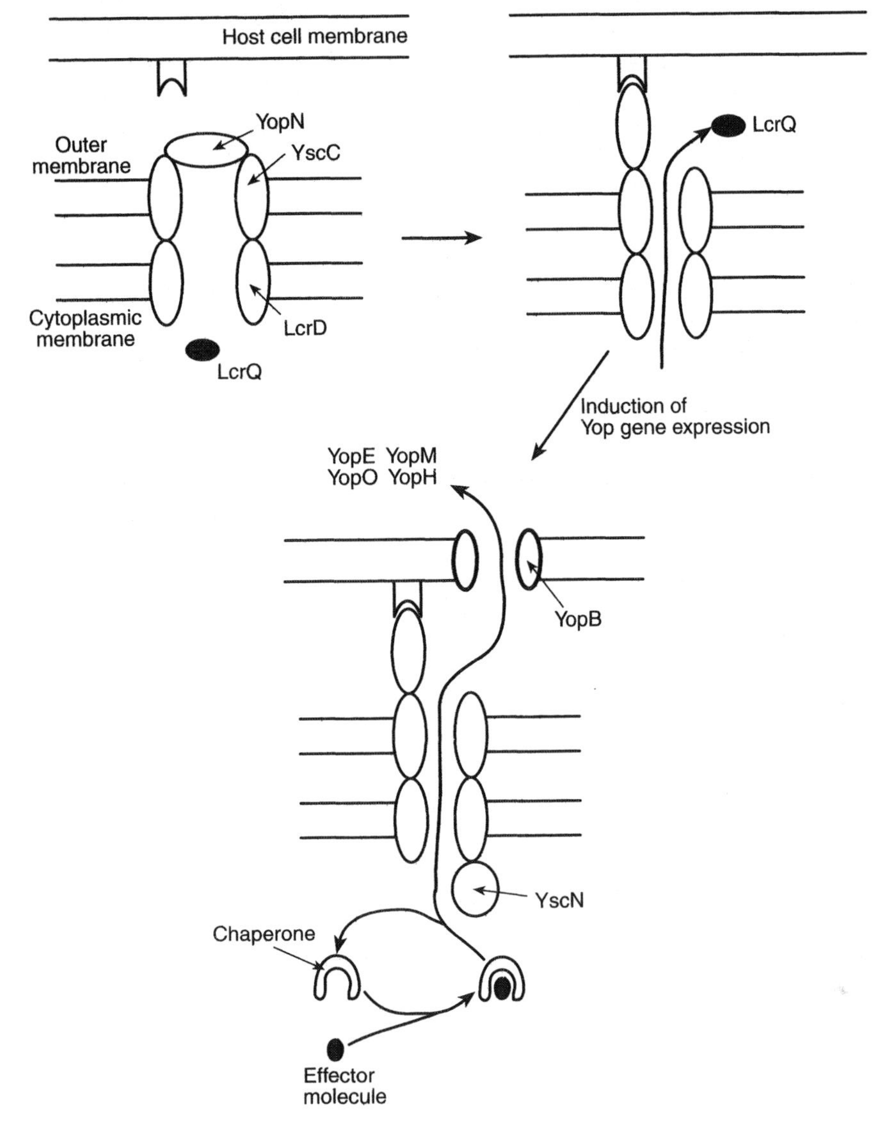

Figure 6.22 Operation of the Type III secretory system in *Yersinia* spp.

IN VITRO AND *IN VIVO* MODELS OF TISSUE INVASION

While examination of biopsies from patients has contributed to our knowledge of bacterial invasion of host tissues, most of what is known about this phenomenon has been derived from studies employing animals and laboratory cultures of eukaryotic cells. It is apposite, therefore, to introduce some of these systems that have been used and point out the inherent drawbacks. Experiments with animals, in particular the target animal, have the advantage of relevance, since the pathogen has to challenge intact multicellular organs and cope with the host immune response, but this, of course, complicates our ability to interpret the results obtained. Such experiments are also costly. Conversely, the use of cells cultured *in vitro* is cheaper and much easier to interpret, but its relevance to the *in vivo* situation is more open to criticism. Within these extremes, various other experimental systems have been devised.

Animals have long been used in the study of infectious diseases of man and the most commonly used species are mice, rats, rabbits and guinea-pigs. These particular mammals are used because of their ease of handling, fast rates of breeding, availability of inbred lines (to reduce experimental variability) and comparatively low maintenance costs. However, there are also a number of important scientific limitations regarding the use of such animals. Their physiology may differ from that of man or the target animal, the pathophysiology of the disease process may differ from that occuring in man and they may not be susceptible to the same strain of bacterium responsible for human disease. Some of these problems may be overcome due to advances in biotechnology which will enable the production of animals with some human characteristics. For example, the SCID-hu mouse (SCID stands for severe combined immunodeficiency) is a strain which has a non-functional immune system and can therefore accept human immune cells which will survive and provide a human-like immune system. For some diseases, however, good animal models exist, and have been responsible for advancing our knowledge of various aspects of pathogenesis, including invasion.

Invasion has also been examined using isolated organs. For example, invasion due to *Salmonella* spp. has been examined using ligated ileal loops in pigs and calves. Here the small intestine of a terminally anaesthetized animal is sutured into several sections that can be injected with bacteria, so that attachment, invasion and cellular damage can be followed over a period of hours. An alternative is to remove sections of intestine from killed animals, open the sections lengthways and clamp them in an apparatus to allow bacteria to be added to one side while the gut wall is under observation. Other similar *in vitro* organ culture (IVOC) methods can also be used, such as tracheal rings to study the attachment of respiratory pathogens. These methods do not permit the full range of *in vivo* responses to take place and, in consequence, give less information but are easier to interpret.

During the past decade, the majority of investigations of bacterial invasion have employed tissue culture models. Although these are easier to work with and facilitate studies of invasion at the cellular and molecular levels, they can also have drawbacks in that: (i) many studies utilize immortalized cell lines that differ from primary (i.e. freshly isolated) cell cultures, for example macrophage cell lines; (ii) the cells are in an undifferentiated state; (iii) the cells are in a monoculture which is artificial; (iv) the cells are often used in a non-polarized state so that there is no differentiation into apical and basolateral surfaces and; (v) environmental conditions under which the cells are grown differ considerably from those encountered *in vivo*, thereby affecting gene expression. In more recent years there has been a trend towards using cells that are more likely to have a relevant phenotype for the pathogen being investigated, for example gut-derived cells (such as Caco-2) for intestinal pathogens and orally derived cells (e.g. KB cells) for oral pathogens. In addition, it is possible to culture some cells (e.g. MDCK (Madin–Darby canine kidney) and Caco-2) cells on filters so that they grow as a monolayer with tight junctions and are polarized with distinct apical and basolateral surfaces. Such cells have been used to study *Salmonella* transcytosis, where it was shown that cells only attached and penetrated from the apical surface. Another trick, which aims to mirror the *in vivo* situation, is to infect animals and then remove cells for *in vitro* culture. This is known as the *ex vivo* technique and has been used to examine invasion of macrophages by *Salmonella* species.

It is fair to say that enormous amounts of information can be derived from cell culture studies but care must be taken not to extrapolate the data obtained from these *in vitro* studies too far.

CONCLUSIONS

What should be apparent from the above survey of bacterial invasion is the, perhaps, surprising extent to which the cell being invaded is actively involved in the process. The term invasion, as it is used in the non-scientific sense, usually implies that the invadee either tries to repel the invader or is so helpless that no resistance can be offered. Bacterial invasion of host cells rarely appears to occur by a process analogous to either of these. In contrast, bacterial invasion invariably involves the active participation of the host cell – the bacterium appears to redirect the activities of the host cell to ensure its uptake. Subversion, or manipulation, of host cell signalling pathways and cytoskeletal organization are phrases which more accurately characterize the invasive process. So, the 'how'? of invasion is becoming clearer and we have a different perspective of the process – no longer should we think of bacteria as being 'little devils' evily spewing out highly toxic molecules at random, thereby killing everything in sight. Instead, we can view these fascinating creatures as being more like the sophisticated confidence trickster who knows what makes people tick and thereby knows how to manipulate them to achieve his own ends. The bacterium does not have to knock down doors – the host willingly opens the door for it!

But we still have no clear impression of the 'why'? of invasion. Invasion of its host will certainly offer some advantages to an organism, once it has developed means of both accomplishing this daunting task and avoiding the internal host defence systems which will be marshalled against it. Hence, an invasive organism will escape the intense competition which exists among the large number of different organisms colonizing mucosal surfaces, all of which are vying for space, nutrients etc. The microbe-free environment within the host would be the microbe's equivalent of the Elysian fields – an enormous supply of nutrients and an ideal temperature for growth. Furthermore, those most adept invaders, *M. tuberculosis, Brucella* spp. and *Coxiella* spp., which maintain an essentially intracellular lifestyle and so avoid many aspects of the host defence system, can happily graze in these fields without constantly having to remain on the alert for the watchdogs of the immune system. The invasive lifestyle certainly appears to have a lot to offer its practitioners.

FURTHER READING

Books

McCrae, MA, Saunders JR, Smyth CJ, Stow ND (1997) *Molecular Aspects of Host–Pathogen Interactions.* Cambridge University Press, Cambridge, UK.

Paradise LJ, Bendinelli M, Friedman H (1996) *Enteric Infections and Immunity.* Plenum Press, New York.

Sussman M (1997) *Escherichia coli: Mechanisms of Virulence.* Cambridge University Press, Cambridge, UK.

Reviews

Cornelis GR, Wolf-Watz H (1997) The *Yersinia* Yop virulon: a bacterial system for subverting eukaryotic cells. *Mol Microbiol* 23: 861–867.

Cossart P (1997) Subversion of the mammalian cell cytoskeleton by invasive bacteria. *J Clin Invest* 100; S33–S37.

Cotter PA, Miller JF (1998) *In vivo* and *ex vivo* regulation of bacterial virulence gene expression. *Curr Opin Microbiol* 1: 17–26.

Falkow S (1997) Invasion and intracellular sorting of bacteria: searching for bacterial genes expressed during host/pathogen interactions. *J Clin Invest* 100: S57–S61.

Fallman M, Persson C, Wolf-Watz H (1997) Yersinia proteins that target host cell signalling pathways. *J Clin Invest* 100; S15–S18.

Finlay BB, Cossart P (1997) Exploitation of mammalian host cell functions by bacterial pathogens. *Science* 276: 718–725

Finlay BB, Falkow S (1997) Common themes in microbial pathogenicity revisited. *Microbiol Mol Biol Rev* 61: 136–169.

Garcia-del Portillo F, Finlay BB (1995) The varied lifestyles of intracellular pathogens within eukaryotic vacuolar compartments. *Trends Microbiol* 3: 373–380.

Guiney DG (1997) Regulation of bacterial virulence gene expression by the host environment. *J Clin Invest* 100: S7–S10.

Jones BD, Falkow S (1996) Salmonellosis: host immune responses and bacterial virulence determinants. *Annu Rev Immunol* 14: 533–561.

Kuhn M, Goebel W (1998) Host cell signalling during *Listeria monocytogenes* infection. *Trends Microbiol* 6: 11–14.

Menard R, Dehio C, Sansonetti PJ (1996) Bacterial entry into epithelial cells: the paradigm of Shigella. *Trends Microbiol* 4: 220–226.

Miller VL (1995) Tissue-culture invasion: fact or artefact? *Trends Microbiol* 3: 69–71.

Zychlinsky A, Sansonetti P (1997) Apoptosis in bacterial pathogenesis. *J Clin Invest* 100: S63–S65.

CHAPTER 7

Bacterial Protein Toxins: Agents of Disease and Probes of Eukaryotic Cell Behaviour

INTRODUCTION

It is part of the human condition to want to put everything into neat and separate categories. It is an approach that can be oversimplified. For example, liking Mozart or Beethoven, or the Beatles or the Rolling Stones (a give-away of the authors' ages), are not mutually exclusive pastimes. Categorization to identify common themes is a large part of scientific analysis, and science itself is divided into ever more specific sub-topics. However, not all subjects can be easily compartmentalized. In biology, the study of bacterial protein toxins is probably the most difficult to fit within one discipline.

Bacterial protein toxins are responsible for the damage caused by many pathogens in disease. There are many stages to this process, which involve several scientific fields of endeavour. Toxin expression is often tightly regulated. Toxins are then transported from the bacterial cell, often using a complex mixture of proteins to achieve export. Some toxins effect direct bacterium-to-host entry; many others have to survive in the unknown and potentially harsh extracellular world, before they latch onto a eukaryotic cell receptor. Certain toxins act on the cell surface, while others enter the cell, usually in a membrane-bound vesicle. Intracellularly-acting toxins are often constructed to exploit the severe conditions there (e.g. low pH and the presence of proteases) in order to expose previously buried residues which can penetrate into the membrane, and also to effect cleavage to release the fragment with the catalytic toxic activity. The active part of the toxin is taken across the vesicle membrane into the cytoplasm, where it finds its target and modifies it. The final stage – that of target recognition and modification – provides a key example of cellular microbiology because toxins have very specific targets, which are crucially important cellular constituents. So toxin research not only can explain disease and lead to antibacterial therapies, but can also identify new and important cellular components. Moreover, the potent and highly specific mechanisms by which toxins work have also led to exploitation of their activity – for example, coupling the cell-killing activity of diphtheria toxin to a protein which targets cancer cells results in an anticancer agent.

Is toxin research really about microbial pathogenesis, microbial genetics, receptors, membrane translocation, structural biology, catalysis, cell biology or cancer? The truth is that it can be about and touch on all of these. Toxins cover a wide sweep of biology and that is one of their fascinations. Over 300 toxins have been characterized, produced by many different genera of bacteria, and they are found in both Gram-positive and Gram-negative bacteria. The range of activities identified is also wide and it is clear that new toxins with novel mechanisms are still there to be discovered.

In this chapter we will look at toxins, their mode of action and their cellular effects, both from the bacterial and target cell viewpoint, and show how analysis of these remarkable molecules lies at the heart of cellular microbiology.

BACTERIAL TOXINS: PAST AND PRESENT KILLERS

Bacteria that produce toxins remain responsible for many of the world's diseases. Improvements in hygiene, the advent of antibiotics and advances in vaccines have eliminated many of the great bacterial killers in the developed world, but these toxin-producing bacteria remain formidable killers in the developing world. For example, tetanus kills more than 500 000 babies a year, whooping cough an enormous number of people, and cholera continues to cause deadly epidemics when conditions allow. Even in the West, outbreaks of *Escherichia coli* serotype O157 and the rise in methicillin-resistant *Staphylococcus aureus* (MRSA) give cause for concern.

Historically, the most infamous toxin-related diseases have been the plague, caused by *Yersinia pestis*, and cholera, caused by *Vibrio cholerae*. The plague swept across Europe from China in the fourteenth century wiping out whole cities in its path, and was again a notorious killer three centuries later in the Great Plague of London. Cholera was endemic in India for many centuries, but epidemics appear to be relatively recent worldwide scourges and have afflicted both Europe and the Americas in the nineteenth century. There have been major epidemics in Asia within the last 50 years. Indeed, cholera continues to cause alarm when the crucial condition of contaminated water supplies combine with extreme stress, like famine or war. Thousands died of cholera in the civil war in Bangladesh in 1971, and thousands more in the refugee camps of Rwanda in 1994.

HISTORY OF RESEARCH

Diseases caused by toxin-producing bacteria have been described since ancient times. Over 2000 years ago Hippocrates described cholera and outlined in detail the symptoms of tetanus. Records from about the same time describe the signs of diphtheria and the poet Virgil wrote an account of anthrax.

The history of research on bacterial protein toxins is just over 100 years old (Figure 7.1). A flurry of intense exploration around the 1880s by the leading bacteriologists of the day (Pasteur, Koch, Roux, Kitasato and von Behring) identified many of the bacteria producing toxins and the poisonous nature of toxin action. These pioneering studies led to some highly effective therapies. In the last 20 years progress has again been particularly rapid as the powerful tools of molecular genetics, and structural and cellular biology have been applied to the biology of toxins. This has led to the identification of new toxins. Moreover the deeper understanding of the relationship between toxin structure and function has facilitated the development of more rationally designed vaccines and other novel strategies to combat disease.

Both *Corynebacterium diphtheriae*, the bacterium that produces diphtheria toxin (DT), and *V. cholerae*, which produces cholera toxin (CT), were cultured in 1883. In the case of diphtheria, progress which turned basic knowledge into effective therapy was remarkably rapid. An antitoxin had been developed in animals for therapeutic administration by Roux and von Behring by the following year. Yersin and Roux speculated that the action of diphtheria toxin was enzymatic in 1888, and the basis of antitoxin therapy, i.e. the production of substances to

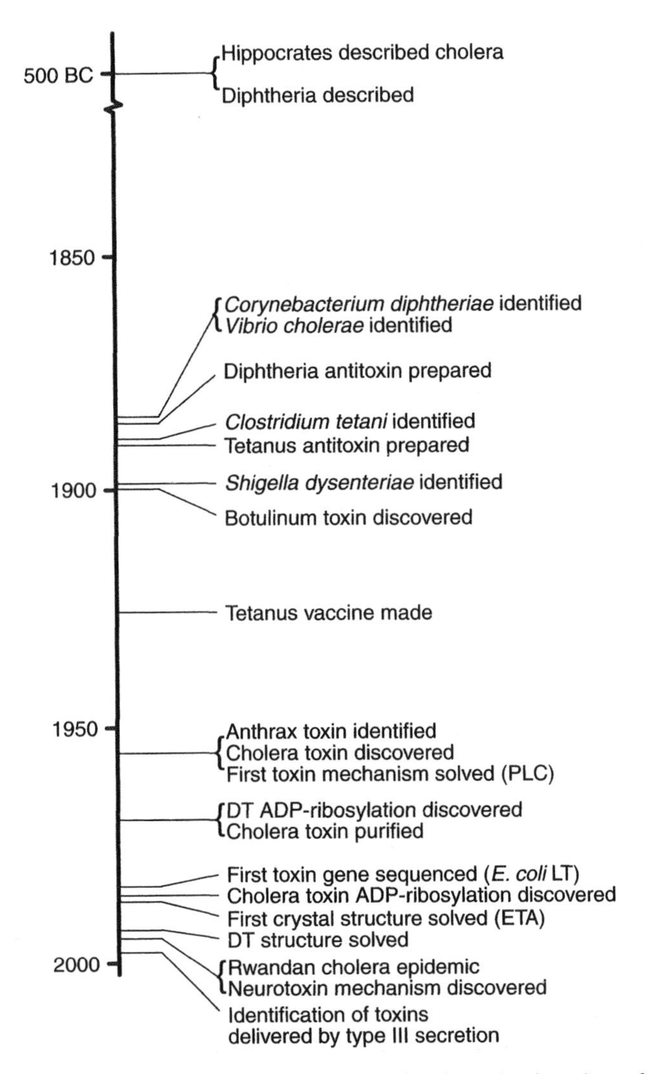

Figure 7.1 Toxin time line. This shows a brief and of necessity selective view of the history of toxins and the diseases they cause

combat diphtheria toxin by challenged animals, was recognized by 1890. The cause of tetanus was also identified about the same time. The transmissibility of the disease was realized in 1884, and *Clostridium tetani* was isolated by Kitasato in 1889. von Behring and Kitasato prepared antitoxin in 1890, and some years later (in 1925) immunization was developed using formaldehyde treatment to produce a non-toxic antigenic vaccine. This treatment has been so effective that it is still used today, and in the developed world reported cases of tetanus occur at the annual rate of one per 3 million of the population. Koch, Pasteur and others were instrumental at this time in discovering the agent of anthrax, *Bacillus anthracis*. However, with anthrax, as with cholera, the evidence that a toxin was involved was only discovered in the 1950s. Similarly, although *Shigella* was identified at the very end of the last century, convincing evidence for a potent toxin took several decades to accumulate.

The first toxin whose mechanism of action was identified was that of the *Clostridium perfringens* α-toxin, which was found to be a phospholipase C – an enzyme capable of damaging animal cell membranes. However, the understanding of intracellular toxin action began with diphtheria toxin in the 1960s, with the sequential discoveries that showed first the role of the co-enzyme nicotinamide adenine dinucleotide (NAD) in the action of diphtheria toxin, and then later the target of diphtheria toxin. The successful conclusion to this part of the story came with the discovery that diphtheria toxin catalysed ADP-ribosylation of the translocation factor, EF2, a protein that interacts with the protein synthesis machinery and that is essential for the elongation of the nascent protein chain. Diphtheria toxin was then the subject of intense study, with the production of numerous described mutants, so that when its crystal structure was eventually solved in 1992 it was possible to ascribe in some detail the role of individual amino acids in the process of intoxication. Despite the spectacular advances in understanding structure–function relationships in other toxins, diphtheria toxin remains a major paradigm, and the detailed knowledge about its structure–function relationship has found broad applications. Nowadays it is easier to identify the mode of action of new toxins partly because of the technology available and partly because the hard-won knowledge acquired over the past hundred years has provided models for how toxins work.

CLASSIFICATION OF TOXINS BY THEIR ACTIVITY

There are essentially three classes of toxin: (i) those that act at the cell membrane, type I; (ii) those that attack the membrane, type II; and (iii) those that penetrate the membrane to act inside the cell, type III (Figure 7.2). We will also discuss the new class of toxins that are directly transported from the bacterium into the target eukaryotic cell by type 3 secretion, as has been introduced in Chapters 2 and 6.

Toxins that damage membranes

Membrane-damaging toxins, sometimes called type II toxins, attack the cell by a direct bombardment on its frontier, the cell membrane, punching holes into it and thus leading to cell death. Over a hundred such toxins have been identified and categorized into groups dependent on function and structural arrangement. One group operates by insertion into the

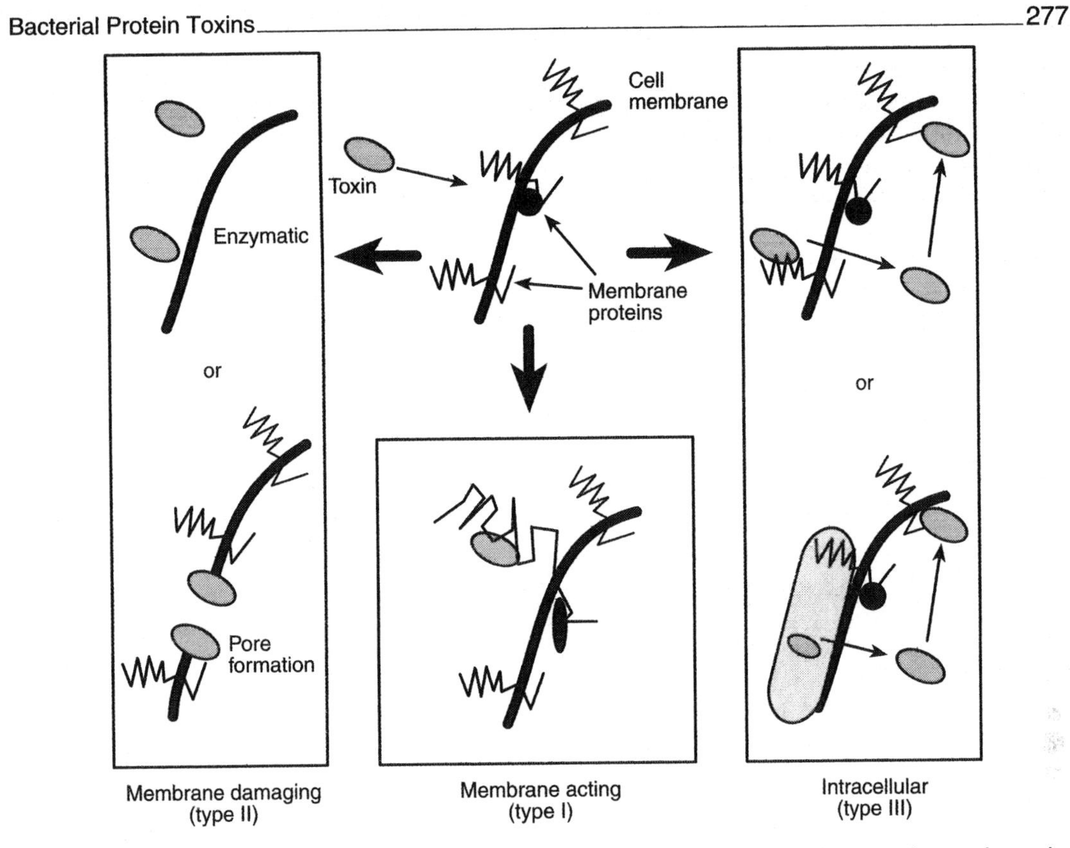

Figure 7.2 Major classes of toxin. There are three major types of toxin. The membrane-damaging toxins either enzymatically degrade membrane proteins or lipids, or insert to form pores. Membrane-acting toxins bind to surface proteins to alter their function. Intracellular toxins gain entry to the cell either by receptor binding and uptake, or via type III secretion directly from the host bacterium

membrane to form pores, and further subdivisions are recognized, depending on both function and structure. The other group are enzymatic and attack components of the cell wall.

Pore forming toxins

These proteins insert as oligomers into the plasma membrane, thus forming a pore. Different toxins produce different-sized pores. The toxins have structural features that enable them to operate both as water-soluble monomers and also as hydrophobic membrane proteins. They do this by sequestration of hydrophobic regions, which are exposed upon membrane insertion (Figure 7.3). There appear to be two basic structural arrangements that can bring this about. One type of toxin has hydrophobic and amphipathic helices, long enough to be membrane-spanning, that can be organized to hide the hydrophobic characteristics in solution; the other consists of soluble sheet structures that oligomerize to display sufficient hydrophobic character for membrane insertion to occur. In the simplest interpretation, the pore allows the efflux of small molecules and the breakdown of the membrane electrical potential. Many experimental approaches have used mammalian erythrocytes, which is why many of these toxins have been referred to as haemolysins. The situation with nucleated cells may be more complex. Such cells attempt to repair the damage, but cell death via apoptosis often still occurs (see Chapters 2 and 3). This suggests either that the temporary breach in the membrane can trigger a specific signal to induce apoptosis, or even that some of these toxins have an intracellular function and may enter the cell to modify the activity of a target. In this regard it is interesting that some pore-forming toxins have been shown to have cellular activity, for example invoking cytokine production, not apparently related to pore formation. Once again, it is difficult to put toxins into absolutely separate categories. It is also relevant to note that many intracellularly acting toxins have been shown to induce pores, as a consequence of translocating their catalytic domains across the membrane into the cytosol (see 'Membrane attachment and toxin translocation', below).

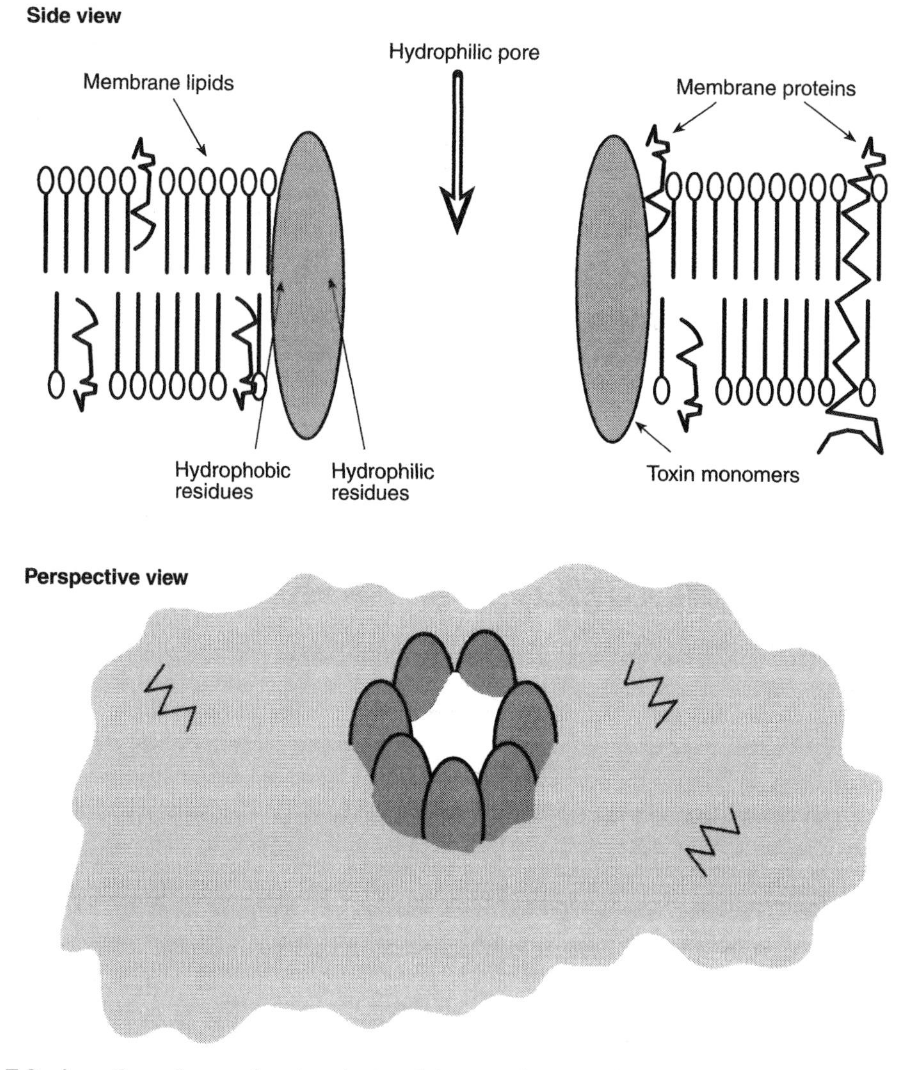

Figure 7.3 Insertion of pore-forming toxins into membranes. Pore-forming toxins organize into oligomers in the membrane and expose hydrophobic residues that interact with the hydrophobic phase of the membrane

Thiol-activated cholesterol binding cytolysins

This subfamily of toxins is produced by Gram-positive bacteria from four genera (Table 7.1) and includes, among others, streptolysin O from *Streptococcus pyogenes*, pneumolysin from *Streptococcus pneumoniae*, listeriolysin from *Listeria monocytogenes*, perfringolysin O from *Cl. perfringens*, and cereolysin from *Bacillus cereus*. This group of antigenically related toxins has had a number of names, many of which are now known to be inaccurate. As with other toxins that attack red blood cells, they have been labelled 'haemolysins', though it is now recognized that this is irrelevant to their *in vivo* activity. Their designation as 'oxygen labile' and 'sulphydryl activated' also turns out to be misleading. Although prolonged exposure of preparations of these toxins to air led to loss of activity which could be reversed by reducing agents, purified toxins are not susceptible to this treatment. Similarly, activation by thiols was thought to be necessary for their action, but mutagenesis of the single cysteine residue in several of these toxins has shown that this is not the case.

Table 7.1 The family of thiol-activated cholesterol-binding toxins

Genus	Species	Toxin
Streptococcus	*Strep. pyogenes*	Streptolysin O
	Strep. pneumoniae	Pneumolysin
Listeria	*Lis. monocytogenes*	Listeriolysin O
	Lis. seeligeri	Seeligerolysin
	Lis. ivanovii	Ivanolysin
Clostridium	*Cl. perfringens*	Perfringolysin O
	Cl. tetani	Tetanolysin
	Cl. sordellii	Sordellilysin
	Cl. caproicum	Caproiciolysin
	Cl. bifermentans	Bifermentolysin
	Cl. septicum	Septicolysin O
	Cl. histolyticum	Histolyticolysin O
	Cl. oedematiens	Oedematolysin O
	Cl. chauvoei	Chauveolysin
Bacillus	*B. cereus*	Cereolysin O
	B. alvei	Alveolysin
	B. thuringiensis	Thuringolysin O
	B. laterosporus	Laterosporolysin

These toxins attack cells with cholesterol in their membranes to form pores of about 30–40 nm comprising about 30 monomers. The oligomerization is believed to occur in the membrane. The monomer size of different toxins varies from 52 kDa to 60 kDa. Toxins within this group show some degree of sequence homology and antigenic relatedness, which is greatest at the C-terminus around the single cysteine. Although mutation of this cysteine residue to various other amino acids shows that it is not essential, other amino acids in this region have been shown to have an important function in cell lysis and binding to cholesterol, which is believed to be the membrane binding site.

Such toxins play a significant role in human disease. This has been shown by toxin gene mutagenesis in conjunction with animal infection and cell studies, and also the appearance of antitoxin antibodies in natural disease. For example, streptolysin O is a cardiotoxin, and is thought to be important in rheumatic heart disease. With the exception of pneumolysin, all of these toxins have an N-terminal signal sequence and are secreted from the bacteria synthesizing them. However, pneumolysin is released by dying bacterial cells, and indeed has been shown to cause morbidity and mortality after antibiotic treatment has eliminated live bacteria. The evolutionary significance of this behaviour is not clear.

The effect of these toxins is not simply limited to pore formation. For example, at sublytic concentrations they have been shown to inhibit B cell function, induce cytokine release and stimulate signal transduction pathways (inositol phosphate release and MAP kinase tyrosine phosphorylation).

RTX toxins

This group of toxins produced by Gram-negative bacteria comprises *Escherichia coli* haemolysin, several toxins from *Actinobacillus* species (notably *Act. pleuropneumoniae, Act. suis* and *Act. actinomycetemcomitans*), and the leukotoxins from *Pasteurella haemolytica, Proteus* and *Morganella* species. The *Bordetella* adenylate cyclase is also included in this group. They have a common structural arrangement (Figure 7.4) that includes a series of glycine-rich nonapeptides, from which their name is derived (repeat in toxin). The number of these repeat units varies between 6 and 47 in different toxins. This repeat region is involved in calcium-dependent binding to membranes. Other characteristic features of the structure of these toxins is an N-terminal amphipathic helical region, followed by several strongly hydrophobic domains, the nonapeptide repeats and a C-terminus that contains the transport signal involved in secretion. The genetic organization of these toxins and the mechanism of secretion and activation are also related (Figure 7.4). The gene for the structural protein (usually named *xxxA*) is flanked by an upstream gene (*xxxC*) and two downstream genes (*xxxB* and *xxxD*) involved in transport of the toxin across the bacterial cell wall. For example, *hlyA* is the structural gene for *E. coli* haemolysin. HlyA is inactive until modified by the *hlyC* gene product in an acylation reaction. HlyB and HlyD are involved in export of HlyA, and both have been shown to be membrane proteins. The organization of the *Act. pleuropneumoniae* toxins is unusual since there are three related *apx* operons, producing three Apx toxins that have slightly different properties, and some serotypes carry and express two RTX toxins. However, in this instance, two complete operons have never been found in the same isolate, and there appears to be sufficient similarity among the Apx toxins for the activation and export machinery of one to function with another. Indeed the *apxII* operon only ever consists of *apxA* and *apxC*. The advantage of such complexity is unclear.

These toxins have been shown to form pores of about 1–2 nm comprising seven monomers. In the case of the *Bordetella* adenylate cyclase toxin, pore formation may not in itself be part of the pathological mechanism, but may be an intermediary stage of toxin entry into the cytoplasm, since the adenylate cyclase activity appears to be the main function. The adenylate cyclase toxin is larger than the other members of the RTX group, which may reflect its additional enzymatic ability.

The toxins are important in both human and animal diseases. The prototype toxin, *E. coli* haemolysin, is associated predominantly with extraintestinal infections. The *Pasteurella* leukotoxin and the *Act. pleuropneumonia* toxins are virulence factors in bovine pneumonic pasteurellosis and porcine pleuropneumoniae respectively. As with the thiol-activated toxins, the nomenclature is misleading, since these haemolysins have been shown to attack other cell types. The *Bordetella* adenylate cyclase is toxic for macrophages during whooping cough. The

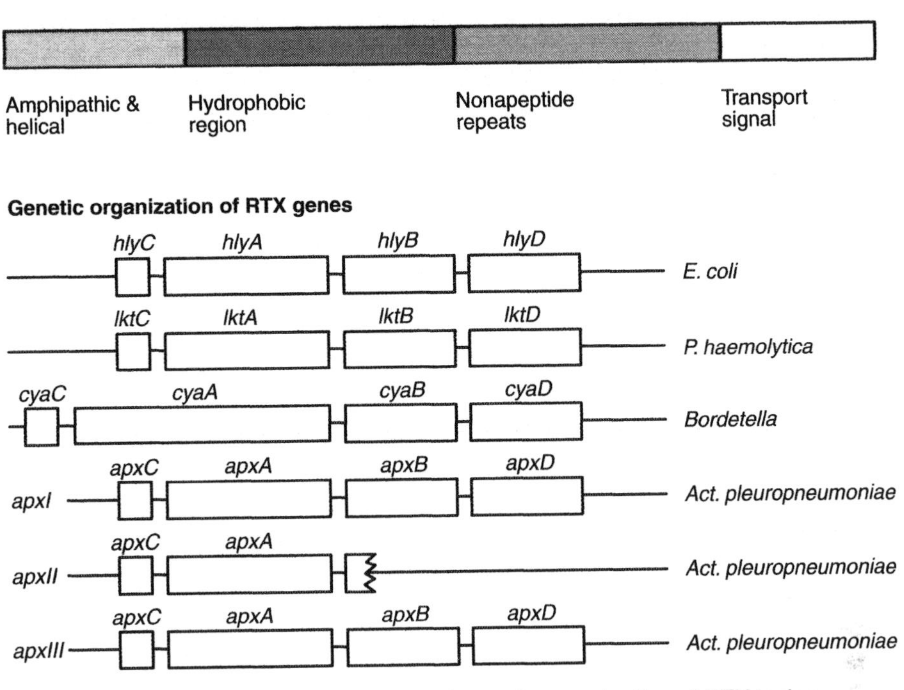

Figure 7.4 Domain structure and genetic organization of RTX toxins

toxins cause more subtle cellular effects than mere pore formation; for example, *E. coli* haemolysin stimulates production of nitric oxide (NO), prostaglandins and inositol phosphates. Thus these toxins attack the host in a number of ways to damage its normal functions.

Staphylococcus aureus toxin

The *Staph. aureus* α-toxin represents another type of pore-forming toxin. Partial unfolding, which can be promoted by a change in pH, exposes hydrophobic patches on the protein and precedes insertion. The α-toxin pore of about 2 nm comprises a hydrophobic β-barrel produced by seven interacting monomers. A central 15-amino acid region has been implicated in insertion, though other regions of the protein are also involved in the process. This contrasts with the RTX toxins and other toxins that contain amphipathic and hydrophobic helices that are sheltered from the exterior of the toxin in the soluble form.

Staph. aureus is an important opportunistic human pathogen that produces many different toxins. The α-toxin is viewed as a major virulence factor and is involved in endocarditis. Cell death after pore formation proceeds via apoptosis, but as with other pore-forming toxins additional cellular events have been noted.

Others

Other pore-forming toxins either do not fall into the categories listed above, or are currently poorly characterized. One of the best understood in the former category is aerolysin, produced by *Aeromonas* species, in particular *Aer. hydrophila,* which causes gastroenteritis and severe wound infections. Aerolysin is synthesized as a preprotoxin. The protoxin dimerizes upon secretion from the bacterial cell. This sequesters hydrophobic patches of β structure and confers water solubility and stability. Proteolytic cleavage of this protoxin does not cause obvious structural changes, but does enable oligomerization and thus pore formation. Aerolysin is one of the toxins whose structure has been determined.

The *Staph. aureus* δ-haemolysin is a 26-amino acid peptide that forms membrane pores; it also induces a cytokine response. The *Staph. aureus* γ-haemolysin is less well defined but also forms pores. The enterotoxins of *B. cereus* cause membrane damage, but although some of the genes have been cloned and sequenced, little is known about their mode of action.

The *Bacillus thuringiensis* insecticidal δ-endotoxins insert a helical domain into the membrane to form pores. These toxins have received considerable attention because of their commercial potential for protecting crops from damage caused by insects, and the crystal structures of several of them have been solved. The toxins are produced as insoluble crystals that are digested in the gut of the insect.

Many bacteria also make pore-forming toxins, called bacteriocins, that are active against other bacteria. Colicin from *E. coli* has been most analysed.

Toxins that damage membranes enzymatically

Phospholipases

Phospholipase C produced by *Cl. perfringens* has, as described, the historical distinction of being the first toxin whose mechanism of action was elucidated (in the 1950s). A number of Gram-positive and Gram-negative bacteria produce similar phospholipases. These include *Lis. monocytogenes, Staph. aureus* (the β-haemolysin), *Pseudomonas aeruginosa, B. cereus* and *Aeromonas* species. Some of these enzymes are zinc phospholipases, but the bacterial phospholipases do not act on polyphosphorylated inositols, and therefore do not produce the important signalling molecule, inositol 1,4,5-triphosphate (IP_3) (see Chapter 3).

The *Cl. perfringens* PLC, or α-toxin, is the most toxic PLC identified and is known to have necrotic and cytolytic activity. *Lis. monocytogenes* has two PLCs a phosphoinositol-specific PLC that aids escape from the primary vacuole, and a broad-range PLC involved in cell-to-cell spread. The *Ps. aeruginosa* PLC may damage lung surfactant.

Proteases

Bacteria release proteases that can attack eukaryotic membrane proteins. While these destructive enzymes are clearly virulence determinants, they have not usually been classed as toxins, perhaps because they were perceived to lack the 'single-minded' precision of other toxins. Two observations challenge that view. The first is that intracellularly acting proteases, such as the tetanus and botulinum toxins, are clearly toxins. The second is the recent discovery that proteases produced by *Porphyromonas gingivalis,* implicated in periodontal (gum) disease, comprise a family of related members, each of which has specificity for a particular integrin (Chapter 3). By specifically attacking integrins these proteases affect eukaryotic signalling and thus really could be said to belong to the family of type I toxins discussed below.

Membrane transducing toxins

These type I toxins (Figure 7.2) disrupt the cell by subtle means, by inappropriate activation of cellular receptors, thus sending the wrong message into the cell – like an intelligence operation that confuses the normal routes of communication.

ST family

The stable toxin (ST) of *E. coli* is only one of a number of toxins produced by that versatile organism. ST is a peptide of 18 or 19 amino acids whose structure is held together by three disulphide bonds. This high level of disulphide linkage is common in small peptides and results in the resistance of ST to heat denaturation, and hence its name. ST binds to membrane receptors to stimulate guanylate cyclase, giving rise to the intracellular message cGMP (see Chapter 3, Figure 3.2), which in turn activates protein kinase G and thus modulates several signalling pathways. The ST receptor is expressed in intestinal cells and its real purpose appears to be to bind to the endogenous ligand guanylin, a peptide hormone that regulates salt and water homeostasis in the gut and kidney. It is therefore hardly surprising that ST causes diarrhoea. The receptor has been proposed as a marker for intestinal cells, and thus ST may be used to direct therapeutic agents against malignant cells of intestinal origin. Several other bacteria including *Yersinia* species make toxins that are either identical or closely related to *E. coli* ST.

The *B. cereus* emetic toxin (also termed cereulide) is a ring of three repeats of four amino acids. This toxin is believed to bind to the 5-HT_3 receptor to stimulate the vagus afferent nerve, leading to vomiting. No details are currently known about how this toxin is produced. The toxin is resistant to heat and other insults. The extent of food poisoning due to *B. cereus*-contaminated food is not known.

Superantigens

Many pathogens stimulate the host to produce an inappropriate immune response that is responsible for the pathological damage produced. This is often a massive release of pro-inflammatory cytokines (see Chapter 9).

Superantigens stimulate the immune response directly by acting as mitogens for a subset of T cells. They bind to the T cell receptor and MHC class II antigen directly outside the antigenic peptide-binding groove without the need for presentation of an antigen, and so activate a whole subset or several subsets of T cells of a particular vβ phenotype. Thus they activate between 5% and 20% of the T cell population instead of the 0.0001 – 0.01% expected with a conventional antigen. This leads to massive release of proinflammatory cytokines such as IL-1, TNF, IL-6 and IL-8, which induces a severe shock response.

Superantigens are produced by *Staph. aureus* and *Strep. pyogenes.* Several such superantigens have been described, of which the most notorious is TSST-1, a *Staph. aureus* toxin, responsible for toxic shock syndrome, that is often associated with the use of tampons (Table 7.2).

Intracellular toxins

These type III toxins (Figure 7.2) are perhaps the most subtle in their action, a Trojan horse that the cell takes up only to release a deadly attack that homes in on a crucial target at the

Table 7.2 Bacterial superantigenic toxins

Organism		Toxin	Mass (kDa)	Disease
Staph. aureus	Enterotoxins	A, B, C1, C2, C3, D, E	~ 27	Food poisoning, etc.
	Exotoxins	A, B	~ 12	
		TSST-1	~ 22	Toxic shock
Strep. pyogenes	Erythrogenic toxins	A, C	~ 25	
		B	~ 40	
		D	~ 13	

heart of the cell's communication system. The intracellularly acting toxins have more stages in their operation than other toxins (Figure 7.5). Receptor binding is not enough. They have to gain intracellular access, survive attack by proteases and protons, and then trick the cell into taking them to their target, which they destroy with enzymatic precision. All the toxins analysed so far have the so-called AB structure (Figure 7.6), where A is the catalytic domain or subunit, and B deals with membrane attachment and translocation of the catalytic domain to the cytoplasm. There are several variations on this general theme. In some toxins, e.g. DT, the AB functions are joined in one molecule, though nicking occurs to yield a molecule held together by a disulphide bridge. Another common variation is the group of AB_5 toxins, where there are five B subunits that form a doughnut ring; the single A subunit is believed to enter the cell through the hole in the centre of the doughnut.

Despite the complexity of this process, these toxins are highly effective at reaching their target and comprise the most deadly group of toxins. Botulinum and tetanus neurotoxins are lethal to humans at a dose of about 0.1 ng, reflecting the crucial requirement of a functional nerve relay system. DT and the *Pasteurella multocida* toxin, PMT, affect cells in almost opposite ways (diphtheria toxin kills cells, whereas PMT promotes cell growth and division), but both kill animals at dose of about 0.1 μg/kg body weight.

Intracellular toxins are usually grouped on the basis of their enzymatic mode of attack. Like other enzymes, they can either catalyse cleavage of molecules or transfer of molecular moieties between molecules, but within these generic groups many mechanisms are found.

Lytic activity

Cleavage of a target has the advantage over other target modifications that it is likely to be a process that requires less energy. Several toxins have successfully adopted this strategy.

The neurotoxins

Although it was known for years that the large *Clostridium botulinum* and *Cl. tetani* neurotoxins targetted the synapse, it was only in the mid-1990s that their mode of action was solved. An amino acid motif common to proteases that have zinc at the active site was found in these toxins, and this led to the identification of both the mode of action and the targets. The tetanus toxin and all the variants of the botulinum toxins (with the notable exceptions of C2 and C3 – see 'Modifying activity', below) act as proteases. Each toxin has a single target protein; the group as a whole attack four proteins (syntaxin, VAMP/synaptobrevin 1 and 2, and SNAP-25; Table 7.3). These targets are involved in the docking of vesicles containing

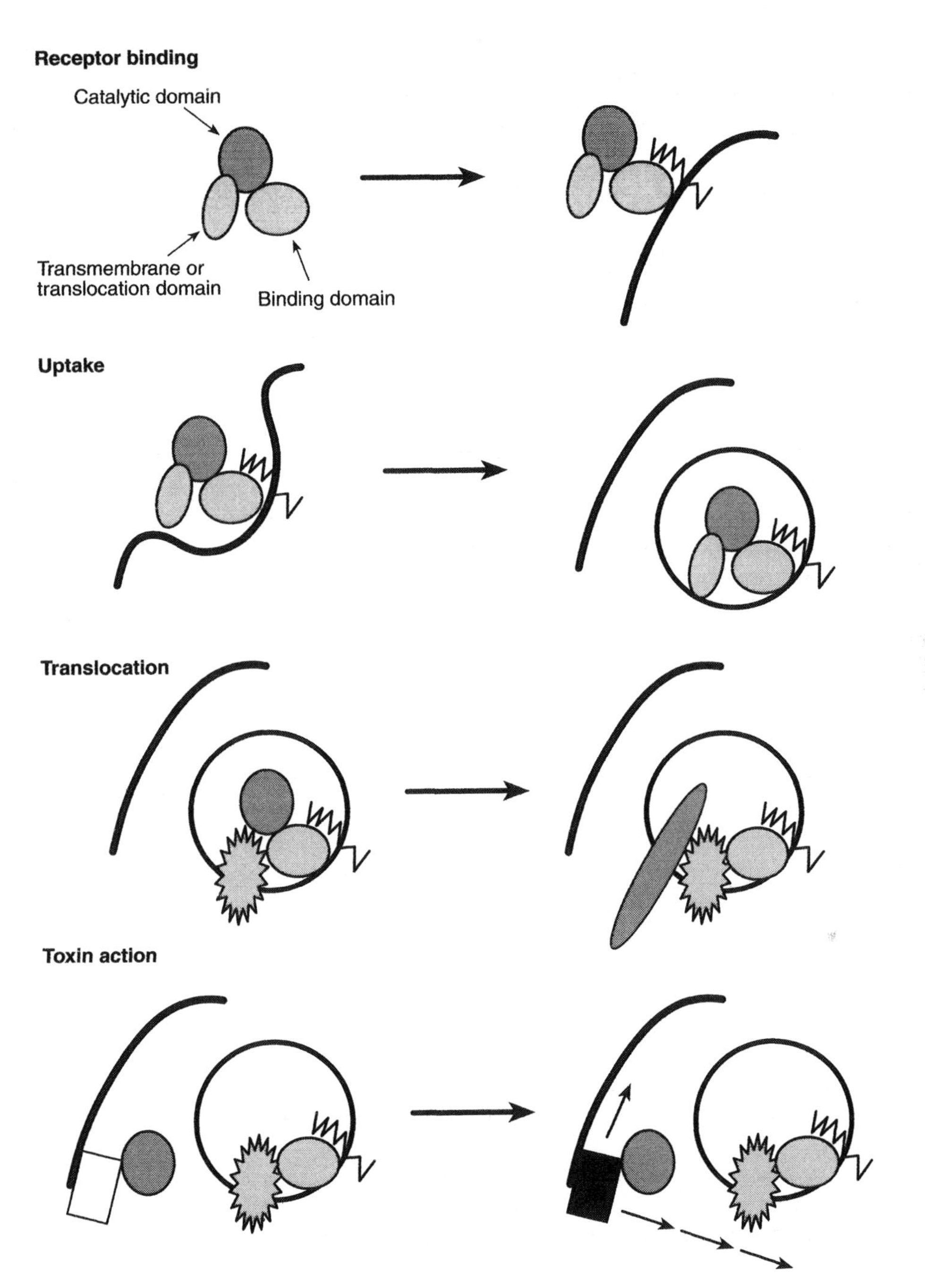

Figure 7.5 The different stages of entry for intracellular toxins. The catalytic domain is also called the A fragment and the translocation and binding domains make up the B fragment that binds to cells and mediates entry of the catalytic domain

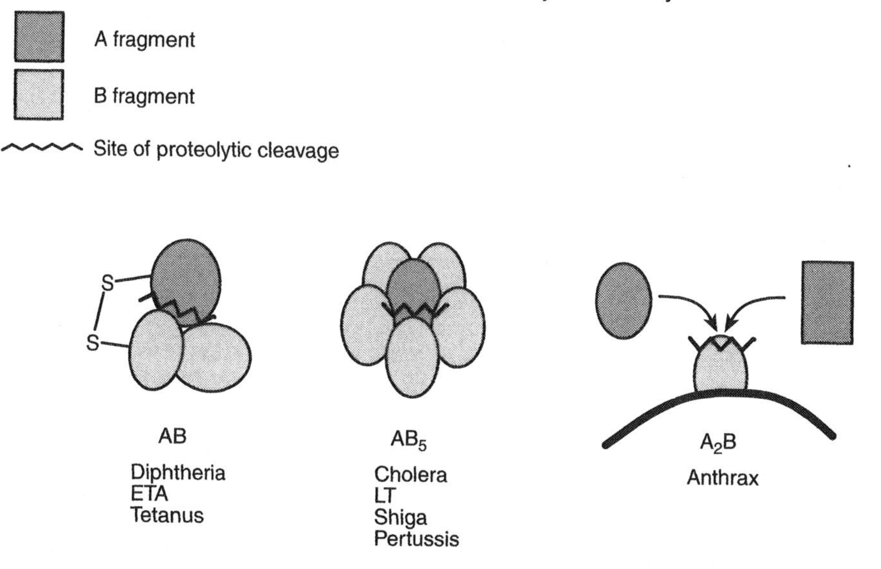

Figure 7.6 Structural architecture of the AB toxins. The two most common arrangements are AB, where the B fragment can be further divided into two domains involved in receptor binding and translocation of the active A fragment

Table 7.3 The targets of the clostridial neurotoxins

Bacterium	Toxin	Target
Cl. botulinum	A	SNAP25
	B	Synaptobrevin/VAMP
	C1	Syntaxin
	D	Synaprobrevin/VAMP
	E	SNAP25
	F	Synaptobrevin/VAMP
Cl. tetani		Syntaxin

neurotransmitter to the synaptosomal membrane which enables membrane fusion and release of the neurotransmitter to take place (see Chapter 3 for detail of endosomal trafficking).

The neurotoxins are all large single-chain toxins of over 150 kDa that are proteolytically cleaved during activation and cellular entry; although the catalytic site of action has been identified, detailed domain structure is not currently available. The different types of botulinum toxins have widely varying potency that may be dependent on the target, enzymatic activity and/or receptor binding. Both the botulinum and tetanus toxins have several cysteines with two disulphide bonds, one of which binds the two chains together (as with diphtheria) and an intrachain disulphide that is also required for activity.

Botulinum toxin is so potent that botulism is caused, not by infection, but by ingestion of food that has become contaminated with the toxin. Botulinum toxins cause a flaccid paralysis by blocking the function of peripheral nerves. Tetanus toxin attacks nerve cells in the CNS and its effects are more dramatic, leading to muscle spasms and rigid paralysis. *Cl. botulinum* can also cause an intestinal infection in infants (infant botulism), which causes a mild self-limiting disease. Botulism in adults is often food-borne and frequently fatal. Tetanus is usually associated with soil-associated wounds, but this deadly disease is controlled by routine vaccination using formaldehyde-inactivated toxin in the developed world.

The potency and specificity of the botulinum toxins have led to their therapeutic use in some nerve disorders and other conditions.

Anthrax lethal factor

Anthrax lethal factor (LF) is a zinc protease that cleaves the N-terminus of MAP kinase to inactivate it. The target is a crucial component of the MAP kinase signalling pathway that regulates proliferation (see Chapter 3). The anthrax toxin complex is an unusual variant on the AB structural theme (Figure 7.6). The receptor binding and translocation domain, called protective antigen (PA), is expressed separately from the catalytic domains. PA behaves as a membrane inserting β-barrel structure, and is unusual in that it serves two catalytic masters: LF, and the oedema factor (EF), which is an adenylate cyclase (see below). This system obviously saves on genome size, but does not appear to be have been adopted elsewhere. The binding of the catalytic domain to PA has been mapped to the N-terminal 250 amino acids of these proteins and this system has been exploited to use PA to carry recombinant foreign proteins into the cell. The deadly disease of anthrax is dependent on the expression of the toxins, since non-toxic strains are avirulent.

Other proteolytic toxins

The *Staph. aureus* epidermolytic, or exfoliative, toxin has motifs implying that it may be a zinc protease, but this has not been confirmed and the target remains unknown. The toxin causes scalded skin syndrome, in which skin layers peel off to cause an inflamed lesion similar to that seen in scalding.

Deamidation

A recently identified mechanism of toxin action is glutamine deamidation. The cytotoxic necrotizing factor (CNF) of *E. coli* deamidates glutamine (Gln)63 in the Rho protein, changing this to glutamic acid (Glu). Rho is a small G-protein that plays a key role in intracellular signalling pathways (see Chapter 3). Gln63 has a crucial role in GTP hydrolysis and thus Rho-Glu63 remains in a GTP-bound and therefore activated state; in particular the ability of RhoGAP (Rho GTPase-activating protein) to stimulate the Rho GTPase is abolished. This has several effects, notably to stimulate tyrosine phosphorylation of the focal adhesion kinase, $p125^{FAK}$, and paxillin, leading to large-scale cytoskeletal rearrangements. Resting cells are also driven into the cell cycle, leading to a round of DNA synthesis; nuclear division takes place but cytokinesis is blocked, resulting in the production of multinucleated cells. The eventual fate of such cells is not known. The signalling events leading from activation of Rho proteins to these effects is not clear; in particular, it is uncertain whether they are related to modification of particular members of the Rho subfamily or due to the lesser degree of cdc42 modification that has been reported. In this regard it is relevant that microinjection of mutant RhoA leads to alterations in stress fibrils, but does not produce multinucleated cells. Whether deamidation is the primary effect of CNF is unclear since this toxin also catalyses the transglutamination of Rho, which presumably also causes activation.

It has been reported that the dermonecrotic toxin (DNT) of *Bordetella* species also deamidates RhoA. DNT shows significant homology to CNF at the C-terminus, which is believed to be the site of the catalytic domain, and its action, though subtly different from CNF, has many similarities, including massive reorganization of actin filaments (Figure 7.7). Putative receptor- binding and translocation domains have also been located on both these toxins.

CNF is one of a number of toxins produced by *E. coli* strains. CNF has been found in isolates associated with both intestinal and extraintestinal origin, and frequently with *E. coli*

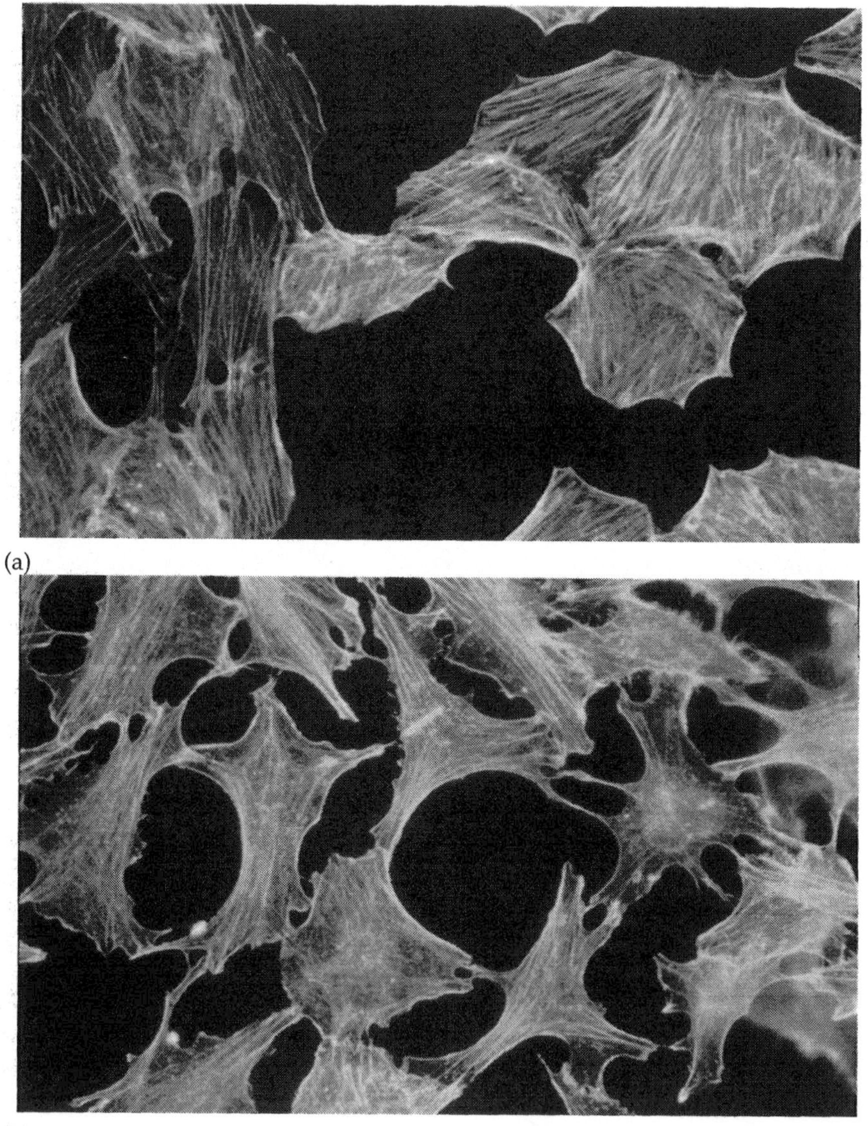

Figure 7.7 The effect of *Bordetalla* dermonecrotic toxin (DNT) on the actin rearrangements within cells. The upper panel (a) shows untreated cells. The lower panel (b) shows the effect of 0.5 ng/ml DNT. Striking reorganization of actin filaments can be seen. Reproduced with the permission of Dr Toni Adams

strains involved in urinary tract infections (see Chapter 5). Nevertheless its exact role and importance in pathogenesis have not been elucidated in detail. The role of the *Bordetella* DNT is more controversial. Although it appears to be highly conserved and is produced by all species of *Bordetella*, experimental infections in mice with *Bordetella pertussis* fail to identify a pathogenic role for it, although it should be noted that the mouse infection is a notoriously poor model for human whooping cough. DNT clearly has a role in animal infections, where it directly affects bone metabolism and enhances colonization by other pathogens.

N-glycosidases

The Shiga toxin family is a group of toxins that attack 28S rRNA to depurinate adenine 4324 which is involved in elongation factor 1 (EF1)-mediated binding of tRNA to the ribosomal complex. This modification inhibits protein synthesis. These toxins are found in *Shigella* species and some *E. coli* serotypes, notably *E. coli* O157 (see Chapter 5), where they have also been referred to as verotoxins because of their toxicity to Vero cells. Shiga toxin has an AB_5 structure (Figure 7.6). The B subunits form a pentameric ring that binds to globotriosylceramide glycolipid receptors. The catalytic A subunit that sits on 'top' of this structure shares sequence and structural homology with the ricin family of plant toxins that carry out an identical reaction. Like many other toxins, the A subunit is activated by proteolytic cleavage on cell entry to yield two fragments joined by a disulphide bridge that is subsequently reduced.

The Shiga toxins are important in the pathogenesis of the bacteria that produce them. The toxins have come to prominence recently with the rise in food poisoning caused by *E. coli* O157 expressing Shiga toxin. The *E. coli* Shiga toxin is associated with haemolytic uraemic syndrome and can lead to kidney failure and haemorrhagic colitis. Other cellular effects, for example NO production, have also been noted. With *Shigella dysenteriae* the pathogenic role is not entirely clear since *Sh. dysenteriae* is primarily an intracellular pathogen.

The specificity of Shiga toxin binding has led to potential uses in cancer therapy (see Chapter 6, 'Invasion of endothelial cells' and 'Chimaeric immunotoxins' (later)).

Modifying activity

Adenylate cyclases

Most toxins that perturb eukaryotic signalling pathways do so by modifying key proteins of these pathways, thereby serving to lock them in either the 'on' or 'off' position. The adenylate cyclase toxins of *B. anthracis* and *Bordetella* species act to increase the concentration of the secondary messenger, the nucleotide cAMP, by direct catalytic action on ATP (see Chapter 3, Figure 3.18). The *B. anthracis* toxin is the adenylate cyclase catalytic domain, and like the lethal factor it uses PA to gain entry to target cells to cause the oedematous reaction. The *Bordetella* cyclase toxin also has haemolytic activity and forms pores, but this is thought to be a by-product of its entry mechanism. Indeed this toxin is also classed as an RTX toxin. Most fascinatingly, both EF and the *Bordetella* cyclase are stimulated by calmodulin (CaM), a eukaryotic calcium-binding protein that acts as a sensor for calcium concentration and works by binding to and modifying the activity of effector proteins. CaM is believed to be restricted to eukaryotic systems, so this is yet another example of the intimacy of the interaction between eukaryotic and prokaryotic systems. It is interesting to speculate on the advantage of such an arrangement and the likely effect of its existence, since it would be predicted that these adenylate cyclases would be active in cells with increased calcium and/or CaM levels and that this selectivity would be advantageous to the aggressor bacterium. Furthermore, factors that stimulate signalling pathways would be predicted to act synergistically with adenylate cyclase. It is known that other pathogens can cause a rise in intracellular calcium. Alternatively, the requirement for stimulation by CaM may prevent these toxins from poisoning the bacteria that produce them with high cAMP; under CaM stimulation the *Bordetella* adenylate cyclase has the highest activity known for an adenylate cyclase. The anthrax and *Bordetella* adenylate cyclases show some homology to each other but not to either the bacterial or mammalian enzymes, which does not help us to define evolutionary relationships.

The *B. anthracis* adenylate cyclase is the cause of the extensive fatal oedema in anthrax. It may also inactivate phagocytic cells. The *Bordetella* cyclase has been shown to inactivate macrophage function.

ADP-ribosylation

A number of toxins attach the ADP-ribose moiety (Figure 7.8) from NAD to their target, which in every case is a nucleotide-binding protein. In most cases, the target is a G protein which binds and hydrolyses GTP, but some toxins modify targets that bind ATP. Presumably ADP-ribosylation arose from an ancestral precursor to these toxins and over the years was modified to arrive in different toxin structures, designed to attack different targets. ADP-ribosylation takes place in eukaryotic cells as part of their normal function, so one possibility for the source of the bacterial motif is that it was originally of eukaryotic origin.

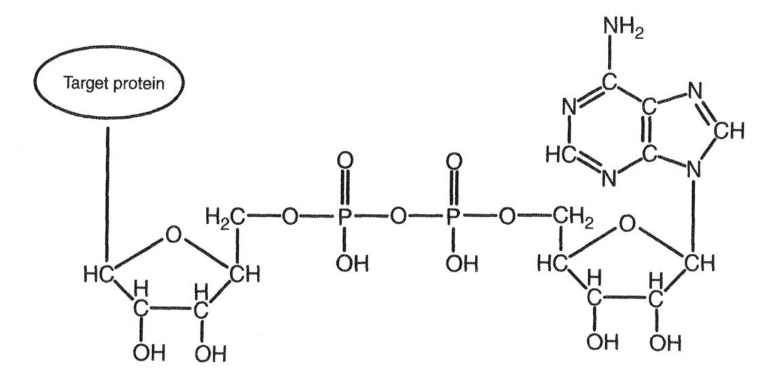

Figure 7.8 ADP-ribose. ADP-ribose produced from NAD is attached to the target protein via the ribose moiety. This method of attack has been adopted by a wide variety of toxins

These toxins fall into several categories, depending on the target attacked and the structural arrangement of the toxin.

Diphtheria toxin (DT) and Ps. aeruginosa exotoxin A (ETA). These are both single chain polypeptides. The catalytic activity transfers ADP-ribose to a modified histidine, diphthamide, on elongation factor 2 (EF2), an essential co-factor that binds to the ribosomal complex during protein synthesis. ADP-ribosylation stops protein synthesis and leads to cell death.

The toxins have no sequence homology. They have a similar structure with the exception that the organization of the domains is different. Thus the C-terminus of diphtheria toxin contains the catalytic site, the N-terminus the receptor binding site and in between these is the transmembrane or translocation domain; in ETA the catalytic domain is at the N-terminus. Unlike diphtheria toxin, the receptor-binding domain in ETA does not comprise a single stretch of polypeptide chain. The route of entry for diphtheria toxin, in particular, has been well worked out. After toxin binding, diphtheria toxin is taken up via a clathrin-coated pit and is thus internalized inside the cell in a membrane-bound vesicle. Proteolytic nicking cleaves the catalytic domain from the rest of the toxin, but it remains attached via a disulphide link. The conformation of the translocation domain changes because of the low pH of the vesicle and inserts deeper into the membrane and in doing so drags the catalytic, domain across the membrane into the cytoplasm where the disulphide linkage is reduced.

Diphtheria is a respiratory disease that is caused by the toxin action. The disease is characterized by production of a pseudomembrane in the throat, comprising bacteria and exudate from necrotic tissue. Severe systemic damage, including cardiac damage, can also occur. Diphtheria is a rare disease in the West and a good vaccine is available. *Pseudomonas aeruginosa* is a prominent cause of infection in immunocompromised subjects and burns patients.

Cholera (CT), heat-labile (LT) and pertussis (PT) toxins. These toxins attack heterotrimeric G proteins (see Chapter 3) and all have an AB_5 structure that is similar to that of Shiga toxin (Figure 7.6). CT and the almost identical LT from *E. coli*, so called to distinguish it from the unrelated heat-stable toxin (ST), both ADP-ribosylate G_s, a heterotrimeric G protein involved in stimulation (hence the $_s$) of adenylate cyclase. The α subunit of G_s is modified at Arg201 which inhibits the GTPase activity, to lock the modified G_s in the 'on' position, leading to permanent activation of adenylate cyclase. Pertussis toxin (PT) ADP-ribosylates the α subunit of G_i at Cys352. G_i is another protein that interacts with adenylate cyclase, but which acts to inhibit its function. Here the ADP-ribosylated form cannot interact with the adenylate cyclase to inhibit its activated state, so again high cAMP ensues.

The PT subunits have a similar ring structure to other AB_5 toxins, even though they are not identical (the B subunit comprises one molecule of proteins named S2, S3, and S5 and two of S4). Such an arrangement enables PT to bind to different glycolipid binding sites (see Chapters 5 and 6).

In the gut epithelial cells, the high concentration of cAMP induced by the action of CT or LT causes massive fluid accumulation in the gut lumen, watery diarrhoea and, particularly in the case of cholera, death if untreated. The biological significance of PT is less clear. *Bord. pertussis* pathogenesis is complex, partly because *Bordetella* species produce a range of toxins, and partly because several host cell types are involved in whooping cough. Nevertheless, it appears that PT causes a variety of effects, including effects on immune function, and it may be important for the characteristic cough: in addition PT is postulated to play a role in attachment.

Other ADP-ribosylating toxins. The *Cl. botulinum* C3 toxin is a small protein that ADP-ribosylates RhoA, B and C at Asn41, and indeed its activity on RhoC led to the identification of the Rho family (described in detail in Chapter 3). The modified Rho has an unaffected GTPase activity, but is unable to signal to its effector proteins. C3 does not have a domain for cellular uptake and there are many reports that C3 cannot enter cells without the use of various experimental tricks *in vitro* (e.g. microinjection), though some researchers claim that it can enter some cells. C3 causes rounding of cells. The contribution of C3 to *Cl. botulinum* pathogenicity is not clear.

The epidermal differentiation inhibitory factor (EDIN) produced by *Staph. aureus* has high homology to C3 and an identical biochemical activity. Application of EDIN or C3 to cultured keratinocytes blocks differentiation and stimulates growth; injection of either toxin into skin leads to hyperplasia of the epidermis.

The botulinum C2 toxin ADP-ribosylates monomeric G actin, an ATP-binding protein, and inhibits its ability to polymerize into F-actin. The modified G actin (at Arg177) also blocks polymerization reactions on F-actin. This leads to a complete loss of the microfilament network and rounding of cells. Other clostridial species produce similar toxins: the iota toxin from *Cl. perfringens*, and *Cl. spiroforme* and *Cl. difficile* toxins. The pathological effects of these toxins include fluid accumulation, hypotonic effects and death.

Glucosylation

A number of toxins have recently been found to glucosylate their target, in other words, attach a glucose moiety. Such toxins use UDP-glucose as a substrate for the reaction. In all cases so far identified, these toxins modify proteins of the Ras superfamily, in particular the Rho subfamily. The *Cl. difficile* toxins A and B both glucosylate Rho on Thr37 to cause a breakdown of the F-actin cytoskeleton and cell rounding. Toxin B has a higher catalytic activity than toxin A. These are large toxins (300 kDa) and the domains involved in the various stages of activity are only just being described. They have been referred to as AB^x toxins, since they have a multivalent receptor-binding domain. This comprises a series of repeats at the C-terminus that are involved in receptor binding. A central hydrophobic domain, found in all the toxins, is postulated to be the translocation domain, and it is assumed that the toxins undergo some form of processing before reaching their targets, but neither this nor the trafficking route is known. There is evidence that the catalytic domain is at the N-terminus. The *Cl. difficile* toxins attack all members of the Rho subfamily (Rho, Rac and cdc42). Related glucosylating toxins modify the same amino acid, but either attack a different subset of the Ras superfamily or add a different sugar (Table 7.4). The *Cl. sordellii* toxin is the first toxin to be found to modify Ras *in vivo*.

Table 7.4 Large clostridial toxins that target the Rho family

Toxin	Size (kDa)	Target	Co-factor
Cl. difficile	LTA 308 LTB 270	Rho, Rac and cdc42	UDP-glucose
Cl. sordellii	300	Ras, Rac and Rap	UDP-glucose
Cl. novyi	250	Rho, Rac and cdc42	UDP-*N*-acetylglucosamine

Cl. difficile can be carried as a commensal, but is the cause of antibiotic-associated pseudomembranous colitis which occurs when competing commensal bacteria are removed by the drug.

Unknown activity

Many interesting toxins exist for which the mode of action is yet to be discovered. Although the cellular effects of some of these toxins have been described, this does not necessarily suggest either the nature of the target or the mode of action. We will highlight two of these. *Helicobacter pylori* took bacterial research into a new area when it was shown to be causally linked not only to gastric ulceration, but also to gastric cancers. The VacA toxin has attracted most attention in the analysis of *Helicobacter* species. The *P. multocida* toxin, PMT, is equally novel as its modus operandi is as a highly potent mitogen.

The *Hel. pylori* vacuolating toxin, VacA, leads to the formation of large acidic vacuoles that contain vacuolar ATPase. It has recently been suggested that VacA might operate through signal transduction pathways that control vesicle trafficking along the endocytic pathway. VacA is thought to be an AB toxin, and has been shown to undergo a pH-related structural transition that leads to increased stability at low pH. The presence of the *vacA* gene in human isolates of *Hel. pylori* is associated closely with ulceration, and both mucosa-associated lymphoid tissue (MALT) lymphoma and gastric adenocarcinoma. However, its role is unclear for several reasons. First, *vacA* and several other *Hel. pylori* genes are encoded on a pathogenicity island, so that *vacA* presence correlates with the presence of many other genes. Secondly, vacuoles are not found in gastric biopsies, which may indicate that the vacuolation is an *in vitro* artefact produced by an unnaturally high concentration of toxin. Finally it has been reported that the related *Hel. mustelae*, a ferret pathogen, does not have the *vac*A gene, yet causes ulceration and leads to gastric cancer. Ferret and mouse infection models display similar pathological features to the human diseases and thereby facilitate study of host–pathogen relationships in disease. A related *Helicobacter* species, *Hel. hepaticus*, causes hepatitis and hepatic tumours in mice.

Most toxin groups are defined by similar biochemical functions: one novel group of toxins differ from this principle. These toxins are defined by their effect on cell cycle progression and cell growth control. The interest in these toxins is spurred on by the belief that they modify key components involved in either cell cycle or cell growth regulation, and thus will prove useful tools for cell biology and cancer research. Some of these toxins are listed in Table 7.5, and they will be discussed in more detail when we consider their effects on cells. Three of the toxins are related and have been analysed in more detail than the others. The cytotoxic necrotizing factor (CNF) from *E. coli* has domains homologous to both the

Table 7.5 Toxins that affect cell growth

Organism	Toxin	Cellular effect	Target	Mode of action
P. multocida	PMT	Intracellular signalling leading to mitogenicity	Unknown	Unknown
Bordetella	DNT	Actin reorganization, DNA synthesis and binucleation	Rho	Gln63 → Glu
E. coli strains	CNF	Actin reorganization, DNA synthesis and multinucleation	Rho	Gln63 → Glu
Cl. botulinum	C3	Actin reorganization	Rho	Asn41 ADPR
Staph. aureus	EDIN	Actin reorganization	Rho	Asn41 ADPR
Por. gingivalis	FAF	Cell cycle progression	Unknown	Unknown
Act. actinomycetem-comitans	Gapstatin	Cell cycle progression	Unknown	Unknown

dermonecrotic toxin (DNT) of *Bordetella* species and PMT, though DNT and PMT show no similarity to each other. PMT is unique among toxins in its action as a potent mitogen. Neither its target nor its enzymatic mode of action has been identified, though several indirect lines of evidence suggest that it might modify G_q, a heterotrimeric G protein intimately involved in the control of growth (see Chapter 2 for details of the eukaryotic cell cycle and Chapter 3 for signalling mechanisms). The creation of an inactive mutant in PMT has helped to identify its catalytic domain. In addition to necrotic effects, PMT and DNT cause bone loss in affected animals.

Membrane attachment and toxin translocation

Although the enzymatic activity of toxins has attracted most attention and excitement, the rest of the toxin molecule that enables it to reach its target is as important to its action and specificity. The *Cl. botulinum* C3 toxin is an enigma that lacks binding and uptake domains and appears to be inactive *in vitro* unless it is injected into the cytoplasm or linked to protein domains which can internalize it. Similarly, toxins with the same enzymatic function and target do not necessarily induce the same cellular consequences. Toxins A and B from *Cl. difficile* each glucosylate Rho on Thr37, but induce different biological sequelae; toxin A is described as an enterotoxin, whereas toxin B is described as a cytotoxin. This is partly due to different receptor binding preferences between the two toxins and also the much higher enzymatic activity of toxin B. The tetanus and botulinum neurotoxins exhibit a similar phenomenon. Tetanus induces a rigid paralysis, whereas botulinum causes flaccid paralysis, yet the tetanus toxin modifies vesicle-associated membrane protein (VAMP)/synaptobrevin at exactly the some location as one of the botulinum toxins. Conversely, the seven botulinum neurotoxins have different targets but are grouped together in terms of the effect they induce. These specific functions are likely to be linked to the targetting of the toxins to different cells, and reflects the importance of the binding and uptake domains.

A variety of receptor types have been subverted by different toxins. Some attach to glycolipids. For example, cholera toxin binds to ganglioside GM1. Other toxins bind to cell surface proteins, e.g. diphtheria toxin binds to the heparin-binding EGF-like growth factor. Binding, uptake and trafficking have been extensively studied in a few model toxins. These include Shiga, diphtheria and cholera toxins.

Cholera and tetanus toxins are examples of toxins that bind to glycolipid ganglioside receptors. Because such glycolipids do not cross into the inner membrane, they are unable to interact with clathrin and so these toxins are taken into the cell via non-clathrin-coated vesicles. Many toxins, including Shiga which also binds to glycolipids, and diphtheria that binds to a protein receptor, initiate vesicle formation via clathrin-coated pits. In either case the toxin is internalized in a membrane-bound vesicle, from which the catalytically active part has to escape (Figure 7.5). The subsequent intracellular processing pathway can be determined by the nature of the receptor and its trafficking route, but also by motifs on the toxin itself. For example, the A subunits of both the cholera and *E. coli* LT toxins have a C-terminal motif that mediates retention of proteins in the ER. Mutational analysis shows that this is essential for activity.

A vast amount is known about protein trafficking from the membrane through the endosomal and lysosomal pathways and retrograde through the Golgi apparatus to the rough ER. Investigation of toxin trafficking has, like much of toxin research, helped to illuminate not only the toxin action, but also the trafficking processes. The addition of weak bases blocks the action of some toxins (e.g. diphtheria toxin), implying a crucial role of the endosomal stage. The fungal toxin brefeldin A interferes with Golgi function and blocks the action of some toxins, implying that such toxins have to be transported to the Golgi before release into the cytoplasm. Most notably, diphtheria toxin action is not affected by brefeldin A, and this taken together with the potent effects of weak bases suggests that the diphtheria toxin A fragment leaves its vesicle at the endosomal stage, while other toxins, for example Shiga toxin, need to progress further, through the Golgi and to the ER before membrane translocation takes place. Extensive analysis of diphtheria translocation suggests that the transmembrane or translocation domain undergoes a conformational change at the low pH of the endosome and becomes more hydrophobic. It penetrates into the membrane, and in the process forms an ion-permeable pore. The attached A fragment is dragged into the cytosol as part of this process. A considerable degree of unfolding of the A fragment is believed to occur. Attachment of peptides with disulphide bridges blocks translocation, and it is pertinent that disulphide bridges have not been found in toxin A fragments. The catalytic domains of other toxins follow a similar type of process.

Intracellular toxins that gain intracellular access via type III secretion

Toxins in this new category are not conventionally toxic. When purified from bacteria they display no toxic activity. Their 'novelty' relies not on the molecules themselves, but their mechanism of cell entry. They avoid the potential dangers of the extracellular space and the difficulties of penetrating a membrane by being carried directly from the bacterial cytoplasm to the eukaryotic cytoplasm by a complex array of proteins that bridge not only the inner and outer membranes of the Gram-negative bacterial cell, but also the eukaryotic membrane. This is the type III secretion system which is often encoded on pathogenicity islands (see Chapter 2). and has been described in detail in Chapter 6. Thus the intracellular 'cruise missile' is carried inside the target cell following a direct breach of its defences.

The discovery of such toxins solved the conundrum that many pathogenic bacteria appeared not to produce toxins, but were known to induce cellular damage that was not related to endotoxin. The bacteria currently known to produce such toxins include *Yersinia*, *Shigella* and *Salmonella* species. This list is likely to expand (see Chapter 6) and it will be interesting to see if Gram-positive bacteria have a related mechanism.

Yersinia. The pathogenicity island responsible for virulence in *Yersinia* has been analysed in some detail (Figure 7.9). There are four elements to the system: (i) a contact secretion system called Ysc that comprises about 22 genes and which builds a complex structure across the bacterial inner and outer membranes; (ii) machinery to deliver the bacterial proteins into the eukaryotic cell that comprises about five proteins (iii) a control protein that effectively closes the channel until contact is made (YopN); and (iv) translocated proteins that are toxic for the eukaryotic target cell. When LcrG contacts the eukaryotic cell the YopB and YopD proteins move to extend the channel into the eukaryotic cell. Selection of proteins for translocation depends on a 15-amino acid N-terminal sequence, which is not cleaved. The *Yersinia* toxins that are translocated into the target cell include a phosphotyrosine phosphatase (YopH) that dephosphorylates the focal adhesion kinase $p125^{FAK}$ (see Chapter 3), a threonine/serine kinase (YopO/YpkA), a cytotoxin that disrupts the actin cytoskeleton (YopE), a protein postulated to bind to thrombin (YopM) and a protein that induces apoptosis (YopP).

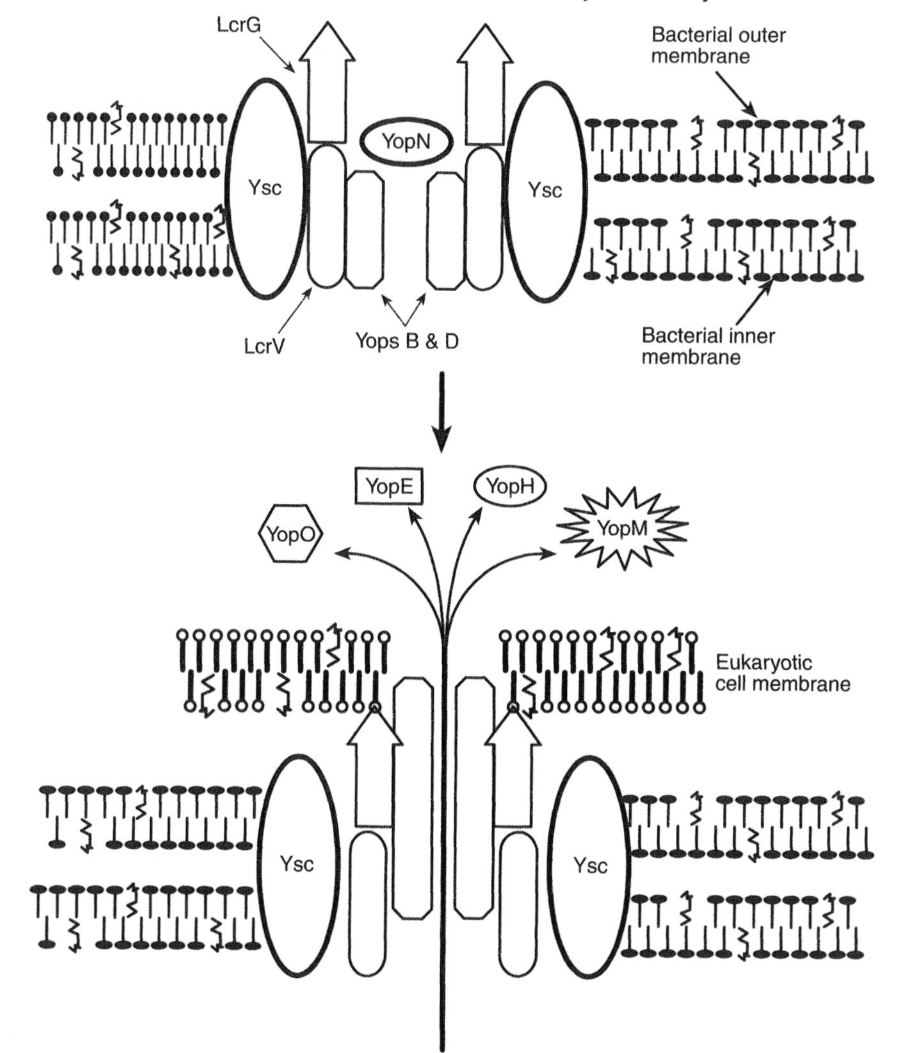

Figure 7.9 Type III secretion delivers toxic proteins from *Yersinia* into target cells

STRUCTURE EXPLAINS FUNCTION

Despite the wealth of biological and mutational information about toxin action, such action has often only been properly understood at the molecular level when structural information was available. The main technical approach has been to use X-ray crystallography, but the large size of some toxins coupled with the increased precision of electron microscopic techniques has allowed this technique to contribute as well. The first toxin structure solved was that of *Ps. aeruginosa,* in 1986, though crystals of some toxins had been made earlier. At the time of writing 19 structures have been obtained (Table 7.6) and each one has been highly instructive in explaining the toxin action. We will consider three structures where good biological evidence is available to correlate with the structure.

Table 7.6 Toxins whose crystal structure has been solved

Bacterium	Toxin
Ps. aeruginosa	Exotoxin A (ETA)
Cor. diphtheria	Diphtheria toxin (DT)
V. cholerae	Cholera toxin (CT)
Bord. pertussis	Pertussis toxin (PT)
Escherichia coli	Labile toxin (LT) Stable toxin (ST) Colicin Verotoxin (VT)
Sh. flexneri	Shiga toxin (ST)
B. thuringiensis	δ-Endotoxin
Staph. aureus	Enterotoxin A (ETA) Enterotoxin B (ETB) Enterotoxin C2 (ETC) Enterotoxin D (ETD) Epidermolytic toxin α-toxin Toxic shock syndrome toxin (TSST-1)
Strep. pyogenes	Pyrogenic exotoxin C
Aer. hydrophila	Aerolysin

Diphtheria toxin

Even before the structure of diphtheria toxin was solved, mutagenesis studies had provided a detailed analysis of the location and role of individual domains. Nevertheless the publication of its structure helped to explain many aspects of its action.

The three domains of diphtheria toxin, which form a Y shape, can be easily discerned in the X-ray structure (Figure 7.10), and each domain is encoded by one continuous stretch of peptide chain. The receptor binding domain comprises a β-sandwich, and the catalytic domain has an α+β structure common to many enzymatic domains. The cleavage site between the catalytic A fragment and the rest of the protein is on an exposed loop. The transmembrane or translocation domain is perhaps of greatest structural interest (Figure 7.10). It comprises nine helices arranged in three layers, an arrangement found in other membrane insertion domains (bacterial pore-forming bacteriocins and the *B. thuringiensis* insecticidal δ-toxin). Several of the helices are hydrophilic, while others are extremely hydrophobic. It is of whimsical interest that the existence of such helices had been predicted, but the exact locations were not precisely as had been suggested. One of the most interesting features is a concentration of acidic residues in the loops between the tips of hydrophobic helices. These give the structure a polar characteristic at neutral pH. However, at acidic pH, when the translocation domain carries out its function to insert into the vesicular membrane, these residues will become protonated, and the increased hydrophobic characteristic assists insertion into the membrane.

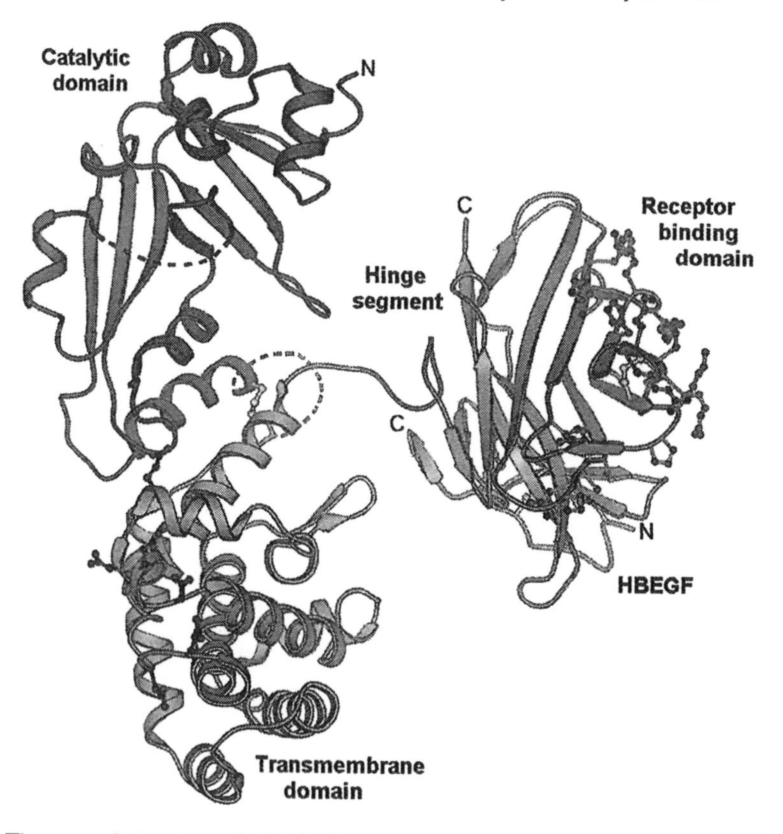

Figure 7.10 The crystal structure of diphtheria toxin bound to a fragment of its cell surface receptor, the precursor of heparin-binding epidermal growth factor-like growth factor (HBEGF). The three individual domains involved in receptor binding, membrane translocation and catalysis can be visualized. Reproduced with permission of Dr Senyon Choe/*Molecular Cell*, by permission of Cell Press

The AB_5 toxin structure

All the AB_5 toxins (Cholera, LT, Shiga and pertussis) have a similar structure, with the A subunit on 'top' of the pentameric arrangement of the B subunits, with part of A projecting through the hole in the B doughnut. Perhaps the most surprising of these structures was that of pertussis toxin, since it would not have been predicted that four different types of B subunit would form a pentameric ring.

The B oligomers all have a similar structure, predominantly of β-sheet with a helix which faces the middle of the ring, and which interacts with an antiparallel helix on the A subunit . The β-sheet on the surface of the ring forms antiparallel sheets containing strands from adjacent subunits. Such an arrangement is seen in some pore-forming toxins, and the *B. anthracis* PA membrane-binding protein has a similar β structure.

The two parts of the A fragment of Shiga toxin interact via a β-sheet that lies on top of the B ring; in LT the interaction is via two helices. The fold of the A subunits largely reflects the encoded enzymatic activity. CT and PT show great similarity to other ADP-ribosylating toxins, whereas Shiga toxin closely resembles the ricin active site.

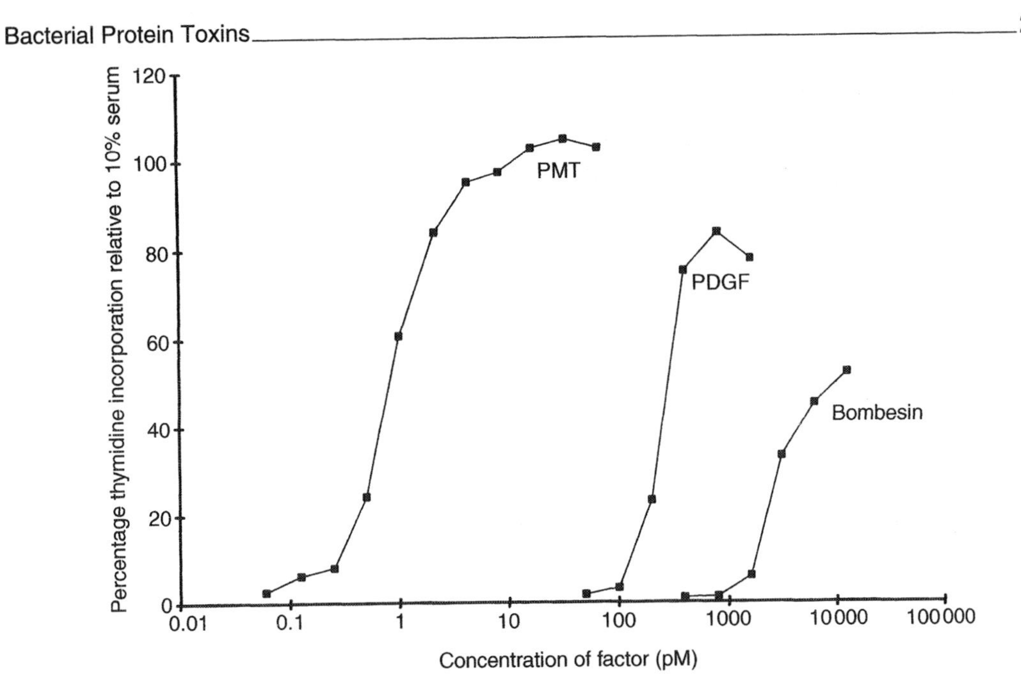

Figure 7.11 The *Pasteurella multocida* toxin is a potent mitogen. The action of PMT in inducing thymidine incorporation (i.e. DNA synthesis) is compared to that of the growth factor, platelet-derived growth factor (PDGF) and the neuropeptide, bombesin. Reproduced with the permission of Professor Enrique Rozengurt

Aerolysin

Aerolysin is expressed as a preproprotein. The expressed proaerolysin is a four domain, predominantly β-sheet, protein that appears as an L shape. The long part of the molecule contains some record-breaking long β strands (93Å) that pass through three of the domains. Aerolysin exists as a dimer of one 'L' and one inverted 'L', that together sequester large patches of hydrophobic residues. Cleavage releases a small C-terminal peptide that covers a large hydrophobic patch, but the exact mechanism that triggers oligomerization is not known. It appears that the dimer binds to the cell receptor and that removal of the peptide enables oligomerization to form a seven member ring with a β-barrel structure inserted in the membrane.

BIOLOGICAL EFFECTS OF TOXIN ACTION

Another way to classify toxins is from the target cell's perspective whereby toxins are grouped depending on the outcome of toxin action. It is pertinent to note that it is usual *in vitro* to focus on the effects of acute toxicity. However, *in vivo* many cells may experience sub maximal doses of toxins and this may lead to a very different set of effects, both immediately and in the longer term, and this is an issue that has not been considered in any detail in the literature. We will try to address this issue as we discuss the cellular effects of these toxins.

Cell death

Many toxins cause cell death, either by complete inhibition of protein synthesis or by pore formation that leads to loss of membrane potential and leakage of cellular components. Some toxins cause cellular and tissue necrosis, suggesting that massive loss of cellular function makes the cell rapidly non-viable. There are also many reports of toxins inducing apoptosis, suggesting that sufficient damage has been sustained for the cell to decide to kill itself. Diphtheria toxin, Shiga toxin, *Staph. aureus* α-toxin, toxins that inactivate Rho proteins and the *Bordetella* cyclase toxin are among these. It is possible that a lower level of damage that does not induce apoptosis may have unforeseen effects.

Nerve transmission

Classically, many toxins were labelled neurotoxins (including diphtheria and Shiga toxins), although now this group is limited to the tetanus and botulinum toxins (except C2 and C3). These toxins attack a very specific aspect of cellular function and although, by interfering with transmission of nerve impulses, they have a highly potent activity on whole organisms, they appear not to have wide-ranging cellular effects. In this regard sublethal effects can be beneficial and the botulinum toxins appear to be safe to use in therapeutic doses.

Signal transduction

Many type I and type III toxins interfere with signalling mechanisms (see Table 7.7). Because of the different roles that signalling pathways play in different cell types, the outcome of toxin action can vary from cell to cell. Some of these toxins do not appear to lead to cell death — indeed there are reports that cholera and pertussis toxins act to block the induction of apoptosis in some cell types. Moreover the effects of sublethal concentrations of these toxins is most likely to cause harmful long-term sequelae.

One group of toxins affect the cytoskeleton whose integrity is essential for most cell processes to take place, and which has increasingly been shown to be an integrated part of the cell's internal communication system (See chapter 3). Some of these toxins (e.g. C2) directly attack actin to prevent its polymerization, while others stimulate or inhibit members of the Rho family, which are closely linked to regulation of the cytoskeleton. This latter group include toxins from *Cl. difficile* and related bacteria and the ADP-ribosylating toxins C3 and EDIN. All these inactivate Rho by interfering with its ability to interact successfully with its effector proteins. Although a construct of C3 that enabled it to be taken into cells induced apoptosis, C3 and EDIN are reported to induce proliferative effects *in vivo*. Is this due to the low level of these toxins that enters cells because of their absence of uptake domains? The *E. coli* cytotoxic necrotizing factors (CNF) and the *Bordetella* DNT both modify Rho but activate it by inhibiting the GTPase activity. Other toxins (most notably the *P. multocida* toxin, PMT) stimulate more distant signalling events which also activate the Rho pathway.

Several toxins result in an elevated concentration of the second messenger, cAMP. This stimulates protein kinase A, which phosphorylates Raf and in general leads to downregulation of the Raf kinase/MAP kinase pathway (see Chapter 3). However, these events produce different sequelae at the whole animal level, reflecting the different sites of action. Thus

Table 7.7 Toxins that interfere with cell signalling mechanisms

Toxin	Target	Molecular events	Cellular effects
Cholera and *E. coli* LTs	G_s	Activated G_s stimulates adenylate cyclase, cAMP produced binds to and activates protein kinase A	Chloride loss into intestine, osmotically followed by H_2O and other signalling events induced by raised cAMP
Pertussis toxin (PT)	G_i	Modified G_i cannot inhibit adenylate cyclase, cAMP produced binds and activates protein kinase A	As above
E. coli ST	Guanylin receptor	Activated receptor stimulates guanylate cyclase, cGMP produced activates protein kinase G	Similar to above toxins
Bordetella DNT CNF	Rho	Rho GTPase inhibited, activated Rho stimulates reorganisation of actin and phosphorylation of $p125^{FAK}$	Cytoskeletal reorganization and stimulation of DNA synthesis and inhibition of cell division
Cl. botulinum C3 and *Staph. aureus* EDIN	Rho	Rho interaction with effector proteins blocked	Cell rounding, cell proliferation *in vivo* stimulated
Cl. difficile toxins	Rho, Rac, cdc42	Rho interaction with effector proteins blocked	Cell rounding. Diarrhoea, via cellular necrosis and cytokine release?
P. multocida toxin (PMT)	Unknown	Not completely ascertained but stimulates phospholipase C and protein kinase C, inositol phosphate release and tyrosine phosphorylation of $p125^{FAK}$	Mitogenesis both in cultured cells and *in vivo*

cholera and LT cause fluid accumulation and diarrhoea, by acting on intestinal epithelial cells, while raised cAMP directly induced by the *Bordetella* adenylate cyclase toxin serves to impair the function of macrophages. The small *E. coli* ST toxin raises cellular cGMP levels, which also causes diarrhoea.

Some of these toxins interact synergistically to stimulate mitogenesis. For example, by itself cholera toxin does not increase mitosis, but potentiates the signal when other growth factors are added. The main mode of action does not appear to be to stimulate growth, and this is only seen when the system is manipulated experimentally. However, transgenic expression of cholera toxin was shown to induce hyperplastic growth. There is one toxin where the main toxin action seems to be to act as a growth promoter or mitogen. The *P. multocida* toxin is highly mitogenic in the absence of any co-stimulatory agents, although it can also potentiate the action of other growth-promoting substances. Its effects on signalling pathways have been described in some detail and will be discussed briefly here.

Pasteurella multocida toxin (PMT)

PMT is unique in its ability to stimulate mitogenesis. It is the most potent mitogen identified for cultured cells and at picomolar concentrations is comparable to 10% serum in stimulating

DNA synthesis and increase in cell number (Figure 7.11). PMT also potently induces anchorage independence in Rat-1 cells, and leads to hyperplasia in *Pasteurella*-infected animals. PMT action stimulates phospholipase C activity, which produces inositol phosphate leading to Ca^{2+} mobilization, and diacylglycerol that activates protein kinase C. In addition, PMT stimulates the pathway that causes marked reorganization of the actin cytoskeleton, and tyrosine phosphorylation of the focal adhesion kinase, $p125^{FAK}$, and paxillin–two proteins that link the cytoskeleton to signalling. These pathways have multiple effects that ultimately lead to mitogenesis (see Chapter 3 for a detailed discussion). There is convincing evidence that PMT enters the cytosol from an acidic compartment (e.g. endosome or lysosome) after low pH processing to yield a biologically active conformation, or fragment, of the toxin. Neither the target nor the modification is known. However, the use of non-hydrolysable nucleotides shows that GTP-binding proteins are involved. Moreover, PMT potentiates the action of bombesin, a neuropeptide that stimulates signalling pathways by binding to a receptor that is coupled to G_q. This suggests that G_q is activated and the cytoskeletal effects seen also clearly implicate small G proteins of the Rho family in its action.

This raises several important issues. First, do the proliferative effects induced by such a potent growth promoter lead to cancer? Secondly, is PMT unique, or will it turn out to be the paradigm for a new group of toxins? In this regard it is interesting that several other bacterial products directly affect the regulation of growth. Many of these are relatively uncharacterized and we will introduce them in Chapter 10.

Bacterial protein toxins and cancer?

As discussed in Chapter 3, carcinogenesis is a multi-stage process: as a cell accumulates mutations so it becomes increasingly unable to restrict its own growth or respond appropriately to externally received signals. Although viruses have long been implicated in this process and indeed led to the first identification of proto-oncogenes, the recent evidence linking *Hel. pylori* infection to some stomach cancers is the most convincing to connect bacteria to carcinogenesis. It is already clear that *Hel. pylori* is not the only example, and it is therefore pertinent to consider characteristics that would be predicted to mark out such bacteria.

Clearly toxins that affect proteins involved in signal transduction are strong candidates. Cholera toxin, which raises intracellular cAMP, has been shown to cause hyperplasia when expressed transgenically. The bacterial toxins that directly affect the regulation of cell growth are also strong candidates for such a role. The *P. multocida* toxin, PMT, very effectively imitates the action of growth factors by inducing growth, even in contact-inhibited cells that would be expected to resist growth signals. It also mimics the action of some tumour promoters in its stimulatory effects on signal transduction pathways. Other bacteria that lead to hyperplastic changes *in vivo*, or appear to stimulate mitogenesis, have been described but not characterized in detail. These include *Bartinella henselae, Citrobacter freundii* and several factors expressed by bacteria associated with oral disease such as *Por. gingivalis*.

Two other members of this grouping of toxins, the DNT of *Bordetella* species and the *E. coli* CNF, both stimulate DNA synthesis but block cytokinesis, leading to the appearance of bi- or multi-nucleated cells, which are presumed to die. These latter toxins operate by activation of Rho, which has been implicated as an essential requirement in oncogenic transformation. Toxin-induced modification of Rho at sublethal levels might also play a role in cancer.

Conversely, it is interesting that Rho inactivation by C3 and EDIN is reported to cause proliferative effects that could perhaps also predispose towards cancer promotion.

Interactions with cytokines

Toxins of all types have been shown to induce cytokine formation. These include pore-forming toxins, cell surface-acting toxins and intracellularly acting toxins. For many toxins such effects are part of a wider array of induced cellular changes; for others, they may represent the primary action of the toxin. In either case, cytokine changes will be a significant aspect of a toxin's effects, both on individual cells and on the whole organism.

THE BACTERIAL PERSPECTIVE

The advantage to the bacterium

Why do such 'simple creatures' as bacteria produce such complex molecules, 'designed' for operation in an alien environment? This is part of the wider question about whether it is advantageous for a bacterium to behave as a pathogen, as opposed to forging the less aggressive role of living in comparative harmony with its host. Pathogenicity is potentially a risky lifestyle. The host being attacked may fight back and kill the aggressor before it has a chance to spread to other hosts. Similarly, a highly aggressive attack may kill the host before bacterial transmission has occurred. On the other hand, successful pathogenicity may enable large numbers of bacteria to grow, compete for nutrients with other pathogens and colonize a large proportion of susceptible hosts. With some toxins, we can identify a clear advantage from the bacterial perspective (Figure 7.12), while with other toxins this is less obvious. For example, CT produces vast quantities of diarrhoea containing *V. cholerae* that contaminate the environment, leading to increased colonization of epidemic proportions. Death of the host is potentially of advantage to strict anaerobes like the clostridial species. It is less clear why perturbation of cell signalling or the cell cycle should be of evolutionary advantage to a toxin-encoding bacterium. Nevertheless, the potent mitogen PMT does aid colonization of its carrier bacterium *P. multocida,* and it is probably true that the apparent subtlety of any particular toxin action is an irrelevance from the bacterial 'viewpoint'. Any gene that bestows an evolutionary advantage will be retained.

Origin and evolution of toxin genes

Toxins were originally defined by their effects on multicellular organisms. This does not necessarily mean that toxins would not give bacteria an advantage against single-celled eukaryotes, since toxins that kill cells would protect bacteria against being engulfed and destroyed. Clearly this would not apply to some toxins; for example, neurotoxins are only effective as toxins against multicellular organisms. Some of the basic principles of toxin action, e.g. membrane insertion and pore formation, are intrinsic to both eukaryotic and bacterial life. There would be many opportunities for genetic exchange by bacterial

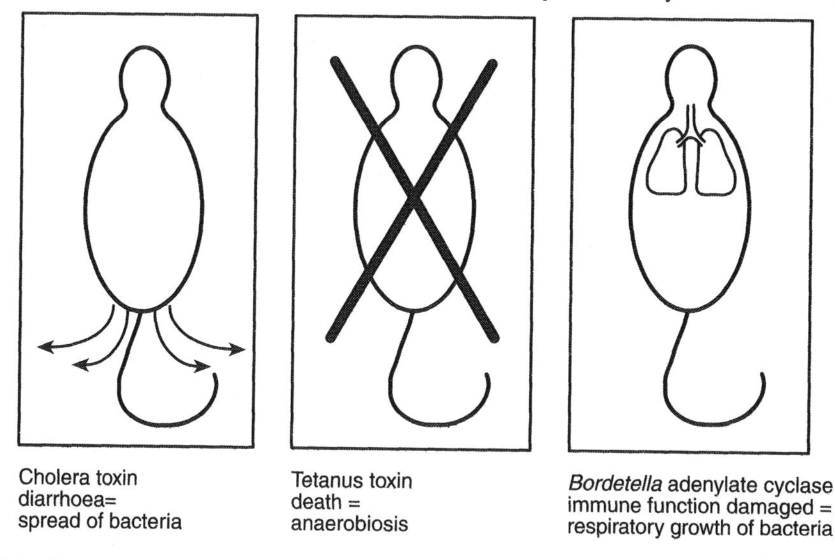

Figure 7.12 Some possible advantages of toxin expression to the host bacterium. Cholera toxin causes large volumes of diarrhoea to be excreted which spreads infected organisms into the environment. Tetanus toxin leads to death of the host and an anaerobic environment which favours bacterial growth. The *Bordetella* adenylate cyclase inhibits macrophage function that would otherwise attack the host bacterium

transformation during the prolonged co-evolution of pro- and eukaryotic organisms. It is highly likely that during this time bacteria acquired several ancestral motifs for the limited number of target types that exist. It is also possible, but distinctly less likely, that such motifs were developed independently by bacteria.

Whatever the origin, there is abundant evidence that toxins are encoded by relatively newly acquired DNA frequently encoded on mobile genetic elements. Many toxin genes are encoded either on plasmids or by phage. A brief list is shown in Table 7.8. There are examples where highly related toxins are encoded on mobile elements in one organism and on the chromosome in another. For example, cholera toxin is chromosomally encoded on a phage, while the highly related *E. coli* LT is plasmid encoded. Indeed many of the chromosomally encoded toxins are on characterized or postulated pathogenicity islands. The availability within the bacterial gene pool of domains that could usefully attack eukaryotic targets (including domains for membrane insertion that are found in many different types of toxins) could then be exploited by evolution to devise subtle variations on a theme. For example, many toxins attack G-protein targets. Subtle changes at the active site would enable different members of the same family to be attacked by the same enzymatic modification (exemplified by the ADP-ribosylating toxins). Different types of changes would enable the same target to be attacked by a different enzymatic mechanism (as illustrated by members of the large clostridial toxin family). In addition, the assignment of different aspects of toxic function to separate domains would enable catalytic domains to be targetted to different cell types expressing other receptors.

This may explain why some proteins are 'popular targets', e.g. Rho family members, while other targets that potentially could result in an equally dramatic outcome (e.g. Raf) appear to have been ignored. However, the targets of all of the known toxins have not yet been

Table 7.8 Location of some toxin genes

Toxin	Location of gene		
	Phage	Plasmid	Chromosome
Streptolysin O and related toxins			+
E. coli haemolysin		+	+
Staph. aureus α-toxin			+
B. thuringiensis β-endotoxin		+	
Staph. aureus superantigenic toxins			
SEA	+		+
SEB		+	+
TSST-1			+
Bordetella toxins (PT, cyclase, DNT, TCT)			+
Diphtheria toxin (DT)	+		
P. aeruginosa ETA			+
Cholera toxin (CT)	+		
E. coli labile toxin (LT)		+	
E. coli stable toxin (ST)		+	
E. coli cytotoxic necrotizing factors (CNF)			
CNF1			+
CNF2		+	
B. anthracis toxin complex(LF, EF and PA)		+	
P. multocida toxin (PMT)			+
Botulinum toxins	+	+	
Tetanus toxin		+	
Shiga toxin			+
Cl. difficile toxins			+
Yersinia YOP proteins		+	

discovered and new toxins are still being recognized, so that further classes of target are bound to be identified. The main classes of target are the membrane, the translation apparatus, GTP-binding proteins and synaptosomal proteins (Table 7.9). This grouping is not, of course, exhaustive and other targets have been identified.

THERAPEUTIC USES OF TOXINS

The powerful nature of toxin action historically made them the first line of attack in the search for effective vaccines, and this process still goes on today but with the advantage of a greater knowledge and the ability to manipulate their toxicity genetically. In addition, their potent actions have found widespread uses in other aspects of biology.

Vaccines

As the major effector of bacterially induced host damage, toxins play a prominent part in both conventional empirical vaccines and the new generation of rationally designed vaccines. It was recognized early that inactivation of the toxic activity could produce highly

Table 7.9 Common targets attacked by several toxins

Target	Example
Membrane	Pore-forming toxins Phospholipases Proteases Superantigens *E. coli* stable toxin (ST)
Translation apparatus	Diphtheria toxin (DT) *Ps. aeruginosa* exotoxin A (ETA) Shiga toxin
GTP-binding proteins	Diphtheria toxin *Ps. aeruginosa* exotoxin A (ETA) Cholera toxin (CT) *E. coli* labile toxin (LT) *E. coli* cytotoxic necrotizing toxin (CNF) *Bord. pertussis* toxin (PT) *Bordetella* dermonecrotic toxin (DNT) *Cl. difficile* and related toxins *Cl. botulinum* C3 *Staph. aureus* EDIN
Synaptosomal proteins	*Cl. botulinum* toxins (except C2 and C3) *Cl. tetani* toxin

effective immunogens because inactivation did not destroy epitopic structure. Such chemically attenuated vaccines have served well in the protection against some diseases (e.g. tetanus and diphtheria). However, this has not been a universally successful approach, either because the induced immunity was poor or short lived (e.g. cholera), or because the vaccines are perceived to cause unacceptable side effects. A good example in this latter case is whooping cough, where the public acceptance of the whole cell inactivated vaccine has not always been high. Several advances in vaccine technology have contributed to the new approaches being adopted to produce more effective and safer vaccines.

Knowledge about toxin structure has enabled scientists to identify which amino acids are involved at the catalytically active site (of intracellular toxins). These can be changed by genetic engineering to produce a protein that has only one or two amino acid changes but is completely devoid of toxin activity. Such a mutant protein is more likely for several reasons to be effective as a vaccine than a toxin with gross alterations. First, it is likely to be correctly folded into the native structure, and so display the epitopes that will trigger an immune reaction that will recognize the active toxin. Secondly, a correctly folded molecule is more likely to be stable and resistant to proteolytic attack in the host. Thirdly, in the case of intracellular toxins, a toxin that is only mutated in its enzymatic function will be able to carry out the first steps in intoxication, i.e. binding and cellular entry. This enables the immune system to process the protein more efficiently and better immunity is raised by vaccination with whole toxin than just the active domain. In this regard it is interesting that intracellular toxins as a group appear to be very effective adjuvants.

This approach relies on fundamental knowledge about the toxins and other virulence determinants (e.g. adhesins) a bacterium makes. It is therefore possible to concentrate on only those proteins important in pathogenesis and thus potentially avoid the side effects of other extraneous bacterial products. This is being applied to pertussis toxin, where a further advantage of such an approach over chemical modification was identified. Formaldehyde inactivation of the toxin, which essentially acts to cross-link the protein, was shown to affect its structure and potentially mask or inactivate immunogenic sites, since the untreated protein was more immunogenic than the chemically inactivated one.

Genetic manipulation of toxin genes is also being coupled to the newer delivery systems, using metabolically attenuated bacteria that can be given orally, e.g. *aro* strains of *Salmonella* that can only survive for a few generations in a host. Such systems have the potential advantages that they have a greater likelihood of inducing protection at mucosal surfaces by administration via the oral route and also are more likely to be of use in the Third World, since an orally administered live vaccine will be cheaper and will not require continual refrigeration.

New uses

The potency and versatility of toxins have found new applications in various fields, from basic science to clinical applications.

New research uses

The value of toxin action in studying signalling processes in cells has already been described, and this section will briefly allude to newer uses. The superantigens have proved to be useful probes of signalling systems in the immune system. Some pore-forming toxins have been used to create controlled permeabilization of cells in order to study cell signalling and membrane transport systems. Expression of the diphtheria toxin catalytic domain under the control of developmentally regulated promoters has provided a powerful method of knocking out a lineage of cells and thereby examining loss of function, particularly in embryogenesis, by removal of a whole class of cells that activate the chosen promoter.

Chimaeric immunotoxins

The potent cell-killing ability of diphtheria, *Ps. aeruginosa* exotoxin A (ETA) and Shiga toxins (as well as the plant toxin ricin) has been exploited to make chimaeric toxins that can be targetted to kill a particular set of cells. A major aspiration is to make a molecule that can specifically attack cancer cells (Figure 7.13) without in any way affecting normal cells. Even before the precise structure of DT was elucidated, chimaeric toxins where the enzymatic function of DT was chemically coupled to other cell binding domains had been constructed and were being tested in clinical trials. Knowledge of the structure of such toxins as DT, ETA and ricin has enabled the design of more precise genetic constructs, where the cell binding domain is replaced by an antibody that will bind to surface epitopes that are specific to malignant cells and that are trafficked and therefore able to direct uptake of the chimaeric toxin to kill only cancer cells. Several such molecules have been tested in clinical trials (e.g. targetted to cells expressing the IL-2 receptor) and have shown great promise for the therapy of lymphomas and leukaemias; they have been less successful in the treatment of solid tumours which the immunotoxin cannot penetrate. This can be circumvented by an immunotoxin that targets the tumour vasculature, by binding to and killing endothelial cells, which express different surface antigens in many tumours as a result of cytokines released by the tumour cells. Another use for such toxins is *ex vivo*, to purge cancer cells from bone marrow to be used in autologous bone marrow transplants. This has been used experimentally with Shiga toxin, whose natural receptor is CD77, which is restricted to activated B cells and tumours derived from these. The potential of such an approach is tremendous, though there is more to be done before this becomes routine in the clinic.

Immunotoxins have also been targetted to other cell types, in particular as a therapy for diabetes.

Muscle spasms

The potent nerve-blocking function of the botulinum toxins has been exploited to tackle a variety of rare but distressing conditions where particular muscles remain permanently

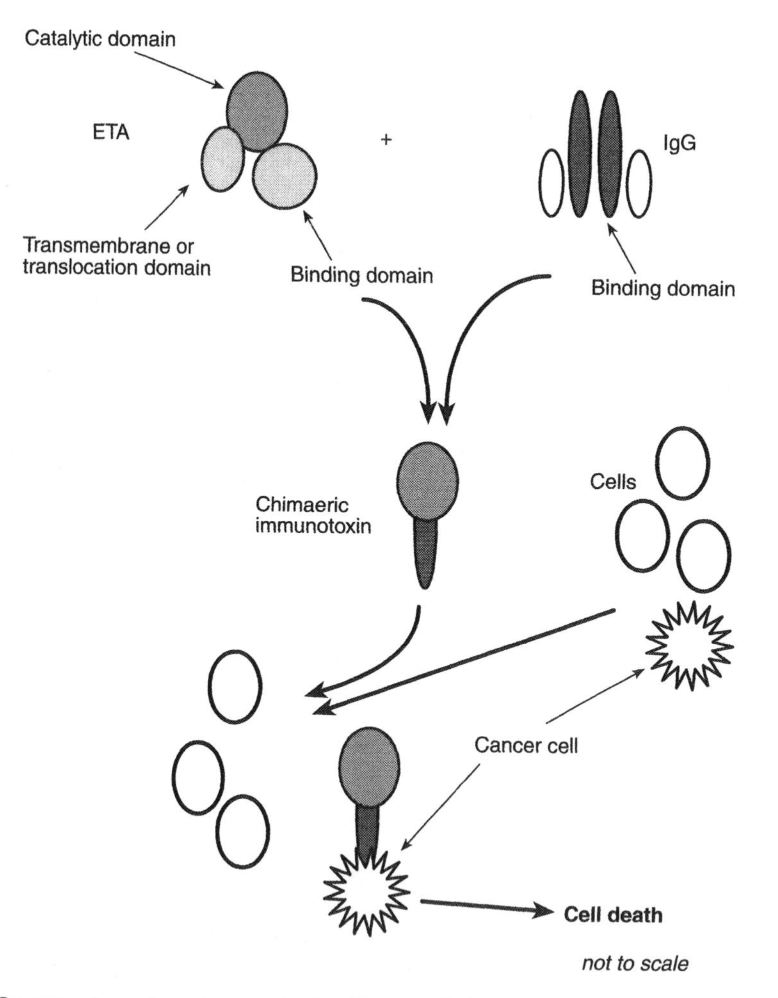

Figure 7.13 Construction of an immunotoxin. The catalytic domain of the *Ps. aeruginosa* ETA is coupled to an antibody binding site that directs the chimaeric toxin only to cancer cells with the receptor that the antibody recognizes. The targetted cell takes up the immunotoxin and is killed

activated. Injection of very low doses of a botulinum toxin relaxes the errant muscle and can give relief for several months. Immunity to the toxins appears to be slow to build up since the dose given is so low and is applied locally. Potentially this could be overcome if other botulinum toxins were commercially available. This technique is also being used by the cosmetic industry as an anti-wrinkle therapy.

SUMMARY AND THE FUTURE

Toxins represent the ultimate in prokaryote/eukaryote interactions. Not only can bacteria interfere with eukaryotic regulatory processes by triggering cell surface receptors, but they can also reach right inside the eukaryotic cell to tweak particular functions. As more is learned about the way these fascinating molecules are constructed, it has become possible to exploit the precision of their action not only to understand cell biology but also as novel therapies for a wide variety of diseases.

The future of toxin research and its applications is likely to be 'more of the same'. However, that should not imply a boring and predictable itinerary! The history of toxins so far has been exciting, unpredictable and varied. We can expect new toxins and new mechanisms of action, leading to novel therapies and fresh insights into cellular microbiology.

FURTHER READING

Books

Aktories A (1997) *Bacterial Toxins: Tools in Cell Biology and Pharmacology* Chapman & Hall: London

Alouf JE, Freer JH (eds) (1991) *Sourcebook of Bacterial Protein Toxins.* Academic Press, London

Frandsen PL, Alouf JE, Falmagne P, Fehrenback FJ, Freer JH, Montecucco C, Olsnes S, Rappuoli R, Wadström T (eds) (1996) *Bacterial Protein Toxins: Seventh European Workshop.* Gustav Fischer Verlag, Stuttgart

Moss J, Vaughan M, Iglewski B, Tu A (1995) *Bacterial Toxins, Virulence Factors and Disease.* Marcell Dekker, New York

Rappuoli R, Montecucco C (1997) *Guidebook to Protein Toxins and their Use in Cell Biology.* Oxford University Press, Oxford

Reviews

Aktories K (1997) Rho proteins: targets for bacterial toxins. *Trends Microbiol* 5: 282–288

Boquet P, Munro P, Fiorentini C, Just I (1998) Toxins from anaerobic bacteria: specificity and molecular mechanisms of action. *Curr Opin Microbiol* 1: 66–74

Ghete V, Vitetta E (1994) Immunotoxins in the therapy of cancer: from bench to clinic. *Pharmacol Ther* 63: 209–234

Henderson B, Wilson M, Wren B (1997) Are bacterial toxins cytokine network regulators? *Trends Microbiol* 5: 454–458

Henderson B, Wilson M, Hyams J (1998) Cellular microbiology: cycling into the millennium. *Trends Cell Biol* 8: 384–387

Kotb M (1998) Superantigens of Gram positive bacteria: structure–function analyses and their implications for biological activity. *Curr Opin Microbiol* 1: 56–65

Lesieur C, Vecsey- Semjen B, Abrami L, Fivaz M, Gisou van der Goot F (1997) Membrane insertion: the strategies of toxins. *Membrane Biol* 14: 45–64
Sears CL, Kaper JB (1996) Enteric bacterial toxins: mechanisms of action and linkage to intestinal secretion. *Microbiol Rev* 60: 167–215
Songer J (1997) Bacterial phospholipases and their role in virulence. *Trends Microbiol* 5: 156–161

CHAPTER 8

The Innate Immune Response and Bacterial Infections

INTRODUCTION

Cellular microbiology focuses on the interactions of microorganisms with host cells. One of the major host cellular systems with which bacteria interact is the immune system. In this, and in Chapter 9, the nature of the mammalian immune system will be explained and the interplay between bacteria and immunity will be delineated.

The complex of cells and humoral factors which are known collectively as 'immunity' can be divided into two interrelated systems. The first such system is known as innate (sometimes natural) immunity and is utilized by all living creatures. The second system, acquired or adaptive immunity, evolved around 400 million years ago and only operates in vertebrates. For most readers the term immunity means acquired immunity with its antibodies and T and B lymphocyte populations. However, 95% of the multicellular animals on our planet are invertebrates and rely solely on innate immune responses for defence against microorganisms. The subject of this chapter will be innate immunity and will describe the cells, soluble factors, cytokines, receptors and so on which constitute and control this complex protective system. Surprisingly, although discovered first, we know less about innate immune mechanisms than we do about the system of acquired immunity.

Innate immunity involves a large number of different cell populations such as epithelial cells, monocytes, macrophages, dendritic cells, polymorphonuclear leukocytes (PMNs), natural killer (NK) cells and various lymphocyte subpopulations (for example CD5-positive B lymphocytes and $\gamma\delta$ T lymphocytes – whose definition and derivation will be described later in this chapter) which bridge the divide between innate and acquired immunity (Figure 8.1). These cells generally arise from precursor cell populations in the bone marrow. Humoral systems are also important and include a large number of cytokines, acute-phase proteins, certain enzymes (e.g. lysozyme), metal-binding proteins, integral membrane ion transporters, complex carbohydrates, antibacterial peptides and the complement pathways. The epithelial cells of the body also form physical barriers to bacterial entry and mechanical systems such as eye blinking and ciliary action are also integral to the innate mechanism. It is important that the reader understands from the very beginning that innate and acquired immune mechanisms are intimately linked, although such links are only now being understood in cellular and molecular terms. It should also be understood that we are only now beginning to recognize the capacity of microorganisms to evade these complex systems of immunity. The nature of the immune evasion mechanisms will be described throughout this chapter.

This chapter will start its consideration of innate immunity by asking the question – how does the system of innate immunity recognize infecting microorganisms? Such recognition is central to the ability of the host to defeat infecting microorganisms.

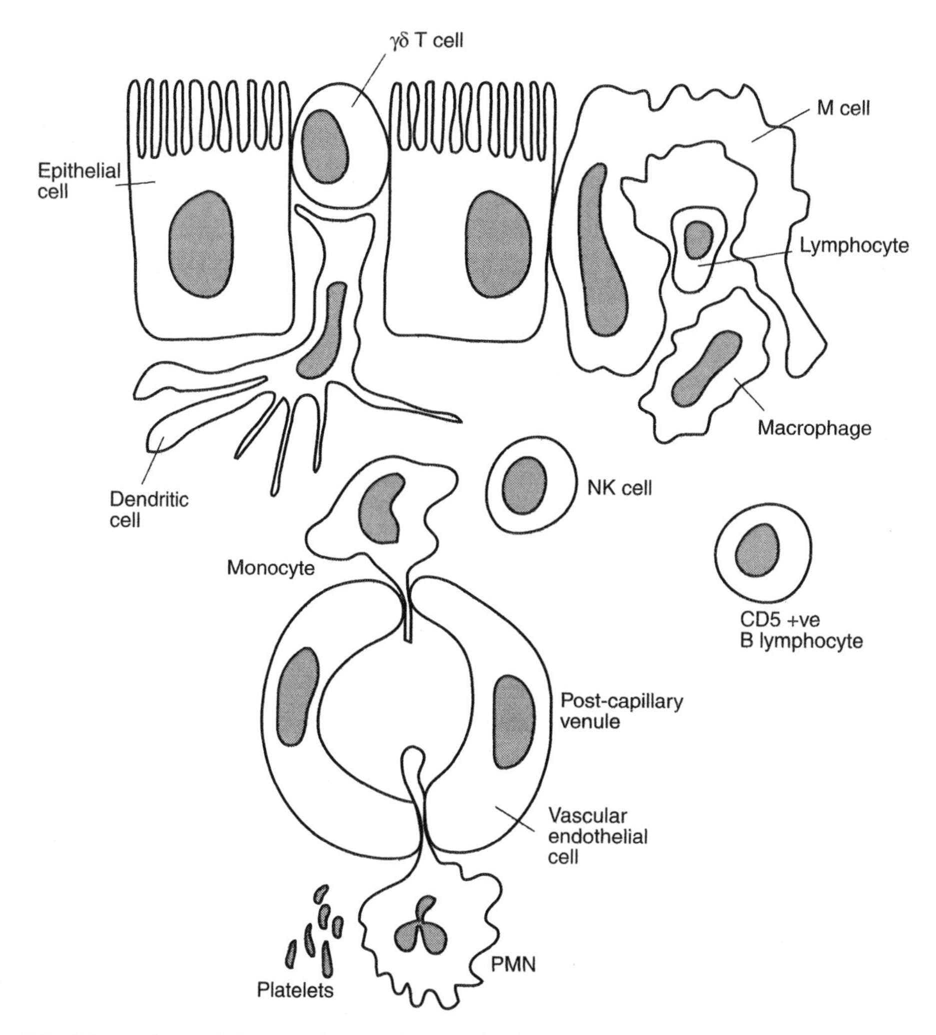

Figure 8.1 The cell populations which are involved in innate immunity. These include epithelial cells which produce a range of antibacterial substances, the most important of which are the antibiotic peptides. Associated with these epithelial cells at the luminal surface are intraepithelial lymphocytes (IELs), which can be thought of as bridging the gap between innate and acquired immunity. Lying underneath the epithelial barrier are the variety of other cells involved in innate immunity. These can either be resident cells (macrophages, dendritic cells and lymphocytes) or capable of being attracted into the area by appropriate signals (PMNs, monocytes, NK cells). The platelet, which enters inflammatory sites via blood vessel leakage is now known to be a source of antibiotic peptides as well as having the capacity of walling-off infectious sites

HOW DOES INNATE IMMUNITY RECOGNIZE BACTERIA AND BACTERIAL INFECTION?

Most readers will have some familiarity with immunology and will know that lymphocytes recognize foreign 'antigens' by making antibodies and/or by producing effector T lymphocytes. However, it can take up to several weeks to make sufficient antibody or produce enough lymphocytes to 'deal' with an infectious microorganism. If this was the only defence mechanism available to vertebrates then they would have disappeared long ago, as a result of infection. Fortunately, innate immune mechanisms defend us against most forms of infectious agents. These innate systems are available, and operative, from the very first instant of the infection. Indeed, much of the activity of innate immunity at epithelial surfaces is invisible to us and it is only when this barrier is breached by a bacterium that we become aware of the functioning of innate immunity. Now while immunologists have largely solved the problem of how acquired immunity recognizes infectious organisms such as bacteria (through systems able to make almost an infinite number of molecular 'shapes' – the antibodies and T cell receptors which will be discussed in Chapter 9) we are only now beginning to address the question of how the system of innate immunity can recognize that the host has been invaded by infectious microorganisms. The data issuing from such studies are revealing the key role played by innate immunity in the control of acquired immunity.

To restate the problem faced by the system of innate immunity – it must immediately recognize the presence of infectious bacteria (and other microbial pathogens), no matter what organism is invading. Over the last decade the immunologist Charles Janeway and his collaborators have argued that the innate immune system recognizes key molecular structures of pathogens – molecules that are essential for bacterial survival and therefore unlikely to evolve their structures, as such mutation would be lethal. Such molecules would include structural components like lipopolysaccharide (LPS) and peptidoglycan and possibly also DNA, particularly that with high CpG content where the cytosine is unmethylated. Although DNA obviously undergoes mutational change it will always contain some CpG residues. One of the authors (BH) has also proposed that molecular chaperones play a role in bacterial recognition. These molecules would be present in all bacteria and Janeway has proposed that the key feature of the innate immune recognition system is that it recognizes, not individual molecules, but patterns of molecules, for which the term pathogen-associated molecular patterns (PAMPs) has been coined. Janeway and his co-workers predict that host organisms will have developed a group of receptors which recognize these PAMPs and these have been referred to as pattern recognition receptors (PRRs). A number of these putative PRRs have now been identified (Table 8.1). The first PRR to be identified was CD14 (present on monocytes and also in the plasma) which binds to a range of bacterial carbohydrates including LPS, peptidoglycan and lipoarabinomannan and activates cells bearing this receptor. The binding of the PAMPs to their receptors is believed to be an early evolved recognition system which stimulates the production of a range of molecules required for anti-parasite defences. In invertebrates and also in vertebrates, binding results in the upregulation of genes producing antibacterial peptides. The nature and activity of these peptides will be reviewed later in this chapter. In vertebrates, binding of PAMPs to their PRRs is proposed to induce, in addition to antibacterial peptides, two other groups of molecules. The first is the cytokines (see Chapter 3). As will be discussed later in this chapter, cytokines are the controlling and integrating signals of innate immunity. They play similar functions in acquired immunity and one of the most important roles of cytokines is in determining the type of acquired immune responses that are produced. In simple terms, the immune response needs to be able to discriminate between bacteria which are extracellular or intracellular. Extracellular bacteria require that antibodies be produced to enable neutralization of toxins and opsonization of bacteria. The T cell population stimulating antibody formation is the so-called the Th_2 cell. Conversely, to deal with intracellular bacteria, Th_1 lymphocytes are required. These cells can activate bacteria-bearing macrophages enabling the killing of the intracellular organisms. The induction of these two different T cells requires different patterns of cytokines and this, in turn, is thought to be dependent on the signals emanating from bacteria. However, to date the nature of such bacterial signals has not been identified. This topic will be discussed in more detail in Chapter 9. The induction of an immune response is a potentially dangerous business as inappropriate activation could lead to chronic inflammation or self-recognition and autoimmunity. A fail-safe mechanism built into lymphocyte activation is the need for what are termed co-stimulatory signals,that is, signals in addition to the ones conferred by binding of the lymphocyte receptor to the bacterial component (immunogen). One of the main co-stimulatory signals are proteins found on the surfaces of antigen-presenting cells. The first such molecule to be discovered was termed B7–1 (CD80) and a second protein discovered later was named B7–2 (CD86). These proteins are also given CD numbers (CD standing for cluster of differentiation) and the use of this nomenclature is described in Table 9.1. It is now recognized that the signals inducing these co-stimulatory proteins are the PAMPs and thus not only do molecules from pathogens define the nature of the immune response, they also are responsible for ensuring that immune responses are only directed at external agents – the pathogens (Figure 8.2).

The PRRs can be divided into two groups: those that are found in the plasma and those that are present on the surface of cells such as leukocytes, epithelial cells and vascular endothelial cells. The majority of those in the plasma belong to the class of molecules known as acute-phase proteins. These are produced during the integrated physiological response to infection known as the acute-phase response, which will be described in a later section. In this response, cytokines (such as interleukin (IL)-1, IL-6, IL-11 and tumour necrosis factor (TNF)α) produced locally at the site of inflammation enter the bloodstream and, by interacting with the brain and liver, induce the liver to produce a number of specific proteins – the acute – phase proteins. Plasma levels of these proteins can rise by up to 1000-fold during infections. Some have evolved to inhibit the deleterious effects of the antibacterial response and are protease inhibitors and free radical scavengers. Others have the capacity to recognize and bind to microbial polysaccharides and form part of the body's portfolio of innate immune recognition systems (PRRs). These include C-reactive protein (CRP) and serum amyloid P (SAP), which are members of the pentraxin protein family, formed of five identical subunits. These proteins bind to the phospholipid phosphorylcholine in bacterial polysaccharides and act as opsonins. The binding of these proteins can also activate the complement cascade. A recently recognized group of carbohydrate-binding proteins which contain collagen triple-helical domains and show structural similarity to the first component of the classical complement pathway, C1q, have been shown to play a role in innate immunity. These proteins are collectively known as the 'collectins' (from *co*llagen-like *lectins*), have both collagen and lectin domains, and include three plasma proteins (mannose-binding protein (MBP), conglutinin and collectin 43) and two lung surfactant proteins (SP-A and SP-D). MBP, which binds to mannose-containing bacterial components, has recently been shown to activate a third pathway of complement activation termed the lectin pathway (the other two

Table 8.1 Molecules acting as recognition 'receptors' in innate immunity

Protein	Location	Ligand	Function
Humoral			
C-reactive protein (CRP)	Plasma	Bacterial polysaccharides	Opsonin, activates complement
Serum amyloid P (SAP)	Plasma	Bacterial polysaccharides	Opsonin
Mannose-binding protein (MBP)	Plasma	Bacterial polysaccharides	Opsonin, activates complement, modulates CD14-dependent signalling
sCD14	Plasma	LPS and other bacterial cell wall components	Enhances sensitivity to LPS and other bacterial components
LBP[a]	Plasma	LPS	Modulates LPS activity
C3	Plasma	Carbohydrates and proteins	Attachment of ligand to receptors such as CD21 and CD35
Cellular receptors			
Scavenger receptors, types I and II	MØ, hepatic endothelium, high endothelial venules	Bacterial cell walls	Clearance of LPS and bacteria
MARCO	MØ, medullary lymph node	Bacterial cell walls	Bacterial clearance
Mannose receptors (e.g. DEC-205)	MØ, dendritic cells, endothelial cells, thymic epithelium	Bacterial carbohydrates	Targets antigens to class II-loading compartment
mCD14	MØ, neutrophils	LPS and other bacterial componenets	Induces cytokine induction, LPS sensitivity, bacterial clearance
CD35 (CR1)	MØ, neutrophils, lymphocytes	C3b, C4b	Enhances C3b and C4b cleavage
CD21 (CR2), CD11b, CD18 (CR3)	B lymphocytes, follicular MØ, neutrophils, NK cells	iC3b, C3dg, C3d, iC3b, LPS, fibrinogen	Augments B cell activation by antigen adhesion, LPS clearance
Bactericidal/ permeability-increasing protein (BPI)	Neutrophils	LPS	Antibacterial
Bacterial DNA	B lymphocytes	Unmethylated CpG dinucleotides	Receptor, if it exists, is undefined

[a] LBP, lipopolysaccharide-binding protein.

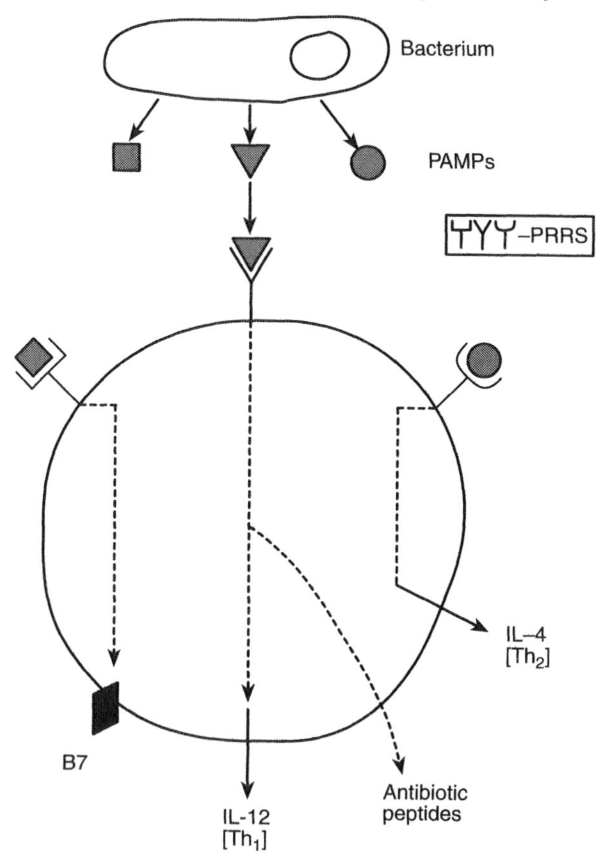

Figure 8.2 How does the innate immune response recognize bacteria? It is proposed that invertebrates and vertebrates have evolved a range of receptors (PRRs) which are cell bound and may also be soluble, able to recognize bacterial molecules (PAMPs) which are essential for cell function and therefore are unlikely to change through evolution. Examples of such bacterial molecules would include LPS, peptidoglycan and CpG DNA. Binding of the PAMPs to the PRRs on vertebrate cells results in the induction of production of antibacterial peptides, appropriate cytokines (e.g. IL-4 and IL-12) for the induction of Th_1 and Th_2 responses and, if the cell is an antigen-presenting cell (macrophage, dendritic cell) the co-stimulatory molecules B7–1, B7–2

pathways – the classical and alternative – will be described in a later section). The other collectins do not activate complement but act as opsonins.

A number of plasma proteins interact with the complex Gram-negative bacterial outer cell wall component LPS. The best studied have been the interactions of LPS with lipopolysaccharide-binding protein (LBP), an acute-phase reactant, and soluble CD14 (sCD14), which is constitutively present in plasma at a concentration of 3 μg/ml. Other plasma proteins which can bind to LPS are cholesterol ester transfer protein (CETP), phospholipid transfer protein (PLTP) and low- and high-density lipoproteins. The interaction of LPS with these various proteins will be described a later section. However, it is clear that these various proteins can both activate and inactivate LPS. CD14 exists both as a plasma protein and as a glycosyl phosphatidylinositol-linked plasma membrane receptor. As described earlier this protein, CD14, has been has been shown to recognize a large number of distinct bacterial carbohydrate-containing components including LPS, peptidoglycan, lipoarabinamannan and polysaccharides.

A number of cell surface receptors, present on macrophages and dendritic cells, which recognize bacteria have been found in recent years. One group with broad ligand specificity are the so-called scavenger receptors. The type I and II receptors (scavenger receptor class A or SR-A) are trimeric membrane proteins containing collagenous domains which have been shown to bind a wide variety of ligands including chemically modified lipoproteins, polyribonucleotides, polysaccharides, phospholipids and even asbestos. The third type of scavenger receptor, which resembles the type I receptor, is referred to as MARCO, and binds both Gram negative and Gram positive bacteria. SR-A is also involved in the endocytic clearance of LPS, and knockout mice lacking SR-A have an increased sensitivity to LPS (see Chapter 4 for a definition of gene knockouts). A second class of bacterial-binding receptor found on phagocytic cells, dendritic cells and certain epithelial cells is the C-type lectin family. This includes the macrophage mannose receptor and a related molecule which has been termed DEC-205. These proteins contain between eight and 10 lectin-binding domains which recognize monosaccharides such as mannose and fucose. These receptors are involved in the binding and internalization of a variety of bacteria, protozoa and fungi.

In addition to receptors which directly recognize bacterial constituents, leukocytes contain a family of receptors which recognize complement products present on the surface of bacteria. These are members of what is termed the *regulator of complement activation* (RCA) gene family which encode the protein: CR1 (CD35), CR2 (CD21), membrane co-factor protein (MCP; CD46) and decay-accelerating factor (DAF; CD55).

As has been stated earlier in this chapter, the systems of innate and acquired immunity are fully interactive and these receptors are an important part of this interactivity. An interesting example of this is the recent finding that measles virus, which is one of the many viruses known to cause immunosuppression in infected individuals, binds to the RCA receptor CD46 on macrophages and in doing so inhibits the ability of these cells to synthesize the cytokine IL-12. IL-12, as has been stated, and will be discussed in Chapter 9, is a key mediator in the development of Th_1 lymphocytes, a cell that functions to activate mononuclear phagocytes to deal with both intracellular and extracellular bacteria.

Another bacterial constituent recognized by host cells is a group of peptides known as bacterial formylated peptides. Many bacterial proteins start with an *N*-formyl group. These formylated peptides are removed and secreted from bacteria and are recognized by specific receptors on monocytes, macrophages and PMNs. The *N*-formyl peptide receptor is a member of the rhodopsin family of G-protein-coupled receptors and has similarity to the chemokine IL-8 receptor (see Chapter 3).

It must also be emphasized that there is remarkable overlap between the inflammatory and wound-healing response and inhibition of inflammation, e.g. by depleting animals of macrophages, inhibits wound healing. One of the key attributes of wound healing is blood coagulation and the reverse process, fibrinolysis. Like the complement system, the processes of coagulation and fibrinolysis can take place on surfaces and involve cascades of enzymic reactions. Bacteria can bind to components of both the coagulation and fibrinolysis pathways and can utilize these pathways to the cells' advantage. It has recently been demonstrated that infusion of *E. coli* into experimental animals stimulates the coagulation pathway, contributing to the pathology of septic shock – an often fatal condition which often accompanies Gram-negative bacterial infections. Thus this is another example of how the defence systems of the body can recognize bacteria. In the case of the coagulation pathway, it is not clear what bacterial surface 'receptors' stimulate coagulation.

Recognition of Lipopolysaccharide

Recognition of bacterial constituents is obviously vital for survival. However, certain bacterial components can, if recognized, produce severe, often fatal, responses. It is estimated that 200 000 people die annually in Europe and North America from septic shock, caused by the release of LPS from Gram-negative bacteria in foci of infections. Causative organisms include *Salmonella typhimurium*, *Neisseria meningitidis* and *Escherichia coli*. Patients die due to the interaction of LPS with leukocytes and vascular endothelial cells. Such interaction generates a range of cytokines, particularly TNFα and IL-1 (see Chapter 3) and other mediators which inhibit the action of the heart, lower systemic blood pressure and clog up the blood vessels of the major organs with activated leukocytes – a process known as disseminated intravascular coagulation (DIC). Collectively, this prevents perfusion of the major organs (lungs, liver, kidneys) and the patient dies. Humans are exquisitely sensitive to LPS. The administration of only nanogram per kilogram quantities of LPS to human volunteers can cause serious physiological effects. In contrast, mice and rats are much less sensitive to this molecule. It is not clear what controls this responsiveness.

The structure of the Gram-negative cell wall has been described in Chapter 2 and it can be seen that the LPS molecule forms an outer layer of long complex polysaccharide chains which are linked to the membrane by a specialized diglucosamine structure known as lipid A. It is this lipid A which is the causative agent of septic shock. LPS is a major danger signal for eukaryotic cells but is, in itself, not particularly 'toxic'. The secret of the activity of LPS lies in its interaction with a small number of host proteins which are not strictly LPS receptors. The first of these proteins is lipopolysaccharide-binding protein (LBP), which binds to the lipid A region of LPS and can increase the specific biological activity of LPS by up to 1000-fold. LBP is a constitutive protein but its concentration increases in plasma during inflammation, demonstrating that it is also an acute-phase protein. Surprisingly, the complex of LPS and LBP is not any more active than LPS alone. The secret to the activity of LBP is that it acts to increase the rate of transfer of LPS, which is normally present in the form of micelles (i.e. aggregates of LPS molecules), to a second LPS-binding protein, CD14 (Figure 8.3). CD14, which can be found either as a soluble or as a membrane-bound molecule, acts as a 'receptor' for LPS and is an example of a PRR. However, to confuse matters, CD14 does not have an intracellular signalling domain (see Chapter 3 for a description of intracellular signalling) and therefore CD14, in turn, must pass LPS onto the final LPS receptor which may be a Toll-like receptor (TLR). The role of CD14 in the response to LPS is clearly seen in mice, which are normally poorly- responsive to LPS. Production of transgenic mice which express high levels of human CD14 on their leukocytes results in the animals being hypersensitive to LPS and knockout of the receptor renders mice refractory to LPS stimulation.

It is not yet clear what the advantage is to vertebrates in amplifying so significantly the biological activity of LPS. The answer may lie in the fact that vertebrates have many systems for inactivating LPS. For example, addition of LPS to serum rapidly results in its binding to a number of serum proteins such as CETP and PLTP, and the end result is that the LPS binds to the high-density lipoproteins (HDLs), in which form it is inactive. A system may therefore have evolved in which there is only a short time-window in which LPS can be recognized

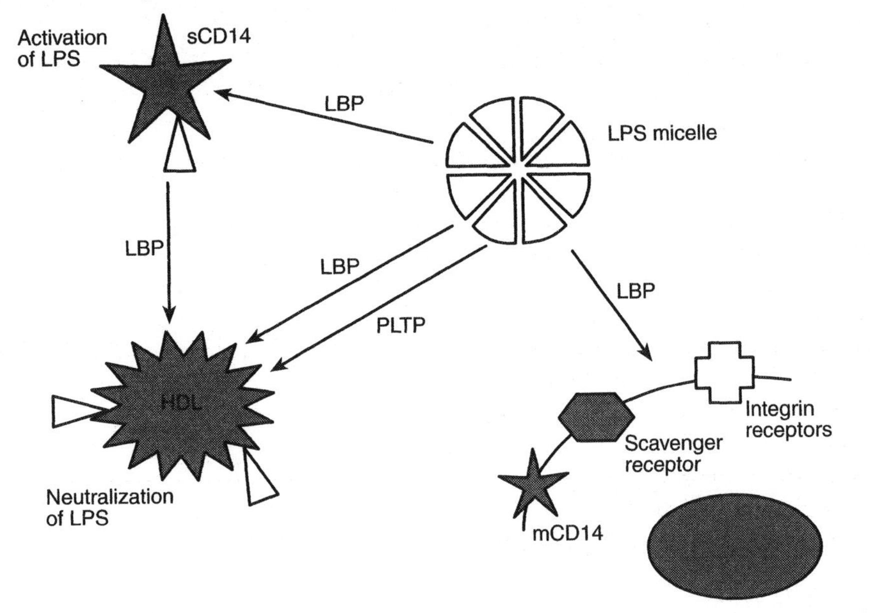

Figure 8.3 The interaction of LPS with host proteins is required to activate and deactivate this key bacterial component

before it becomes inactivated. Rapid amplification of the message of the presence of LPS (and of other bacterial molecules which bind to CD14) may be a mechanism to recognize very low levels of LPS and therefore to recognize the very early signs of bacterial invasion. The cost of this early warning system in the human appears to be a life-threatening condition which occurs if exposed to too much LPS.

Many diferent forms of microorganisms can cause pathology. Thus far the system that has been described can recognize pathogens but does not seem able to discriminate between different pathogens. However, recent studies are suggesting that the recognition systems of innate immunity are able to discriminate between different pathogens and, surprisingly, the genes enabling such discrimination are also involved in the process of development.

Discrimination between different pathogens

Many different types of microorganisms can infect invertebrates and vertebrates, including viruses, bacteria, protozoans and fungi. It has been a tacit assumption that the innate immune system responded equally to all these classes of infectious agents. However, evidence is now accumulating to suggest that innate immunity may be able to discriminate between different classes of pathogens and produce appropriate responses at the genomic level. The fruit fly *Drosophila,* which is much beloved of geneticists, produces seven peptides with antibacterial and anti-fungal properties. Five of these have antibacterial properties, drosomycin has antifungal activity and metchnikowin has both antifungal and antibacterial actions. Analysis of the genes which control the synthesis of these peptides is revealing a fascinating story. The synthesis of the antifungal peptide drosomycin is under the control of a set of genes involved in dorsoventral regulation in *Drosophila.* The insect body, like our own, has two main axes. There is the anterior-posterior (head to tail/feet) axis and the dorsal-ventral (back-front) axis. Control of the dorsoventral patterning is believed to involve the synthesis and proteolytic processing of an inductive protein called späetzle which binds to Toll, a cell surface receptor, inducing intracellular signalling via a pathway which resembles the NF-κB pathway described in Chapter 3. The intracellular domain of the Toll receptor resembles the type I IL-1 receptor (see Chapter 3). This NF-κB signalling pathway is now recognized as being very important for the production of inflammatory responses. The NF-κB analogue in *Drosophila* is called Dorsal (which is part of a family known as the Rel proteins) and the regulatory protein in this insect system I-κB (which binds and inactivates NF-κB) is called Cactus. Mutations in the Toll signalling pathway results in insects which show defective responses when exposed to fungi, revealing that these dorsoventral patterning genes are also involved in host defence responses. However, these mutations do not affect the insect's response to bacterial infections. Moreover, exposure of *Drosophila* to entomopathogenic fungi only caused the production of peptides with antifungal properties but did not switch on antibacterial peptide synthesis (Figure 8.4).

These findings clearly establish that, at least in insects, the innate immune response can discriminate between fungi and bacteria and switch on the correct antifungal response. In addition, this work clearly indicates the deep foundations of the innate antimicrobial defences of the body.

BACTERIA–HOST INTERACTIONS AT BODY SURFACES

Having discussed the mechanisms which have evolved to enable host cells to recognize bacteria it is now time to widen the discussion to examine the totality of the mechanisms which constitute our innate immunity to infection. Given the astronomical numbers of bacteria that exist in our environment it is perhaps surprising that infection is such a relatively rare occurrence. This is, of course, due to the multifunctional systems of innate immunity, which can be divided into those systems which (i) act at the body (epithelial or mucosal) surfaces and (ii) those which are present in the submucosal tissues (Figure 8.5). It needs to be appreciated that our immune system can be further subdivided into those activities which function external to the body, i.e. on the surface of the epithelial cells of the mucosa (lining the mouth, lungs, intestines etc.), and those which function within the body. We are only beginning to appreciate the importance of the former. This is the 'invisible' immune system because we only know of its existence when it stops working and we become infected. The only portal of entry of a bacterium into the body is through the epithelial (also called, where appropriate, the mucosal) surfaces of the skin, eyes, oral cavity, upper respiratory tract, gastrointestinal (GI) tract or genital tract. These various mucosal surfaces have evolved a whole set of mechanisms to prevent the growth and invasion of pathogens (Figure 8.6). The clever tricks that bacteria employ to adhere to and enter into cells, including epithelial cells, have been described in detail in Chapters 5 and 6. Common and site-specific mechanisms exist at the different externally exposed epithelial sites to prevent infection, including: (i) the synthesis and release of antibacterial agents such as the enzyme lysozyme which can degrade the peptidoglycan of bacterial cell walls or antibacterial peptides which can kill bacteria; (ii) the natural tendency of epithelial surfaces to slough off, thus constantly removing the bacteria present on the epithelium; and (iii) the inhibitory effect of the normal microflora on

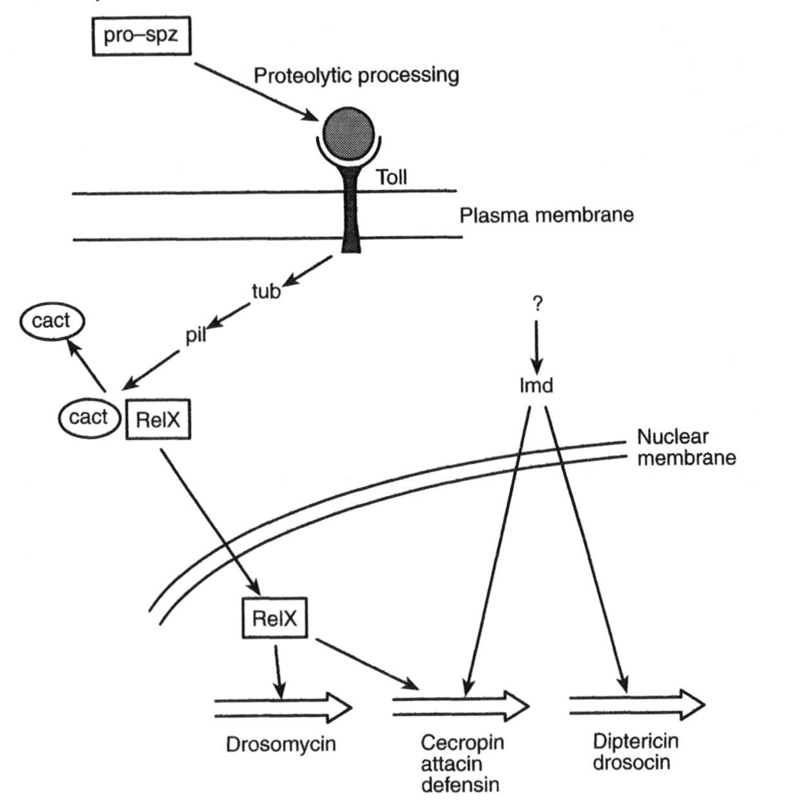

Figure 8.4 A proposed model for the control of the expression of the genes encoding antimicrobial peptides in *Drosophila*. Infection in *Drosophila* is thought to stimulate coagulation of phenoloxidase pathways which results in the proteolytic processing of the protein Spatzle, resulting in the production of a ligand able to bind and activate Toll. This results in intracellular signalling pathways which results in the dissociation of the NF-κB-like protein Rel from its inhibitor cactus. Rel can then translocate to the nucleus and stimulate the transcription of drosomycin. Other antimicrobial peptides are believed to be under the control of the transcription factor Imd (immune deficiency gene), whose precise mechanism has not been established

the growth of exogenous bacteria. If we examine the antibacterial mechanisms used by one of the largest organs in our body – the skin, a tissue constantly exposed to bacteria – these various strategies can be seen in operation. The surface of skin is generally dry, acidic and contains relatively high levels of salt – conditions normally inhibitory to bacterial growth. In addition, the outer layer of cells which constitute skin – the keratinocytes – are dead and are constantly being shed, thus removing bacteria. Skin has many openings such as pores, hairs and sweat glands which can be sites of bacterial colonization. However, these openings are protected by the production of lysozyme, toxic lipids and, as has recently been discovered, antibacterial peptides. The skin also has a relatively sparse normal microflora which will compete with exogenous bacteria for nutrients and colonization sites. Minor damage to the skin (scratches, grazes etc.) generally does not lead to infection, presumably due to a combination of the epithelial mechanisms described and the leukocytes present in the dermis (which underlies the epithelial cells of the epidermis). These leukocytes form what is termed the skin-associated lymphoid tissue (SALT). This is, of course, a misnomer as the SALT does not constitute an actual tissue but is an interacting network of leukocytes acting to prevent access of bacteria into the blood. The other major body epithelia have similar systems, which are generically termed the mucosa-associated lymphoid tissue (MALT).

The body contains many epithelial surfaces which are exposed to the external environment. These include the epithelial surfaces of the eye, mouth (tongue, gums or gingivae and the buccal mucosa on the inside of the cheeks, roof and floor of mouth), trachea, lungs, stomach, gut and genital tract. Most of these epithelial surfaces are, unlike the skin,

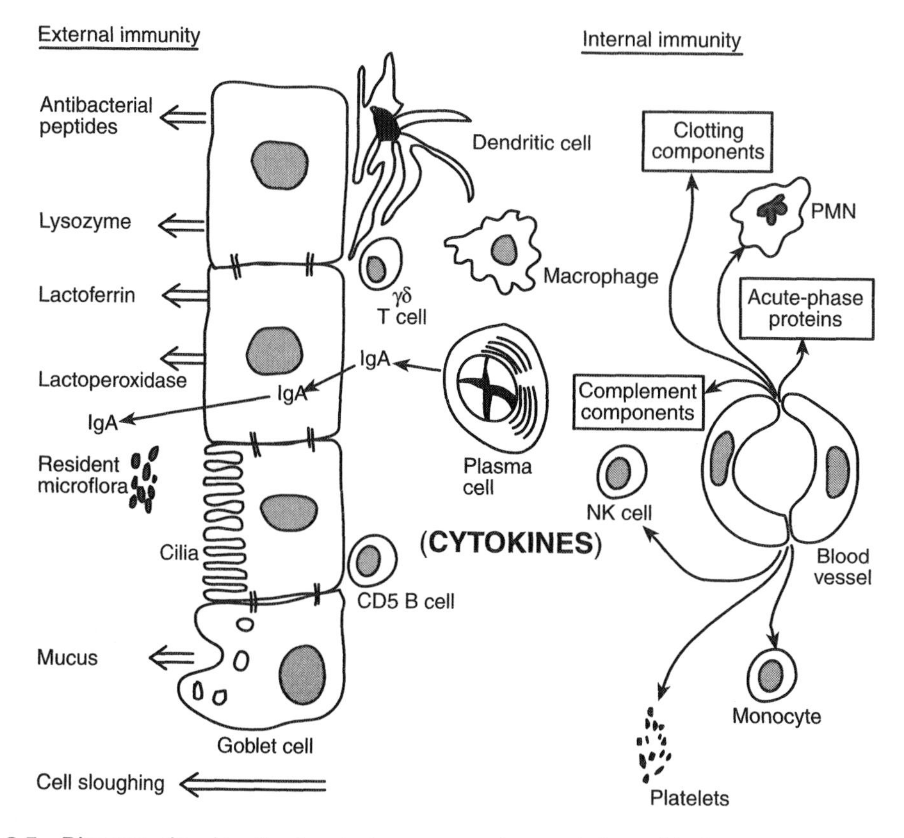

Figure 8.5 Diagram showing the innate immune systems of the epithelia and of the submucosal tissues. The epithelial cells produce a variety of soluble components which are antimicrobial, including antimicrobial peptides, lysozyme, lactoferrin and lactoperoxidase. Mucus is produced by goblet cells. In certain sites the epithelial cells are ciliated and can sweep up bacteria and remove them. IgA is also able to transcytose through epithelial cells to end up on the mucosal surface. These defences can be thought of as an external immune system. Included in this system must be the commensal microflora whose presence acts to compete for nutrients, adhesion sites etc. with any foreign organism, including pathogens. Another defence system which is not often considered, but which is targetted by bacteria, is the tight junctions which exist between epithelial cells to form a continuous barrier to bacterial entry. Epithelial cells on mucosal surfaces generally undergo a maturation process and are shed, removing with them any adherent bacteria. The internal or submucosal defence mechanisms involve a variety of resident cells (CD5 B lymphocytes, γδ T lymphocytes, macrophages and dendritic cells) and cells which can be attracted into the site once infection has occurred (PMNs, monocytes, NK cells, platelets). In addition vascular leakage will bring into the site of infection complement components, acute-phase proteins and components of the clotting cascade. Once an infection is under way many different cytokines would be produced to control the inflammatory response

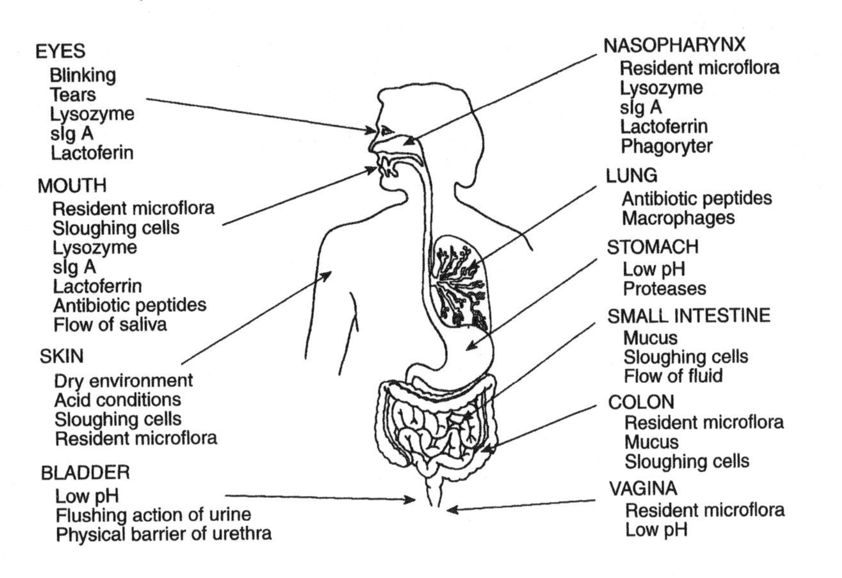

Figure 8.6 Schematic diagram showing the epithelial/mucosal sites in the body and the mechanisms utilized in these various sites to deal with pathogenic microorganisms

composed of only one layer of cells and exist in a moist, warm environment. These are conditions obviously conducive to bacterial growth. Such epithelia are known as mucous membranes and the immune mechanism associated with them is known generically as mucosal immunity. Due to the increased susceptibility of the mucous membranes to colonization with exogenous bacteria, evolution has added further mechanisms for antibacterial defence in addition to those used by the skin (Table 8.2). These mechanisms include the production of mucin, a complex of polysaccharides and proteins, produced by goblet cells in the epithelium, and which has a lubricant function. Mucin also acts as a mechanical barrier and can trap bacteria and prevent them making contact with the epithelium. Of course the possibility exists that the mucin could also act to protect bacteria from their hostile host environment. This is exemplified by the recent discovery that gastric ulceration is caused by a bacterium, *Helicobacter pylori.* This bacterium lives in the acidic environment of the stomach, where it colonizes the mucus layer, thus gaining protection from the low pH.

Many key cellular enzymes contain iron (e.g. the cytochromes of the electron transport chain), and thus this element is an essential nutrient for bacterial growth. Mucus contains the iron-binding protein lactoferrin, which limits the free iron in the environment of the epithelium. This lack of iron has a bacteriostatic action. Another host mechanism for controlling divalent cation concentrations (the *Nramp* gene system) will be described at the end of this chapter. However, it is now clear that bacteria can respond to this low-iron environment by upregulating the production of low molecular mass iron-binding molecules called siderophores. For example, uropathogenic *Escherichia coli* binds to epithelial cells in the urinary tract, an environment low in iron. This binding event triggers the transcription of bacterial genes which produce siderophores and enable the bacterium to take up iron. Lactoperoxidase, also found in mucin, is an enzyme capable of producing superoxide, an oxygen-derived free radical (ODFR), which is damaging to many bacteria. Such free radicals are also produced by neutrophils and macrophages as part of their antibacterial armamentarium.

Table 8.2 Antibacterial mechanisms employed by external epithelia

Tissue site	Mechanism	Action
Skin	Dry, low pH, <37°C	All limit bacterial growth
	Epidermal cell sloughing	Removal of bacteria
	Resident microflora	Competition with exogenous bacteria for colonization
	Resident microflora produce anti-inflammatory proteins	Inhibits inflammatory response to bacteria
	Antibacterial peptides	Kill bacteria/inactivate LPS
Sweat gland/hair follicle	Lysozyme and toxic lipids	Lethal to bacteria
Mucosal epithelia	Mucus	Sticky material which traps bacteria
	Lysozyme	Kills bacteria by digesting cell walls
	Lactoferrin	Binds iron, which is essential for bacterial growth
	Lactoperoxidase	Generates free radicals to kill bacteria
	sIgA	Blocks bacterial attachment to mucosal cells
	Antibacterial peptides	Kill bacteria/inactivate LPS
	Resident microflora	Competition with exogenous bacteria
	Resident microflora produce anti-inflammatory proteins	Inhibits inflammatory response to bacteria
	Ciliated epithelium	Removes a lot of bacteria
	Epidermal cell sloughing	Removal of bacteria
	Epithelial cell tight junctions	Prevents bacteria from invading between epithelial cells

The immunoglobulin molecule is produced in five different forms (termed antibody classes, see Chapter 9) and has a number of functions in immunity. On external epithelial surfaces the only antibody class found is IgA. This is produced locally by plasma cells in the submucosa and is then transported across the epithelium, where its main function is to block the adhesion of bacteria to epithelial cells.

Two additional systems which contribute to antibacterial defence and to the regulation of bacterially induced inflammation are controlled by distinct groups of peptides and protein. The first are antibacterial peptides produced by most cells in the body but particularly by epithelial cells, neutrophils and macrophages. The second are a group of proteins which have been hypothesized to exist by two of the authors (BH and MW). These proteins, termed microkines, are proposed to act to inhibit inflammation in response to the normal microflora which exist at mucosal surfaces.

Epithelial cells are tightly coupled by specialized protein complexes termed tight junctions, which interact with the cell cytoskeleton. It is the formation of such junctions that turns the epithelium into a barrier and prevents bacteria entering into the underlying submucosa. It is of interest that certain bacteria target these tight junctions. For example, *Vibrio cholerae* produces a toxin known as zonula occludens toxin (ZOT), which disrupts epithelial cell-to-cell junctions and increases tissue permeability (see Chapter 7 for more details).

Natural 'antibiotics' and host defence: a newly-r

Salvarsan was first used for the treatment of syph
introduced in the mid-1930s. Since then many ch
for treating bacterial infections. Most of these h
metabolites of soil bacteria and fungi and have
that we use clinically were initially used by
another. With hindsight, the fact that micro
have suggested to those workers developin
would produce similar compounds. Howev
studying insect immunity, that Boman and colleag
tics' from *Hyalophora cecropia* and also from *Drosophila*.
antibacterial peptides was the studies of the granules of mammalia
number of what have been termed cationic antimicrobial proteins (CAP
range of actions, including the intracellular killing of bacteria. The term antibiotic
molecules produced by bacteria and which kill other bacteria. Bacteria make antibacterial peptides which kill other bacteria and thus, in truth, these molecules are antibiotics. However, as it is now clear that the production of such peptides is probably universal the term antibacterial peptide will be used in preference to antibiotic.

The insect antibacterial compounds were short peptides and were called cecropins. Cecropins and related analogues have been identified in many insect species (Figure 8.7). The first mammalian cecropin was isolated from the small intestine of the pig in 1989. By 1998, at the time of writing, more than 100 antibacterial peptides have been described in organisms ranging from bacteria (the polymyxins) to man (e.g. human defensins). Plants produce antibacterial peptides. A bewildering variety of names have been coined for these peptides, including and magainins, protegrins, brevinins, apidaecins indolicidin. They can be subdivided into a number of groups depending on their size, conformational structure and predominant amino acids (Table 8.3). Most cells appear to have the capacity to produce antibacterial peptides, including the non-nucleated platelet. An important finding has been that mucosal epithelial cells in all species studied produce antibacterial peptides. In aggregate, these peptides are active against Gram negative and Gram positive bacteria, fungi and certain eukaryotic microorganisms such as the protozoan *Giardia lamblia*. Antibacterial activity is demonstrated over the concentration range 1–100 μg/ml (approximately 0.3–30 μM – based on an average molecular mass of 3 kDa). The peptides appear to interact with the bacterial cell wall and inner membrane resulting in the bacterial membrane being permeabilized with subsequent cell lysis. Some of the antibacterial peptides are cytotoxic to mammalian cells at high concentrations.

Current textbooks of microbiology and immunology focus on the actions of what can be termed the 'internal' immune system – that is, leukocytes in blood and in the major lymphoid organs: spleen, bone marrow, thymus and lymph nodes. Very little attention is focused on mucosal immunity apart from a brief synopsis of secretory IgA. However, a new paradigm is emerging in which the key defence against infectious bacteria is viewed as a continuous process acting at epithelial surfaces and involving the release of antibacterial peptides. Indeed, as described earlier, antibacterial peptides are the major defence mechanism used by invertebrates against bacteria and fungi. It is possible that these peptides are also responsible for controlling the growth of the commensal bacteria which populate all the external epithelial surfaces of the body. Are these peptides themselves capable of being regulated?

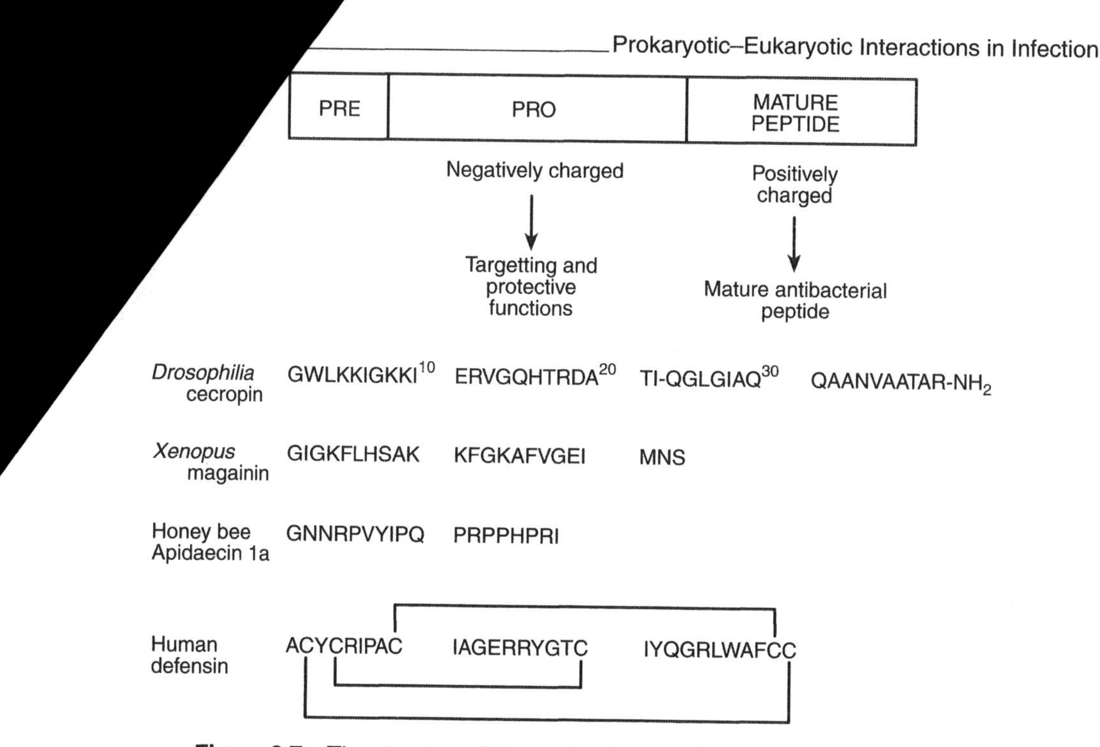

Figure 8.7 The structure of the major classes of antibacterial peptides

The answer is yes, and it has been shown that epithelial cells exposed in culture to the Gram-negative bacterial outer cell wall component, LPS, produce more antibiotic peptides. This suggests that the epithelia of the body may monitor their external environment for the presence of bacterial components and modify their synthesis of antibacterial peptides and other antibacterial activities, accordingly. It is interesting that some of the antibacterial peptides have been shown to have chemokine-like activity (the actions of chemokines has been discussed in Chapter 3) and thus these peptides may have roles in controlling the inflammatory response.

The discovery of antibacterial peptides could not have come at a better time for a world which is running out of antibiotics. In the past five years or so a number of companies with names such as Magainin Pharmaceuticals and IntraBiotics have been set up and are now testing antibacterial peptides in clinical trials for diverse conditions such as foot ulcer infections, oral mucositis and stomach ulceration due to *Hel. pylori*. It has even been proposed that transgenic plants and animals could be produced which would have elevated resistance to infections by incorporating genes for these antimicrobial peptides. If these natural peptides do not produce significant clinical effects, due to the fact that peptides are easily broken down in the body, it may be possible to produce non-peptidic forms of these molecules in which the conventional amide bond is changed to one which is resistant to proteases. Of course, it is possible that once we understand the mechanisms by which the antibiotic peptides are transcriptionally controlled (see the earlier section on Toll) it might be possible to selectively switch on peptide production at sites of infections and utilize the body's ability to make antibiotics to treat infection.

As has been described, there is a co-evolutionary war going on between bacteria and the multicellular organisms they are trying to colonize. The defence systems of innate immunity

Table 8.3 Antibiotic peptides

Name	Source	Size (no. of amino acids)	Active against
CECROPINS			
Cecropins A–D	Insect haemolymph	30–45	Gram –ve bacteria
Cecropin P1	Pig intestine		Gram –ve bacteria
MAGAININS	Frog skin, intestine	20–27	Gram +/–ve bacteria/fungi
BREVININS	Frog skin	24–30	Gram +/–ve bacteria
Ranalexin	Frog skin	20	Gram +/–ve bacteria
APIDAECINS	Honey bee haemolymph	18	Gram –ve bacteria
Abaecin	Honey bee	34	Gram –ve bacteria
Hymeonotaecin	Honey bee	93	Gram –ve bacteria
Drosocin	*Drosophila*	19	Gram –ve bacteria
Cyclic dodecapeptide	Bovine	12	Gram +/–ve bacteria
TACHYPLESINS	Horseshoe crab	16–18	Gram +/–ve bacteria/fungi/ viruses
Protegrin	Pig		Gram +/–ve bacteria/fungi/ viruses
Bac5	Bovine	42	Gram –ve bacteria
Indolicidin	Bovine	13	Gram +/–ve bacteria
DEFENSINS			
Human neutrophil peptide	Human		Gram +/–ve bacteria
Cryptidin	Mouse	20	Gram +/–ve bacteria
Tracheal antimicrobial peptide (TAP)	Bovine		Gram +/–ve bacteria

are constantly being challenged by bacteria and we predict that it will not be long before it is shown that the antibacterial peptides can be defused by one or other bacterial mechanism, such as the action of specific proteases or the use of soluble or cell surface specific-binding proteins.

Before leaving this subject of antibacterial peptides, recent studies have shown how important they are in the pathology of cystic fibrosis, an autosomal recessive disease and the most common genetic lesion in Caucasians. The condition is caused by mutations in the gene encoding the cystic fibrosis transmembrane conductance regulator (CFTR), a phosphorylation-regulated Cl^- channel located in the apical membrane of involved epithelia. The major phenotype of these mutations is the formation of mucin at epithelial surfaces which has an increased NaCl content and is unusually thick. The thickness of the mucin apparently makes the clearing of bacteria from the lungs and airways much less efficient and renders sufferers susceptible to colonization with the bacterium *Pseudomonas aeruginosa*. Recurrent infections with this organism produces progressive damage to the lungs until the patients die in early adulthood. It should be noted that despite the continuous exposure of human airways to aspirated bacteria they remain sterile in healthy individuals. This suggests some problem in dealing with aspirated bacteria in sufferers from cystic fibrosis. An explanation for the relationship between the genetic defect in cystic fibrosis patients and the colonization of the lungs by *Ps. aeruginosa* has recently been suggested and related to the sensitivity of some of the antibiotic peptides to conditions of high ionic strength. Addition of *Ps. aeruginosa* to normal human airway epithelia resulted in the rapid killing of these organisms, with no obvious changes in the epithelial cells. In contrast, airway epithelia from patients with cystic fibrosis failed to kill bacteria and indeed the bacteria survived and multiplied. However, when CFTR was transfected into cystic fibrosis airway epithelia, correcting the fault in the NaCl balance, the bacteria were killed. This suggested that the failure to kill bacteria was related to a salt-dependent defect in epithelial cell antibacterial activity. Human β-defensin 1 (hBD-1), produced by lung epithelia, is now recognized to be important in innate immunity in the lungs. This antibiotic peptide is active against Gram negative organisms, including *Ps. aeruginosa*. It is non-functional at the NaCl concentrations found in the lungs of cystic fibrosis patients. Thus it is suggested that the inhibition of activity of hBD-1 by the high salt concentration in the lung extracellular fluid is the cause of the life-threatening airway infections in patients with cystic fibrosis.

INNATE IMMUNITY AND THE ACUTE-PHASE RESPONSE

The previous sections have attempted to highlight aspects of innate immunity often ignored in textbooks of microbiology and immunology. As described in previous chapters, bacteria such as *Salmonella typhimurium*, *Yersinia enterocolitica* and enteropathogenic strains of *E. coli* have developed very sophisticated strategies to disable or control epithelial cells. Much of the interplay between bacteria and the host must occur on our external epithelial surfaces and it must be supposed, given the relative rarity of bacterial infections, that most of this interplay is invisible. This is presumably why so little is known about the microbiology of the mucosal surface. However, in the circumstance in which an infectious bacterium defeats the systems of epithelial immunity, the body's backup systems swing into operation. These systems are those classically described as innate immunity and the totality of the short-term programmed response to infection is called the acute-phase response. This is a whole body response designed to activate multiple organ systems such as the brain and liver in order to stimulate immunity and get rid of the offending bacteria. The system comprises cellular elements and a vast range of proteins which can be divided into groups that have distinct but interacting activities. These include the acute-phase proteins (some of which were described in an earlier section), involved in the opsonization of bacteria and the activation of the complement pathway. The complement pathway itself is a complex multiprotein cascade which activates inflammatory systems and kills bacteria. The third, and possibly most important group of soluble proteins are the cytokines, which act to control the working of the acute-phase response.

Cell populations of the acute-phase response

The response of the body to the entry of bacteria past the epithelial barrier is shown diagrammatically in Figure 8.8. Epithelial cells will play an initial role by producing particular patterns of proinflammatory cytokines which will signal to nearby blood vessels and fixed tissue macrophages. The blood vessels and macrophages will in turn produce cytokines and other signals which will accelerate the local inflammatory process. Having signalled the presence of bacteria, what are the processes needed by the body to remove these potentially dangerous organisms? The requirement of the acute-phase response is to recognize where an infection is occurring and to get to that site certain *matériel*. This includes cells with the capacity to phagocytose and kill bacteria (cells given the generic name – phagocytes), molecules called acute phase proteins which can bind to the invading bacteria and increase their capacity to be ingested by phagocytes (the process known as opsonization) and components of the complement pathway which promote opsonization and can kill (some) bacteria directly.

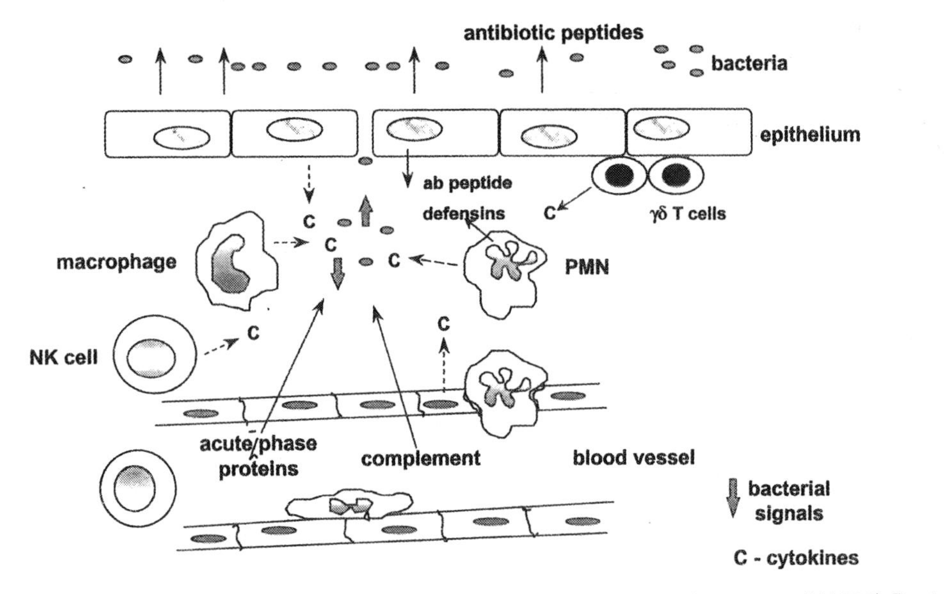

Figure 8.8 Schematic diagram of the acute-phase response (from Henderson *et al* (1998) Bacteria-Cytokine Interactions in Health and Disease, Portland Press, London, with permission)

Unfortunately, a major problem exists in dealing with infections. The cells and soluble proteins required to deal with the invading organisms are sequestered within the blood vessels which permeate all tissues. The leukocytes flow down the centre of the blood vessels and are therefore not normally available. What solution has nature evolved to deal with this situation? The answer is shown in Figure 8.9. Bacterial components released locally at the site of invasion stimulate cells to produce and release a small number of very active proinflammatory cytokines such as IL-1, TNFα and IL-8 which stimulate the endothelial cells lining particular regions of the vasculature (the so-called post-capillary venules) to synthesize and express cell surface receptors for leukocytes and induce the production of chemotactic cytokines called chemokines. These endothelial cell receptors are proteins and include endothelial (E) and platelet (P)-selectins, intercellular adhesion molecules (ICAMs) and vascular cell adhesion molecule (VCAM). The function of these endothelial cell surface proteins is to slow down and then bind strongly (capture) neutrophils and monocytes. E- and P-selectin recognizes certain oligosaccharides (the best-known example being sialyl-Lewis x [s-Le^x]) on the plasma membranes of leukocytes, while the ICAMs recognize heterodimeric counter-receptors known as β_2-integrins. Of interest is the finding that pertussis toxin contains domains which mimic the selectins and can inhibit neutrophil binding to E-selectin. Certain bacterial carbohydrates have also been shown to prevent cell binding to E-selectin. How important such bacterial mechanisms are in inhibiting antibacterial responses is not established. The integrins are present on circulating neutrophils and monocytes in an inactive form but on exposure to chemokines such as IL-8 in activated blood vessels they become phosphorylated and thus able to bind to the counter-receptors (ICAMs) on the blood vessel endothelium. Once the leukocytes have been stopped, the process known as diapedesis can take place. This involves the movement of the leukocyte across the vascular endothelial cell barrier by passage between adjacent endothelial cells and into the site of infection. Leukocytes can then move by chemotaxis along the gradient of signals until they reach the bacteria. This passage of cells accross the vascular wall can also render the local blood vessels leaky and thus introduce plasma factors (including acute-phase proteins and complement components) into the infected site. There are a small number of studies suggesting that bacteria can inhibit this process of leukocyte recruitment. As leukocyte recruitment is a central antibacterial process it is to be expected that bacteria would have evolved systems to inhibit the process. The consequences of the failure to recruit leuokocytes is seen in conditions like leukocyte adhesion deficiency (LAD), an autosomal recessive condition in which leukocyte β_2-integrin expression is markedly diminished or totally absent. With total loss of integrin expression patients are susceptible to life-threatening infection. Partial loss results in recurrent necrotic infections of skin, soft tissues and mucous membranes, revealing the importance of correct leukocyte trafficking in our defence against bacteria.

Bacterial killing

The workhorses of bacterial killing are the polymorphonuclear neutrophil (PMN or simply neutrophil) and to a lesser extent the monocyte/macrophage (Figure 8. 10). These cells arise from a common progenitor in the bone marrow, where they develop and mature before being released into the blood. The neutrophil is a short-lived cell, surviving only a few days at most in the blood. It is estimated that each day 10^{11} neutrophils are lost from the human body. The monocyte is a longer-lived cell and a proportion of circulating monocytes enter tissues under normal circumstances and mature into fixed tissue macrophages which are given a range of names depending on the tissue (Kuppfer's cells in liver, microglia in brain etc.). These fixed tissue macrophages act as part of the early warning system for bacterial infections. Monocytes entering sites of infection have to mature into macrophages before they can efficiently ingest and kill bacteria. The first process in bacterial killing by neutrophils and macrophages is binding and phagocytosis. Both cells are believed to have similar mechanisms of phagocytosis and this process is dependent on the bacteria being opsonized by host molecules. Opsonins include antibodies, complement components (C3b and iC3b) and acute-phase proteins (serum amyloid A -SAA etc.) and the phagocytes have receptors for these various host proteins. They also have receptors which can recognize bacterial carbohydrates and these have been discussed. Thus function of the opsonin is to immobilize the bacteria on the surface of the phagocyte. Ingestion of bacteria is followed by their killing, within an intracellular body called a phagolysosome, by a combination of mechanisms (Table 8.4; see also Chapter 6). Not all the mechanisms described in Table 8.4 are

Figure 8.9 Schematic diagram outlining the mechanism by which leukocytes are attracted into sites of infection. (A) Under normal circumstances circulating leukocytes do not interact with the vascular endothelial cells that make up the conduits in which these cells circulate. Leukocytes contain two major ligands for interacting with vascular endothelial cells. These are carbohydrate moieties (e.g. sialyl-Lewis x, also called CD15), which bind to selectins, and a group of homodimeric proteins termed the β_2-integrins, which in resting cells are in a non-phosphorylated low-affinity conformation which does not bind to their cognate receptors on the endothelium. (B) However, infection within tissues releases a range of

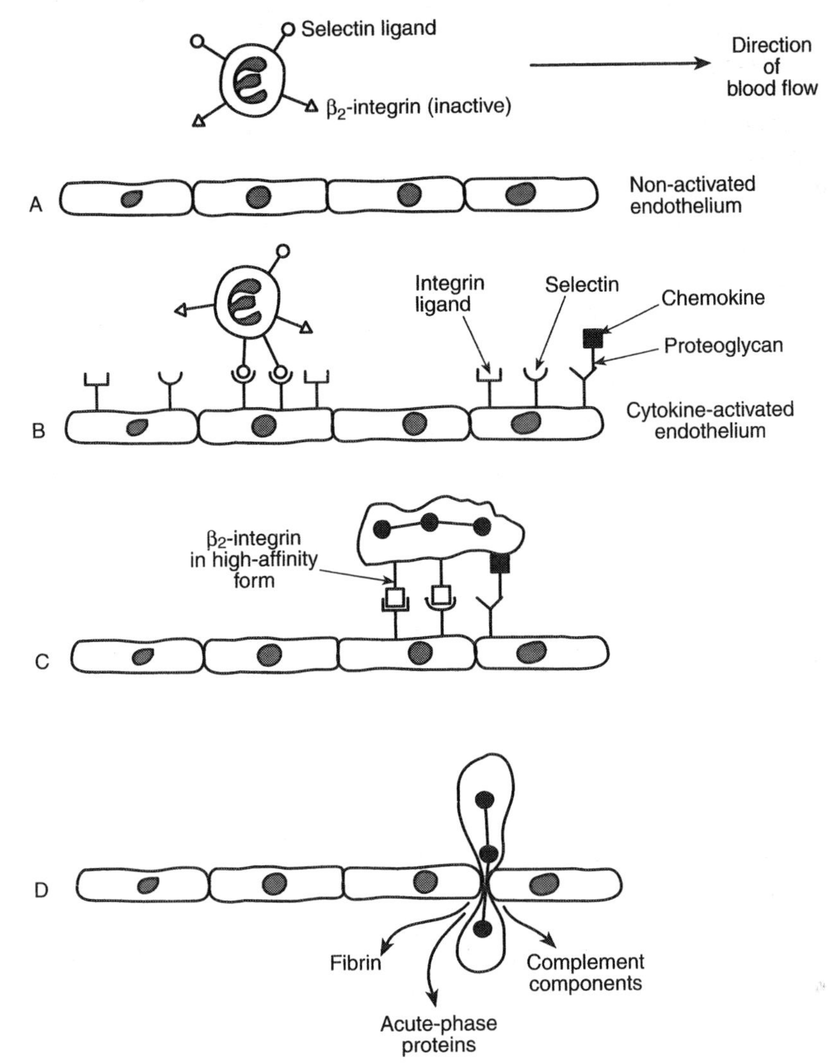

signals, including bacterial components (LPS, peptidoglycan etc.) and the cytokines that these components stimulate (IL-1, TNF), which interact with the vascular endothelial cells and stimulate the production of a range of cell surface receptors – the selectins and the receptors for the integrins. Vascular endothelial cells also express chemokines produced locally on their cell surfaces bound onto proteoglycans. The first step in this process is that the leukocytes flowing down the centre of the post-capillary venule begin to detach and start rolling down the vessel wall. This process involves the interaction of CD15 on the leukocyte with the selectins (P and E). This phase is called tethering and allows the leukocytes to be stimulated by factors produced by the endothelial cells such as the surface-expressed chemokines. (C) Activation of the leukocytes causes phosphorylation of the integrins which now bind to their cognate receptors (ICAM-1, ICAM-2) and make firm contact with the endothelial cells. (D) The activated, firmly bound, leukocytes now crawl along the endothelial surface until they find the interendothelial cell junction. The process known as diapedesis can now occur in which the cell crawls between the endothelial cells and emerges into the extravascular space. This process is facilitated by another adhesion molecule platelet–endothelial adhesion molecule 1 (PECAM-1) or CD31. Leukocytes entering the extravascular space can now chemotact to the site of the bacteria and begin to kill them. In the process of diapedesis plasma components may also leak into the site of infection used to kill all organisms. Killing of microorganisms can be divided into those mechanisms which are oxygen-dependent and those which are oxygen-independent. Oxygen-dependent pathways produce reactive species such as superoxide, hydroxyl radical, singlet oxygen and hypohalite. Genetic deficiency of certain of these pathways are found, such as chronic granulomatous disease (CGD) and congenital myeloperoxidase deficiency. Sufferers have deficient antimicrobial activity and have recurrent infections. A number of bacteria have evolved protective mechanisms against oxidative killing that may involve modulation of endosomal pathways (see Chapter 6 for more details) or the production of anti-oxidant systems. *Listeria monocytogenes* produces large amounts of catalase which has a protective function.

Table 8.4 Mechanisms employed in killing ingested bacteria in phagocytes

Acidification of the phagosome
Lysosomal enzymes (proteases, glycosidases etc.) active at acidic pH
Reactive oxgen species (hydroxyl radical, superoxide, singlet oxygen and hydrogen peroxide)
Toxic oxidants such as HOCl produced by interaction of hydrogen peroxide and Cl^- and utilizing enzymes such as myeloperoxidase
Cationic proteins and defensins (antibiotic peptides) active at neutral pH
Nitric oxide produced by inducible nitric oxide synthase (iNOS)
Growth inhibitor (e.g. lactoferrin and arginase)

A recently discovered antimicrobial mediator is the gas nitric oxide. This is produced by an inducible enzyme known as nitric oxide synthase (NOS) from arginine and is now recognized as being important in defence against intracellular pathogens such as the bacteria *Mycobacteria* and *Salmonella*, and *Leishmania*, a protozoan. This statement must, however, be qualified. Nitric oxide production is clearly of importance in the mouse, and murine macrophages can be stimulated to make large quantities of this gas. However, it has proved difficult to demonstrate that human monocytes or macrophages produce significant quantities of this fascinating mediator. The mechanism by which NO kills bacteria is not established but is probably due to production of a range of intermediates such as peroxynitrite ($ONOO^-$), S-nitrosothiols (RSNO), nitrogen dioxide (NO_2), dinitrogen trioxide (N_2O_3), dinitrogen tetroxide (N_2O_4) and dinitrosyl iron complexes. The production of these various products requires interactions with the reactive oxygen-producing systems in the phagocyte. DNA appears to be an important target of these compounds, as do a range of key cellular proteins. It should be noted that the nitric oxide synthase system can also operate against cytosolic pathogens. As expected, microorganisms have developed defence mechanisms against NO. These include the actions of low molecular mass thiols such as reduced glutathione (GSH) and homocysteine. Again it is expected that we will uncover a range of defences against NO intermediates.

Non-oxidative defences are increasingly seen to be important in killing bacteria. This belief is based on the finding that leukocytes from patients with chronic granulomatous disease (who cannot produce reactive oxygen radicals) can still kill many microbial pathogens. The presence of antibiotic peptides (defensins) and cationic proteins has been described and these have potent antimicrobial properties. Other proteins such as lactoferrin are found in phagocytes. The acidification of the phagolysosome containing ingested bacteria can also have effects of bacterial viability. These various mechanisms of bacterial killing are shown schematically in Figure 8.10.

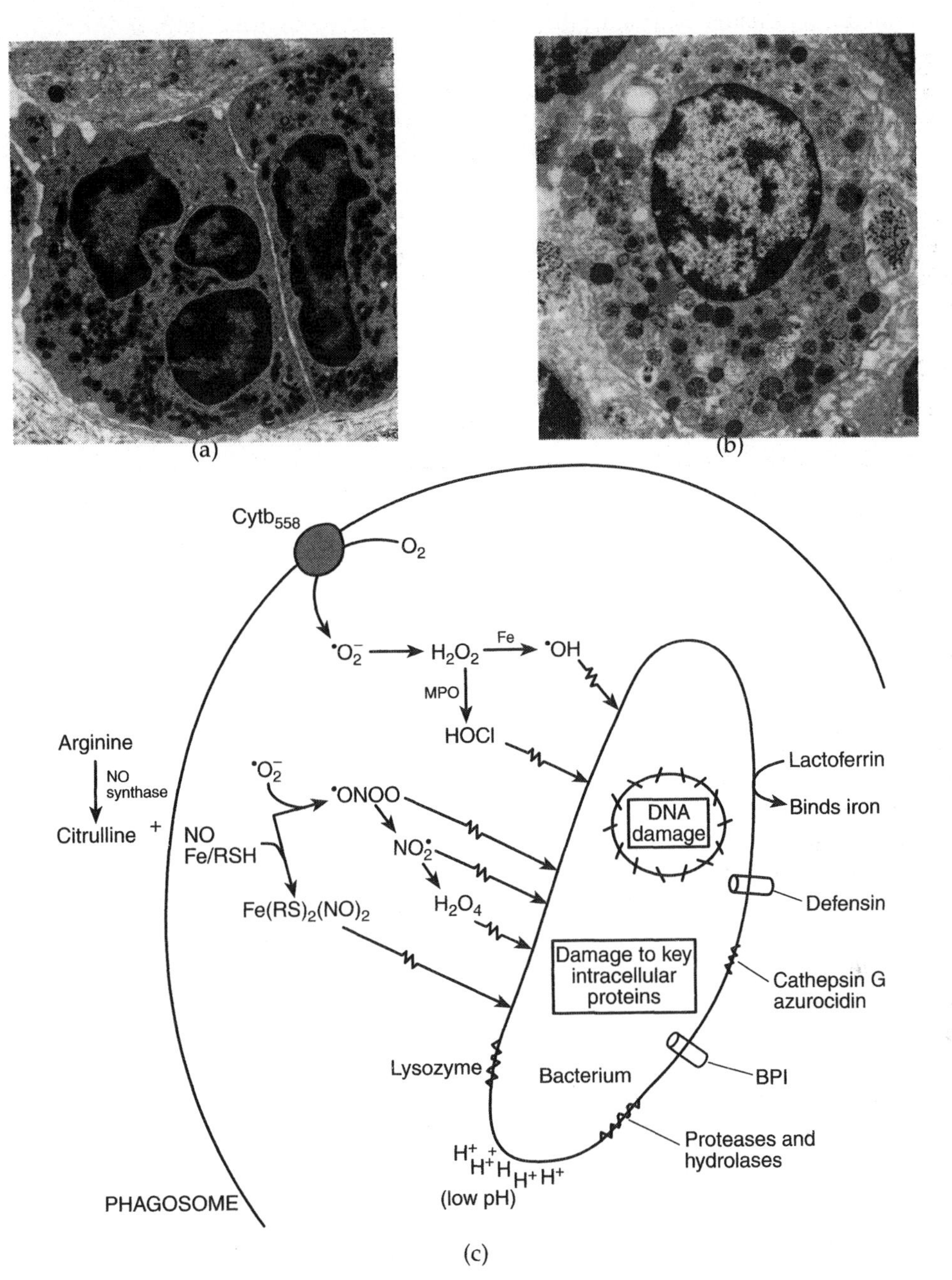

Figure 8.10 Bacterial killing is the province of the two major phagocytic leukocytes – the PMN (a) and the macrophage (b). In (c) a schematic diagram shows the various mechanisms acting within the phagosome to kill bacteria. This can be divided, as stated in the text, into those mechanisms that utilize reactive oxygen species, produced by the cytochrome b in the phagosome membrane, or by interaction of NO with these reactive oxygen species. These free radicals can damage both the DNA of the cell and key intracellular and cell surface proteins. A range of non-oxidative mechanisms are also operational. These include the low pH within the phagosome, lytic enzymes, proteins which can bind and permeabilize the membrane (e.g. defensins and BPI) etc.

A cell which, to some extent, bridges the gap between innate and acquired immunity is the natural killer (NK) cell, which is part of the T lymphocyte lineage. However, unlike classic T lymphocytes, the NK cell does not express the T cell receptor which is required for conventional T cells to be activated (see Chapter 9 for more details about these receptors). This cell appears to act as a first line of defence in certain infections with particular viruses and intracellular bacteria. For example, mice lacking T or B lymphocytes are resistant to infection with the intracellular bacterium *Lis. monocytogenes*. However, depletion of NK cells renders animals highly susceptible to infection with this organism. The antimicrobial actions of NK cells seem to stem from three distinct activities. Firstly, although NK cells are poorly phagocytic, they do appear able to bind to certain pathogens and have direct microbicidal activity. Secondly, they can lyse host cells infected with intracellular pathogens, releasing organisms and exposing them to additional immune antibacterial mechanisms. Thirdly, NK cells can be stimulated to release cytokines including the important stimulatory cytokine IFNγ.

The neutrophil is the main defence against many bacteria, and this is clearly shown by the consequences of neutrophil dysfunction, which is found in a range of diseases such as severe congenital neutropaenia, chronic granulomatous disease (CGD), LAD and Chediak–Higashi syndrome. In severe congenital neutropenia (also known as Kostmann's syndrome) there is a defect in neutrophil production and sufferers have frequent and severe bacterial infections (pneumonia, otitis media, gingivitis, perineal and urinary tract infections) which often kill them in early life. In CGD – a group of inherited disorders – the defect is in the oxygen-dependent microbicidal activity of neutrophils and other phagocytic cells. The major clinical problem is abscess formation in the skin, lungs and the external-facing mucous membranes. Given the importance of the phagocyte in antibacterial defences it would be expected that bacteria would have evolved mechanisms to try and foil such cells.

Antiphagocytic mechanisms of bacteria

As has been described, three processes are required for a professional phagocyte (the term refers to neutrophils and macrophages) to kill a bacterium: (i) opsonization; (ii) phagocytosis; and (iii) oxidative and non-oxidative killing. Many bacteria have evolved defences against being opsonized and can thus escape being phagocytosed by neutrophils and macrophages. The best-known defence is the production of a bacterial capsule and many bacteria including *Haemophilus influenzae, Neisseria meningitidis, Streptococcus pneumoniae, Klebsiella pneumoniae, E. coli* and group B streptococci have polysaccharide capsules which essentially shield the bacterial surface. It is well recognized that non-encapsulated forms of these various bacteria are much less virulent. The exact mechanism of action of bacterial capsules is still not fully explained. Capsules may either prevent the attachment of antibodies and complement components to the bacterium's surface or, having bound these molecules, the capsule may act to mask their interaction with the appropriate receptors on the surface of phagocytes (Figure 8.11). It is also important to note that the components of the capsule are generally poorly immunogenic and so do not raise an antibody or cell-mediated response. A related mechanism is utilized by *Streptococcus pyogenes*. This bacterium produces the M-protein, which forms hair-like projections that are linked to the bacterial surface. This protein is largely composed of repeating amino acid sequences. The key to the actions of M-protein is that it binds the complement inhibitory factor – factor H – which prevents the production and binding of C3b to the bacterial surface. The binding to factor H also acts to make the bacterium appear as self. A second bacterial strategy to prevent opsonization, and one probably most successfully used by the African trypanosome, the cause of sleeping sickness, is to vary the surface antigenic structure of the parasite. This process is termed antigenic variation. The best bacterial 'costume changer' is *Neisseria gonorrhoeae*. This bacterium has two major variable surface proteins: a pilin and an outer membrane protein termed P.II (or Opa). By a variety of mechanisms *N. gonorrhoeae* can create up to 1 million pilin antigenic variants and a smaller number of variants of the P.II protein. This prevents the binding of pre-existing antigonorrhoeal antibodies and thus prevents opsonization. A related mechanism of immune avoidance is called phase variation and involves the environmental regulation of the synthesis of a particular bacterial component, e.g. pili or LPS. A more detailed description of phase variation is to be found in Chapter 9.

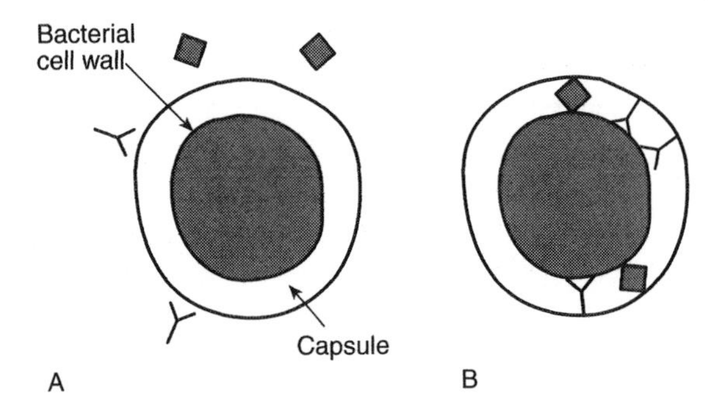

Figure 8.11 Mechanism of action of bacterial capsules in preventing opsonization by antibacterial antibodies or by C3b. In (A) the capsule acts to prevent the binding of antibody to the cell surface or the binding of C3b. In (B) the capsule is of looser texture and allows some binding of these host components. However, the capsule, because of its physical nature, prevents the antibody or C3b making contact with the appropriate receptors on the phagocyte surface

Another fascinating example of the evolution of a bacterial defence mechanism is the finding of numerous Fc-binding proteins in bacteria. The best-known examples of such proteins are protein A of *Staphylococcus aureus* and protein G of *Strep. pyogenes*. These proteins bind to the constant region of the antibody molecule, that is, that part of the antibody molecule recognized by Fc receptors on phagocytes – receptors used to recognize antibody-opsonized bacteria. The presence of such Fc-binding proteins on the surface of bacteria results in the organism being coated by antibody and thus becoming invisible, as far as phagocytes are concerned. Staphylococci and streptococci also have receptors for other host proteins such as collagen, fibronectin and α_2-macroglobulin which presumably act to mask the bacteria and thus prevent opsonization.

If the bacterium is unable to prevent opsonization, all is not lost. The process of phagocytosis starts with the binding of the bacterium to phagocyte cell surface receptors, thus triggering the cells' intracellular microfilament system, resulting in pseudopodia formation; these surround the bacterium producing a phagocytic vacuole (phagosome), which can then fuse with lysosomes containing the lytic machinery to form a phagolysosome. Some bacteria, such as the intracellular organism causing tuberculosis, *Mycobacterium tuberculosis*, can inhibit phagolysosome formation by the release of stongly acidic sulphatides. *N. gonorrhoeae* uses an outer membrane protein, a porin called P.I, to inhibit phagolysosome formation. Unusually, this protein can translocate from the bacterium into human cells, where it can insert into membranes making pores and collapsing the transmembrane potential. This prevents proper maturation of phagolysosomes. The detailed ability of bacteria to frustrate phagolysosome formation and bacterial killing is presented in Chapter 6.

The mechanisms of bacterial killing have been briefly described, as have pathogen defences. If all else fails the bacterium can simply kill the phagocyte. A number of bacteria including *E. coli*, *Staph. aureus* and the oral organism *Actinobacillus actinomycetemcomitans* produce leukotoxins which kill neutrophils and macrophages by a variety of mechanisms, including the induction of apoptosis (see Chapter 2).

COMPLEMENT AND ANTIBACTERIAL DEFENCE

The presence of an antibacterial system in the blood was first demonstrated in France in the 1890s. This was described as a heat-labile bactericidal activity in serum able to be triggered by the binding of antibodies to bacteria. The activity was said to 'complement' antibody. We now recognize complement as a multicomponent amplificatory system, utilizing a cascade of proteases, and acting to: (i) induce and control inflammation; (ii) kill bacteria; and (iii) remove immune complexes. The importance of the complement system is highlighted by the fact that it constitutes 10% of the total protein content of serum. A number of cell receptors for complement breakdown components are now recognized and their activity is understood. The main changes in our understanding of complement in recent years has been the realization of the importance of non-antibody-mediated complement activation. The three pathways of complement activation (classical, alternative and lectin) are now recognized (Figure 8.12) and the functions of the various proteins involved are detailed in Table 8.5. In the classical pathway, the binding of the first complement component C1q to IgG or IgM immune complexes on the surface of a bacterium activates a serine protease active site (C1s) associated with the complex. This protease then cleaves C4 into two fragments, C4a and C4b, the latter covalently attaching to the bacterial surface. C4b then binds C2 to form a complex also able to be proteolysed by C1s to form C4b2b which, in turn, is a protease that can cleave the most abundant complement component C3 (present at a concentration of 1.3 mg/ml in serum). Cleavage of C3 is the major amplificatory step in the pathway and, for example, one molecule of C4bC2b can cleave up to 1000 molecules of C3. When C4 and C3 are cleaved the large cleavage products (C4b, C3b) produced contain a reactive thioester group which has a very short half-life and reacts with adjacent hydroxyl or amino groups in its vicinity. This is a mechanism for ensuring that the complement components are linked to the bacterial surface and that there is no spillover of reactive molecules onto host cells, causing self-damage. In the lectin pathway the C1s is replaced by another serine protease – MBP-associated serine protease (MASP).

The complex C4bC2bC3b has an altered protease specificity and is now termed a C5 convertase, able to cleave C5 into C5a and C5b. We now find that there is a different mode of behaviour with the terminal complement components. Rather than a series of proteolytic cleavages the proteins C5a, C6 and C7 bind sequentially. C7 on binding exposes a hydrophobic domain which allows the complex C5aC6C7 to bind into the membrane. C8 binds and allows the complex to insert into the membrane and this complex promotes the polymerization of the final component C9 into a ring-like structure called the membrane attack complex

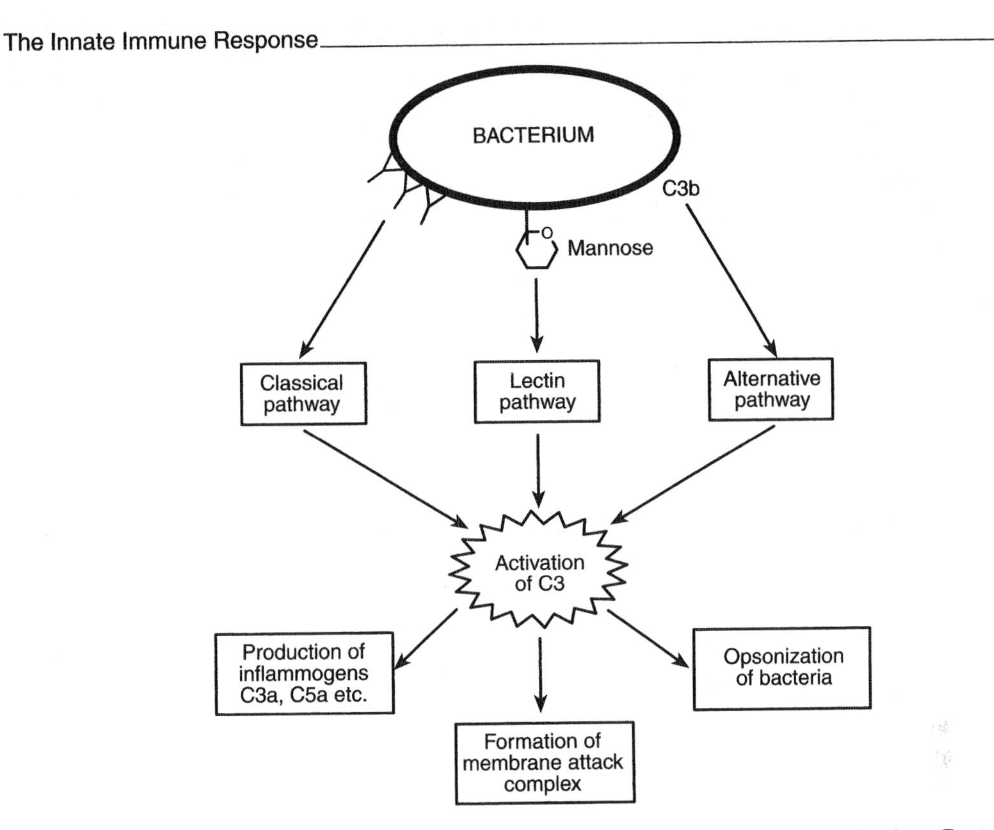

Figure 8.12 The relationship between bacterial infection and complement activation. Bacteria can trigger complement activation by three mechanisms. Antibody binding triggers the classical pathway of complement activation. The presence of certain carbohydrates on the bacterial surface can be recognized by the collectin, mannan-binding lectin, which triggers the lectin pathway. Bacterial surfaces also provide the capacity to overcome the complement inhibitors I and H, allowing the cleavage of C3. This process is termed the alternative pathway of complement activation. The key attribute of these three pathways is the activation of the pivotal component of the complement cascade C3. The activation of this protein allows the production of various proinflammatory molecules. Perhaps the most important is C5a, which is a potent chemoattractant for leukocytes. Cleavage of C3 results in the production of C3b which binds to bacteria, thus opsonizing these organisms. The complement components C5 to C9 produce a macromolecular complex called the membrane attack complex which forms pores in certain bacteria, killing them

(MAC), which acts as a pore allowing free passage of solutes from the cell and loss of cellular integrity (Figure 8.13). The complement pore is now recognized to be very similar to that used by cytotoxic T cells to kill target cells (see Chapter 9).

The alternative pathway of complement activation has a distinct initiation mechanism but shares the terminal sequence of pore formation. Under normal circumstances and with the large amounts of C3 in plasma, there is a significant rate of spontaneous hydrolysis of C3 to C3b and C3a. C3b can bind to factor B and this complex can be cleaved by a plasma protease, factor D, to produce C3bBb, which is the alternative pathway C3 convertase. Deposition of C3bBb on host cells results in the inhibition of further C3 convertase activity because of the presence on the cell surface of the proteins CR1, decay-accelerating factor (DAF) and membrane co-factor protein (MCP) and the plasma protein factor H. CR1, DAF and factor H bind to C3b, displacing Bb. In addition, factor H, CR1 and MCP render C3b susceptible to

Table 8.5 Functions of the complement proteins

Function	Proteins
Binding to antigen–antibody complexes	C1q
Binding to bacterial carbohydrates	MBP
Proteases activating other molecules	C1r C1s C2b Bb D MASP
Proteins regulating the complement pathway	C1 INH C4bp CR1 MCP DAF H I P CD59
Membrane-binding proteins and opsonins	C4b C3b
Peptides which induce inflammation	C4a C3a C5a
Membrane attack complex proteins	C5b C6 C7 C8 C9
Complement receptors	CR1(CD35) CR2 (CD21) CR3 (CD11b, CD18) CR4 (CD11c, CD18) C1qR C3aR C5aR (CD88) MCP (CD46) DAF (CD55)

cleavage by factor I, an active circulating serine protease. Thus host cells are protected from complement activation and deposition on their surfaces. In contrast, microbial surfaces do not contain these control proteins and therefore the C3bBb complexes formed on bacterial surfaces function as an active C3 convertase. Furthermore, these complexes are stabilized by an additional protein, properdin. Binding of C3b to the C3bBb complex forms $C3b_2Bb$, which acts as the C5 convertase of the alternative pathway leading to the production of the membrane attack complex.

What are the consequences of triggering complement activation? Cleavage of C4, C3 and C5 produces a series of small protein breakdown fragments, C4a, C3a and C5a, which have

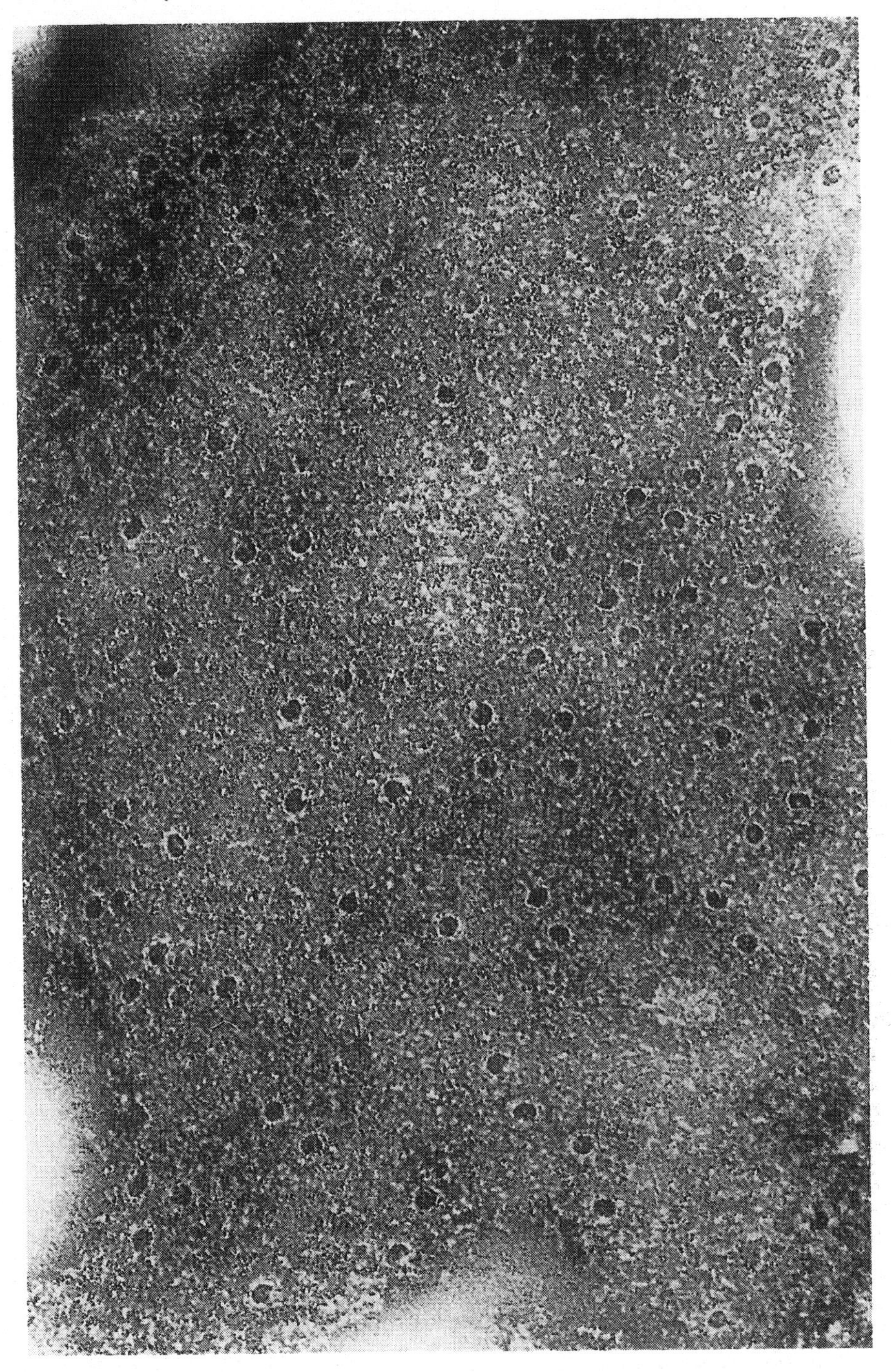

Figure 8.13 Electron micrograph showing the effect of the formation of the membrane attack complex in an erythrocyte plasma membrane. This electron micrograph was kindly provided by Dr E.A. Munn and reproduced with permission of Oxford University Press

proinflammatory actions inducing vasodilatation, oedema formation and leukocyte accumulation. C3a and C5a are called anaphylotoxins, and the latter is an important chemoattractant for neutrophils, acting via a specific C5a receptor. Using homologous recombination to knock out the murine C5a receptor suggests that this receptor, and thus the activity of C5a, is of particular importance in mucosal host defence in the lung.

Probably the most important role of the complement cascade is to promote the uptake and destruction of pathogens by phagocytic cells. This occurs by the binding of activated complement components to the bacterial surface (the process known as opsonization) and their recognition by cell surface receptors (complement receptors) on phagocytes. The major complement opsonin is C3b and inactive forms of C3b (iC3b, C3d, C3dg) can bind to the four receptors on phagocytes called complement receptors (CR 1 to 4). Of interest is the finding that fusion of C3dg to the protein hen egg lysozyme (HEL) resulted in this complex producing an antibody response some 10 000 times greater than that of HEL itself. This suggests that complement components may play a role in potentiating antibody responsiveness to the components they bind to, such as bacterial cell surface molecules.

The clearest indication of the importance of the complement cascade, and of the individual actions of this multifunctional pathway, has come from clinical research on complement deficiencies. As expected, deficiency in C3 causes recurrent infections in affected individuals who also show signs of immune complex disease. One of the important functions of the complement pathway is the clearance of immune complexes. Deficiencies in the classical complement components (C1, C4, C2) are associated with immune complex disease and also recurrent infections. Deficiency in MBP is associated with infections in children. Deficiency in alternative pathway components is associated with recurrent upper respiratory infections (factor D) or severe meningitis (properdin). Surprisingly, in light of the potential damage that the membrane attack complex could do, deficiency in the terminal components of the complement cascade has only been associated with susceptibility to *Neisseria* species – the causative agent of gonorrhoea.

Bacterial defences against complement

Bacteria, viruses and parasites have developed defence mechanisms against the activated complement cascade. Such defences can be passive. For example, Gram positive bacteria have very thick cell walls which are resistant to complement attack. Some Gram negative bacteria lack sites for the binding of C3b. The role of the bacterial capsule in preventing binding of opsonic complement components (and also the MAC) has already been explained. The bacterial S layer (see Chapter 2) has recently been reported to prevent binding of C3b to bacteria. In other bacteria C3b is bound on long external chains of peptidoglycan which prevents the build-up of the MAC on the bacteria cell membrane. Bacterial proteinases probably inhibit complement activity by inactivating complement components (e.g. C3b, C4b and the MAC) on the bacterial surface. Bacteria proteases can also inactivate the proinflammatory anaphylatoxins such as C5a.

The binding of C3b and C4b to bacteria can also be prevented by so-called decoy molecules which bind to these opsonins. Viruses also encode proteins with similar activity. For example, poxviruses encode a protein (termed vaccinia complement control protein, VCP) which has sequence similarity to the regulator of complement activation (RCA) superfamily. VCP binds C3b and C4b and blocks activation of the classical and alternative pathways of complement activation. A second protein with sequence similarity to the RCA superfamily has also been found but has not been demonstrated to bind complement proteins. *Candida albicans* produces an integrin-like protein which binds iC3b.

Thus the complement pathways represent another defence system which has co-evolved with the microbial world and which shows certain 'chinks in its armour'.

NATURAL RESISTANCE-ASSOCIATED MACROPHAGE PROTEIN (Nramp)

There is increasing interest in determining the genetics of susceptibility and resistance to infections. For example, it has recently been established that the ability of certain individuals to remain uninfected by human immunodeficiency virus (HIV) is due to a homozygous 32 bp deletion in the gene encoding the chemokine receptor CCR5. This receptor, along with other receptors for chemokines, is involved in the uptake of HIV into CD4 T cells and the mutation presumably inhibits viral entry into cells.

It has been known for some years that in the mouse resistance/susceptibility to infection with several intracellular pathogens, including mycobacterial species, *Salmonella typhimurium* and the protozoan *Leishmania donovani*, was under genetic control and the locus was identified on chromosome 1 and was termed *Bcg*. The *Bcg* gene is present in two allelic forms: Bcg^r which is dominant and confers resistance and Bcg^s which confers susceptibility to infection and is recessive. Positional cloning was used to identify this gene, which was named *Nramp1* (natural resistance-associated macrophage protein 1), that encodes a highly hydrophobic macrophage membrane-associated protein with 12 predicted transmembrane domains and a sequence motif found in several prokaryotic and eukaryotic transport proteins. A second *Nramp* gene has been cloned and this highly related protein has been demonstrated to be a divalent cation transporter for Fe^{2+}, Mn^{2+} and Zn^{2+}. *Nramp* homologues have been identified in *Drosophila*, the worm *Caenorhabditis elegans* and in plants. A homologous gene has even been found in several bacterial species, including *M. leprae*.

What function do these proteins play in protection agains intracellular pathogens? Both Nramp1 and Nramp2 proteins are believed to be membrane-associated and involved in the transport of divalent cations, including iron, manganese, magnesium, copper and zinc. One suggested mechanism for these proteins proposes that they function within the phagosomes containing living pathogens and act to remove cations required for the growth of these parasites. The finding that several bacterial species have homologous systems suggests that there has been parallel evolution of this transport system and that both may act in competition within the bacteria-containing phagosome. In addition, it has been found that in animals in which the *Nramp1* gene had been inactivated, the pH of the phagosomes containing live *M. bovis* failed to show the expected reduction in pH. This suggests that *Nramp1* plays a key role in modulating phagosomal pH with the consequent effects on the replication of internalised pathogens.

The human *Nramp1* gene has been cloned and a number of polymorphic variants have been identified, including three polymorphisms that localize to potential regulatory sequences in either the promoter or 3' untranslated regions. Preliminary population studies suggest that allelic variants in the human *Nramp1* locus are associated with susceptibility to leprosy and tuberculosis.

CYTOKINES: THE INTEGRATORS OF IMMUNITY

It is clear that innate immunity requires the participation of many cell populations, including epithelial cells, various fixed tissue and circulating leukocytes, vascular endothelial cells and probably a range of other cells such as fibroblast populations, which will be involved in tissue repair should it be necessary. The activities of these various cells have to be controlled both in time and space. As described in Chapter 3 one, if not the major, group of integrating biological signals is the very large 'family' of proteins known as cytokines. These proteins have the ability to bind to high-affinity receptors on all the cells present in our bodies and to produce a vast range of cellular effects. Cytokines were discovered by biologists studying inflammation and are clearly the major controlling signals in both acute and chronic inflammation and in innate and acquired immunity. As described in Chapter 3, the major level of control by cytokines is the cytokine network. We know very little about how such networks are controlled and it is likely that such knowledge will only become available when it is possible to mathematically model the immensely complex interactions of cytokines with cells.

Cytokines in the control of the innate response to infection

Determining the role of individual cytokines in inflammation is often difficult due to the enormous number of cytokines that can participate in particular inflammatory responses. An obvious way to define their role would be to try and ablate their activity. This can be done with antibodies, but the results can be difficult to interpret as the effects may only be short lived due to immune responses to the injected antibody. A much more satisfactory method would be to inactivate the gene for the cytokine under study in an appropriate experimental animal. This is now possible using a genetic technique called homologous recombination in which a segment of DNA is inserted within the gene of interest, leading to the synthesis of an inactive protein.

Cytokine gene knockouts

Homologous recombination produces animals in which a particular gene is 'knocked out' and the animals, in consequence, are known as gene knockouts. The methodology to produce such knockouts has been described in Chapter 4. A large number of cytokine gene knockouts have now been reported. The consequences of knocking out individual cytokines is difficult to predict and certain unusual results have been reported in the literature. For example, the ablation of the activity of the growth-promoting cytokine transforming growth factor (TGF)α produces mice in which the major phenotypic change is that they have wavy hair and whiskers and a propensity to develop ulceration of the cornea. However, the expected changes – failure of embryonic development or impairment of wound healing in the adult – failed to materialize. The ability of these TGFα knockout mice to develop normally may be due to the action of an overlapping cytokine, epidermal growth factor (EGF), which can bind to TGFα receptors. The role of EGF as a growth factor for *M. tuberculosis* is discussed in Chapter 10. This is an example of functional redundancy in the cytokine network. In contrast, knockout of the cytokine transforming growth factor (TGF)β, which is one of the cytokines showing significant anti-inflammatory activity, results in animals which survive to term and then rapidly develop a diffuse, lethal, inflammatory syndrome characterized by massive mononuclear infiltration of the heart, lungs and gastrointestinal tract. This is in spite of the fact that there are about 30 proteins in the TGFβ superfamily.

Certain cytokine or cytokine receptor knockouts result in susceptibility to microbial infections. For example, mice deficient in active IFNγ remained healthy in the absence of pathogens but were susceptible to normally sublethal doses of the intracellular bacterium *Mycobacterium bovis*. Knockout of NF-IL-6, the major transcriptional control element of the IL-6 gene, renders mice highly susceptible to infection with *Lis. monocytogenes*. Mice lacking the 55 kDa TNFα receptor (TNF-R55, TNFR1) showed the same response to this intracellular bacterium. Knockout of the TNF-R55 receptor or of the cysteine protease ICE (caspase 1, see Chapter 2), which converts the inactive IL-1β precursor to the active form, renders mice insensitive to the effects of endotoxin. However, if the gene for IL-1β is ablated, animals respond normally to LPS. This discrepancy between the IL-1β knockout mice and the ICE knockouts may be due to the fact that in the latter there is also a deficiency in production of IL-1α. IL-1β knockout mice, however, do show impaired inflammatory responses to turpentine and an impaired contact hypersensitivity response to trinitrochlorobenzene. Knockout of the gene encoding IL-6 rendered mice deficient in the local inflammatory response to a tissue-damaging agent – turpentine – but failed to alter the pathological responses animals could mount to TNFα or LPS. As described, IL-1, IL-6 and TNFα are the major endogenous pyrogens responsible for the fever response to bacterial products. Thus to further complicate the issue of the role of proinflammatory cytokines in the acute-phase response, the injection of LPS or IL-1β into IL-6-deficient mice failed to produce fever, although the same treatment caused fever in wild-type mice. Intracerebroventricular injection of IL-6 into IL-6 knockouts induced fever but similar injection of IL-1 was without effect. This suggests that centrally produced IL-6 is an obligatory factor in the fever response and that IL-6 acts downstream from both peripheral and central IL-1β.

Cytokines in the acute-phase response

The entry of bacteria into tissues releases a range of signals, some of which have already been discussed and others will be discussed below. It is likely that all cells in a tissue can recognize such bacterial signals, but certain cells, fixed tissue macrophages and dendritic cells may be the most responsive. Apart from LPS, we know very little about the bacterial molecules which stimulate cytokine synthesis, and yet many components of bacteria are as potent or are more potent than LPS in stimulating cytokine synthesis. These bacterial components induce local cells to synthesize and secrete two major proinflammatory cytokines: IL-1 and TNFα. These two proteins appear to be the pivotal cytokines on which the acute-phase response balances and their production results in: (i) the local synthesis of chemokines (see Chapter 3); (ii) the activation of vascular endothelial cells; (iii) the release of cytokines into the circulation (e.g. IL-6, IL-11, leukaemia inhibitor factor (LIF) and oncostatin); and (iv) the activation of both the brain and the liver to produce fever and the upregulation of the synthesis of acute-phase proteins (Figure 8.14). In addition to opsonization, the acute-phase proteins can decrease the plasma levels of iron and zinc (required by bacteria for growth) and increase the levels of copper, which is bacteriostatic. It is not clear how this modulation of cation concentrations meshes with the *Nramp* system described in the previous section. The increase in body temperature is believed to inhibit the growth of the invading bacteria

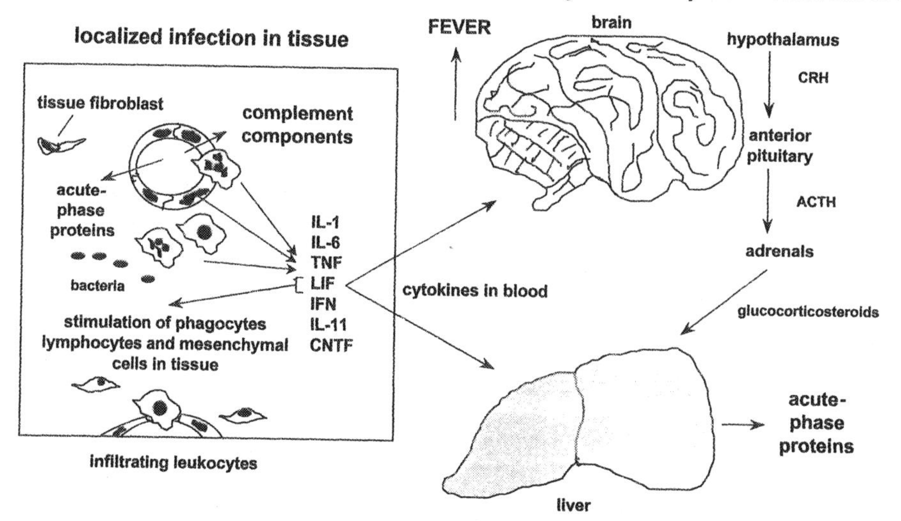

Figure 8.14 Systemic effects in the acute-phase response (reproduced from Henderson *et al.* (1998) Bacteria-Cytokine Interactions in Health and Disease, Portland Press, London, with permission)

and to increase the efficiency of the body's immune defences. The local production of cytokines can also activate coagulation pathways, causing the local production of fibrin which seals off the site of infection. This is a mechanism used by primitive organisms to deal with infections.

The attraction of leukocytes, complement components and acute-phase proteins to the site of infection (plus antibodies if this is a recurring infection) allows the invading bacteria to be opsonized, phagocytosed by leukocytes and directly killed by the complement systems. During this process there is the constant production of cytokines, including additional chemokines and interferons. Once the infection begins to be brought under control it is envisaged that the pattern of cytokines produced locally changes and that there is an increase in the synthesis of anti-inflammatory cytokines which begin to bring the activity of infiltrating leukocytes under control and promote their apoptosis. The change in cytokine networks in inflamed tissues may be stereotyped or it may be dependent on changes in the amount or composition of the molecules being released from bacteria. In recent years a number of cytokines have been discovered which act to inhibit the actions of inflammatory cells such as macrophages. Examples of such cytokines are interleukin 1 receptor antagonist (IL-1a), IL-4, IL-10, IL-13 and TGFβ_1. These cytokines can act to inhibit the actions of macrophages and dendritic cells. Tissue repair is also an important process in inflammation and the local production of growth factors must occur during the final stages of inflammation (Table 8.6). However, the reader should be warned that while we have a reasonable understanding of what induces inflammation we know very little about the mechanisms which result in the resolution of inflammation. An understanding of the latter would allow much better control of the many chronic inflammatory conditions which afflict mankind.

Bacterial components and their role in controlling inflammation

The role of evolutionarily stable bacterial molecules such as LPS and peptidoglycan have already been discussed as part of the innate immune system's recognition system. The

Table 8.6 Cytokines in acute phase response to simple infection

Initiation of inflammation
IL-1α
IL-1β
TNFα
Cytokines interact and may synergize with each other and with bacterial factors such as LPS
Perpetuation of inflammation
Chemokines
IL-8
Groα
Groβ
MIP-1α
MIP-1β
MCP-1,2,3
IL-6
IFNs
Cessation of inflammation/controlling cytokines
IL-1ra
IL-4
IL-10
IL-13
TGFβ
Cytokines involved in growth and repair
Colony-stimulating factors
EGF
TGFα
FGF
PDGF
TGFβ family

molecules have been termed PAMPs and have been proposed to interact with host cells, and the major outcome of this interaction is the synthesis of cytokines. LPS has already been discussed as a major bacterial proinflammatory signal, acting via its ability to stimulate virtually every cell in the body to produce cytokines. For many years LPS was believed to be the key inflammatory signal of Gram negative bacteria. It was known that Gram positive bacteria also caused severe inflammation but it was not known what components were responsible. However, since the beginning of the 1990s it has become apparent that Gram negative and Gram positive bacteria produce many components of every biochemical class (proteins, carbohydrates, lipids, nucleic acids) which are able to stimulate mammalian cells to produce proinflammatory cytokines and are therefore potentially able to contribute to the inflammatory actions of bacteria. The finding of such a large collection of bacterial molecules able to induce host cells to release cytokines has suggested to the authors that, in addition to acting to warn the host that it has been infected, these bacterial molecules may also be virulence factors in their own right.

Modulins: a new bacterial virulence factor

One plausible hypothesis to explain the evolution of the many cytokine-inducing bacterial molecules discovered in the past few years is that they function as virulence factors. The authors have named these molecules modulins by analogy with the other known classes of virulence factors (adhesins, impedins, aggressins and invasins: see Chapters 1, 2 and 5). The term modulin implies that by inducing cytokine synthesis these bacterial molecules *modulate* the activity of the cytokine-producing cells and of bystander cells. This concept is shown schematically in Figure 8.15. LPS would be the prototypic modulin. In addition to LPS, a very large number of bacterial-derived molecules have the capacity to induce a range of mammalian cells (myleoid, lymphoid, mesenchymal, epithelial) to produce various cytokines (Table 8.7). Some of these components are well known and include peptidoglycan, teichoic acids and lipoarabinomannan which are all PAMPs. Others are unexpected inducers of proinflammatory cytokine synthesis. For example, bacterial molecular chaperones (also known as heat shock or stress proteins) are potent inducers of cytokine synthesis. The molecular chaperones that have received most attention are the chaperonins (cpns), which are oligomeric proteins composed of subunits of 10 (cpn 10) and 60 (cpn 60) kDa molecular mass. Two of the authors (BH and MW) have recently demonstrated that complete proteolysis of the *E. coli* chaperonin 60, a molecule known as groEL, by trypsin, has no affect on the ability of this protein to stimulate human peripheral blood mononuclear cells to secrete IL-1 and IL-6. Bacteria contain large amounts of these chaperonins and the death of bacteria and release of these proteins (even if they are broken down by proteases) would be a major proinflammatory stimulus, which has previously been unrecognized.

A surprising recent finding is that bacterial DNA can act as a potent inducer of cytokine synthesis and can even provoke a septic shock-like state in experimental animals. Perhaps the most surprising finding is that all bacterial exotoxins tested are able to induce cytokine synthesis and that the most potent bacterial cytokine inducers are bacterial exotoxins (see also Chapter 7). Indeed, the most potent inducer of cytokine synthesis reported is the lethal

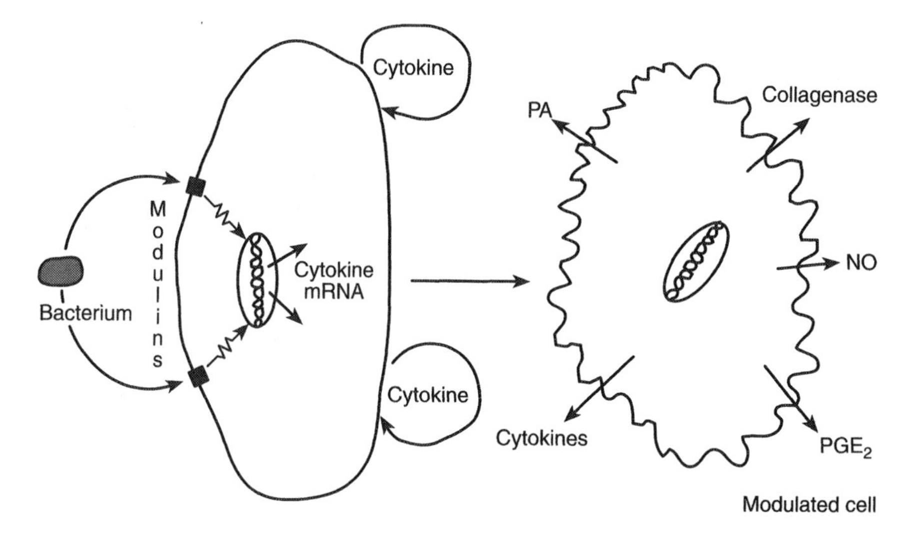

Figure 8.15 The modulin concept. Bacteria contain and release a wide range of molecules which can activate host cells to produce cytokines. It is proposed that host cells have receptors for these various components (either PRRs or other receptors) and that binding to these receptors switches on the production of various cytokines. Now cytokines have the propensity for binding in an autocrine manner. That is, they can bind and activate the cell which has produced them. In this manner the cytokines modulate the function of the cell. One obvious way of doing this is to activate the cell such that it produces proinflammatory mediators such as the prostanoids, nitric oxide (NO), metalloproteinases or other cytokines. This cell activation is proposed to be a virulence mechanism

Table 8.7 Modulins produced by bacteria

Bacterial molecule	Lowest concentration inducing cytokine synthesis (ng/ml)	Mechanism of action
Cell surface polysaccharides	2500	CD14-d
Teichoic acid	1000	CD14-d
Protein A	1000	NCD14-d
Lipoarabinomannan	100	CD14-d
Peptidoglycan	100	CD14-d
Peptidoglycan fragments	50	CD14-d
Porins	10	NCD14-d
Fimbrial proteins	10	NCD14-d
Superantigens	1–10	NCD14-d
Molecular chaperones	1–10	NCD14-d
Lipid A-associated proteins	1	NCD14-d
Lipoproteins	0.5	?
Lipids	0.1	?
Glycoproteins	0.05	?
Exotoxins	1 to >0.0001	NCD14-d
DNA	?	?

CD14-d–CD14-dependent NCD14-d – not CD14-dependent

factor protein of the anthrax exotoxin triple complex. This molecule has been reported to be active at attomolar (10^{-18} M) concentrations.

Many of the carbohydrates, proteins, lipoproteins, glycoproteins, nucleic acids and lipids of bacteria can stimulate cytokine synthesis. It is unlikely that all of these components are also PAMPs as many of them are not evolutionarily stable, a major prerequisite for these bacterial components. Could the evolution of these modulins have been a bacterial evolutionary strategy to confuse the immune system and prevent the generation of properly co-ordinated immunological responses to bacterial infection? The answer to this question is currently not available as we do not know what pattern of cytokines each of the many bacterial modulins produce. The ability of these various bacterial molecules to induce cytokine synthesis also implies that eukaryotic cells have many receptors for bacterial components.

One of the consequences of the realization that all bacteria produce a range of cytokine-inducing molecules was that it resulted in a reappraisal of the interaction which must be going on between the normal microflora and the 'external' epithelial surfaces of the body. On mucosal surfaces such as in the colon, large numbers of bacteria appear to live harmoniously in contact with the epithelial surface. Given that the bacteria constituting the normal microflora contain, and must release, proinflammatory cytokine-inducing molecules and that the intestinal epithelial cell is known to be an active producer of cytokines, the expectation is that all epithelial surfaces should be inflamed, due to the response of the epithelial cells to the normal microflora. However, this is patently not the case. A possible explanation for this paradox is provided in Chapter 10, which looks at the future trends in cellular microbiology.

Modulation of cytokines as an immune evasion mechanism

Having indicated the enormous importance of cytokines in innate (and, as will be described in Chapter 9, acquired) immunity it would be sensible to predict that microorganisms would have evolved mechanisms to manipulate cytokine networks to their own advantage. This is exactly what has been found during the past decade with viruses, and evidence is now emerging that bacteria can produce proteins which alter cytokine synthesis and activity.

Anti-cytokine actions of viruses

During the last decade viruses, paticularly the double-stranded DNA viruses with large genomes, have been found to encode a growing number of proteins which are able to combat host immune responses by inhibiting cytokine networks. These proteins have three general mechanisms of action: (i) inhibition of the synthesis or release of cytokines; (ii) antagonism of cytokine–receptor interactions; and (iii) functional antagonism of cytokine networks through homologues of anti-inflammatory cytokines. Viruses have evolved a number of specific mechanisms for dealing with the antiviral cytokines – the interferons. It is also emerging that viruses can encode certain chemokine receptors which are expressed on the surface of the infected cell and can respond to the respective ligands.

As has been described in other chapters, monocytes contain a unique cysteine proteinase which cleaves the 31 kDa inactive precursor form of IL-1β at Asp^{116}–Ala^{117} to produce the 17 kDa active form of this cytokine. This proteinase has been termed pro-IL-1β-converting enzyme (ICE) and the cloned active form of the enzyme is a cysteine proteinase composed of two subunits of molecular masses 10 and 20 kDa, with the larger subunit containing the active site. ICE is homologous to the *ced*-3 gene of the nematode worm *Caenorhabditis elegans*, a gene encoding a cysteine proteinase essential for the control of apoptosis in this organism and in the mammal, and is one of at least 12 proteinases termed caspases which are now recognized to induce apoptosis when overexpressed in cells (see Chapter 2). Cowpox virus contains the gene *crm*A and vaccinia virus the gene *B13R* (the protein produced is known collectively as SPI-2) which encodes a serpin inhibitor (a member of a class of proteins best known for inhibiting serine proteases). This viral protein inhibits ICE by forming a tight complex with an equilibrium constant of inhibition of greater than 4×10^{-12} M. Thus cells infected with vaccinia virus cannot produce the active IL-1β. However, it is not clear whether this inhibition has significant effects on IL-1β-driven pathology and it may be more concerned in the blocking of apoptosis. Baculoviruses, which infect insects, encode two apoptosis inhibitors: p35 and IAP. The former is believed to act, like SPI-2, as a direct protease inhibitor, whereas IAP may act, not directly on ICE but to prevent activation of this, and other, apoptotic proteases. In addition to inhibiting IL-1β these ICE-inhibiting viral proteins will also block the formation of active IL-18 (also known as IL-1γ) which is produced in a like manner to IL-1β. IL-18 is a potent inducer of IFNγ and is involved in the production of Th_1 cells (see Chapters 3 and 9).

The second set of gene products expressed by these double-stranded DNA viruses are soluble forms of cytokine receptors. Proteins with the capacity to bind IL-1, TNFα, IFN-γ and various chemokines (IL-8, MIP-1α, RANTES, SDF-1) have been described. In addition, hepatitis B virus contains an envelope protein with no homology to the IL-6 receptor but which binds IL-6. Vaccinia virus contains another gene that encodes the protein B15R, which has 30% identity with the type II non-signalling IL-1 receptor. This protein binds with high affinity to IL-1β and competes with this cytokine for binding to the cell surface IL-1 receptor. However, this viral protein did not not bind to IL-1α, giving an evolutionary view of how this virus regards the IL-1 family members. It is of interest that the type II IL-1 receptor does not transduce a signal and has been termed a decoy receptor and probably acts to regulate the action of IL-1 at the cell surface. Vaccinia virus produces large amounts of this IL-1 receptor, with more than 10^5 molecules being produced every 24 hours by infected cells. This soluble receptor blocks the biological effects of IL-1β and is therefore anti-inflammatory. When the gene encoding B15R was disrupted, the mutant virus produced increased morbidity. Strikingly, viruses in which B15R was deleted caused an elevation in body temperature which was not seen with the wild-type virus. Thus B15R appeared to be acting to inhibit the febrile response. These findings support the hypothesis that proteins of this type have a role in ensuring that the host survives its own inflammatory response to the infection. Evolution is a mechanism to ensure survival of DNA and thus the evolution of viral genes which allow the host to survive viral infection, and as a consequence enhance viral survival, would be expected to be favoured. This co-evolution of genes which favours survival of both the parasite and the host has not been studied in any detail but will, we predict, be a common theme in host–parasite interactions.

As has been referred to in this, and other, chapters, TNFα is a major proinflammatory cytokine. Thus it is not surprising that viruses have evolved mechanisms for neutralizing this cytokine. For example, adenoviruses encode at least four genes that inhibit the effects of TNF. Epstein–Barr virus (EBV) encodes a soluble receptor for macrophage colony-stimulating factor (M-CSF).

One of the most striking findings in the study of viral 'cytokine genes' has been the number of receptor-like proteins they produce that can inhibit the activity of chemokines. The biological activity of chemokines has been described in Chapter 3 and, to reiterate, this cytokine 'family', which is one of the largest and now includes about 30 proteins, are promoters of leukocyte movement to inflammatory sites. Thus they are fundamental in the induction of localized inflammation and their inhibition would have serious consequences for the host. Viruses such as human cytomegalovirus and herpesvirus saimiri contain genes which encode soluble forms of receptors for a range of chemokines, including IL-8. An interesting response to the knockout of a chemokine gene (MIP-1α – chemotactic for monocytes, basophils and eosinophils) has recently been reported. Mice infected with coxsackievirus develop a serious, sometimes fatal, inflammation of the heart and those animals infected with influenza virus develop pneumonia. Transgenic mice lacking the gene for MIP-1α, when infected with these viruses, failed to show serious inflammatory pathology of the heart or lungs, yet were still able to cope with the viruses and resolve infections. This again raises the question of whether the viral inhibition of a host defence molecule benefits the virus or the host.

The third method developed by viruses to, apparently, combat host defence systems is the production of anti-inflammatory cytokines. The only example of this strategy, to date, is the presence of genes in (EBV) and equine herpesvirus type 2 (EHV2) for the anti-inflammatory cytokine IL-10. This molecule has a major inhibitory effect on monocytes and Th_1 lymphocytes and its administration to animals can inhibit a number of experimental conditions, including lethal endotoxin shock. In the early 1990s it was discovered that the human (h) and murine (m) IL-10 genes exhibited marked homology to an open reading frame in the EBV genome, BCRF-1. Surprisingly, the protein sequences of hIL-10 and BCRF-1 are 84% identical. Comparison of the mouse, human and viral sequences show that the human and viral sequences are the most closely related at the protein level, while mIL-10 and hIL-10 are significantly more homologous in their DNA sequences. Such sequence relationships suggest that the mIL-10 and the hIL-10 genes evolved from a common ancestor. The amino acid sequence of BCRF-1, showing such identity with the human protein, suggests that the ancestor of BCRF-1 may have been a captured, processed host IL-10 gene. The equine virus EHV2 also encodes a viral homologue of IL-10. Viral IL-10 appears to play a crucial role during the early transformation of B cells by EBV and transformed B cells produce large amounts of human IL-10 The viral IL-10 product of the BCRF1 gene shares the cytokine synthesis inhibitory activity of murine IL-10 although it has a lower specific activity and this may be important in suppressing the antiviral activities of IFNγ and macrophages. However, viral IL-10 lacks the activities of mammalian IL-10s in co-stimulating thymocyte and mast cell proliferation and inducing class II MHC expression on B cells. These are obviously activities which may militate against virus survival and, if BCRF1 is a captured human gene, have presumably been evolved out of the original gene structure.

The interferons (IFNs) are potent antiviral proteins and viruses have evolved a number of mechanisms for overcoming their biological effects. Induction and release of IFNs leads to cell binding and the transcription of IFN-responsive genes. Viruses encode proteins which inhibit such IFN-responsive genes. They also produce soluble forms of interferon receptors.

The mechanisms by which viruses can ablate the actions of cytokine are shown in Figure 8.16. and the range of viral anti-cytokine genes is shown in Chapter 10, Table 10.3.

Anti-cytokine actions of bacteria

The long-term neglect of bacteria and the much larger genome sizes of these cells, compared to that of viruses, has meant that we know very little about how bacteria control the cytokine networks that are being used in the defence of the host. The finding that bacteria produce so many different cytokine-inducing components may offer a clue to the likelihood that bacteria can inhibit host defences by controlling the synthesis of cytokines. In Chapter 10 the argument is forwarded that the commensal bacteria must produce proteins which inhibit mucosal proinflammatory cytokine networks. In support of this argument is the finding that a number of bacteria have been reported to produce proteins which inhibit the production of one or other cytokine (Table 10.4). However, few of the bacteria shown in this table are actually members of the commensal microflora. The other bacteria may therefore utilize these cytokine-inhibiting proteins in order to overcome host inflammatory defence mechanisms. A number of bacterial exotoxins are able to inhibit the synthesis of TNFα but not that of IL-6, suggesting that those bacteria producing these toxins may require the upregulation of IL-6 (which has negative immunomodulatory actions) and the downregulation of TNF to evade lethal inflammatory host responses.

One organism which targets TNFα is *Y. enterocolitica,* a bacterium already discussed in other chapters and which utilizes a 70 kb virulence plasmid (pYV). This plasmid encodes the *Yersinia* outer proteins (Yops) which are secreted by a plasmid-encoded type III secretion system. It has been shown that *Yersinia* inhibits the production of TNFα and IFNγ. The mechanism by which this bacterium does this has recently been partially elucidated. We discussed at the begining of this chapter the role of NF-κB in host defence and cytokine induction. The active NF-κB, which is a heterodimer (p50/p65), plays a central role in immunological processes controlling the expression of various cytokine genes: IL-1, TNF, IL-8, GM-CSF and, as has been discussed earlier, antibacterial peptides. NF-κB activity is controlled by inhibitory proteins I-κBα and I-κBβ which bind to NF-κB and prevent nuclear translocation of the complex. Degradation of these inhibitory proteins is necessary to enable NF-κB to enter the nucleus and induce cytokine transcription. *Y. enterocolitica* acts to inhibit the breakdown of these inhibitory proteins and thus prevent the transcriptional upregulation of the genes for cytokines like TNFα.

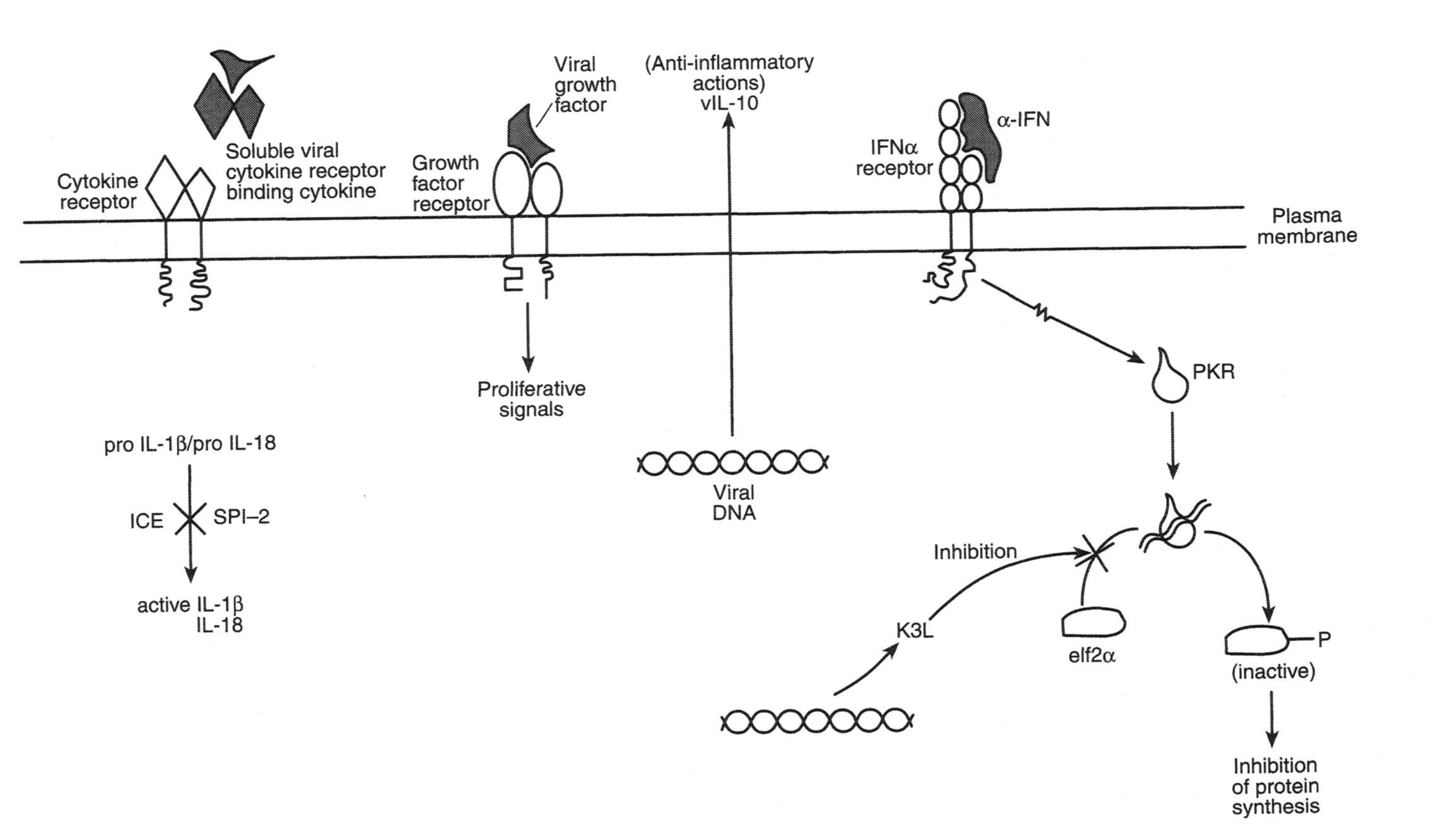

Figure 8.16 Schematic diagram showing some of the major mechanisms used by viruses to inhibit cytokine-driven host defences. Viruses encode a wide range of soluble cytokine receptors which are released by the infected cell and which can bind and inactivate proinflammatory cytokines (IL-1, TNFα, IFNs, chemokines) in the vicinity of the viral-laden cell. Certain viruses encode the anti-inflammatory cytokine IL-10, which is secreted from the virally infected cell and can downregulate the actions of nearby macrophages and dendritic cells. A few viruses encode inhibitors of ICE (caspase 1). The major effect of such proteins is probably inhibition of apoptosis although it could also inhibit the production of active IL-1β and IL-18. Other viruses encode growth factors which can stimulate the proliferation of virally infected cells. Viruses have a number of means of evading the activity of the interferons. One mechanism is described. Binding of IFNα to cells stimulates the production of a 68 kDa protein called PKR (RNA-activated protein kinase). PKR is inactive until it binds double-stranded DNA, which is often found in virally infected cells. Once activated, PKR phosphorylates (and thus inactivates) a cellular factor, eIF2α, which is required by ribosomes for the initiation of protein translation, and thus stops cellular protein synthesis. Vaccinia virus encodes a protein, K3L, which binds double-stranded DNA and prevents the activation of PKR

Other bacteria have apparently evolved proteins which can inhibit the synthesis of a variety of lymphocyte-activating cytokines – the lymphokines. Thus certain enteropathogenic strains of *E. coli* produce a protein (lymphocyte inhibitory protein) which inhibits the synthesis of IL-2, IL-4, IL-5 and IFNγ. *Act. actinomycetemcomitans*, an oral Gram negative opportunistic pathogen, produces a 14 kDa protein which inhibits the synthesis of the same lymphokines, posing the question – are these two bacteria producing the same protein? Much more work needs to be done before we will know the full answer to the question – how effective are bacteria at controlling host cytokine networks.

CONCLUSIONS

The system of cells and soluble mediators which constitutes innate immuity is all that stands between invertebrates and death from infection. The success of invertebrate colonization of the planet is a tribute to the effectiveness of these innate defence mechanisms. In mammals, the innate mechanisms have evolved into complex networks of interacting cellular and humoral systems which are closely integrated by the array of cell-to-cell signals known as cytokines. The evolution of these systems has obviously been shaped by the nature and activities of the microbes that they are defending against, but we have little knowledge of the dynamics of this co-evolutionary battle in time and space. While innate immunity can defend us it is also clear that bacteria can evolve strategies for neutralizing aspects of this defence. An excellent example of bacterial evolution in our own century is the development of antibiotic resistance which, in the case of penicillin, appeared within weeks of the first publication of its clinical use and has increased relentlessly since. It is therefore of great interest that the recently discovered innate defence mechanism relying on the production of antibacterial peptides still appears to show limited signs of resistance. Is this due to the mechanism of action of these peptides or do multicellular organisms have some mechanism for producing novel peptides? In this context the report that the coccinellid beetle *Epilachna borealis* uses combinatorial chemistry to produce a diverse array of defensive polyamines is of interest. Combinatorial chemistry is a process for producing random diverse chemical structures and is now widely used by the world's drug industry to produce novel pharmaceuticals. The possibility that similar combinatorial approaches are being used to generate antibacterial compounds would be worth exploring.

In mammals the innate immune response forms a continuum with the more recently evolved acquired system. In the next chapter this system and the mechanisms used by bacteria to defeat acquired immunity will be discussed.

FURTHER READING

Books

Abraham JS, Wheeler JG (eds) (1993) *The Neutrophil*. Oxford University Press, Oxford.

Aggarwal BB, Puri RK (eds) (1995) *Human Cytokines: Their Role in Disease and Therapy*. Blackwell, Cambridge, MA.

Goode J (ed.) (1994) *CIBA Foundation Symposium No. 186. Antimicrobial Peptides*. Wiley, Chichester.

Henderson B, Poole S, Wilson M (1998) *Bacteria/Cytokine Interactions in Health and Disease*. Portland Press, London.
Ibelgaufts H (1995) *Dictionary of Cytokines*. VCH, Weinheim.
Janeway CA, Travers P (1997) *Immunobiology: The Immune System in Health and Disease* (3rd edn). Current Biology/Garland Publishing.
Law SKA, Reid KBM (1995) *Complement* (2nd edn). Oxford University Press, Oxford.
Male D, Cooke A, Owen M, Trowsdale J, Champion B (1996) *Advanced Immunology*. Mosby, London.
Meager T, (1998) *The Molecular Biology of Cytokines*. Wiley, Chichester.
Mims CA, Dimmock NA, Nash A, Stephen J (1995) *Mims' Pathogensis of Infectious Diseases*. Academic Press, London.
Paradise LJ, Bendinelli M, Friedman H (1996) *Enteric Infections and Immunity*. Plenum Press, New York.
Playfair J (1995) *Infection and Immunity*. Oxford University Press, Oxford.
Poole RK (ed.). (1995) *Advances in Microbial Physiology*. Academic Press, London.
Salyers AA, Whitt DD (1994) *Bacterial Pathogenesis: A Molecular Approach*. ASM Press, Washington, DC.
Thomas L (1974) *The Lives of a Cell: Notes of a Biology Watcher*. Viking Press, New York.
Volk WA, Gebhardt BM, Hammarskjöld M-L, Kadner RJ (1996) *Essentials of Medical Microbiology* (5th edn.) Lippincott-Raven, Philadelphia.

Reviews

Boman HG (1995) Peptide antibiotics and their role in innate immunity. *Annu Rev Immunol* 13: 61–92.
Dinarello CA (1989) Was the original endogenous pyrogen interleukin-1? In *Interleukin-1, Inflammation and Disease*. (eds Bomford R, Henderson B), pp17–28. Elsevier, Amsterdam.
Govoni G, Gros P (1998) Macrophage NRAMP1 and its role in resistance to microbial infections. *Inflamm Res* 47: 277–284.
Henderson B, Poole S, Wilson M (1996) Bacterial modulins: a novel class of virulence factors which cause host tissue pathology by inducing cytokine synthesis. *Microbiol Rev* 60: 316–341.
Henderson B, Poole S, Wilson M (1996) Microbial/host interactions in health and disease: who controls the cytokine network? *Immunopharmacology* 35: 1–21.
Martin E, Ganz T, Lehrer RI (1995) Defensins and other endogenous peptide antibiotics of vertebrates. *J Leucoc Biol* 58: 128–136.
Medzhitov R, Janeway CA (1997) Innate immunity: the virtues of a nonclonal system of recognition. *Cell* 91: 295–298.
Smith GL, Symons JA, Khanna A, Vanderplasschen A, Alcami A (1997) Vaccinia virus immune evasion. *Immunol Rev* 159: 137–154.

Papers

Goldman MJ, Anderson GM, Stolzenberg ED *et al.* (1997) Human β-defensin-1 is a salt-sensitive antibiotic in lung that is inactivated in cystic fibrosis. *Cell* 88: 553–560.
Hackam D, Rotstein OD, Zhang W-J, Gruenheid S, Gros P, Grinstein S (1998) Host resistance to intracellular infection: mutation of natural resistance-associated macrophage protein 1 (*Nramp*1) impairs phagosomal acidification. *J Exp Med* 188: 351–364.
Kirshning CJ, Wesche H, Ayres TM, Rothe M (1998) Human Toll-like receptor 2 confers responsiveness to bacterial lipopolysaccharide. *J Exp Med* 188: 2091–2097.
Lemaitre B, Nicolas E, Michaut L, Reichhart J-M, Hoffman JA (1996) The dorsoventral regulatory gene cassette spatzle/Toll/cactus controls the potent antifungal response in *Drosophila* adults. *Cell* 86: 973–983.
Lemaitre B, Reichhart J-M, Hoffmann JA (1997) *Drosophila* host defence: differential induction of antimicrobial peptide genes after infection by various classes of microorganisms. *Proc Natl Acad Sci USA* 94: 14614–14619.
Schroder C, Farmer JJ, Attygale AB, Smedley SR, Eisner T, Meinwald J (1998) Combinatorial chemistry in insects: a library of defensive macrocyclic polyamines. *Science* 281: 428–431.
Smith JJ, Travis SM, Greenberg EP, Welsh MJ (1996) Cystic fibrosis airway epithelia fail to kill bacteria because of abnormal airway surface fluid. *Cell* 85: 229–236.

CHAPTER 9

Acquired Immunity in the Defence Against Bacteria

INTRODUCTION

It is estimated that some 400 million years ago there evolved a second system of immunity which has been variously called acquired, adaptive or specific immunity. In this volume the term acquired immunity will be used. Acquired immunity relies on a unique mechanism whereby genetic 'mutation' occurring in two specialized cell populations, B and T lymphocytes, produces an enormous number (estimates vary from 10^{12} to 10^{20}) of molecular 'shapes' in the form of the antibody molecule and the T cell receptor. These 'shapes', if they bind to a structurally related (cognate) 'shape' (termed an antigen), and provided other signals (called co-stimulatory signals) are present, result in the proliferation of antigen-specific lymphocytes and the generation of a specific immune response. The secret of the immune system is this ability to amplify selective immune responses by the so-called clonal proliferation of lymphocytes. Two additional groups of molecules play major roles in acquired immunity: major histocompatibility complex (MHC) gene products and cytokines. The latter, as described in earlier chapters (3 and 8), are essential integrators of the intercellular communication which must occur for the proper functioning of the acquired (as well as the innate) immune response. An important point to understand from the start is that the acquired immune response is 'anticipatory'. That is, the appropriate receptors (antibodies and T cell receptors) exist prior to the organism encountering specific microbial antigens. It should also be understood that the various populations of cells and products constituting acquired immunity have been shaped by the nature and cellular location of the infectious agent. Acquired immunity contains some of the most complex cell-to-cell interactions and the most complex genetic mechanisms in the whole of biology. It is important to realize that the evolutionary driving force behind the development of this system, which pervades the whole of our bodies, is infectious organisms – in particular, bacteria and viruses. Moreover, as has been described, acquired immunity interacts closely with innate immunity and, in terms of responses to pathogens, is largely guided by it (see Chapter 8).

The basic foundations of acquired immunity largely developed in the period between the 1950s and the 1980s, a time when it was believed that the threat of bacterial infections had been eliminated by the use of antibiotics. Thus immunologists tended to focus their research on model systems, for example the immune response to purified antigens, such as hen egg lysozyme (HEL) or sperm whale myoglobin. Paradoxically, for experimental animals to develop immune responses to such proteins required that they were emulsified with sonicated bacteria (e.g. *Mycobacterium tuberculosis*) to act as an adjuvant. It is now believed that the presence of bacterial constituents, termed pathogen-associated molecular patterns (PAMPs), in such adjuvants was responsible for inducing the co-stimulatory molecules on antigen-presenting cells, thus allowing T cells to recognize and respond to the immunizing protein. Immunization with the protein in the absence of an adjuvant normally results in the development of immunological tolerance. The development of transplantation surgery in the 1960s gave immunological research an enormous boost, and at around the same time interest began to focus on autoimmune diseases such as rheumatoid arthritis and on the immune responses to tumours (tumour immunology). During this period, lip service was given to the fact that acquired immunity in the mammal had evolved to deal with infection but very little work was done on the role of this system in dealing with bacterial infections. While the work that was done in the period between the early 1960s and the early 1990s cannot be decried, the rules of immunology which were defined by such studies are now beginning to unravel at the edges as immunologists begin to examine, for the first time, immunity to bacteria. In this chapter the nature of the immune responses required to produce antibacterial immunity will be described and the countermeasures used by micro-organisms will be highlighted. In addition, bacteria produce certain molecules which can have profound effects on immunity, specifically superantigens, molecular chaperones and quorum-sensing compounds. The effects of these bacterial components on acquired immunity will be discussed.

This chapter describes the interactions of bacteria and their constituent components with the acquired immune system and is not meant to be a chapter on immunology. However, it is necessary to provide the reader with some background information. This will only cover the salient points, and the interested reader can find additional information in a range of modern textbooks whose titles have been provided in the reference section.

In Chapter 8 the mechanisms used by bacteria and viruses to combat innate immunity were described. In this chapter similar weight will be given to describing how bacteria and, where appropriate, viruses have evolved mechanisms to defeat acquired immune defences. The reader must be made aware of the limited attention that has been paid to this subject and it is likely that the richness of these bacterial mechanisms will only be appreciated after this textbook has been published.

A BRIEF DESCRIPTION OF ACQUIRED IMMUNITY

There are many points of entry of infectious agents into the human body, and many sites within the body where such organisms accumulate. With a number of bacteria and, obviously, with viruses, this includes entry into cells (see Chapter 6). Thus, any effective immune system must constantly patrol the body's tissues and cells in order to identify intruders. This is precisely what occurs in vertebrates. The immune system can be considered as one integrated organ pervading the body, although it is normal to view it as being divided into specialized tissues (such as the thymus, spleen and lymph nodes), a specialized transport network (the lymphatics) and specialized resident cells in tissues (macrophages, dendritic cells, T and B lymphocytes and their subsets). The major tissues of immunity, known as the lymphoid organs and lymphoid tissues, are shown schematically in Figure 9.1. Most of the cells constituting innate (macrophages, PMNs, NK cells) and acquired (dendritic cells, macrophages, B lymphocytes, T lymphocytes) immunity are produced in the marrow spaces of the bones. Two populations of T lymphocytes exist which are discriminated on the basis of surface proteins known as CD4 and CD8 and thus these cells are known as CD4 and CD8 T cells or T lymphocytes. The average human has 10^{12} lymphoid cells and these are produced in huge numbers daily. For example, it is estimated that the short-lived neutrophil is released into the blood at a rate of 7 million cells per minute. Most of the leukocytes which exit the bone marrow are committed to their genetically defined function. An exception to this are the T lymphocytes, which require to travel to the thymus where they develop. The importance of the thymus in T cell development is shown by the finding that patients with DiGeorge syndrome, in whom the thymus fails to develop, produce B lymphocytes but not T

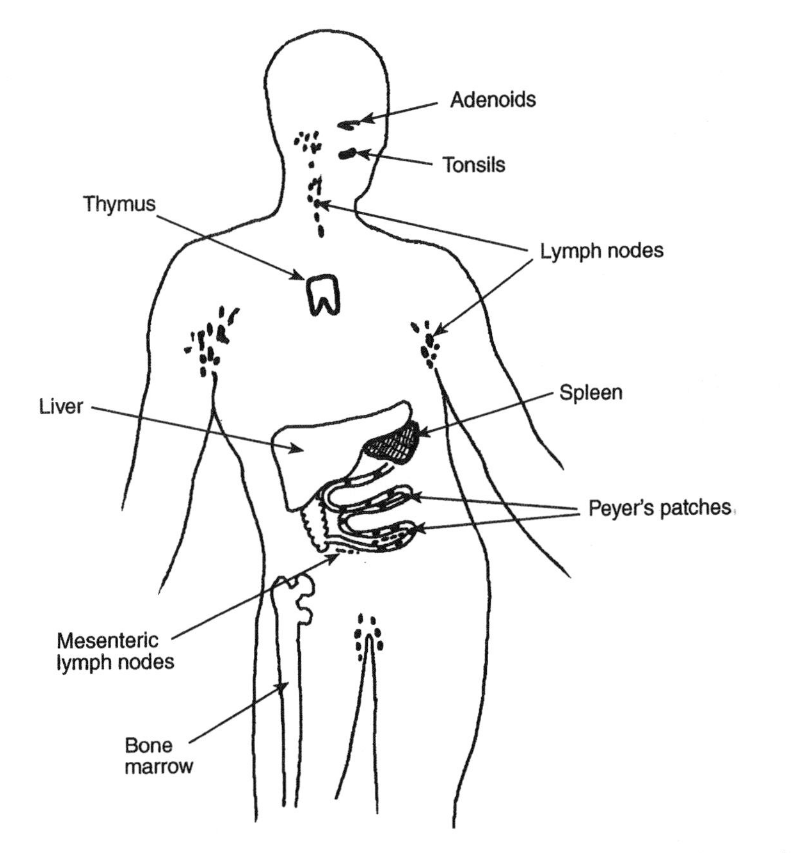

Figure 9.1 Simple schematic diagram showing the locations of the primary and secondary lymphoid tissues in the human body. Mature lymphocytes are produced in the bone marrow and the thymus and these tissues are in consequence called primary, or central, lymphoid tissues. Mature T and B cells enter the blood and can then travel to and through the other lymphoid tissues of the body which are called secondary or peripheral lymphoid tissues. These include the spleen, the large numbers of lymph nodes which pervade our bodies and the mucosal-associated lymphoid tissues (MALT) consisting of Peyer's patches in the gut and the adenoids and tonsils, which together are termed Waldeyer's ring

lymphocytes. These individuals can deal with extracellular bacteria but are unable to fight infections by intracellular organisms. The bone marrow and thymus are known as primary (or central) lymphoid organs, as they are the source of the lymphocytes. A further series of secondary (or peripheral) lymphoid organs and tissues permeate the body. These include the spleen, a large number of lymph nodes (or glands) which are capsulated tissues, and tissues that are more diffuse (i.e. unencapsulated), which associate with the mucosal surfaces of the body. These latter aggregates of lymphoid cells are given the generic term – MALT (mucosa-associated lymphoid tissue) and in different tissues are known as GALT (gut-associated lymphoid tissue), SALT (skin-associated) and even BALT (bronchial-associated). Troublesome tissues such as the tonsils, adenoids and appendix are part of the GALT. The study of these epithelial-associated immune systems is known as mucosal immunity.

T and B lymphocytes that have matured in the thymus and bone marrow respectively and which have not met their cognate antigen are referred to as naive lymphocytes and undergo a continuous circulation from the blood into the secondary lymphoid tissues, entering via specialized areas of the vasculature known as high endothelial venules (HEVs). They then return to the blood via the lymphatic vessels (Figure 9.2). In this way all available lymphocytes can circulate (immunologists like to use the term *traffic*) through all the tissues of the body (with the exception of certain tissues which either are not vascularized, like the anterior chamber of the eye and articular cartilage, or have a modified vasculature, such as the testis and brain) and interact with foreign antigens. Tissues through which lymphocytes do not traffic are called immunologically privileged sites.

The entry of bacteria into the body results in the carriage of bacterial antigens to the local (or draining) lymph nodes. This is generally the result of the uptake of bacterial antigens by specialized antigen-presenting cells (APCs) called dendritic cells which carry the antigens to such lymph nodes (Figure 9.3). It is in the draining lymph node environment where naive lymphocytes encounter their cognate antigens, that is, the antigens which bind with sufficient affinity to the lymphocyte receptors to activate these cells. The key response of any naive lymphocyte to its cognate antigen is that the cell is stimulated into cell cycle (see Chapter 2), producing large numbers of identical cells. This is known as clonal expansion or clonal proliferation (Figure 9.4). Each B lymphocyte, for example, produces only one antigenically specific antibody molecule. This clonal proliferative response is the secret of acquired immunity as it provides large numbers of effector cells (e.g. plasma cells producing antibodies or Th_1 T lymphocytes able to activate macrophages) to cope with the infectious agent and, of enormous importance, also produces memory cells for subsequent encounters with the antigen. These effector lymphocytes are produced relatively rapidly (within days) and the host does not have to carry the heavy cost of supporting large numbers of identical lymphocytes. It is possible to recognize cells which have encountered antigen. For example, the naive human CD4 T lymphocyte strongly expresses CD45RA but only weakly expresses CD29 and CD44 and does not express CD25, CD45RO, CD54 or class II MHC antigens. When activated, the expression of CD45RA is blocked and the cell expresses large amounts of CD25, CD29, CD44, CD45RO, CD54 and MHC II. Memory T cells can be identified by their high expression of CD29, CD44 and CD45RO. Apologies are due to the reader for this rapid decline into CD gobbledegook. The CD nomenclature is one of the major stumbling blocks to understanding immunology, but is also central to our understanding of this discipline. Thus a brief description of what CD stands for and examples of key CD receptors involved in immunology are provided in Table 9.1.

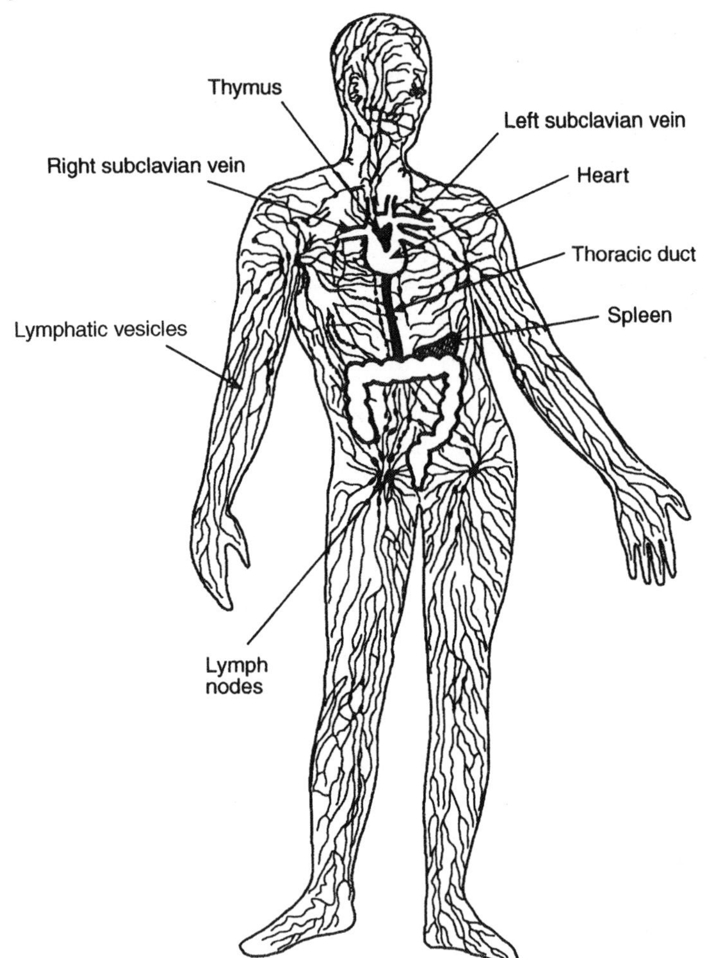

Figure 9.2 The network of lymph glands and vessels (lymphatics – seen as a series of fine lines permeating the diagram) in which the lymphocytes traffic. The lymphatics drain extracellular fluid (lymph) through the the lymph nodes and into the thoracic duct, which returns the lymph to the bloodstream by emptying into the left subclavian vein. Lymphocytes that circulate in the bloodstream enter the peripheral lymphoid organs and thence into the lymph, where they are carried to the thoracic duct to re-enter the bloodstream. In this way the lymphocytes can patrol the whole body

CELL POPULATIONS INVOLVED IN ACQUIRED IMMUNITY

The reader has already been introduced to some of the cell populations (macrophages, dendritic cells, natural killer (NK) cells, neutrophils – also known as PMNs) which play a role in both innate and acquired immunity. However, the two key cell populations which essentially define acquired immunity are the B lymphocyte and the T lymphocyte. The development of these various cell populations from bone marrow precursor cells is shown in Figure 9.5. B and T lymphocytes use distinct mechanisms to combat bacterial infections. The former

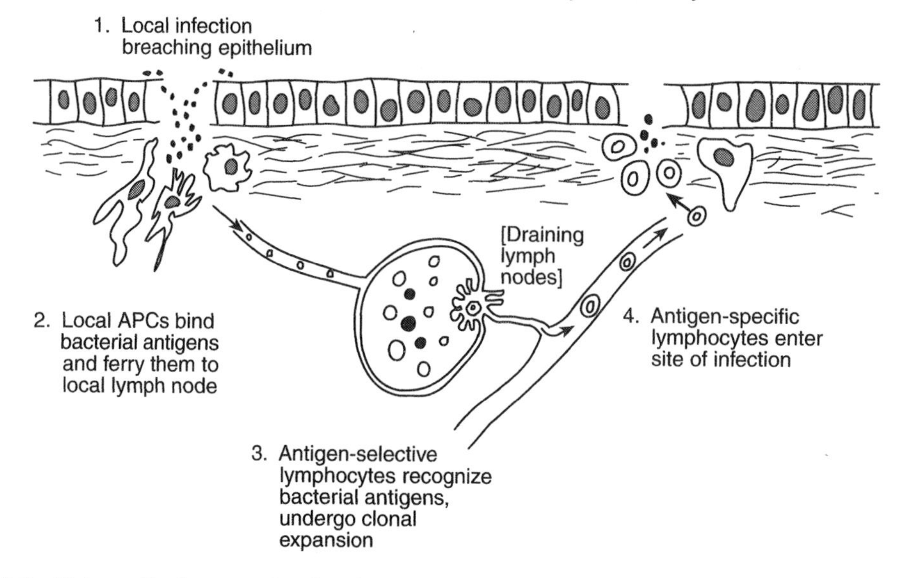

Figure 9.3 Schematic diagram showing the consequences of infection at an epithelial surface in terms of the initiation of an acquired immune response. Once bacteria have overcome the immune barriers described in Chapter 8 and have entered into the submucosa they encounter specialized antigen-presenting cells (APCs) which are given the general title dendritic cells. These cells take up bacterial antigens, become activated (including the upregulation of B-7 expression), and carry them to local or draining lymph nodes, where they present them to the large numbers of lymphocytes that traffic through such secondary lymphoid organs. Encounter of the appropriately expressed antigen with antigen-specific T cells results in clonal proliferation of reactive lymphocytes which can then traffic to the site of infection and promote antigen-specific immunity

cells produce a humoral factor – the antibody molecule – to recognize and 'inactivate' free-living bacteria or bacterial components such as exotoxins. The actions of antibodies can be further amplified by the complement system as described in Chapter 8. Antibody synthesis also depends on the interaction of B cells with one subset of T lymphocytes – the so-called helper T cell. With this exception, the T lymphocyte is largely employed in dealing with intracellular parasites (viruses, intracellular bacteria within phagocytes and intracellular protozoa) and its actions involve direct cell-to-cell interaction, resulting in cell killing, and/or cell-to-cell interaction leading to cellular activation via the synthesis of specific patterns of cytokines. There are also specialized populations of B and T lymphocytes to deal with infectious agents, in particular, bacteria.

Subpopulations of T lymphocytes

The different subpopulations of T cells are recognized largely by their expression of surface proteins (CD markers – see Table 9.1). All T cells express CD3, a hetero-oligomeric protein which is part of the T cell receptor complex, and can be further subdivided into those cells which express CD4 and those which express CD8. The CD4 lymphocyte population can be further functionally subdivided into two subpopulations termed Th_1 and Th_2 lymphocytes. Th_1 and Th_2 cells were initially discriminated, not on the basis of cell surface markers, but on

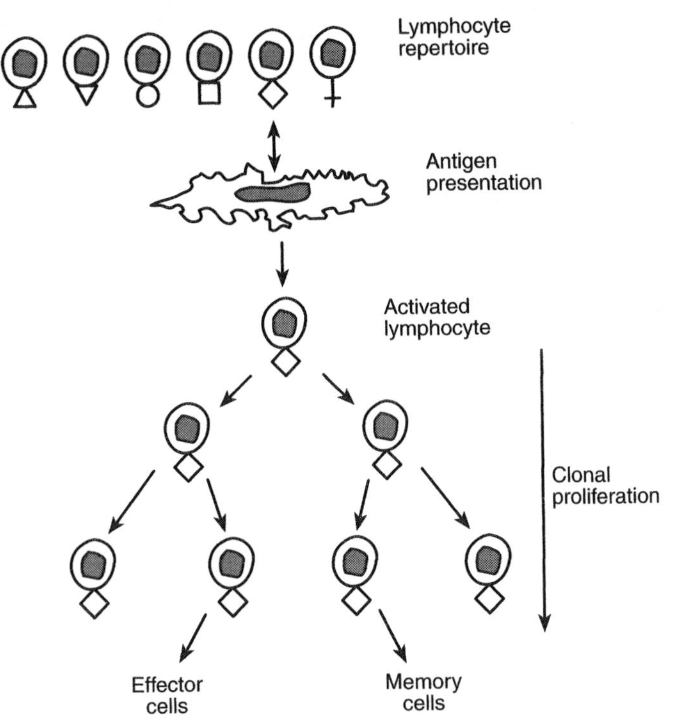

Figure 9.4 Schematic diagram elucidating clonal expansion. When a lymphocyte meets its cognate antigen on the surface of a properly activated antigen-presenting cell (APC) the result is activation of the lymphocyte, cell division and the continued division of progenitor cells. This process is known as clonal proliferation or expansion and leads to the production of effector cells (e.g. antibody-producing plasma cells, CD4+ T lymphocytes or cytolytic CD8+ T lymphocytes) and memory cells which provide for long-term acquired immunity to infections

the basis of the patterns of cytokines they produced. In very recent years it has proved possible to identify these distinct CD4 subpopulations on the basis of their expression of chemokine receptors. The role of Th_1/Th_2 lymphocytes in infection will be described in more detail later in this chapter. Each T cell contains a unique receptor termed the T cell receptor (TCR), which was discovered in 1984 after a number of years of very active research. This receptor is composed of two distinct peptide chains known as the α and β chains and the T cells are now known as αβ T cells. Not long after this long-awaited discovery of the T cell receptor, another subset of T cells was discovered which had a similar heterodimeric receptor but whose individual subunits were the products of different genes. These T lymphocytes were termed γδ T cells and they have an important role in the immune response to infectious organisms. The γδ receptor is thought to have evolved prior to the αβ receptor and is therefore termed TCR1 and the latter is denoted TCR2. These γδ T cells can be CD4 or CD8 positive or, indeed, express neither of these surface antigens. In the human, γδ T cells make up only 0.5–15% of the T cells in peripheral blood. In contrast, they show a much greater predominance in the skin and colon. γδ T cells are increasingly being viewed as important in the protective responses to bacterial infections.

Certain T cell populations are cytotoxic, that is, they can kill other cells. Virally infected cells are a major target of CD8-positive T cells, which kill in an antigen-dependent manner.

Table 9.1 CD nomenclature and examples of major immunological receptors

CD number	Other terminology	Function	Cellular distribution	Ligand
CD1a/1b/1c		Peptide. lipid antigen presentation	Thymocytes, LC, DC	
CD2	LFA-2	T cell adhesion to APC	Thymocytes, T, NK	LFA-3, CD58
CD3		T cell activation	Thymocyte, T	
CD4		Th activation	Thymocyte/T cell subpopulations, M	MHC II
CD5	Leu-1, Ly-1	T cell activation	Thy, T cell, B cell (sub)	CD72
CD8		T cell subpopulation marker	Thy (sub) T cell (sub)	MHC I
CD11a/CD18	LFA-1	Leukocyte adhesion protein	Most leukocytes	ICAM-1,2,3
CD11b/CD18	Mac-1	Leukocyte adhesion protein	Most leukocytes	ICAM-1, iC3b Fibrinogen
CD11c/CD18	CR4m p150,95	Leukocyte adhesion protein	PMN, NK	iC3b, fibrinogen
CD14		LPS binding protein	M , some fibroblasts	LPS
CD16	FcγRIII	Phagocytosis and ADCC	PMN, M , NK	Fcγ
CD19	B4	B cell activation/ proliferation	B	
CD23	FcεRI	B cell activation/IgE regulation	B, M , FDC, Plt	FCε CD21
CD28		Co-stimulatory molecule	T (sub) B	CD80, CD86
CD54	ICAM-1	Adhesion molecule	Leukocytes, endothelium	LFA-1/Mac-1
CD62E	E-selectin	Vascular adhesion molecule	Endothelial cells	Oligosac-charides
CD71	transferrin receptor	Receptor for transferrin	Activated leukocytes	Transferrin
CD80	B7–1	Co-stimulatory molecule	Monocytes/dendritic cells	CD28
CD86	B7–2	Co-stimulatory molecule	Monocytes/dendritic cells	CD28
CD88	C5a receptor	C5a receptor	Myeloid cells	C5a
CD152	CTLA-4	Co-stimulatory molecule	Activated T cells	CD80/CD86

DC – dendritic cell, FDC – follicular dendritic cell, LC – Langerhans cell, M – macrophage, NK – natural killer cell, plt – platelet

The immunology literature abounds with CD this, and CD that, but what do these various CDs refer to? Historically, once Caesar Milstein had developed the technology for producing monoclonal antibodies, there was a worldwide drive to make as many monoclonal antibodies (mAbs) as possible. Many of the mAbs were produced to the cell surfaces of leukocytes and the proteins recognized by the mAbs were initially given a variety of appelations – producing an extremely confusing time in immunology. To bring order to this confusion international workshops were set up in order to enable different laboratories worldwide to compare the specificity of their mAbs. If a number of mAbs react with the same antigen then it is given a CD number – the term CD standing for cluster of differentiation. There are currently about 200 CD specificities defined and they are still growing in number. The CD antigens which will be discussed in this chapter are defined above. As will be seen, a number of these CD antigens are part of cognate pairs which interact with each other (e.g. CD5 and CD72, CD11b/CD18 and CD54 (ICAM)).

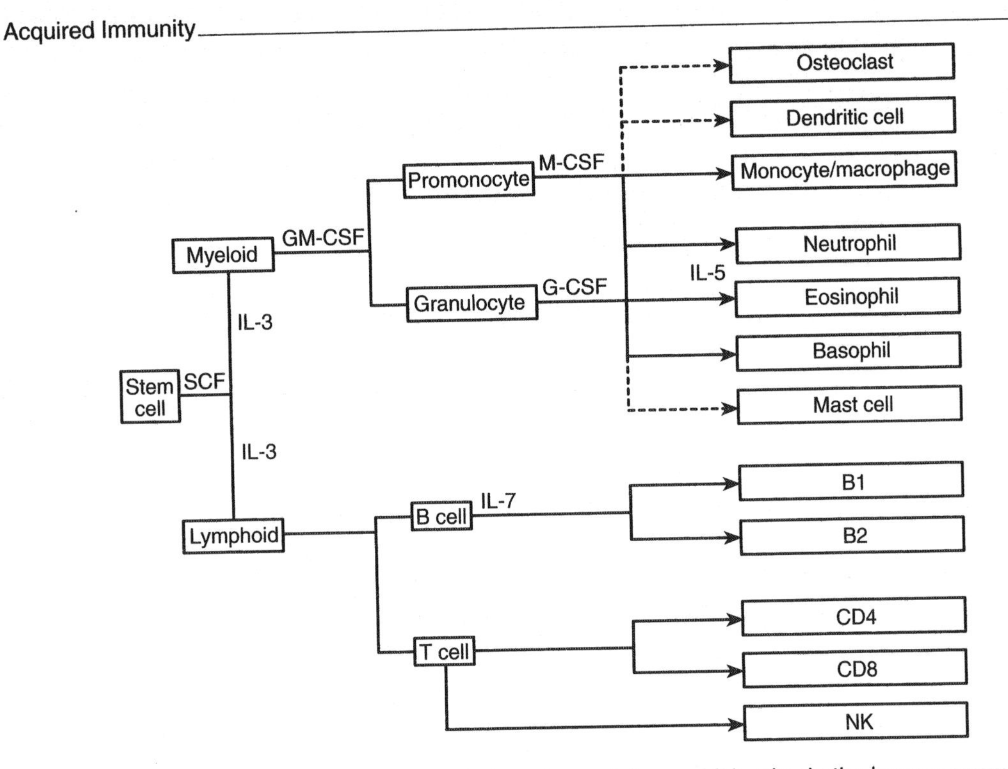

Figure 9.5 The various myeloid and lymphoid cell populations which arise in the bone marrow from a stem cell precursor population. SCF stem cell factor; GM-CSF granulocyte, macrophage colony-stimulating factor; M-CSF, macrophage colony-stimulating factor

There is another class of circulating cytotoxic cell, which derives from a common T cell precursor, called natural killer (NK) cells. These belong to a morphologically distinct population of lymphocytes known as large granular lymphocytes (LGLs), which express the low, affinity FCγ receptor (CD16). Unlike T cells generally, NK cells do not express a TCR. These NK cells can spontaneously kill certain susceptible target cells in a manner which is not antigen-restricted and are believed to act as a first line of defence in a variety of infectious diseases such as listeriosis. The various subpopulations of T cells are shown diagrammatically in Figure 9.6.

Subpopulations of B lymphocytes

The human bone marrow manufactures 10^{10}–10^{11} B lymphocytes each day and, unless activated by antigen, these cells generally have a short lifespan and die via normal apoptotic mechanisms (see Chapter 2). Each B lymphocyte contains on its cell surface membrane-bound immunoglobulin (Ig: normally IgM or IgD) which acts as the cell's antigen receptor. Exposure to antigen results in B cell activation, clonal proliferation and the differentiation of the B cell into an antibody-producing factory known as a plasma cell. The antibody produced is of the same antigenic specificity as the surface-bound antibody. B memory cells are also produced in this process of clonal expansion. There are two distinct B cell subpopulations (Figure 9.7). The most interesting, in terms of microbial immunity, are the CD5-positive B

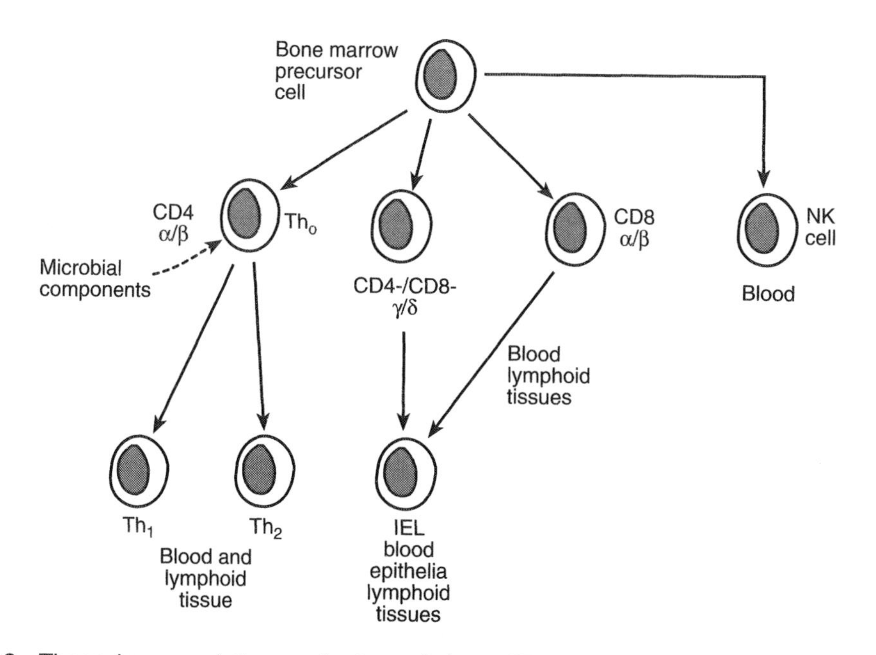

Figure 9.6 The various populations and subpopulations of T cells in the mammal. All cells arise from a precursor cell population in the bone marrow. Circulating blood will contain all these T cell populations, as will lymphoid tissues, due to normal cell trafficking. The γδ T cells and CD8 T cells are also found in the epithelia of the body, where they form the intraepithelial lymphocyte (IEL) cell population. CD4 T cells can also mature into Th_1 and Th_2 cells depending on the nature of the signals they receive from microorganisms

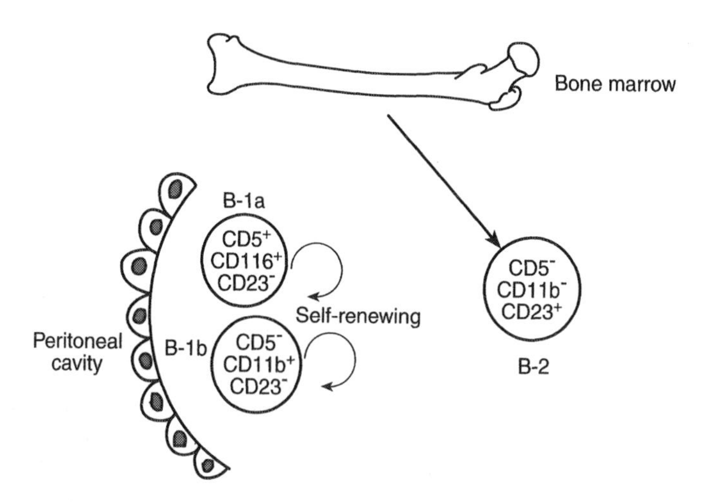

Figure 9.7 There are three subpopulations of B cells. The cells conventionally called B cells are bone marrow-derived cells which are CD23 positive and lack CD5 and CD11b expression. However, two related B cell populations, one CD5 positive and one CD5 negative, are found in the periphery and are self-renewing, although precursor cells are present in bone marrow

cells. These cells may be seen as being analogous to the γδ T cells found in mucosal tissues. Unlike conventional B cells, which arise from the bone marrow, the CD5-positive B cells renew themselves in the periphery and are the predominant lymphocyte in the peritoneal cavity. CD5 B cells make antibodies to bacterial polysaccharides and do so rapidly without the need for T cell help. Mice deficient in CD5 B cells are more susceptible to infection with *Streptococcus pneumoniae* due to the failure to produce antibodies to phosphatidylcholine that protects against this bacterium. Like γδ T cells, the CD5 B lymphocyte is limited in its range of specificities of the antibodies it produces. These B and T cell subpopulations are increasingly being viewed as a transitional phase in the evolution of the acquired immune response.

GENERATION OF DIVERSITY: THE SECRET WEAPON OF ACQUIRED IMMUNITY

Although immunology can be traced back to the pioneering work of Pasteur, Koch and Metchnikoff, it was only in the 1960s that the central role of the T lymphocyte in immunology was discovered. Moreover, it was only in the 1980s that the central problem of the acquired immune system – how can it recognize so many antigens? – was solved. Thus immunology is still very much an emerging science, with many new discoveries, and reinterpretations of old discoveries, being made. In this section the central mechanism of acquired immunity, generation of diversity (often abbreviated to GOD), will be described in a historical context.

The word immunology can be traced to the Latin *immunitas* and *immunis*, which derive from the Roman concept of being exempt from (normally military) service. The modern use of this term was probably first employed by the Roman poet Marcus Annaeus Lucanus (AD 39–65) in his poem *Pharsalia* to describe the resistance of members of the Psylli tribe of North Africa to snakebites. The modern concept of immunization or vaccination is normally ascribed to the pioneering work of the English physician Edward Jenner. It is interesting, at least for the readership of this volume, that it was the work of bacteriologists that established the science of immunology. The studies of the chemically trained bacteriologist Louis Pasteur (1822–1895) can truly be said to have initiated modern immunology by using attenuated strains of bacteria or viruses to produce vaccines. The rival schools of Pasteur in France and the great bacteriologist Robert Koch in Germany were instrumental in the late nineteenth and early twentieth centuries for the discovery of antibodies and of complement. For most of the history of immunology it has been the humoral factors (antibodies, complement) that have been studied. Cellular immunology is a child of the 1960s.

The discovery that immunization of animals, or man, with appropriately 'packaged' antigens gave rise to the production of molecules termed antibodies, which were specific for the immunizing agent, was made early this century. The awesome ability of the immune system of mammals to recognize and respond to antigens was rapidly discovered and put on a solid experimental basis by the pioneering discoveries of Karl Landsteiner, who immunized rabbits with proteins (carriers) conjugated to small organic molecules (haptens) such as aminobenzoic acid. He found, for example, that by immunizing with *ortho- meta-* or *para*-benzoic acid conjugated to a carrier the antibody response could distinguish between these three isomers. More impressively, Landsteiner demonstrated that rabbit antibodies can distinguish between the optical isomers of tartaric acid. These and other findings suggested that mammals could make many different antibodies, possibly tens of thousands to millions (current views are that the immune system may be able to produce 10^{12}–10^{20} distinct antibody molecules). As each antibody molecule is a protein, and as each protein is the product of a gene, the finding of the capacity of the immune system to create so many different antibodies created a major paradox. If each antibody gene was in the genome was there enough genetic material to encode them? If there were not enough genes to encode each antibody molecule, then how were they encoded? We now know that the human genome, for example, contains DNA encoding for only 60–100 000 genes (see Chapter 2). Therefore it is clear that we do not carry around in our cells the genes for each antibody molecule (or for the T cell receptor, which comes into the picture much later in immunological history).

Examination of the structure of the antibody molecule eventually gave the clues needed to solve this paradox. The basic structure of the antibody molecule was elucidated in the early 1960s by the chemical studies of Rodney Porter in Oxford, England, and Gerald Edelman working at the Rockefeller University, New York. These scientists, who won the Nobel Prize for their work, revealed that the antibody molecule consists of two light (L) chains (consisting of around 220 amino acids) and two heavy chains (H) (of about 450 residues). In the human there are in fact two different types of light chain called κ and λ light chains. The two L and two H chains are identical to each other and are joined by disulphide bonds (Figure 9.8). Unfortunately, these findings gave no clues as to the reasons why different antibodies bound to different antigens and it proved impossible at this time (early 1960s) to isolate individual antibody molecules from the heterogeneous collections of antibodies that we and animals have in our sera. Thus it was not possible to analyse the chemical composition

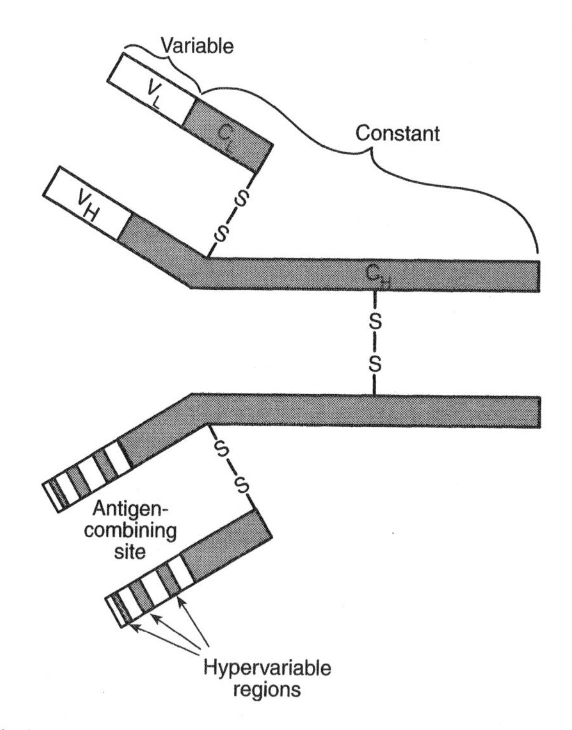

Figure 9.8 The two-chain structure of the antibody molecule showing the variable and constant regions, the antigen-combining site and the hypervariable regions

(amino acid sequence) or physical structure (by crystallization and X-ray diffraction) of the antibody molecule. An escape from this impasse was the discovery of homogeneous populations of antibodies. The three main sources of such proteins have been myeloma proteins (a myeloma is a cancer of a single clone of B cells which gives rise to a homogeneous antibody), monoclonal antibodies and homogeneous antibodies to certain bacterial carbohydrates. Amino acid sequencing of these homogeneous antibodies revealed that the N-terminal segments (the first 110–120 amino acid residues) of both the light and heavy immunoglobulin chains had three regions in which the amino acid sequences proved to be very diverse (or hypervariable) (Figure 9.9). These hypervariable (HV) regions are found at the approximate positions-amino acids 28–35 (HV1), 49–59 (HV2) and 92–103 (HV3). The stretches of amino acids between the HV regions are termed framework regions. Using X-ray crystallography it was found that these hypervariable regions lie on the outside of the antibody molecule and the HV regions of both the antibody light chain and the heavy chain form the antigen-binding site of the antibody. As the hypervariable regions constitute the antibody-binding domain they are more commonly known as the complementarity-determining regions or CDRs (thus CDR1, CDR2, CDR3). Readers should not confuse CDR with cluster of differentiation (CD).

The current concept of the immunoglobulin molecule is of a protein composed of discrete globular domains, called immunoglobulin domains, of approximately 110 amino acids. The antibody light chain is composed of two immunoglobulin domains: the variable (V_L) domain covalently linked to a contiguous constant domain (C_L). The antibody heavy chain consists of one variable domain (V_H) linked, depending on the class of the antibody, to three to five C_H domains.

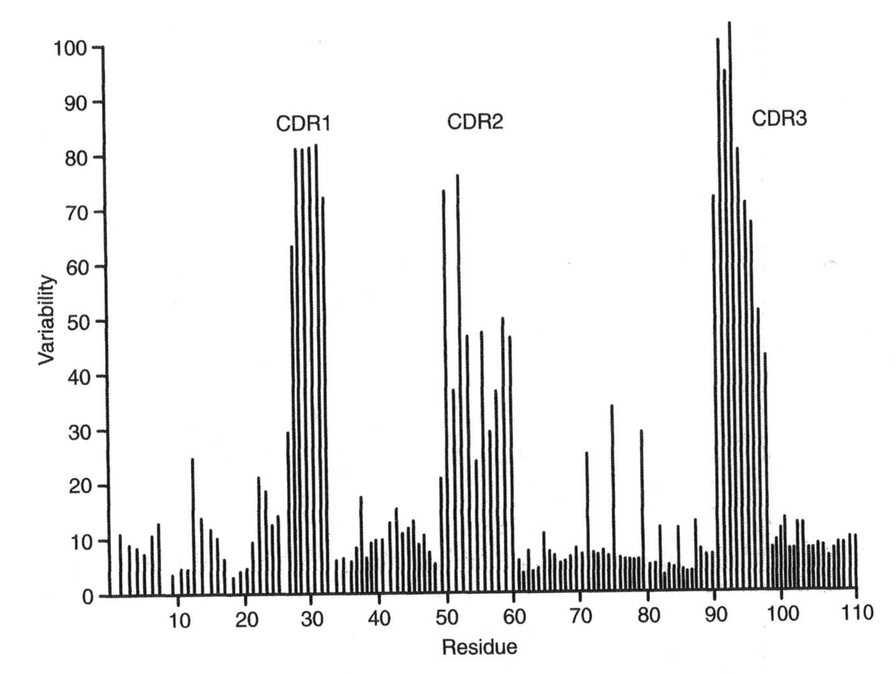

Figure 9.9 The regions of amino acid variability on the immunoglobulin light chain. A large number of Ig molecules have been sequenced. The histogram delineates the extent of the variability of each individual amino acid residue at each position among the sequenced proteins. This analysis shows that there are three 'hypervariable' regions. A similar hypervariability is found in the Ig heavy chain

Genetics of antibody synthesis

The capacity to synthesize the variable domains of the light and heavy immunoglobulin chains provides the acquired immune system with its ability to recognize almost any antigen. The mechanisms responsible for producing this enormous diversity of molecular shapes involve a number of unique genetic manipulations including a combinatorial gene-shuffling mechanism analogous to playing molecular poker. This combinatorial mechanism, plus a system for introducing variation during the combination of DNA sequences (called junctional diversity), is common to the production of both the antibody receptor (i.e. the antibody molecule) and to the production of the T cell receptor. The B cell is also able to induce mutations to occur in the antibody molecule – a process known as somatic hypermutation. In this volume only the mechanisms involved in the production of diversity of the antibody molecule will be discussed in detail. It is fascinating, although counterintuitive, to realize that the genetics of the immune response anticipates the encounter with antigen and is in no way dependent on antigen for the generation of TCRs and antibodies.

All cells in the human body are totipotent. That is, they contain all the genes present in the germ cells (ova and sperm). That is why it is possible to clone animals. However, each distinct cell type in the body only uses a fraction of the total number of available genes, thus accounting for differences in, for example, cell shape, biochemistry and behaviour in the various tissues of the body. It is now established that B and T lymphocytes have the capacity to do what no other cell can do – namely, use gene rearrangement and somatic hypermutation to produce the diversity of molecular shapes that are antibody and TCR. One of the major early findings which led to our understanding of the generation of immunological diversity was that the variable and constant regions of the immunoglobulin molecule were coded for by different genes. It now turns out that the variable domains of antibody molecules and of T cell receptors are encoded for by a number of genes (Figure 9.10). If we concentrate first on the κ light chain, it is encoded for by two genes: a variable region gene (V_κ which encodes for the first 95 residues and contains the first two CDRs of the antibody) and a second gene downstream (3') called J for (joining), encoding for residues 96–108 – the third hypervariable region of the light chain. Now, in fact, there is not simply one V_κ gene in the human genome – there are around 40 functional genes encoding this light chain and there are five genes encoding the J regions. In germ cells of the human, the V_κ genes are arranged in a linear array, each with its own leader sequence and separated by introns (see Chapter 2). The J genes are similarly arranged (but with no leader sequences). Separated by another long intron is the single constant region of the κ chain (C_κ). Early in B cell development the so-called pre-B cell selects one of the 40 V_κ genes and physically joins it to one of the J_κ genes. This joining involves the looping-out of the intervening DNA and the linking of conserved recognition sequences that are found at the 5' and 3' ends of the V and J regions. Thus one V and one J gene segment are brought together in the genome of this one B cell precursor. From this rearranged DNA a primary RNA transcript is made which is then spliced to remove all intervening intronic (non-coding) RNA to bring the V_κ, J_κ and C_κ exons together to produce a mature mRNA encoding the light chain. This can then be translated into the polypeptide chain and can join with the heavy chain to form a functional antibody molecule (Figure 9.10). Thus in the mature B cell the choice of antibody molecule has been made, but other mechanisms can increase the diversity of the final product. Current estimates are that the λ light chain is the product of 35 V_λ genes and 4 J_λ genes.

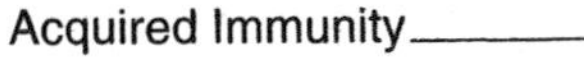

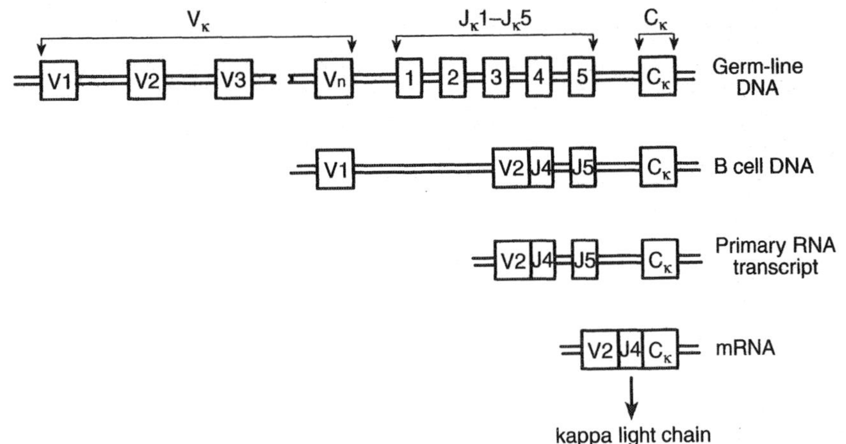

Figure 9.10 Mechanism of the synthesis of the V_κ light chain. During the differentiation of the pre-B cell one of the several V_κ genes in the germ line DNA (denoted V1 to Vn) is recombined and apposed to a Jκ segment (Jκ1–Jκ5). The B cell transcribes a segment of DNA into a primary RNA transcript that contains a long intervening sequence of additional J segments and introns. This transcript is processed into mRNA by splicing the exons together and is then translated by the ribosomes into the κ light chain

With the synthesis of immunoglobulin heavy chains the pre-B cell has to make a choice between three different pools of genes: variable (V), diversity (D) and joining (J) genes. It is estimated that in man there are 51 V genes, approximately 30 D genes and six J genes. The pre-B cell then chooses one of each of these three genes from the genes available and produces the final heavy chain gene in much the same way as was described for the light chain.

This ability to choose between different pools of genes, known as combinatorial diversity, is one of the four mechanisms which produce the enormous repertoire of antibodies and TCRs which constitute the immune response. We can quickly calculate how many different antibody molecules can be made from the 171 genes ($40V_\kappa + 5J_\kappa + 35V_\lambda + 4J_\lambda + 51V_H + 30J_H + 6J_H$) encoding the antibody molecule in *Homo sapiens*. For the human κ light chain there are 40 V and five J gene segments, providing the capacity to make 200 different V_κ proteins. Similarly with the λ light chain the calculation is 35×4, equalling 140 products. So, in total 340 light chains can be produced. For the heavy chain in the human there are 51 functional V_H gene segments, 30 D gene segments and six J_H gene segments, giving the capacity to produce 9180 different proteins. The light and heavy chain variable regions contribute to antibody specificity and each of the 340 light chains could be combined with each of the 9000 odd heavy chains to give approximately 3×10^6 different specificities. This process, which is analogous to choosing hands of cards from a pack of cards, is known as combinatorial diversity. Thus the inheritance of multiple copies of antibody genes and the capacity to rearrange these genes and choose particular sets of genes are two of the basic mechanisms for producing diversity. The third mechanism is known as junctional/insertional diversity. When the VJ, VD or VDJ gene segments are joined, the cutting of the DNA and its rejoining are not done precisely and so nucleotide omissions or insertions can occur, leading to additional variations in the protein's sequences. It should be noted that these additional variations are occurring in the third CDR of the antibody molecule. The final mechanism responsible for producing variation is the most mysterious of all. It has been known for many years that B cells undergo a process known as affinity maturation in which the initial antibody molecule, produced during the primary immune response, shows increased affinity for the antigen during a secondary response. The process responsible for this affinity maturation has been termed somatic hypermutation and involves point mutations in the antibody molecule which can occur in any of the three CDRs and introduces additional variation to the final antibody used by memory cells.

The T cell receptor

When the T cell receptor was isolated in the 1980s it turned out to be a membrane-bound molecule composed of two disulphide-linked chains which were named α and β. Each chain folds into two Ig-like domains, one having a relatively invariant structure and the other showing similar variability to the Ig variable region (V_α/C_α, V_β/C_β, Figure 9.11). As explained earlier, there is a second TCR composed of γ and δ chains. The mechanism of producing the TCR is very similar to that for antibodies, with one exception. The gene segments encoding the T cell receptor β chains follow a similar arrangement of V, D, J and constant gene segments with rearrangements of these segments occurring in early T cell development to produce a continuous VDJ sequence. The α chain pool lacks D gene segments but has a very large number (50) of J gene segments. Joining of these various gene segments introduces additional variation as was described for antibodies. The mechanism of somatic hypermutation does not, however, occur with the TCR. There are 52 V_β gene segments in man and these can be divided into 20-odd 'families'. Certain bacterial proteins called superantigens have the ability to stimulate T cells bearing certain of the V_β families; this will be discussed in a later section and has also been touched upon in Chapter 7.

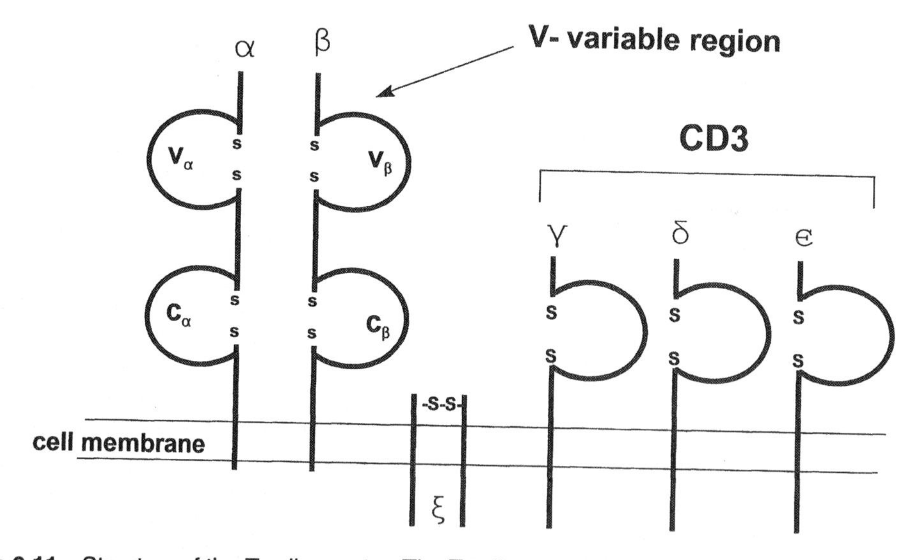

Figure 9.11 Structure of the T cell receptor. The T cell receptor has similarity to the Ig molecule and is composed of two separate proteins chains (α and β) with each chain having a variable and a constant domain

ANTIGEN PRESENTATION AND LYMPHOCYTE ACTIVATION

Having explained how the immune system can generate the proteins which recognize bacterial antigens, what happens if those antigens enter into the body as part of the process of bacterial infection? Antigen recognition occurs by binding to either surface-bound monomeric Ig on B cells or to the TCR on T lymphocytes. The consequence of such binding is normally the activation of the B or T cell clone, the proliferation of that clone, with the consequent generation of an appropriate immunological response (antibody synthesis, cell-to-cell contact with cell killing etc.) and the production of memory cells, so that a repeat of the infection can be more rapidly and effectively dealt with. This memory response can be extremely long lived even in the absence of re-exposure to the antigen. Indeed there are documented cases of individuals having been exposed to an infective agent 60–80 years previously, with no known subsequent exposure, and still exhibiting immune responses to the infectious organism.

Activating a T or B cell clone can be dangerous, as it could possibly produce an autoimmune response or become uncontrolled. Thus during the evolution of the immune response, the system has become hardwired with mechanisms to prevent lymphocyte activation or to shut off activated cells. As will be described, lymphocytes generally do not respond directly to antigen but require two or more signals to become activated. The concept of self-reactive lymphocytes and the mechanisms used to ablate such self-reactive cells lies at the heart of immunology and has been largely driven by the interest in autoimmunity. This topic will not be dealt with in any detail in this chapter and the interested reader should see the bibliography at the end of the chapter.

B cell activation

The antigen recognition receptor on the B lymphocyte is surface-bound monomeric IgM or IgD. The IgM has only a short intracellular domain which does not signal when antigen binds. Transmission of the signal depends instead on at least two other proteins which are called Igα and Igβ. The cytoplasmic tails of these two proteins contain sequences (known as immunoreceptor tyrosine-based activation motifs – ITAMs) that can be phosphorylated by Src-family tyrosine kinases (see Chapters 2, 3 and 7), leading to a cascade of intracellular signals promoting entry into cell cycle. Other cell surface proteins such as CR2 (the receptor for the complement component C3d), CD19, CD40 and the Toll receptor are also involved in B cell activation. In order to activate the B cell, the surface-bound IgM molecules must be cross-linked. B cell activation normally requires the participation of a particular class of T lymphocyte – the Th2 helper T cell. However, certain bacterial constituents such as lipopolysaccharide (LPS), polymeric proteins and polysaccharides can directly activate B cells without the need for T cell participation. Such molecules are known as thymus-independent (TI) antigens. These TI antigens fall into two classes which activate B cells by different mechanisms. The first class – TI-1 antigens, which include molecules like LPS – have the ability to directly induce the activation of B cells independent of their antigenic specificity. This phenomenon is known as polyclonal B cell activation and the molecules able to activate in this manner are called B cell mitogens – a mitogen being any substance that can stimulate cells into cycle and through mitosis. TI-1 antigens can activate both mature and immature B cells. The action of these TI-1 antigens may have some analogy with bacterial superantigens, to be described in a later section, which are polyclonal activators of T cells. The second class of thymus-independent antigens are molecules with highly repetitive structures, good examples being bacterial cell wall polysaccharides. Unlike the TI-1 antigens, these repetitive TI-2 antigens have no intrinsic B cell-activating activity and are only able to activate mature B lymphocytes. The major subpopulation of B cells responding to TI-2 antigens appears to be the CD5 B cell, which has been mentioned previously, it being thought that the major function of this cell population is to make antibodies to bacterial polysaccharides. TI-2 antigens are believed to function by extensively cross-linking the monomeric immunoglobulin molecules on the surface of specific mature B lymphocytes. As with most biological responses, too extensive cross-linking can render mature B cells unresponsive to such TI-2 antigens.

Many bacteria, as described in Chapter 8, have extracellular capsules that contain large amounts of polysaccharides and have the ability to block innate immune responses, including ingestion by macrophages. As will shortly be described, the activation of T cells requires the participation of macrophages and macrophage-like cells. Thus bacterial capsules are unable to activate T cells. It therefore has to be assumed that the relentless struggle between eukaryotes and prokaryotes has resulted in the maintenance of the direct B cell response to bacterial capsules. Binding of the low-affinity IgM antibodies produced by the CD5 B cells can activate complement and help phagocytes ingest capsulated bacteria. *Homo sapiens* exhibits a large number of so-called immunodeficiency diseases which can affect innate B or T cell immunity. In one of these immunodeficiency syndromes, known as Wiskott–Aldrich syndrome, the sufferers respond normally to protein antigens but fail to make antibody to polysaccharide antigens and are, in consequence, highly susceptible to infections with extracellular capsulated pyogenic bacteria.

The third and largest class of molecule which can stimulate antigen-specific B lymphocyte activation is the thymus-dependent (TD) antigen. These are generally proteins which can bind to surface Ig of B cells but do not provide sufficient signal to trigger a response – a safety device built into the immune response, as described earlier. In order for B cells to respond to such antigens the antigen has to be internalized, proteolytically processed and presented on the B cell surface linked to the class II MHC. As will be described in the next section, T cells only respond to antigen linked to MHC proteins. With the antigen present on the surface of the B cell in the groove of the MHC molecule the B cell can act as an APC and can interact with T lymphocytes which have also responded to the same antigen. The T cell provides additional signals such as cytokines and the cognate pairing of CD40 on the B cell with CD40L on the T cell (and also B7.1/7.2 on the B cell with CD28 on the T cell – see next section); the final result is that the B lymphocyte is stimulated into clonal proliferation. This mechanism of B cell activation is called linked recognition and, as described, requires specific antigen-stimulated T cells.

Activated B cells make antibody, and the nature of the antibody can alter by two mechanisms. The first has been described, and is due to somatic hypermutation leading to the synthesis of antibody with higher affinity for the antigen – a process known as affinity maturation, first described in the 1960s. The particular antibody produced by any one B cell can also change with respect to its class or subclass. In primary immune responses the major class of antibody produced is IgM and, in secondary responses, IgG. Cytokines such as interferon γ (IFNγ) or interleukin (IL)-4 can cause what is termed class-switching. This means that the B cell can use one of its other Fc genes to produce the same (idiotypic) antibody but one linked to a different Fc molecule and therefore having different biological functions such as activating the classical complement pathway.

Viral and bacterial infections are often associated with the appearance of circulating rheumatoid factors. These are antibodies which bind to the Fc region of immunoglobulins. They are called rheumatoid factors because they were first discovered in the blood of patients with the chronic autoimmune condition, rheumatoid arthritis. However, they have proved hard to reproduce experimentally. It has recently been shown that immunization of experimental animals with viruses or bacteria coated with antibodies to repetitive surface structures (e.g. LPS) is able to induce rheumatoid factors and this may be the way that they are produced naturally. These findings may shed light on the mechanism of rheumatoid factor production in rheumatoid arthritis.

The T cell: a complex of activation

B lymphocytes can be triggered directly by certain bacterial antigens binding to the surface immunoglobulin receptor. In contrast, the activation of T cells is a much more complex affair requiring the processing of the protein antigen by a so-called antigen-presenting cell (APC) and its binding to another cell surface molecule known as the major histocompatibility complex (MCH) molecule. It is this complex of MHC molecule and processed peptide *on the cell surface* of APCs that is recognized by the TCR. The fact that the T cell only recognizes antigen appropriately packaged on the surface of another cell is an important point which the reader needs to grasp to follow the rest of the chapter. In addition to the binding of the TCR to the peptide–MHC complex, other sets of counter-receptors on the T cell and APC must interact before the T cell can be triggered into cell cycle. This is a very complex process and will be described in turn, as an understanding of this process is vital for comprehending how the immune system deals with different forms of infectious agents. It is also important to realize that the immune response to certain bacteria has recently been shown to break these previously inviolate immunological rules.

The major histocompatibility complex

The MHC gene complex encodes for two different sets of MHC proteins known as class I and class II molecules (Figure 9.12), which can be thought of as the third set of recognition molecules for antigens employed by the immune response (the other two being the immunoglobulin molecule and the TCR). The function of the MHC is to bind peptides which have been produced by the proteolysis of the constituents of infectious organisms and to stimulate antigen-specific T lymphocytes. Each class of MHC protein is recognized by a distinct T cell population. Thus CD4 T cells only recognize antigen in combination with MHC class II molecules and CD8 lymphocytes only recognize antigen on cells which express class I MHC proteins. The elucidation of the biology of the MHC can be traced back to the work in the 1930s of Peter Gorer in England and George Snell in the USA, who were interested in the process of tissue rejection; and the impetus to study the MHC has largely come from transplantation surgery. The two best-studied MHC systems are the murine MHC complex known as H-2 and located on chromosome 17, and the human MHC termed HLA (human leukocyte antigens) on chromosome 6. For space considerations only the human MHC will be described. In the human genome there are three independent MHC class I loci (termed A,

B and C) coding for a transmembrane polypeptide of molecular mass 43 kDa and composed of three immunoglobulin-like extracellular domains (α_1, α_2, α_3). Each class I molecule is expressed on the cell surface in association with a 12 kDa polypeptide, β_2-microglobulin, which has an immunoglobulin domain structure. Class I molecules are expressed on most cells, the main exception being erythrocytes. Levels of expression vary but can be upregulated by α, β and γ interferons. Class II molecules are encoded by three sets of genes known as DP, DQ and DR, each of which encodes for a cell surface glycoprotein consisting of two chains (α, 35 kDa; and β, 28 kDa), both with two immunoglobulin domains ($\alpha_1\alpha_2/\beta_1\beta_2$). In contrast to class I molecules, the MHC class II proteins are normally expressed only on the surfaces of B lymphocytes and APCs (macrophages and dendritic cells). Other cells can be induced to express class II molecules if stimulated by IFNγ and other proinflammatory cytokines.

Two important properties of the MHC contribute to its evolutionary success in dealing with infectious agents. The first is that the system is polygenic and both maternal and paternal genes are expressed (expression is said to be co-dominant). Thus with three class I genes and four class II genes each individual human can express six different class I and

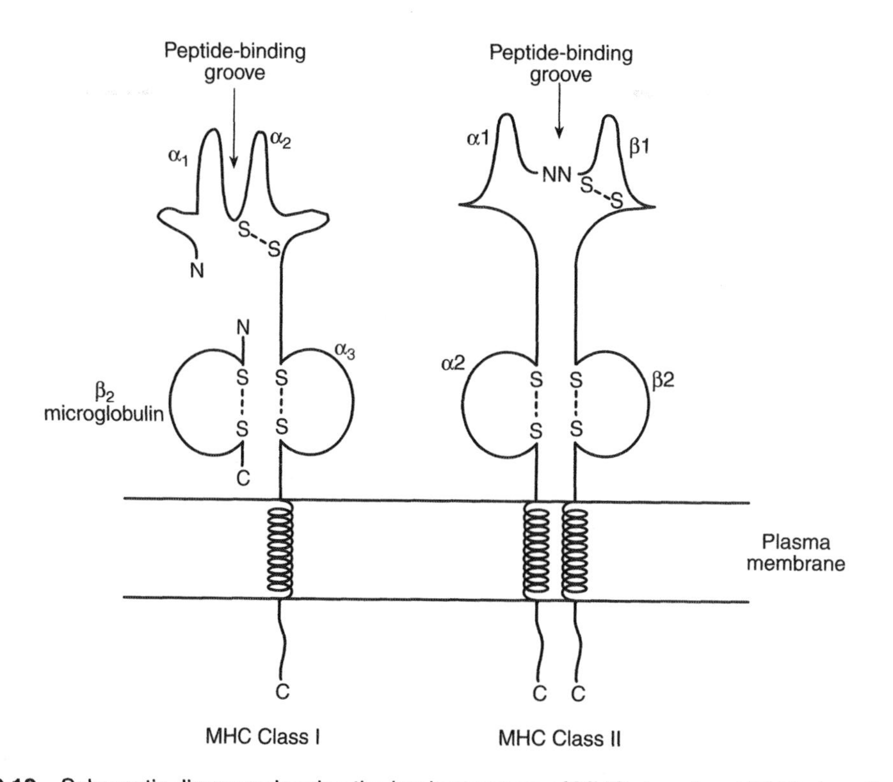

Figure 9.12 Schematic diagram showing the basic structure of MHC class I and MHC class II proteins. The class I molecule consists of an α chain of 44 kDa consisting of three domains and associated with a separate protein of 12 kDa termed β_2 microglobulin. The α_1 and α_2 segments are homologous and interact to form a peptide-binding domain which has the structure of a groove in which the peptide sits. In contrast to the structure of class I molecules, class II proteins are composed of two non-covalently associated polypeptide chains: the 32–34 kDa α chain and the 29–32 kDa β chain. These chains associate to form a peptide-binding domain

eight different class II molecules on his or her cells. In fact, by combining different α and β chains from different chromosomes, each individual human can express between 10 and 20 different class II molecules. The second attribute of the MHC system contributing to its success is that it is polymorphic (from the Greek *poly*, many; *morph*, shape or structure). Thus while each individual has only a relatively small number of MHC molecules, within the population as a whole there are different forms, or alleles, of each MHC gene. For example, there may be up to 100 alleles of some MHC loci. The mechanism responsible for generating and maintaining the MHC polymorphism is not clear, but it is believed that the major selection pressure comes from infectious organisms.

X-ray crystallographic analysis of MHC class I and class II proteins has revealed that the part of the molecule furthest from the membrane contains a deep groove or cleft. This groove has proved to be a binding site for peptides. The groove in the class I molecule can accommodate linear peptides of 8–10 residues. In the case of the class II protein it can bind peptides 12–25 residues in length. It is this MHC groove, with its associated foreign peptide (derived from an infecting organism), that is recognized by the TCR (Figure 9.13). Thus the MHC plays a major role in the recognition of infectious organisms. How the peptides gain access to the MHC proteins will be described in the next section.

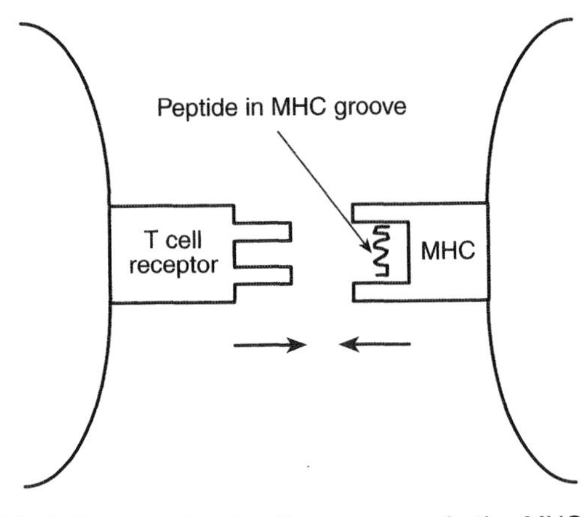

Figure 9.13 Highly simplified diagram showing the presence in the MHC groove of bound peptide to remind the reader that it is the complex of peptide and MHC molecule that is recognized by the CD4 or CD8 T lymphocyte via its T cell receptor

The description of the MHC system given above has derived largely from model systems using simple proteins. However, in recent years, driven by the need to understand the true anti-infectious functions of immunity, attention has focused on the presentation of bacterial molecules to T cells. Genes encoding relatively non-polymorphic MHC class I molecules have been found in man and mouse. One of these molecules, H-2M3 in the mouse, has been found to present bacterial *N*-formylmethionine peptides to T cells. The synthesis of all bacterial (but not mammalian) proteins begins with a formyl-substituted methionine residue which is generally removed post-translationally. These peptides are released from bacteria and are well-known and potent chemoattractants for PMNs (and described in Chapter 3). It now appears that in bacterial infections these peptides can induce the expression of H-2M3. This is now believed to be particularly important in the stimulation of CD8 T cells specific for *Listeria monocytogenes*. An even more exciting finding is that there is another non-polymorphic family of antigen-presenting molecules – the CD1 gene family (in man *CD1 A* to *E*). It has now been recognized that human CD1b and CD1c present, not peptide antigens, as has been the immunological dogma, but lipid and lipoglycan antigens (of *Mycobacterium tuberculosis* and *M. leprae*) to T cells. This opens up a new mechanism for the recognition of bacteria by the acquired immune system (Figure 9.14). As will be described, the presentation of bacterial peptides on MHC I or MHC II requires proteolytic processing of bacterial proteins in either the proteosome or in lysosomes. This obviously does not occur with non-peptide antigens. Most work on CD1 presentation has been done with lipoarabinomannan (LAM), an LPS-like molecule from *M. tuberculosis*. LAM is taken up into APCs via pattern recognition receptors (see Chapter 8) such as CD14 and/or the mannose receptor. It is in the acidified late endosome that changes in the conformation of CD1, due to the low pH, allow the LAM to bind, with surprisingly high affinity to the CD1. The CD1–LAM complex then traffics to the plasma membrane, where it can interact with antigen-specific T lymphocytes. Indeed, it has been reported that γδ T cells can recognize one of the building blocks in lipid synthesis, isopentenyl pyrophosphate.

Is there any evidence for the participation of CD1 in infectious diseases? The best example described thus far is leprosy. Dendritic cells bearing CD1a, -b and -c have been found in the granulomas of the skin lesions of leprosy patients. The frequency of cells bearing CD1 was found to correlate with the level of cell-mediated immunity to *M. leprae*. Patients with the

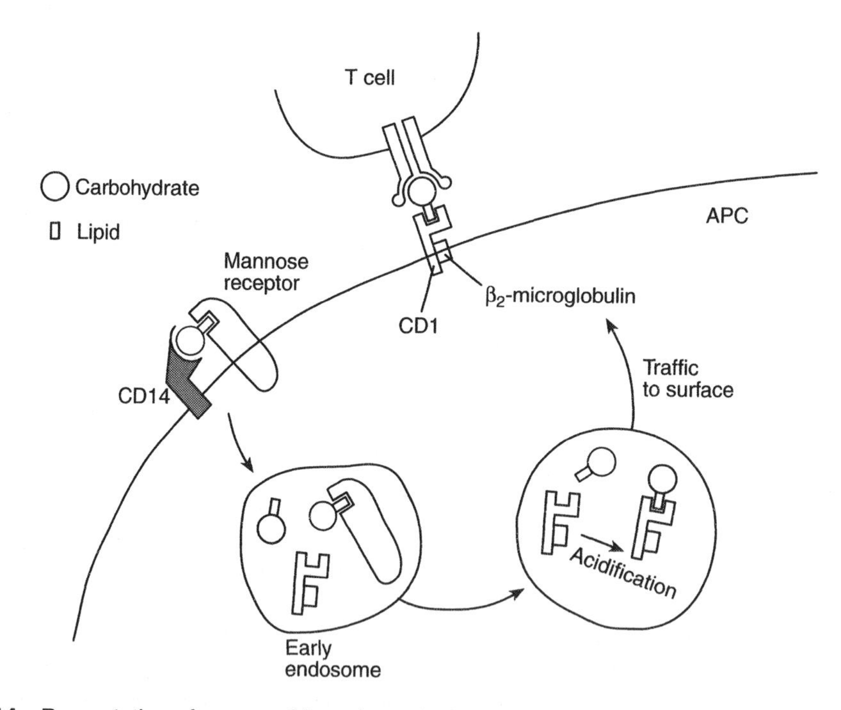

Figure 9.14 Presentation of non-peptide antigen via CD1. A lipoglycan, in this example it is LAM, binds to the surface of an antigen-presenting cell (macrophage or dendritic cell) via pattern recognition receptors such as CD14 or the mannose receptor (see Chapter 8). The mannose receptor is taken up into endocytotic vesicles where at low pH it undergoes a conformational change, allowing it to bind to the lipoglycan. The CD1–LAM complex can then traffic to the cell surface where it is recognized by the appropriate T cell via its T cell receptor

responsive tuberculoid form of disease had 10 times as much CD1 expression as did those with the non-responsive lepromatous form of the disease. These findings are beginning to rewrite the 'laws' by which immunologists believed the acquired immune system works. They also demonstrate another case in which the innate and acquired immune systems are linked. The presentation of LAM requires the participation of pattern recognition receptors which, as described in Chapter 8, are the major recognition and activation antigens in innate responses to infectious agents.

T cell activation

One of the unsolved mysteries of immunity is how the system prevents recognition of self antigens and the development of autoimmune disease. Such disease, which is believed to be orchestrated by self-reactive T cells, includes organ-specific conditions such as myasthenia gravis (muscle), Hashimoto's thyroiditis (thyroid) and diabetes (pancreas), and conditions which appear to affect the connective tissue systems (rheumatoid arthritis, systemic lupus erythematosus). In aggregate, these diseases (of which there may be 30-odd recognized conditions) affect 1–3% of the population. It used to be thought that the immune system was unable to recognize self. However, it is now realized that we all exhibit low-affinity autoantibodies in our blood. Thus reactivity to self antigens can occur, but is potentially dangerous as it could lead to autoimmunity, and therefore the immune system takes great care to ensure that such activation is carefully controlled. This is done by requiring that when a T cell meets an APC activation is only achieved if both cells share the correct cognate signals (Figure 9.15). The first signal is obviously that the T cell receptor recognizes the complex of the MHC and antigenic peptide. However, as stated, this is not sufficient to stimulate the T cell. This may simply reflect the low affinity of the binding between the TCR and the MHC–peptide complex and may require additional binding events to stabilize the cell–cell complex. With CD4 cells the CD4 antigen must bind to a non-polymorphic region of the class II MHC molecule; with CD8 T cells the CD8 binds to MHC class I. Additional binding events include leukocyte function antigen (LFA)-3 and intercellular adhesion molecule (ICAM)-1 on the APC binding to CD2 and LFA-1, respectively, on the CD4 T cell. If these binding events are satisfied, the gene for the T cell growth factor IL-2 is transcribed. However, due to the mRNA for IL-2 (as is the case for most cytokines) being extremely labile, the protein is not produced and, in the absence of additional signals, this T cell will enter a state of unresponsiveness known as T cell anergy. This is a fail-safe mechanism to ensure that we are not continuously producing inappropriately activated T cells. In order for the T cell to be triggered into clonal proliferation additional signals, known as co-stimulatory signals, are required. At present, the best-understood co-stimulatory signals are the B-7 family of molecules (B7–1 also known as CD80 and B7–2 or CD86) on the surface of APCs which interact with CD28 on naive T cells and the related molecule, CTLA-4 (CD 152), on activated T cells. The binding of B7 to CD28 stabilizes IL-2 mRNA transcripts, allowing IL-2 to be produced and act in its capacity as a T cell growth factor. T cell activation involves a complex cascade of interacting kinases and phosphatases including Ras, Rac, calcineurin, the MAP kinase pathway, PI3 kinase, PLC etc. There is insufficient space to describe these specific signalling pathways but the reader is referred to Chapter 3, which contains an overview of these and related signalling pathways.

The induction of the B-7 receptors is a major controlling step in the activation of T lymphocytes. It has been proposed that the PAMPs described in Chapter 8, are important in the upregulation of B-7 receptors on APC, leading to specific activation of lymphocytes

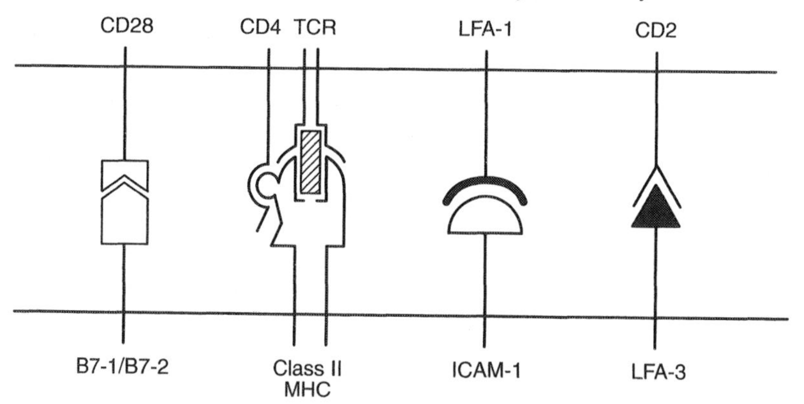

Figure 9.15 Recognition signals required to activate a T cell (in this case a CD4 lymphocyte). The obvious first one is the antigen recognition by the T cell receptor (TCR) of the complex of peptide and MHC II. Further pairing of cognate receptors strengthens this binding event. This includes CD4 binding to MHC II, LFA-1 binding to ICAM-1 and CD2 binding to LFA-3. However, in order to stimulate T cell clonal proliferation the antigen-presenting cell must express B7–1 (CD80) or B7–2 (CD86) and the T cell must express CD28

recognizing such pathogens. This again ties together innate and acquired systems into a continuum and demonstrates that the former system has a major role to play in the control of the latter. Given the importance of such co-stimulatory signals it is likely that this B-7 receptor system will be a target for pathogen immune avoidance and there are reports that certain bacteria such as *M. tuberculosis* can inhibit the synthesis or expression of B-7 and therefore block antibacterial lymphocyte responses. This promises to be a key area of pathogen–immune evolutionary interactions and also an area which could lead to the production of immune response-controlling drugs.

ACQUIRED IMMUNITY IN ANTIBACTERIAL DEFENCE

The reader not familiar with modern immunology will now have sufficient information to understand how acquired immunity has evolved to cope with infectious agents. In the rest of this chapter the focus will be on the ability of acquired immunity to deal with bacterial infections. Much of our understanding of the role of acquired immunity in dealing with infections comes from the related discipline of virology and some of the recent information on how viruses attempt to defeat the actions of acquired immunity will be discussed.

The immune system of the vertebrate (in particular the mammal) has been evolutionarily shaped by the nature of the organisms which infect it. More specifically, the system has been cast in response to the environments within the host in which infectious organisms have evolved to live. The first major division is between organisms which live outside cells (many bacteria, certain protozoa, helminths) and those which live within cells (prions, viruses, certain bacteria, certain protozoa) (Table 9.2). Those organisms which live in cells can do so within two discrete compartments. The first is the cytoplasm of the cell within which one finds viruses and certain bacteria and these organisms can be termed cytosolic pathogens. The other environment is the intravesicular apparatus of the cell – the endosomes and acidified phagolysosomes – which is much beloved of bacteria like *M. tuberculosis* and *M. leprae*, and certain protozoans. The cell biology of the vesicular apparatus and the bacteria which can live there has been reviewed in Chapters 2 and 6. By understanding where infecting organisms live, one can see the logic of acquired immunity, and this is shown in diagrammatic fashion in Figure 9.16. Extracellular pathogens, and the toxins they release, are dealt with by B cells and their product – the antibody molecule which ultimately stimulates macrophages. Intravesicular pathogens such as *M. tuberculosis* are the target of CD4 T lymphocytes (specifically Th_1 cells) and require the participation of MHC class II molecules and the newly discovered CD1 system. Cytosolic parasites are controlled by CD8 T lymphocytes in conjunction with MHC class I molecules. It is now becoming clear that it is the packaging of peptides from intracellular pathogens onto the groove of class I or class II MHC molecules that delivers the appropriate signal to the correct T lymphocyte. In order to understand this more fully, the next section will describe how MHC molecules bind to immunogenic peptides. It should now be appreciated by the reader what a major role the phagocytic macrophage, and the related dendritic cell, plays in acquired immunity.

Table 9.2 The two major subdivisions of pathogens and the diseases they cause

Extracellular	Disease	Intracellular	Disease
Bacteria		**Prions**	
Staphylococcus aureus	Boils, toxic shock etc.	BSE prion	BSE
Streptococcus pneumoniae	Pneumonia		
Protozoa		**Viruses**	
Entamoeba histolytica	Amoebiasis	Variola	Smallpox
		Influenza	Influenza
Fungi		**Bacteria**	
Candida	Candidiasis	*Mycobacterium tuberculosis*	TB
		Listeria monocytogenes	Listeriosis
Helminths		**Protozoa**	
Tapeworms	Hydatid disease	*Trypanosoma* spp.	Trypanosomiasis
Flukes		*Plasmodium* spp.	Malaria
		Helminths	
		Trichinella spiralis	

Antigen processing and presentation on MHC class I/class II proteins

Figure 9.17 provides a schematic overview of the process of antigen presentation to CD4/CD8 lymphocytes. A simple explanation will be given, but more detail can be found in the various textbooks in the reference section of this chapter.

Exogenous antigens taken up by APCs (dendritic cells, macrophages, B cells), or bacteria such as *M. tuberculosis*, end up in lysosomal vesicles which are acidic and contain a population of proteases and peptidases which result in the production of peptidic breakdown products. MHC class II molecules are synthesized on the rough endoplasmic reticulum (ER) and traffic via the Golgi and trans-Golgi network to eventually reach the cell surface. As

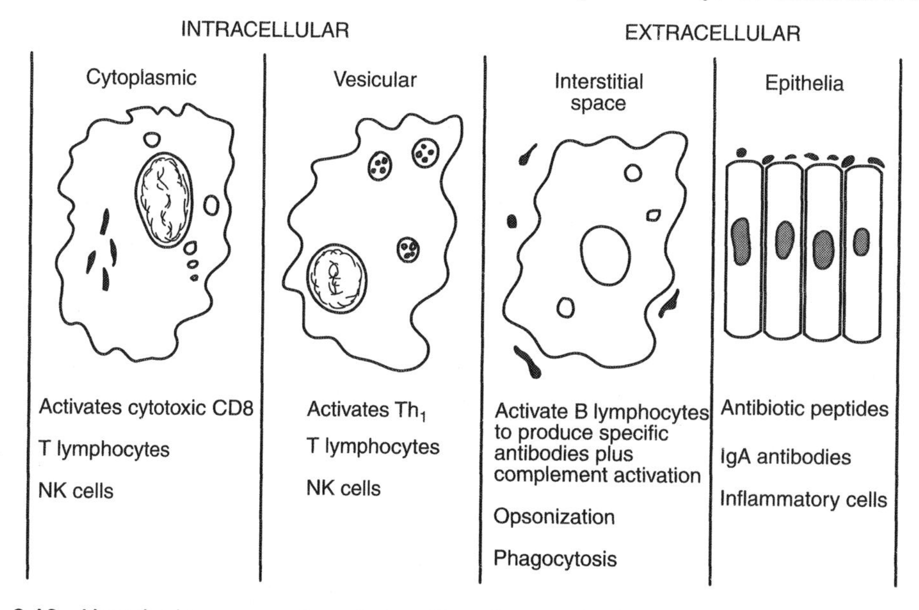

Figure 9.16 How the immune system deals with pathogens in the various cellular and extracellular sites which they infect

these class II molecules are synthesized, they associate with a third molecule known as invariant chain (CD74) to form a complex which prevents the binding of self peptides generated in the ER. Vesicles containing MHC class II–CD74 complexes interact with acidic vesicles at some point on their way to the cell surface. In the acidic environment the invariant chain (CD74) dissociates and the class II proteins can bind to foreign peptides present in the acidic vesicle. These then traffic to the surface of the APC, where they can be sampled by passing CD4 T lymphocytes.

Proteins that produce peptides which bind to MHC class I molecules undergo a distinct processing pathway. Unlike the proteins in the vesicular apparatus, which can really be said to be still outside the cell (very much like the mouth, oesophagus and intestine can be thought of as forming an external and continuous cavity within the body), antigens binding to class I molecules are endogenous in that they are synthesized within the cell. In this case we are dealing with any cell, not just APCs. This is the site of synthesis of viruses and certain bacteria and parasites. Peptide fragments of cytoplasmic proteins are believed to be continually generated by an ancient and ubiquitous intracellular protein complex called a proteosome. The proteosome is a barrel-shaped hollow structure with four layers of rings, each composed of seven subunits. Proteolysis takes place inside this central cavity. Peptide fragments are selectively transported into the ER by two transporter proteins known as TAP-1 and -2 (transporter associated with antigen processing). In the ER these peptides can bind to newly synthesized MHC class I molecules. Complexes of MHC class I (associated with β_2-microglobulin) move to the surface of the cell and interact with antigen-specific CD8 T lymphocytes.

It should now be clear to the reader that although the immune system can generate possibly trillions of TCRs, the only ones that can bind to antigen are the those that also associate with the appropriate MHC protein. In the case of class I MHC molecules the only

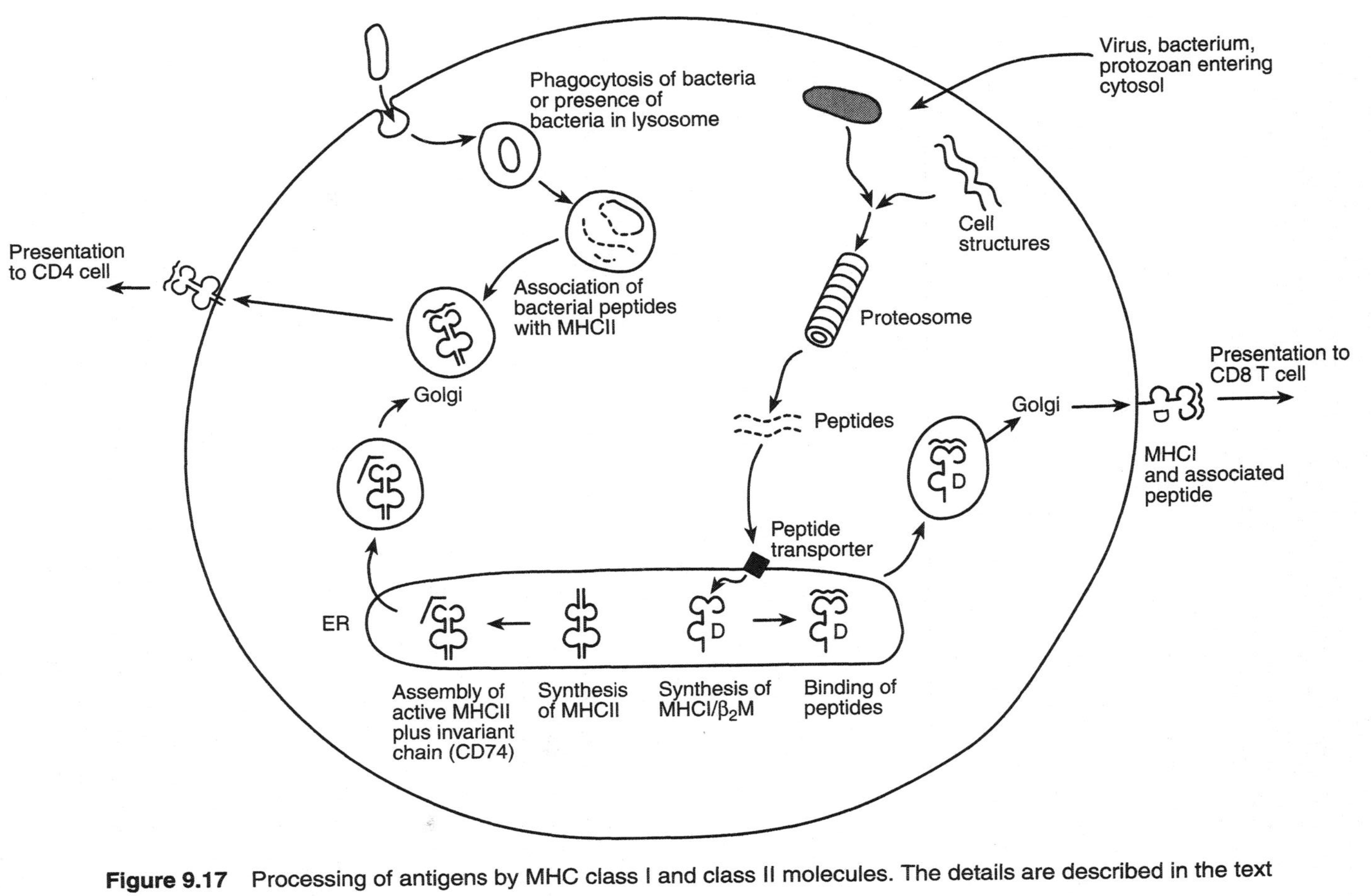

Figure 9.17 Processing of antigens by MHC class I and class II molecules. The details are described in the text

peptides that can bind appear to be those generated by the proteosome. As the proteosome evolved very much earlier than the acquired immune system (indeed it appeared early in bacterial evolution) it is now suggested that proteosomes may have influenced the evolution of MHC class I molecules (see paper by Niedermann *et al.*, 1997).

B cell immunity to bacteria

Antibodies of the IgG, IgM, IgA and IgE isotypes are important in the defence against a range of pathogens including viruses, bacteria, fungi and protozoa. The biological roles of the various antibody isotypes are shown in Table 9.4. Antibodies of the IgG and IgA isotypes are the most efficient at neutralization, be it of bacterial toxins or, by binding to bacteria, of bacterial adherence. As described in Chapter 7, many bacteria make exotoxins which may be the major causes of pathology. Good examples are the exotoxins which produce the pathology of tetanus, anthrax and botulism. In recent years, there has been increasing concern about the toxins of *E. coli*, particularly the Shiga-like toxin produced by serogroup 0157. Two of the most successful vaccines for bacterial infections are dipththeria and tetanus toxoids. A toxoid is an inactivated (normally using formaldehyde) toxin, but one which retains its immunogenic structure. This shows the protective nature of antitoxin antibodies. It is important that such toxoids should produce a memory response to provide long-term protection.

A very important role for one antibody isotype is in the blocking of adherence of bacteria to host cells. This is the function of IgA antibodies at mucosal surfaces. These antibodies are made by plasma cells found in the lamina propria, lying immediately under the basement membrane of mucosal surfaces. Of course, this epithelial layer is a barrier to the antibody getting to the luminal surface. It is now clear that the epithelial cells at mucosal surfaces allow bacteria (and also certain bacterial toxins) to traffic through them (see Chapters 5 and 6). These same cells also enable IgA antibody to move in the opposite direction (Figure 9.18). The IgA molecule is produced as a dimer linked together by a 15 kDa protein called a J chain (not to be confused with the J gene segments described earlier). This complex binds to a receptor called the poly-Ig receptor on the basolateral face of the epithelial cell. The receptor–IgA complex is then internalized and carried in a transport vesicle through the epithelial cell cytoplasm to the apical surface. At this apical surface the poly-Ig receptor is proteolytically cleaved, releasing the extracellular portion of the receptor (which is termed the secretory component) still attached to the IgA dimer. This secretory component may protect the IgA from proteolytic cleavage. Secretory IgA antibodies are believed to be particularly important in preventing infectious bacteria from adhering to epithelial surfaces. However, IgA deficiency is relatively common in the Caucasian population (with one person in 700 demonstrating deficiency), and such individuals do not appear to have a greater susceptibility to infection, although IgA deficiency appears to be commoner in individuals with chronic lung disease, possibly suggesting a protective role for IgA in lung infections.

Antibodies can bind to the surfaces of bacteria but this has little direct effect. Some mechanism is required to link such binding to the killing of the bacteria. Many leukocytes have cell surface receptors for the Fc region of immunoglobulins. On phagocytic cells (PMNs, macrophages) the Fc receptors recognize the IgG bound to bacteria, rather than free IgG which is present in excess, because the IgG on the bacterial surface is aggregated and shows subtle conformational changes which increase its affinity for the Fc receptor. These

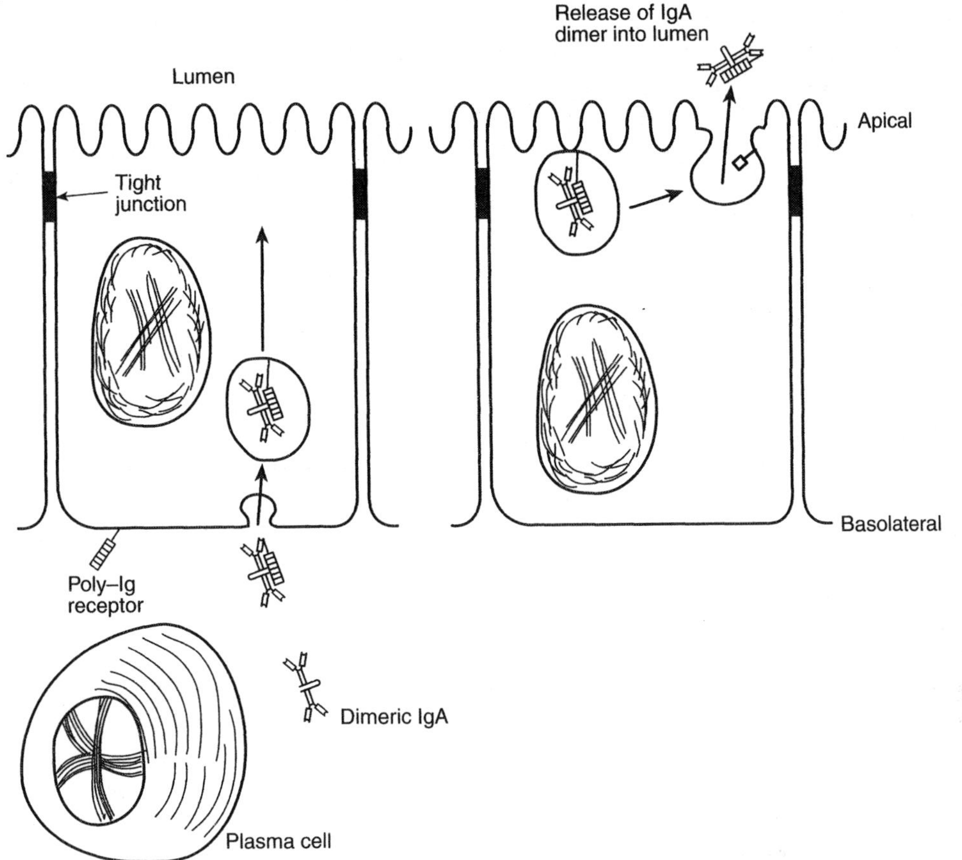

Figure 9.18 Trafficking of IgA antibody at mucosal surfaces. Local plasma cells produce dimeric IgA which diffuses across the basement membrane and binds to the poly-Ig receptor on the basolateral surface of the epithelial cell. The receptor–IgA complex then undergoes transcytosis, a process involving vesicular transport to the apical surface where the poly-Ig receptor is cleaved, leaving the extracellular IgA-binding component bound to the IgA molecule, in which form it is known as the secretory component. This component is eventually degraded, leaving the IgA dimer on the external mucosal surface

antibodies therefore act to opsonize the bacterium and enable phagocytes to ingest and kill the organism. The mechanisms used by phagocytes to kill ingested bacteria have been described in Chapter 8. Other leukocytes, e.g. eosinophils and mast cells, secrete factors when their Fc receptors are engaged. The best-known example of this is the mast cell which binds IgE (reaginic) antibody through high-affinity Fc receptors. When these receptors are cross-linked by antigen, the mast cell is triggered and releases a range of proinflammatory signals which aids in the body's defence against pathogens. Mast cells were believed to be involved mainly in immunity to helminths, but in recent years they have been implicated in antibacterial responses.

An important role played by antibodies, particularly of the IgG and IgM isotypes, is in the activation of the classical pathway of complement. This pathway has been described in some detail in Chapter 8 and will not be discussed in any detail in this chapter. The key recognition element in the classical pathway is the complex molecule C1q,C1r,C1s. The C1q oligomer, which is shaped like a bunch of six tulips, binds to immunoglobulin molecules participating in immune complexes with bacteria. Conformational changes in C1q due to binding to the Fc region of complexed immunoglobulin then activate C1r and C1s and thus triggers the whole pathway. This results in the production of proinflammatory anaphylatoxins such as C5a and the formation of an oligomeric molecule termed the membrane attack complex, which can form pores in certain bacterial cell walls and kill the organism. Deficiency in components of the C1 complex renders sufferers susceptible to recurrent infections with pyogenic bacteria (see Chapter 8).

Bacterial defences against antibodies

B cell activation and the consequent production of antibodies is an important antimicrobial defence and microorganisms have developed means of evading such antibodies. These defences can be divided into those that prevent the microorganism from being immunogenic and those which in some way inactivate the antibody or prevent it binding. The role of the bacterial capsule in preventing recognition of bacteria has been described in Chapter 8. One strategy to prevent recognition is by coating the bacterium with immunoglobulin. By doing this the bacterium would become invisible as far as the immune system is concerned. Protein A, produced by the Gram-positive bacterium *Staph. aureus*, was the first such bacterial protein to be found. This protein binds to the Fc region of immunoglobulins, thus preventing the effector functions of antibodies from coming into play and shielding the organism from immune surveillance. Since the discovery of protein A, a large number of Fc-binding proteins produced by bacteria have been discovered. Some of these proteins are of commercial importance in the isolation of antibodies.

An alternative to using the antibody molecule to shield the organism is to directly attack the antibody. This is a strategy adopted by a number of bacteria on mucosal surfaces where IgA antibodies are believed to function to prevent adherence of bacteria to the epithelial cells. Gonococci, streptococci, certain oral bacteria and *Haemophilus influenzae* produce proteases with specificity for the destruction of IgA and in consequence are known as IgA proteases. It is proposed that by destroying IgA locally the inhibitory effects of this antibody on bacterial adhesion and colonization are neutralized.

Antibodies need to bind to the bacterium to have any effect. Many bacteria release components from their cell surfaces, including LPS, capsular polysaccharides and surface-associated proteins. In pneumococcal pneumonia and meningococcal meningitis, bacterial polyaccharides are detectable in body fluids. If these bacterial components bind to sufficient antibody they could prevent opsonization of the bacteria.

There are increasing examples of bacteria and eukaryotic cells sharing common macromolecules. Bacterial macromolecules which share similarity with eukaryotic molecules are likely not to be immunogenic. Thus in evolutionary terms there should be pressure on bacteria to resemble their hosts. *Bacteroides* spp., the commonest commensal bacteria in the mouse intestine, share antigens with intestinal epithelial cells and mice are known to be unresponsive to *Bacteroides* antigens. Thus it is possible that by evolving to produce surface components that mimic host cell components, bacteria can fool the immune system. Of course, the best-known example of bacterial molecular mimicry is the cross-reactivity between β-haemolytic streptococci and the human myocardium, a response which has severe consequences for the infected human in the form of rheumatic fever and heart valve damage.

The immune responsiveness to another shared series of proteins, the molecular chaperones, will be discussed in a later section of this chapter.

The final example of microbial defences against antibody-mediated immunity are the phenomena of antigenic and phase variation. The former describes a process in which an antigen is qualitatively varied. In contrast, in phase variation, antigens are either expressed or are absent. These mechanisms are particularly important for extracellular pathogens whose surfaces contain the immunodominant antigens responsible for antibody production. There are essentially two mechanisms resulting in antigenic variation. In the first, random mutation in immunodominant epitopes results in a population of antigenically variant bacteria that can be categorized as different serotypes or serovars (serological variety) (these terms arising from the fact that bacteria can be distinguished serologically). Many infectious organisms exist as a variety of antigenic types. As an example, there are more than 80 known types of *Strep. pneumoniae*, the organism responsible for pneumonia. Each type has an an antigenically distinct polysaccharide capsule. Infection with one serotype results in protective immunity only to that serotype, but not to bacteria with a different serological reactivity. The immune system therefore sees each serotype as a separate organism. In the second mechanism, there is an element of control over the change in DNA, either because of a specific DNA rearrangement or a region of DNA that is prone to mutation. These types of trick to avoid the immune system are, of course, found in other types of pathogens. The antigenic drift found in influenza virus proteins is an example of the first mechanism, and antigenic shift in influenza virus is an example of a genome rearrangement that is possible because of the genome arrangement of segmented viruses. The variant-specific glycoprotein of trypanosomes is another example of a programmed DNA rearrangement. DNA rearrangement also occurs in a number of bacteria. The first such example was found in the bacterium *Salmonella typhimurium*, a common cause of food poisoning. This organism can invert a segment of its DNA which contains the promoter region for the bacteria's flagellin genes (termed H1 and H2). In one orientation the promoter drives expression of H2 and also a repressor of the H1 flagella expression. When the promoter is inverted, neither gene is expressed, enabling H1 to be expressed from its own promoter. A more complex example is provided by switching between FimA and FimE expression in *E. coli*, where other linked genes are involved in the regulation of the inversion event. *Neisseria gonorrhoeae*, the causative agent of gonorrhoea, can also alter a surface pilin protein and other proteins as well. There are two mechanisms involved: recombination of silent pilin genes that places them downstream of a promoter, and transformation of DNA from lysed cells followed by recombination. This latter mechanism resembles the antigenic shift found in influenza virus where 'new' DNA is brought into the organism. Antigenic variation has been shown to occur during *Neisseria* infection and is even displayed at different stages of the menstrual cycle. *Borrelia* species, which cause relapsing fever, can express around 30 different genes, called variable major proteins, by transposing silent genes carried on linear plasmids into an expressing site. The disease causes episodes of fever, separated by about two weeks where infected individuals are free of symptoms, this being linked to the appearance of new serotypes. There appears to be some regulation to the order that these serotypes appear, with certain ones occurring at early relapses and others at later stages in the disease. Other variations occur as a result of features of the DNA that promote mistakes during DNA replication either as a result of slipped-strand mispairing or recombination.

Haem. influenzae, a commensal bacterium, but one which can cause meningitis, demonstrates antigenic variation in its LPS structure. Three genetic loci, *lic1*, *lic2* and *lic3*, have been identified as containing genes encoding the phase variation of glucose and galactose residues in the outer core of the LPS. Multiple tandem repeats of the tetrameric oligonucleotide 5'-CAAT-3' are found within the first open reading frame of each of the loci and changes seen in the number of repeats have been implicated as the mechanism for generation of phase variants. The LPS of *Legionella pneumophila* has recently been reported to undergo phase-variable expression which contributes to the virulence of this organism.

Runs of the same nucleotide are obviously prone to error, and a run of about 15 cytidine residues in the fimbrial promoter of *Bordetella* species causes variation. *Bordetella* fimbrial variation is known to be important in the design and efficacy of *Bordetella pertussis* vaccines. A similar run of bases in the *bvg* promoter region causes phase variation, but it is less clear whether this is significant for disease. Other examples are found in *Neisseria* and *Yersinia*.

Thus it is clear that bacteria and other microorganisms have a wide range of mechanisms able to subvert the actions of antibodies and, as described in Chapter 8, certain of the consequences of antibody binding.

T cell defences against bacteria

It is important to reiterate that extracellular parasites, be they bacteria, protozoa, helminths etc., are dealt with by antibodies and the consequences of antibody binding, such as pathogen ingestion by phagocytes. If the infecting organism enters and resides within a cell then antibodies are of little use and the two other T cell subpopulations (CD4 Th_1 and CD8 lymphocytes) are called into play. CD4 lymphocytes recognize antigen presented on MHC class II antigens which has come from the vesicular apparatus of the host cell, normally an APC. This results in appropriate signals that generate Th_1 cells, which can then interact with the bacteria-laden cell and activate it to kill the organisms. This will be dealt with later. CD8 lymphocytes recognize antigen which has come from the cytoplasm of the cell and is presented on MHC class I antigens of almost any cell in the body with the exception of erythrocytes. These cells are the mainstay against viral infections but they are also important in killing cells bearing cytoplasmic bacteria.

Having said this, a role for antibodies in protecting against intracellular pathogens has recently been suggested in a review by Casadevall which the interested reader will find in the references.

Role of CD8 cells and related cytotoxic cells

The preferred strategy for killing viruses, which only survive inside eukaryotic cells, is to kill the virally infected cell. Certain bacteria, such as *Lis. monocytogenes*, live, like viruses, within the cell cytoplasm. It has been established that, in the mouse, CD8 cells can recognize listerial antigens in the context of the non-polymorphic gene product, H-2MR, which recognizes bacterial formylated peptides. However, the role of such cytotoxic CD8 cells in killing *Listeria* is not understood. Although this book is concerned with prokaryotic/eukaryotic interactions, there are lessons to be learned from the mechanisms cytotoxic T cells use to kill virally infected cells and the rich seam of countermeasures employed by viruses. It is likely that bacteria will also use such mechanisms

There appears to be two mechanisms by which CD8 cells kill virally infected cells. CD8 cells need to make intimate contact with antigen-specific target cells in order to kill them.

This cell-to-cell contact results in the release of pre-formed secretory granules containing two distinct forms of cytotoxins. The first is a protein termed perforin which, when released from the CD8 cell, can polymerize to form a pore-like structure, analogous to the membrane attack complex of the complement pathway, thus causing membrane leakage. The other cytotoxic moiety is a collection of at least eight serine proteinases (designated A to H), called granzymes or fragmentins, which can enter target cells and also cause cell damage, although they are not essential for killing.

There are two distinct ways in which a cell can die. It can undergo necrosis, which is a messy way to die and generally results in the induction of inflammation. The other way is via apoptosis, often referred to as programmed cell death (see Chapter 2). In apoptosis, the dying cell regulates its break-up and does not leave an 'inflammatory signature'. CD8 cells are thought to interact with certain target cells and induce apoptosis by two mechanisms. The first is the interaction of a member of the TNF receptor family called Fas on the target cell with a membrane-bound cytokine called Fas ligand on the CD8 lymphocyte, resulting in the activation of two proteins: FADD (Fas-associated death domain, also known as MORT1) and RIP (receptor interacting protein) (the scientists in this field have obviously got into the spirit of things). These proteins ultimately upregulate the group of intracellular proteases (caspases) known to control apoptosis, thus leading to cell death. Details of this signalling pathway are found in Chapters 2 and 3. This mechanism appears to be important in controlling lymphocyte responsiveness generally. In contrast, the killing of cells bearing intracellular microorganisms utilizes the perforin/granzyme system. The granzymes are thought to activate the pro-forms of the intracellular caspases, thus promoting the apoptotic death of infected cells.

Although not MHC restricted, the NK cell population is also involved in killing cells which have been infected with viruses.

Viral strategies to combat immune responses

During the last decade viruses have been found to have evolved a number of mechanisms for combating the host immune responses. Cytokines play an important role in the defence responses of acquired immunity and it is now established that viruses have evolved three major defence mechanisms for inhibiting cytokine networks, which have been described in Chapter 8.

In addition to having genes encoding cytokine-like molecules and cytokine-inhibitory molecules, viruses which are targetted by CD8 lymphocytes and NK cells also contain genes encoding other proteins to evade the attention of these immune cells.

As has been described earlier, the killing of virally infected cells by CD8 lymphocytes requires the recognition of viral peptides bound to MHC class I proteins. Prevention of expression of viral peptides would render infected cells invisible to CD8 lymphocytes, although it would potentially render them susceptible to NK cells. Viruses have developed a number of mechanisms for blocking the expression of MHC I containing viral peptides. The production of viral peptides in the cytosol, which will end up on class I receptors, is due to the action of the proteosome. Human cytomegalovirus (HCMV) and Epstein–Barr virus (EBV) can inhibit the generation of proteosome-derived viral T cell epitopes. Peptides produced by the proteosome have to be transported to the lumen of the ER in order to bind to the MHC class I proteins. Herpes simplex virus (HSV) types 1 and 2 inhibit this peptide transport mechanism. Once peptides have attached to the class I molecules they must be transported and inserted into the plasma membrane. A number of viruses block this process at various stages.

While inhibition of MHC class I expression can limit attack by CD8 T cells, such virally infected cells may still be susceptible to attack by NK cells. NK cells are normally prevented from killing other cells by the presence of MHC class I molecules containing self peptides on the cell surface. If viruses prevent the appearance of MHC class I proteins on the surfaces of infected cells then these cells become susceptible to NK cell attack. However, it has recently been described that human and murine CMV encodes an MHC class I homologue protein which can inhibit attack by NK cells.

A final and fascinating example of evasion is the finding that vaccinia and pox viruses encode a 3β-hydroxysteroid dehydrogenase. It is postulated that this enzyme synthesizes steroids which are immunosuppressive and therefore could inhibit host inflammatory responses.

CD4 T lymphocytes: the Th_1:Th_2 balance

An exciting hypothesis which curently has a strong hold on the immunological community can be traced back to the development of the technology for cloning murine T cells. Two types of cloned CD4 T lymphocytes were discriminated on the basis of the profile of cytokines they released. These two different murine CD4 lymphocyte subsets were termed Th_1 and Th_2 cells. The predominant cytokines made by the former are IFNγ, IL-2 and tumour necrosis factor (TNF)β. In contrast, the Th_2 T lymphocyte predominantly makes IL-4, IL-5, IL-6, IL-10 and IL-13 (Figure 9.19). These profiles of cytokine production have not been so clearly delineated with rat or human CD4 lymphocytes and it is likely that the real-life situation is more complex.

Th_1 CD4 lymphocytes produce the potent effector cytokine IFNγ and are also known as inflammatory T cells. Unlike the cytotoxic CD8 T lymphocytes, which kill infected cells, the CD4 Th_1 T lymphocyte activates macrophages which are infected with intracellular bacteria such as *M. tuberculosis* (Figure 9.20). The importance of this mechanism for killing intracellular pathogens can be seen with the organism *Pneumocystis carinii*, which appears to be a major cause of infections in humans, with an estimated 70–90% of children experiencing a respiratory infection with this organism by the age of four. However, such infections are either mild or asymptomatic. *Pn. carinii* has only come to prominence with the epidemic of AIDS. Up to 70% of AIDS patients have infections with this organism, and in these patients infection can be fatal. The suppression in the numbers of CD4 T lymphocytes is presumably the explanation for the severity of the response to this organism and shows how important this mechanism can be to our survival.

How do CD4 Th_1 T cells activate macrophages? The first requirement is that the T cell recognize the infected macrophage via the MHC class II–bacterial peptide complex. Once antigen specificity has been satisfied the CD4 Th_1 cell releases the potent macrophage-activating cytokine IFNγ. The second signal required is TNFα, either as a membrane-bound molecule on the T cells or, at higher concentrations, as a soluble molecule. These two signals activate the macrophage, allowing it to prime its intracellular mechanisms for killing bacteria. These include production of oxygen-derived free radicals, nitric oxide, defensins etc. (see Chapter 8). Activated macrophages also release IL-12, a key cytokine in the development of CD4 Th_1 T lymphocytes. The importance of both IFNγ and IL-12 in protective immunity to

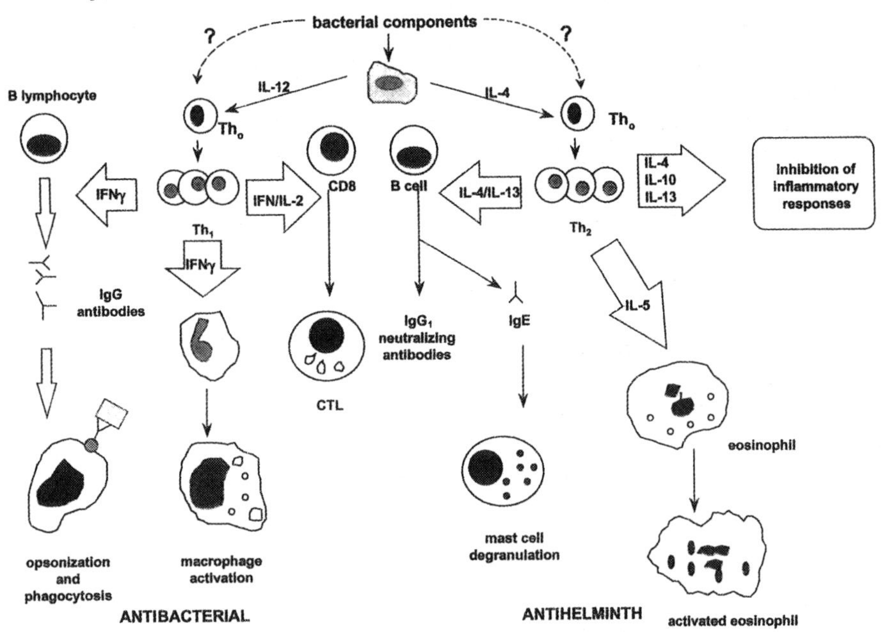

Figure 9.19 Th_1 and Th_2 cells, derivation and cytokine profiles (Reproduced from Henderson *et al* (1998) Bacteria-Cytokine Interactions: In Health and Disease, Portland Press, London, with permission)

M. tuberculosis is clearly seen in mice in which the genes for these cytokines have been inactivated by homologous recombination. This produces a so-called knockout mouse which cannot produce the active cytokine. Such animals are extremely susceptible to infection with *M. tuberculosis*.

Th_1 T lymphocytes can be regarded as cells which interact with macrophages to promote the latter's activity. The direct activation of antigen-specific macrophages has been described. Th_1 cells also can cause activated B cells to undergo class switching, to produce antibody classes or subclasses which are good at opsonizing bacteria (see Table 9.3), thus facilitating the uptake of bacteria into macrophages. The profile of cytokines produced by CD4 Th_2 T lymphocytes allows these cells to have a different range of actions from that of the Th_1 cell. First and foremost, the Th_2 cell is a helper T cell, enabling B cells to make antibody. The IL-4 produced by Th_2 cells is also able to cause B lymphocytes to class switch their immunoglobulin molecules to produce IgG_1 and IgE. IL-4 and IL-5 produced by Th_2 cells are also important for the activation and proliferation of cells involved in dealing with parasitic worms such as helminths (tapeworms, flukes and nematodes) that cause diseases such as schistosomiasis, and which are still major causes of morbidity worldwide (one third of the world's population are believed to have worm infestations). Much of our understanding of the immunology of Th_2 cells in dealing with helminths has used the nematode *Nippostrongylus brasiliensis*. Eosinophils and mast cells activated by IL-4 and IL-5 produce a complex web of interactions which involve B lymphocytes and intestinal epithelial cells in the killing and expulsion of worms. Unfortunately, the Th_2 cells and their products are also involved in the pathology of allergic diseases such as asthma, eczema, hay fever and food allergies and are the topic of intensive investigation. One hypothesis that is receiving much attention is that the two major classes of immunologically based diseases – autoimmune and allergic – are due to inappropriate and mutually exclusive alterations in the Th_1 to Th_2 balance.

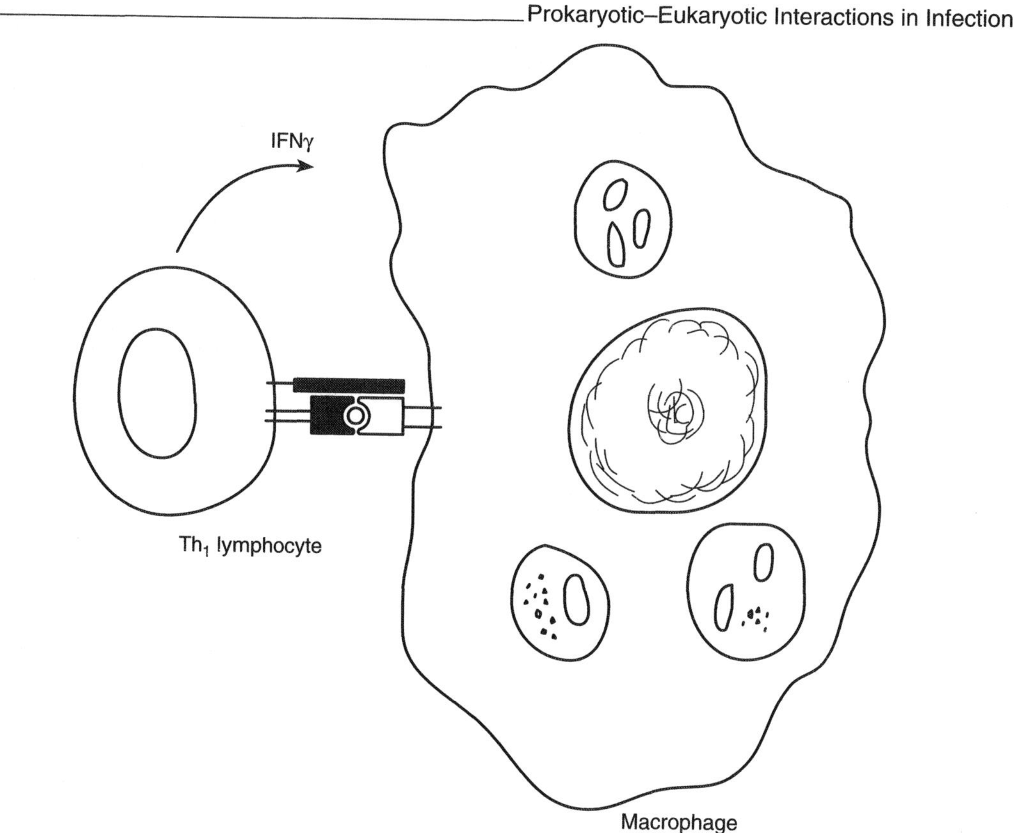

Figure 9.20 Antigen-specific interaction of a macrophage containing intracellular bacteria with a Th_1 lymphocyte results in the lymphocyte producing IFNγ, which stimulates the macrophage to kill its intracellular organisms

The Th_1:Th_2 balance

Th_1 and Th_2 cells are not genetically fixed cells, but are induced to differentiate from a precursor cell by appropriate environmental stimulation. Two hypotheses have been suggested to account for the generation of Th_1/Th_2 cells. They may either arise directly from virgin T cells (those CD4 T cells leaving the thymus to enter the periphery) or from an intermediate cell which has been termed a Th_0 cell. Different signals appear to be able to divert the virgin or Th_0 cells to differentiate into either Th_1 or Th_2 cells. The macrophage cytokine, IL-12, is one of the main signals causing the development of Th_1 cells. IL-4, produced by T cell subsets (such as a $CD8^-$/$CD4^-$/NK1 cell) is the major differentiating factor inducing the development of Th_2 cells. Bacterial LPS, for example, can stimulate macrophages to produce IL-12. It is assumed that many other bacterial components can do likewise, but very little is known about the signals emanating from bacteria that can modulate macrophage function and, in particular, IL-12 production. The same is true for the stimuli inducing IL-4 and the generation of Th_2 cells. The cytokines produced by Th_1 and Th_2 lymphocyte subsets have actions on the other subset. The CD4 Th_1 T lymphocyte, because of its ability to stimulate macrophages, is a potentially dangerous cell and is involved in the pathology of chronic autoimmune diseases. Thus, in any sensible evolved system it would be

Table 9.3 Properties and functions of antibodies

Property	Immunoglobulin								
	IgG_1	IgG_2	IgG_3	IgG_4	IgM	IgA1	IgA2	IgD	IgE
Heavy chain	γ_1	γ_2	γ_3	γ_4	μ	α_1	α_2	δ	ε
Molecular mass (kDa)	146	146	165	146	970	160	160	184	188
Valency for antigen binding	2	2	2	2	5(10)	2(4)	2(4)	2	2
Serum level (mg/ml adult)	9	3	1	0.5	1.5	3	0.5	0.03	0.00005
Half-life (days)	21	20	7	21	10	6	6	3	2
Activation of classical complement pathway	++	+	+++	–	+++	–	–	–	–
Activation of alternative complement pathway	–	–	–	–	–	+	–	–	–
Binding to phagocytes	+	–	+	–	–	–	–	–	+
Binding to mast cells and basophils	–	–	–	–	–	–	–	–	+++
Reactivity with protein A of staphylococci	+	+	–	+	–	–	–	–	–

under tight control. It is now established that IL-10 and IL-13, products of Th_2 lymphocytes, can inhibit the development of Th_1 cells by downregulating the activity of APCs. Likewise IFNγ can prevent the activation of Th_2 cells. IL-6 also has cytokine network-inhibitory properties.

It is likely that infectious organisms can modulate this Th_1:Th_2 balance for their own benefit and this ability may be related to the genetic susceptibility of the host. The correspondence between susceptibility to infectious disease, and the profile of Th_1/Th_2 subsets produced by individuals, was first clearly defined with the causative agent of leishmaniasis, *Leishmania major*, a protozoan resembling trypanosomes, which lives within macrophages. This organism can either grow within the liver and spleen, producing visceral leishmaniasis, or within the skin, producing cutaneous leishmaniasis. BALB/c mice when infected with this parasite fail to make IL-12 and thus fail to generate the correct Th_1 response. Instead, they produce helper Th_2 cells which produce antibody, which is of little use with an intracellular parasite. Thus BALB/c mice are very susceptible to this parasite. In contrast, C57BL/6 mice make IL-12 when infected, generate the correct Th_1 response, and recover from infection. Patients with progressive leishmaniasis are characterized by deficient IFNγ production and increased expression of the cytokine IL-10 which inhibits macrophages and the development of Th1 cells. Patients with leprosy also show two distinct CD4 responses. The earliest evidence for this concept was the recognition of the clinical patterns of leprosy. This disease, which has been so feared for thousands of years, is caused by *M. leprae* and two forms of the disease are recognized. Tuberculoid leprosy is frequently a self-limiting disease and has been known to regress spontaneously. The main features of this disease are the production of skin lesions and the involvement of nerves, producing sites of anaesthesia. The causative organisms are difficult to find in this form of the disease. In contrast, lepromatous leprosy is a progressive disease which, if not treated, is fatal. The causative organism is found in all organs and pathology is seen in the skin and nerves. The explanation for these two forms of the disease becomes clear when the immunological responses of patients are assessed. In tuberculoid leprosy, patients have a strong delayed-type hypersensitivity (DTH) response, but low levels of circulating antibodies. In contrast, in patients with lepromatous leprosy, the immunological pattern is the inverse of what has been described, with minimal DTH reactions and high antibody titres; exactly the immunological profile required to cope not with an intracellular parasite but with an extracellular one. In more recent years the profiles of cytokines in both groups of patients have been studied. In lepromatous leprosy the lesional tissues show a Th_2 profile of cytokines (IL-4, IL-5, IL-10) while in tuberculoid leprosy lesional tissues contain Th_1 cytokines. As described earlier, these patient groups also differed in their expression of CD1. Whether these different patterns of cytokines in the lesional tissues are due to host genetics or bacterial factors (or, more likely, a combination of both) is still not fully established.

Role of specialized B and T cells in antibacterial defence

Three lymphocyte subpopulations are believed to play specific roles in antibacterial defences. These are the CD5-positive B cell, the γδ T cell and a cell population defined largely by its topology – the intraepithelial lymphocyte (IEL).

CD5 B lymphocytes

The CD5 B lymphocytes comprise a cell population found predominantly in the various internal cavities of the body, for example the peritoneal cavity, where it is believed to be able to rapidly respond to bacterial invasion by producing a restricted range of IgM anti-polysaccharide antibodies. These antibodies are made rapidly and thus act as a stop-gap measure between innate immunity and the induction of acquired immunity. These cells have also been implicated in the pathogenesis of autoimmune diseases, but this may simply reflect the growing belief that autoimmune diseases are caused and/or driven by infectious agents (mainly viruses and bacteria).

γδ T Lymphocytes

The γδ T lymphocyte is attracting increasing attention for its possible involvement in antibacterial defence. These cells also appear to be rewriting some of the basic rules of immunology. In animals such as humans, mice and chickens γδ T lymphocytes represent only a small fraction of the circulating T cells. Most of our understanding has come from the study of the mouse and, at present, it is not clear if the findings in this species also pertain to the human. In mice the γδ T lymphocytes are a major cell population in skin and in mucosa. In mouse skin, the γδ T lymphocyte adopts a dendritic morphology allowing it to make cell-to-cell contact with nearby melanocytes and keratinocytes. The γδ T cells are also present in another major population of T cells, which will be discussed later in this section, the IELs.

The mechanisms of joining of the V, D and J segments to produce the diversity of the antibody molecule and of the TCR have already been discussed and the importance of the third hypervariable region in the antigenic specificity of antibodies has been described. The γδ T lymphocytes have smaller numbers of V, D and J gene segments than have αβ receptors, and gene rearrangement only produces a limited T cell repertoire. However, there is generation of a large amount of junctional diversity to form the third hypervariable region and in this respect the TCR of the γδ T lymphocyte resembles the antibody molecule. One of the central tenets of cellular immunology is that T cells only recognize foreign peptides attached to the groove in the MHC receptor. However, it is now recognized that the γδ T cell (and a small subset of αβ T cells which express neither CD4 nor CD8 receptors – so-called double negative cells) do not obey this tenet. These cells can therefore recognize peptides larger than those presented to conventional T cells on MHC molecules. One protein which a subset of γδ T cells recognizes is the heat shock protein, chaperonin 60. As chaperonin 60 is overexpressed on cells which are stressed, it is possible that γδ T cells can monitor cell stress locally. A major upset in immunological thinking occurred when it was demonstrated that γδ T cells could also recognize non-peptide antigens. These molecules were originally isolated from *M. tuberculosis* and have been shown to be the major stimulatory molecules of this and many other bacteria, both pathogenic and non-pathogenic. Phosphate is an essential component of such ligands and the term phospholigand has been introduced to describe them. The first such ligand to be identified was isopentenyl pyrophosphate, which is a key five-carbon building unit for the synthesis of an enormous number of molecules by the three cellular domains. Examples include rubber, plant fragrances, carotenoids, vitamins (e.g. vitamin D, vitamin K), steroids, cholesterol and bacterial lipids. It is thought that isopentenyl pyrophosphate on the membranes of bacteria is recognized directly by γδ T lymphocytes. The recognition of mycobacterial components such as lipoarabinomannan and mycolic acid by αβ T cells in the context not of class II antigens but the non-polymorphic CD1b has been discussed earlier in the chapter. Thus there is increasing evidence for the ability of T cells to recognize non-peptidic bacterial components by a novel mechanism which does not involve the MHC complex. This is a major departure from mainstream immunology and is probably only a taste of what is to come now that immunologists have begun to concentrate on the study of the true nature of immunity – a mechanism to protect the host from infection.

Having described some of the unusual aspects of γδ T lymphocytes, what is the function of these cells? Abalation experiments in which γδ T lymphocytes are inhibited or destroyed by antibodies, or genetic ablation to produce knockouts, exacerbate experimental infections with various bacteria and protozoa, particularly intracellular bacteria. Stimulated γδ T cells generally display cytolytic activity. An interesting recent finding has been that murine γδ T cells produce an epithelial growth factor. Thus, although the biology of these γδ T cells is only just beginning to be understood, they appear to act as a first line of acquired immune defence against bacterial infection at epithelial barriers and appear to have a role in detecting changes in epithelial integrity (possibly by monitoring stress protein synthesis) and thus may act to maintain epithelial integrity by the production of growth factors.

Intraepithelial lymphocytes

Mucosa, particularly that of the intestine, contains lymphocytes interspersed amongst the epithelial cells which are, unusually in immunology, correctly termed intraepithelial lymphocytes (IELs). Collectively the IELs represent a large pool of cells comparable in size to that of all the peripheral lymphocytes in the spleen. In the mouse about 30–40% of the IELs are γδ T cells and about 70% are CD8 positive. There appear to be two populations of IELs one that develops in the thymus (as all good lymphocytes do) and, surprisingly, one that differentiates independently of the thymus. This indicates that the gastrointestinal epithelium must be regarded as a primary lymphoid organ (along with bone marrow and thymus). These cells have some unusual and poorly understood properties. For example, it was recently shown that their generation requires factors from the associated epithelial cells. One of the surprising factors required for the production and maintenance of IEL numbers is thyroid-stimulating hormone (TSH), a peptide hormone normally produced by the thyroid gland. Thus it is clear that at mucosal surfaces the epithelial cells play a major role in the immune mechanisms.

M cells

In the gut the epithelium effectively excludes the uptake of intact antigens by transendothelial or other routes. However, there is a specialized cell population called the M cell which is functionally specialized to enable intact antigens, and ideed microbes, to cross the mucosal barrier. These M cells have a role in local antigen presentation to the lymphoid tissue of Peyer's patches, which are part of the GALT. M cells have specialized structures and functions such as a deeply invaginated basolateral surface which forms a large intraepithelial pocket, shortening the distance between the apical and basolateral surfaces. The M cell makes contact with immune cells such as APCs and lymphocytes. It also has specific glycosylation patterns on its plasma membrane which appears to be related to the ability of these cells to bind bacteria. A number of bacteria, including all members of the Enterobacteriaceae family (*E. coli, Salmonella, Shigella*) and also *Vibrio cholerae*, bind to M cells and can be transported by these cells and can enter into the body by this means. Viruses can also bind to M cells and be transported. These very interesting cells are still poorly understood and the complete range of their activities remains to be elucidated.

BACTERIAL PRODUCTS INFLUENCING ACQUIRED IMMUNITY

It is well known that bacteria, bacterial sonicates and various crude mixtures of bacterial constituents can modulate innate and acquired immune responses. Much of our knowledge of immunity has depended on the ability of either whole *M. tuberculosis* or the sonicated organism in paraffin oil and emulsifier (known as Freund's complete adjuvant) to promote immune responses to antigens. In the absence of this adjuvant it is extremely difficult to induce immune responses to specific antigens or to produce many of the whole animal immune models (of arthritis, thyroiditis, nephritis etc.) used by immunologists to study immunologically-based diseases. However, it is only in relatively recent years that the actions of purified bacterial products on immunity have begun to be examined. In Chapter 8 the concept of pathogen-associated molecular patterns (PAMPs) and pattern recognition receptors (PRRs) in innate immunity was introduced. The PRRs are believed to be cellular receptors, mainly on APCs, which recognize evolutionarily conserved bacterial constituents such as peptidoglycan, LPS and bacterial DNA. It is believed that these interactions are required to instruct and focus the acquired immune response. However, it is also emerging that several bacterial constituents, which may or may not be PAMPs, can have far-reaching effects on acquired immunity. Four of these bacterial components will be discussed: superantigens, exotoxins, heat shock proteins and one group of simple molecules involved in quorum sensing – the acyl homoserine lactones.

Superantigens

The terms antigen and mitogen have been employed and described earlier in this chapter. The term superantigen (SAg) was introduced by Marrack and Kappler in 1989 to describe the capacity of certain bacterial proteins to stimulate a substantial part of the T cell repertoire of mice and humans (Table 9.4). Viruses, including the murine mammary tumour virus (MMTV), which is a vertically transmitted retrovirus, and the rabies virus, also produce SAgs but they will not be discussed in this chapter. The variable genes in the TCR β chain have already been discussed in terms of the gene rearrangement involved in the generation of diversity of the TCR. On the basis of sequence similarities the V_β gene elements in the human are divided into 25 different families. Thus each V_β chain is found on approximately 5% of the available αβ T cells. What superantigens are able to do is to cross-link MHC class II antigens on APCs with selected V_β-containing T cells and to stimulate all T cells bearing such β chain variable regions (Figure 9.21). The discovery of the actions of bacterial superantigens was a major spur to the study of the V_β repertoire.

Table 9.4 Comparison of antigens, superantigens and mitogens

Feature	Antigen	Superantigen	Mitogen
% of responding cells	10^{-4} to 10^{-6}	5–20	80–90
Accessory cell requirement	Yes	Yes	Yes
Requirement for MHC II expression	Yes	Yes	No
MHC restriction in presentation	Yes	No	No
Antigen processing	Yes	No	No
Restricted T cell V_β usage	Possible	Yes	No

A growing number of bacteria have been found to produce superantigens, which are all proteins of molecular mass in the approximate range 20–60 kDa (Table 9.5). *Staph. aureus* wins the prize for producing the greatest number of superantigens – the well-known pyrogenic exotoxins and enterotoxins produced by this organism. *Streptococcus pyogenes* also produces a number of superantigens (see Chapter 7 for a description of bacterial enterotoxins). Other bacteria producing superantigens include group B and group G streptococci, *Mycoplasma arthritidis, Pseudomonas aeruginosa, Clostridium perfringens* and *Yersinia pseudotuberculosis*. It is not known if the production of superantigens is a property of all bacteria. The staphylococcal and streptococcal superantigenic toxins consist of single polypeptide chains of between 27 and 28 kDa, with the exception of toxic shock syndrome toxin (TSST), which is 22 kDa, and all but TSST and streptococcal pyrogenic exotoxin (SPe)C contain a small disulphide loop in the central part of the sequence. These bacterial exotoxins are stable to most forms of denaturation and resistant to all known proteases. However, complete reduction and alkylation of the single disulphide group cause a loss of T cell stimulation,

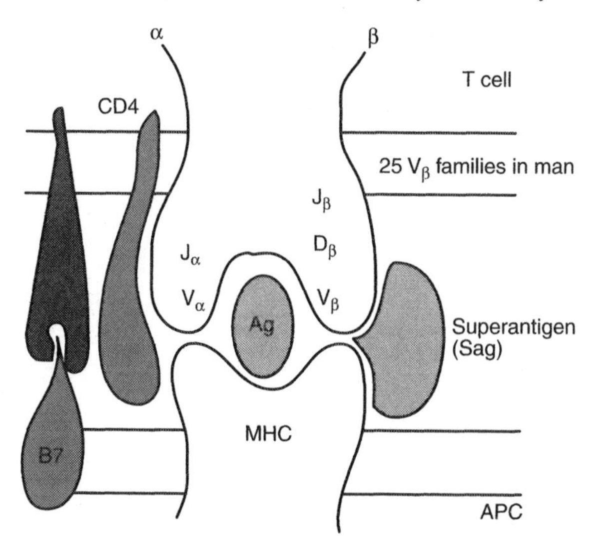

Figure 9.21 Action of superantigens cross-linking $V_β$ TCR protein and MHC class II

although the protein still binds to the MHC class II protein. Sequence alignment of the staphylococcal and streptococcal toxins reveals the presence of three homology groups. SEA, SED and SEE form group 1 with approximately 80% homology. Group 2 consists of SEB, SEC1,2,3, SPeA and SPeC. There is only around 30% homology between groups 1 and 2. Group 3 consists only of TSST, which shares limited homology with the other toxins. It is thought that TSST represents a very early divergence from a primordial toxin gene. The *Cl. perfringens* SAg, in contrast, bears no homology to the staphyococcal pyrogenic exotoxins. The staphylococcal enterotoxins (SEs) bind directly to human class II MHC antigens without prior processing and this is an essential requirement for defining SAg activity. Human MHC class II molecules bind better than do murine and HLA-DR isotypes (H-2E in mice) bind all SEs better than either HLA-DQ (H-2A) or HLA-DP. Group 1 SEs bind to DR with the highest affinity while the group 2 toxins are 100–500-fold less avid. TSST binds with an intermediate affinity. Interestingly, HLA-DR polymorphism has an influence on SAg binding affinity. For example, the dissociation constant for SEA binding to HLA-DR1 and DR6 is 36 nM and 320 nM respectively. Competitive binding experiments suggest that the binding sites on MHC class II for group 2 and group 3 toxins are separate and that group 1 binds to a site which is between these two sites and overlaps them. Binding of SAgs has little effect on the binding of peptides in the MHC groove and vice versa.

What is the consequence of the binding of a SAg? The cross-linking of the MHC class II proteins on the surface of an APC (B lymphocyte, macrophage or dendritic cell) with the TCR of T lymphocytes with particular $V_β$ specificities results in the activation of a significant proportion of the T cell population and the production of large amounts of cytokines produced by both cells being bound by the SAg (Figure 9.22). It should be noted that the profile of cytokines produced by SAgs is different from that induced by LPS. As will be described, it is this production of cytokines that contributes to the pathogenic actions of superantigens. Thus, in the short term, SAgs induce the formation of inflammatory mediators. However, what happens to the activated T lymphocytes which have bound the SAg? These cells are activated and undergo rapid proliferation. However, experimental studies suggest that the

Table 9.5 Superantigens produced by bacteria

Bacteria	Superantigen	Human V_β specificity
Staph. aureus	SEA	1, 5, 6, 7, 9
	SEB	3, 12, 14, 15, 17, 20
	SEC1	3, 6, 12, 15
	SEC2	12, 13.2
	SEC3	3, 5, 12, 13.2
	SED	5, 12
	SEE	5.1, 6, 8, 18
	TSST-1	2
	Exfoliative toxin	2
Strep. pyogenes	SPeA	2, 12, 14, 15
	SPeB	8
	SPeC	1, 2, 5.1, 10
	SPeF	2, 4, 8, 15, 19
	SSA	1, 3, 5.2, 15
Mycoplasma arthritidis	MAM	6, 8.1–8.3
Clostridium perfringens		6.9, 22
Yersinia pseudotuberculosis		3, 9, 13

majority of the activated T cells either enter a state of unresponsiveness termed T cell anergy, or else they become apoptotic. Thus the ultimate response of the host to a superantigen is the loss of T cells bearing particular V_β TCRs, although in the early response to the SAg these particular V_β T cell populations will be present in larger than normal proportions. Much of the interest in superantigens has come from the belief that these molecules are important in a variety of human diseases. The staphylococcal enterotoxins are the cause of food poisoning associated with contaminated food. Symptoms of food poisoning associated with these toxins occur rapidly and are unlikely to be due to activation of T lymphocytes. These toxins may act by stimulating intestinal epithelial cells. Indeed, it has been shown that SEB can bind significantly better to epithelial cells than to B cells. It has also been shown that SEA and TSST-1 can undergo facilitated transport through intestinal epithelial cells and this pathway may deliver these toxins into the blood. The diseases in which the actions of SAgs are best defined are the relatively recently described shock-like syndromes in humans caused by TSST-1 (staphylococcal toxic shock syndrome) and SPeA (streptococcal toxic shock syndrome). While still rare, the incidence of these diseases has increased dramatically in recent years. Both are lethal conditions which are difficult to treat and pathology is due to the rapid and massive production of proinflammatory cytokines from stimulated APCs and T cells. In patients with staphylococcal toxic shock syndrome (TSS) it has been reported that there was expansion of V_β2-bearing T cells in patients' blood. In patients with streptococcal TSS depletions of V_β1-, V_β5.1- and V_β12-bearing T cells in the peripheral blood have been reported. These findings suggest that the SAg activity is related to the disease symptoms.

In addition to the conditions described, there is extensive speculation as to the involvement of bacterial superantigens in various severe human autoimmune and immune diseases, including rheumatoid arthritis, autoimmune diabetes, multiple sclerosis, Grave's disease, psoriasis and Kawasaki disease. These speculations are based on the finding of biased V_β gene usage in T cells from such patients, the suggestion being that such bias is the result of a

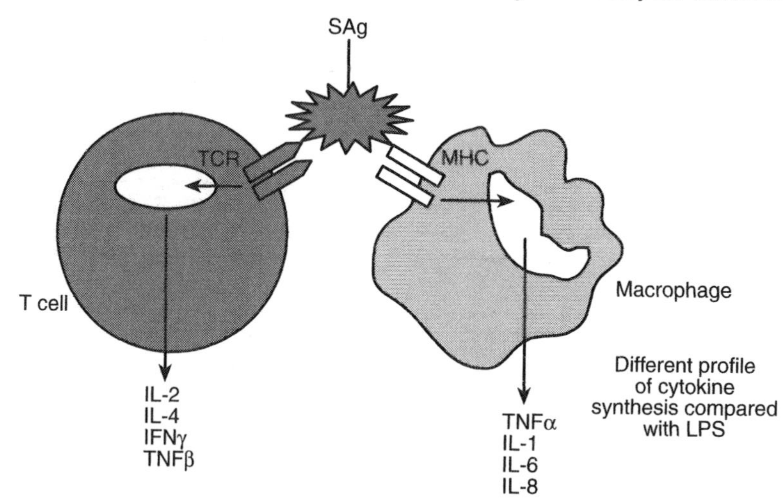

Figure 9.22 Cytokine stimulation by SAgs. The pattern of cytokines produced is different from that induced by LPS

pre-existing infection and action of a bacterial SAg. This hypothesis is the subject of great debate and much confusion in the literature and will not be discussed further.

The obvious question that needs to be asked of superantigens is – what is the evolutionary advantage to the bacterium to produce proteins with such a mode of action? The answer to this important question is still unknown, but has been a spur to the scientific imagination. On the face of it, the ability to produce molecules which stimulate inflammation in the host would seem to be counterproductive. With some of the SAgs the toxicity is so great that it kills a significant proportion of infected individuals. Again, this is not a particularly sensible strategy if one is relying on the host for survival. However, it may be that the massive induction of inflammatory mediators induced by SAgs overwhelms the normal control mechanisms of immunity, allowing bacteria to enter without inducing appropriate localized protective inflammation. In the longer term, the deletion of a large proportion of the T cell repertoire of the host must compromise its ability to deal with pathogens. However, this is likely to benefit all pathogens, not only those producing SAgs. The reader should be clear that although science has not answered the riddle of the SAg, there will be a simple answer to the question of the evolutionary development of these fascinating proteins.

Bacterial exotoxins

The SAgs described in the previous sections are examples of bacterial exotoxins with particular properties *vis-à-vis* T lymphocytes. In recent years other bacterial exotoxins have been shown to have actions on immune cells. For example, it has been established by two of the authors (BH and MW) that many, if not all, exotoxins have the capacity to induce cytokine synthesis. Indeed, some of the most potent cytokine inducers are exotoxins. This suggests that the biological actions of exotoxins rely, to an unknown extent, on their actions on the immune system. Certain toxins have been shown to have powerful immunological actions. For example, cholera toxin is a powerful adjuvant for oral immunization with soluble antigens. It is suggested that cholera toxin may induce specific Th_2 responses in mucosa to promote an IgA response. Tetanus toxoid is also known to have powerful adjuvant activity. In contrast, recent work is revealing that other toxins have the ability to inactivate or inhibit immune responses. For example, the C3 exotoxin from *Cl. botulinum* which acts on Rho, and has been described in some detail in Chapters 3 and 7, blocks macrophage phagocytosis induced by binding to Fcγ receptors. Rho is one of the small GTPases controlling the actin cytoskeleton and other exotoxins which can influence these intracellular signalling proteins may also have effects on macrophage phagocytosis. The vacuolating toxin from *Helicobacter pylori* has inhibitory effects on the synthesis, but not recycling, of MHC proteins. Certain toxins from a range of bacteria act as leukocidins and directly kill leukocytes. This will have the effect of inhibiting leukocyte-driven immune processes although this is an area that has not received significant attention. In addition to these *in vitro* studies, it has been reported that the *E. coli* heat-labile toxin has the ability to inhibit immune responses and to prevent an experimental autoimmune disease, collagen-induced arthritis, in mice. The potential use of bacterial exotoxins in controlling immunological responses is an area which is attracting increasing interest.

Heat shock proteins (Hsps)

If a fertilized hen's egg is placed under a chicken for the requisite time the result is the hatching of a chick. If the same egg is placed in boiling water for 3–4 minutes, then there is no chick, only breakfast. What is the difference? As the reader will be aware, the difference is protein denaturation – the boiled egg consisting of non-functional denatured proteins. In the last decade it has become apparent that protein denaturation is a major problem for all cells as they contain extremely high concentrations of proteins which, if not properly folded, can interact and cause cellular mayhem. The solution to this problem of cellular protein denaturation evolved with the beginning of life on this planet. This was the evolution of classes of proteins known variously as molecular chaperones, heat shock proteins (normally abbreviated to Hsps) or stress proteins (Figure 9.23). These are present in all cells on the planet and have molecular masses ranging from 8 kDa to 100 kDa. The function of these molecular chaperones is to catalyse the correct folding of proteins as they form on the ribosome and to refold proteins where appropriate (see Chapter 2). These proteins were first discovered by scientists examining the response of various organisms to heat stress. Exposure to temperatures greater than normal body temperature was found to induce the production of a range of proteins which were termed heat shock proteins. It was then discovered that many stresses will cause cells to produce such proteins. These include elevated temperature, changes in ionic conditions, free radicals, certain drugs, toxins, ischaemia and mechanical stress. It appears that if cells are exposed to any stress the result is an increase in protein denaturation which is a trigger to the synthesis of Hsps. Of course, the unstressed cell also produces protein-folding proteins normally, as there is a certain obligatory amount of protein folding in every cell. These constitutive proteins are called molecular chaperones. In situations of cell stress the synthesis of these constitutive proteins may also be increased. A list of the major heat shock proteins is provided in Table 9.6.

Bacteria were the first organisms to evolve Hsps (probably around 3.5 billion years ago) and one of the major findings, when the sequences of these proteins were being collected, was the significant homology between them, no matter the source. Thus bacterial and, for

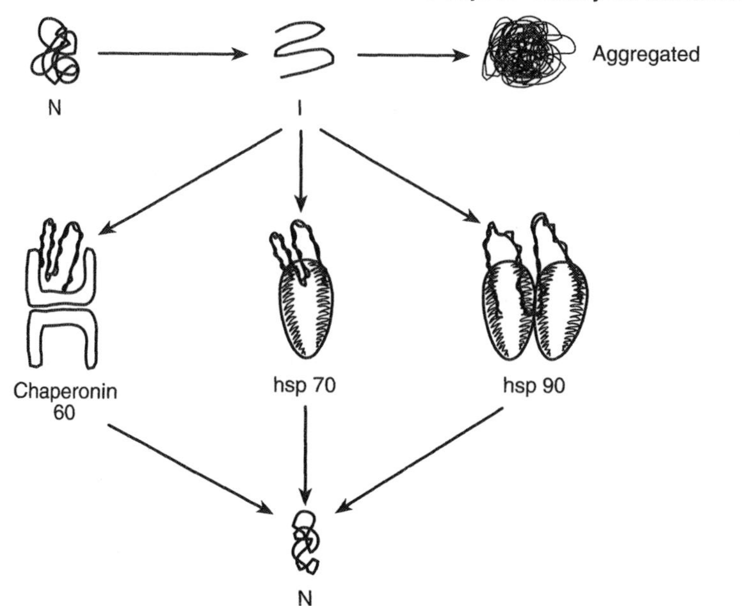

Figure 9.23 Schematic diagram of three of the most important molecular chaperones or heat shock proteins in terms of immune responses. The upper part of the diagram shows what can happen to a normally folded protein (N) in conditions of cell stress. This can lead to the formation of unfolded non-native intermediates (I) which are prone to aggregate. Three of the major intracellular proteins which can act to provent denaturation and promote correct refolding are chaperonin 60 (Hsp 60, an oligomeric protein composed of 14 subunits), Hsp 70 (a monomeric protein) and Hsp 90 (a dimer)

Table 9.6 The heat shock proteins or stress proteins or molecular chaperones

Name	Molecular mass (kDa)
Ubiquitin	8
Chaperonin 10	10–12
Thioredoxin	10–12
Parvulins	10–19
Small heat shock proteins	15–30
Cyclophilins	18–40
HSP40 (DnaJ-related) proteins	40
FK506-binding proteins	12–60
Chaperonin 60	55–65
Hsp 70 family	70
Hsp 90 family	90
Hsp 100 family	100
100 Hsp 110/SSE proteins	110

example, human Hsps, show significant amino acid sequence identity, which is remarkable considering the enormous evolutionary time gap between them. For example, the human and the *E. coli* chaperonin 60 exhibit 40–50% positional identity in their amino acid sequences. Thus these molecular chaperones are probably another example of a PAMP. Another way of viewing the bacterial Hsps is that they contain significant stretches where their sequence is identical to the human Hsp. In other words, these proteins contain what immunologists term self epitopes. Now there is a general rule in immunology that the closer a molecule is to self, the less immunogenic it will be. Thus it comes as a surprise to find that infections with various pathogens, including bacteria, protozoa and helminths, are associated with dominant antibody responses to the Hsps (Table 9.7). This was first shown in the mid-1970s and the antigens, which were not identified at the time, were termed 'common antigens'. The role of antibody to Hsps is not clear. The Hsps are normally found within cells and so antibody is unlikely to have much effect. However, there is increasing evidence that Hsps are expressed on the surfaces of cells (both bacteria and host cells) during infection and thus antibodies could have a protective effect. For example, certain bacteria are now thought to use Hsps as adhesins. Immunization of experimental animals with various Hsps has been shown to confer some degree of resistance to infection with both bacteria and malarial parasites.

Table 9.7 Immunity to heat shock proteins in human infections

HSP	Pathogen	Disease
Chaperonin 10	*Mycobacterium tuberculosis*	Tuberculosis
Small Hsps	*Schistosoma mansoni*	Schistosomiasis
	Mycobacterium leprae	Leprosy
Chaperonin 60[a]	*M. tuberculosis*	TB
	M. leprae	Leprosy
	Treponema pallidum	Syphilis
	Borrelia burgdorferi	Lyme disease
Hsp 70	*M. tuberculosis*	TB
	M. leprae	Leprosy
	Chlamydia trachomatis	Trachoma, urethritis
	Trypanosoma cruzi	Chagas' disease
	Plasmodium falciparum	Malaria
Hsp 90	*Plasmodium falciparum*	Malaria
	Trypanosoma cruzi	Chagas' disease
	Schistosoma mansoni	Schistosomiasis

[a] Also known as Hsp 60 or, in *Escherichia coli*, groEL.

In addition to inducing antibody synthesis, certain of the Hsps are extremely potent inducers of T cell immunity. Most attention has been focused on an interacting pair of molecular chaperones termed chaperonins, abbbreviated Cpns. One Cpn is an oligomeric molecule consisting of 60 kDa subunits (Cpn 60, the molecule from *E. coli* being known as groEL). The other is composed of 10 kDa subunits (Cpn 10). The ability of these molecules to activate T cells is shown by the finding that as much as 20% of the mycobacteria-reactive T cells in mice immunized with this bacterium recognize Cpn 60. In a leprosy patient one third of the T lymphocytes responsive to this bacterium recognized Cpn 10. Given that both bacteria probably contain between 2000 and 3000 proteins, many of which will be very different from murine or human proteins, it seem surprising that a major T cell response is to a protein with such sequence similarity to the host Cpn 60 or Cpn 10. Another surprise is the finding that the blood of newborn babies, who have not been exposed to any bacterial infections, contains significant numbers of lymphocytes responsive to mycobacterial Cpn 60. Lymphocytes reactive with other mycobacterial components were absent, thus demonstrating the special nature of this anti-Cpn 60 response. The finding that a proportion of γδ T lymphocytes recognize Cpn 60 has been mentioned in an earlier section of this chapter. Newborn mice contain similar numbers of Cpn 60-reactive T cells in their thymus gland as are found in the peripheral lymphoid organs of adult animals. These findings of Cpn-reactive cells in the newborn, who have not been exposed to bacteria, suggest that the immune response is in some way already primed to recognize this particular bacterial antigen.

In the previous section on SAgs the suggestion that these T cell-activating proteins were involved in autoimmunity was briefly touched on. Autoimmunity is an immunological condition in which the immune response is directed to self antigens. A good example of such an autoimmune disease is myasthenia gravis. In this condition patients suffer from severe muscle weakness. The transmission of nerve impulses from the brain to the muscle is dependent on the release of acetylcholine from the nerve endplate, the diffusion of this neurotransmitter across the neuromuscular junction, and the binding of acetylcholine to the acetylcholine receptor. Patients with myasthenia gravis have antibodies (so-called autoantibodies) to the muscle acetylcholine receptor. These prevent binding of the acetylcholine and so the muscles become progressively weaker. The basis of autoimmune diseases has been a mystery since their discovery. There has been a prevailing view that the immune system should be barred from recognizing self. Paul Ehrlich, a pioneer of immunology, coined the term *horror autotoxicus* for the inability of the immune system to recognize self. However, we now recognize that there is some degree of self-reactivity (autoimmunity) in all of us. The key question is what prevents this from becoming full-blown autoimmunity. In the past 20 years there has been increasing belief in the hypothesis that autoimmune diseases are initiated by some form of infection. The finding that bacterial Hsps had large amounts of self epitopes suggested the hypothesis that certain autoimmune diseases, for example rheumatoid arthritis (which affects the major joints of the body), were caused by cross-reactivity to bacterial Hsps. The idea was that an infection would cause an immune response to the bacterial Hsp. This response would then cross-react with the human Hsp homologue, which is found in the mitochondria, and would give rise to a chronic autoimmune disease. During the 1980s and early 1990s an enormous amount of circumstantial evidence accumulated to support this idea. However, the latest evidence suggests that at least in one form of autoimmune disease – juvenile rheumatoid arthritis – responsiveness to Cpn 60 may be protective. The protective nature of an immunological response to Cpn 60 or Cpn 10 has also been shown in animal models of arthritis in rodents. It is, at present, not clear what the implications of these finding are.

One fascinating speculation on the role of Hsps in immunity has been suggested by Irun Cohen, a well-known immunologist. He argues that the inherent result of the enormous capacity of the immune system to recognize antigens is chaos. Remember, the key problem in immunity is to recognize foreignness in a sea of self antigens. The answer to this chaos, it is argued, is for the immune system to recognize only a small group of dominant antigens in each organ system of the body. These dominant, potentially self antigens, blind immunity to the myriad of competing self antigens and thus reduce the requirement to be continually tolerizing the system. Moreover, the self molecules chosen for and attracting autoimmune responses (the Hsps, for example) are those for which regulatory control networks exist. However, as Cohen admits, there is no free lunch and a few per cent of the population pay the price of this system by developing autoimmune disease due to the misdirection of antimicrobial immunity from the microbe to self.

Another very intriguing current immunological speculation, which is attracting an enormous amount of attention, is Polly Matzinger's Danger Model. This model or hypothesis claims that the immune system has not evolved to allow the discrimination of self from non-self. Rather the system has evolved to detect and protect against danger. Danger, in this context, would be the induction of aberrant cell behaviour and cell necrosis (rather than apoptosis). One of the key responses of cells to infections is the induction of a stress response which means the increased production of Hsps. Matzinger suggests that the Hsps may be one of the key danger signals which the immune system recognizes. The ideas of Cohen and Matzinger may go some way to account for the large number of $\gamma\delta$ T lymphocytes able to respond to Cpn 60. Perhaps the overexpression of Cpn 60 by, say, epithelial cells exposed to infectious bacteria is the first trigger to inducing a protective immune response.

In Chapter 8 the modulin concept was introduced. Modulins are the fifth and most recent form of bacterial virulence factor to be described. Bacteria produce a large and growing variety of molecules which can interact with host cells and induce cytokine formation. These cytokines feed back onto the producing cell and 'modulate' its behaviour. In the last five years it has become clear that molecular chaperones are modulins, in addition to being immunogens. Cpn 60 from *E. coli*, for example, is both a potent inducer of human monocyte cytokine synthesis and a growth factor for rodent osteoclasts. These latter cells are myeloid cells involved in bone remodelling and their overproduction can lead to excessive skeletal destruction, suggesting that the chaperonins may play a role in bone diseases. An interesting finding is that even when broken down by proteases the *E. coli* Cpn 60 still retains its potent cytokine-inducing activity. Thus these proteins, by activating myeloid cell function, are likely to be involved in the activation and control of immunity.

Acyl homoserine lactones

Quorum sensing (from *quorum* – 'the minimum number of people required to transact business', Chambers Dictionary) has been described in detail in Chapter 3. To reiterate, this is a cell-to-cell signalling system in bacteria, in which a low molecular mass diffusible molecule, released by bacteria, signals to the bacterial colony when the cell density has reached a set limit. At this cell density the amount of quorum-sensing agent is sufficient to switch on gene transcription. One of the main quorum-sensing molecules are a group of acyl homoserine lactones (AHLs). With the increasing realization that signals used by one set of cells (e.g. bacteria) can be 'read' by other cells (e.g. eukaryotic cells) the effects of the AHLs on host cells have been examined. Initial studies suggested that the AHLs could stimulate cytokine synthesis. However, in a detailed investigation Pritchard and co-workers (see references) have revealed that relatively high concentrations of one particular AHL, that produced by *Ps. aeruginosa*, has a modulatory effect on various measures of immunity (see Figure 3.13).

CONCLUSIONS

Innate and acquired immunity have evolved to protect large multicellular organisms from colonization by certain microorganisms. We are beginning to appreciate the deployment of the cells constituting acquired immunity and how the evolution of these cells has been driven by the lifestyles of the organisms that they are meant to defend us from. Much of our understanding of the role of acquired immunity in infections has been derived from virology and it is now clear that viruses have evolved multiple strategies for inactivating immune protection. We know much less about the interactions of bacteria with the immune system. For example, it is now becoming clear that the normal rules of engagement between T cells and antigens are being rewritten by bacteria. The dogma is that T cells only recognize peptide antigens. However, in the past few years it has been found that non-peptidic bacterial antigens can be recognized by T cells. How do bacteria deal with the immune system? Again, this is an area that is poorly understood. We have known about the actions of capsules on complement and about antigenic variation for some years. However, it is only in recent years that bacteria have been shown to inhibit the actions of cells involved in acquired immune responses. For example, a number of bacterial exotoxins, including superantigens, are now known to have profound effects on immunity. However, it is likely that these findings are only the 'tip of the iceberg' of immune-deviation mechanisms utilized by bacteria. In the final chapter of this book a number of newly discovered areas of the interaction between bacteria and host cells will be discussed.

Further Reading

Books

Abbas AK, Lichtman AH, Pober JS (1997) *Cellular and Molecular Immunology* (3rd edn). Saunders, Philadelphia.

Barclay AN, Brown MH, Law SKA, McKnight AJ, Tomlinson MG, van der Merwe PA (eds) (1997) *The Leukocyte Antigen Facts Book* (2nd edn). Academic Press, London.

Bell JI, Owens MJ, Simpson E (eds) (1995) *T Cell Receptors*. Oxford University Press, Oxford.

Benjamini E, Sunshine G, Leskowitz S (1996) *Immunology: A Short Course* (3rd edn). Wiley–Liss, New York.

Clarke WR (1989) *The Experimental Foundation of Modern Immunology* (4th edn). Wiley, New York.

Gething M-J (1997) *Guidebook to Molecular Chaperones and Protein-Folding Catalysts*. Oxford University Press, Oxford.

Janeway CA, Travers P (1998) *Immunobiology: The Immune System in Health and Disease* (4th edn). Churchill Livingstone, Edinburgh.

Kuby J (1996) *Immunology* (2nd edn). Freeman, New York.

Male D, Cooke A, Owen M, Trowsdale J, Champion B (1996) *Advanced Immunology* (3rd edn). Mosby, Turin.

Roitt I (1997) *Essential Immunology* (9th edn). Blackwell, Oxford.

Roitt IM, Brostoff J, Male DK (eds) (1998) *Immunology* (5th edn). Mosby, London.

Silverstein AM (1991) *A History of Immunology*. Academic Press, San Diego.

Reviews

Banchereau J, Steinman RM (1998) Dendritic cells and the control of immunity. *Nature* 392: 245–251.

Casadevall A (1998) Antibody-mediated protection against intracellular pathogens. *Trends Microbiol* 6: 102–107.

Cohen IR, Young DB (1991) Autoimmunity, microbial immunity and the immunological homunculus. *Immunol Today* 12: 105–110.

Gray KM (1997) Intercellular communication and group behaviour in bacteria. *Trends Microbiol* 5: 184–188.

Horton J, Ratcliffe N (1998) Evolution of immunity. In *Immunology* (5th edn) (eds Roitt IM, Brostoff J, Male DK), pp. 15.1–15.22. Mosby, London.
Kotb M (1995) Bacterial pyrogenic exotoxins as superantigens *Clin Microbiol Rev* 8: 411–426.
Matzinger P (1994) Tolerance, danger and the extended family. *Annu Rev Immunol* 12: 991–1045.
Mitchie CA, Cohen J (1998) The clinical significance of T-cell superantigens. *Trends Microbiol* 6: 61–65.
Slifka MK, Ahmed R (1996) Long-term humoral immunity against viruses: revisiting the issue of plasma cell longevity. *Trends Microbiol* 4: 394–400.
Takahashi A, Earnshaw W (1996) ICE-related proteases in apoptosis. *Curr Opin Gen Dev* 6: 50–55.

Papers

Fehr T, Bachmann MF, Bucher E, Kalinke U, DiPadova FE, Lang AB, Hengartnet H, Zinkernagel RM (1997) Role of repetitive antigen patterns for induction of antibodies against antibodies. *J Exp Med* 185: 1785–1792.
Jullien D, Stenger S, Ernst WA, Modlin RL (1997) CD1 presentation of microbial nonpeptide antigens to T cells. *J Clin Invest* 99: 2071–2074.
Niedermann G, Grimm R, Geier E *et al.* (1997) Potential immunocompetence of proteolytic fragments produced by proteosomes before evolution of the vertebrate immune system. *J Exp Med* 185: 209–220.
Telford G, Wheeler D, Williams P, Tomkins PT, Appleby P, Sewell H, Stewart GSAB, Bycroft BW, Pritchard DI (1998) The *Pseudomonas aeruginosa* quorum-sensing signal molecule *N*-(3-oxododecanoyl)-L-homoserine lactone has immunomodulatory activity. *Infect Immun* 66: 36–42.

CHAPTER 10

Future Developments in Cellular Microbiology

INTRODUCTION

Science is not a collection of facts and figures but is the resultant of the accepted hypotheses that workers in the field adhere to at any one time. The controversial philosopher of science, Thomas Kuhn, suggested that there were two kinds of science. Firstly, there is the everyday science in which workers in a particular field accept a common set of principles (the framework) within which they work. Kuhn called this common set of principles, ideas or framework – the paradigm. In Kuhn's words the paradigm is 'the entire constellation of beliefs, values, techniques, and so on, shared by members of a given community'. However, if the paradigm no longer works, that is, if experimental results cannot be interpreted within the commonly accepted paradigm, then a paradigm revolution occurs and a new paradigm has to be produced within which scientists can work and interpret their hypotheses. It is this paradigm revolution that constitutes the second kind of science. Examples of paradigm revolutions are Einstein's theory of gravitation replacing (at least for relativistic velocities) Newton's theory (or laws). In biology, a good example would be Peter Mitchell's chemiosmotic hypothesis replacing earlier hypotheses about how mitochondria produce ATP.

The past decade appears to have been one in which microbiology has gone through, and continues to go through, a paradigm revolution. This revolution has been fuelled by revolutionary findings in a number of disciplines and by cross-fertilization between disciplines. Molecular phylogenetics has revolutionized our views on species diversity and has introduced the concept that there are only three domains of life (Bacteria, Archaea and Eukarya) and the finding that the major diversity of the life-forms on our planet lies in the single-celled (i.e. microbial) organisms. Most (estimated at 90–99.99%) of these single-celled organisms, revealed by cloning and determination of the sequence of 16S ribosomal DNA, are not culturable under laboratory conditions and thus we know nothing about them other than what can be deduced from the sequences of their genomes. This failure to culture the vast majority of the microorganisms which have been revealed by these 16S rRNA safaris suggests that we have many surprises in store. For example, the estimates of the numbers of bacterial species which cause disease in humans or which live in harmony with us (our normal microflora) may be completely awry. If estimates of culturable organisms defined in other surveys are applied to man, then perhaps rather than 1000 different bacterial species constituting our normal microflora, as is currently estimated, it may be 10 000 or even 100 000. Microbiologists, for obvious reasons, have spent the last century growing bacteria in isolation, a technique called axenic culture, in which only a single bacterial species is grown. This was obviously necessary for taxonomic and other reasons and was realized to be rather artificial. It was also relatively unproductive as we have only identified 5–10 000 microbial species. However, within the last decade or so we have begun to realize just how interactive bacteria are – both with themselves and with eukaryotic cells. In most environments bacteria exist as competing species and it is to this competition, for example, that we owe the evolution of antibiotics and of antibiotic resistance.

Bacteria can also exist in the form of single species and multi-species biofilms, in which competition is more limited and positive interactions between organisms must occur (Figure 10.1). Bacterial biofilms are increasingly seen to be important in medicine and dentistry. For example, they form on indwelling medical devices and can be a major clinical problem, as bacteria within a biofilm can be relatively resistant to the antibiotics to which planktonic organisms are susceptible. Mixed bacterial biofilms also exist on the mucosal surfaces of our bodies and we will return to the paradox that such biofilms confront us with later in this chapter.

The key discoveries which are central to the belief that we are in a paradigm revolution in microbiology are those which reveal just how diverse and intimate are the interactions which bacteria can make with eukaryotic cells. The literature describing how various disease-causing bacteria such as *Escherichia coli, Yersinia enterocolitica, Shigella* spp., *Salmonella* spp., *Mycobacterium tuberculosis* and *Listeria monocytogenes* (among others) interact and control the functions of eukaryotic cells has been described in detail in earlier chapters. The mechanisms

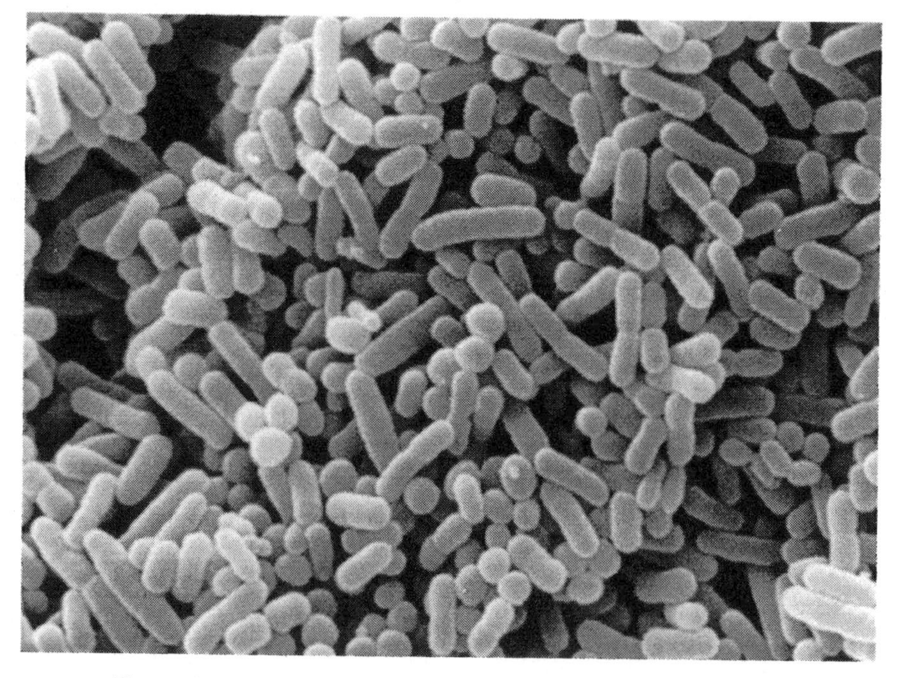

Figure 10.1 Scanning electron micrograph of a bacterial biofilm

by which these various bacteria produce their cellular effects, through genes encoded by pathogenicity islands, the use of type III secretion systems, the synthesis of cytokine-modulating molecules, etc., are being rapidly elucidated and the information is creating a new paradigm for studying bacteria–host cell interations. It is such interactions that we have taken as being the subject of this textbook on cellular microbiology.

In this final chapter some newer areas of cellular microbiology which are still in their formative stages will be described and we will suggest what the future might hold for the science of cellular microbiology.

PROKARYOTIC/EUKARYOTIC INTERACTIONS IN BACTERIAL GROWTH

Self-replication is the defining characteristic of living organisms. The process of replicating the constituents of a cell in order to form two daughter cells has been described for both bacteria and for eukaryotic cells in Chapter 2. Both bacteria and eukaryotic cells have evolved a complex set of mechanisms in order to allow then to replicate and partition their intracellular constituents prior to the process of cell division. These mechanisms, in totality, are called the cell cycle. In this, and the next section, the ability of bacteria and eukaryotic cells to interact to control each other's cell cycles will be discussed. This is a poorly studied area, but one which holds great promise in terms of scientific understanding of disease mechanisms and the possible development of novel therapeutics based on such knowledge.

Bacterial growth is a major problem for cell biologists who face the constant problem of bacteria growing in their eukaryotic cell cultures. Thus, intuitively it is felt that bacteria grow readily. However, it is now clear that many bacteria which are recognized in environmental samples by16S rRNA gene sequencing prove to be unculturable. It is not clear if such unculturable organisms fail to grow simply because they are dead, or because the culture media being used lack essential nutrients, gases etc. Good examples of bacteria which scientists have been attempting to grow in culture for decades are *Mycobacterium leprae*, the causative organism of leprosy, and *Treponema pallidum*, the pathogen responsible for syphilis. Propagation of these bacteria requires the use of animals. For example, the animal used to cultivate *M. leprae* is the nine-banded armadillo. Bacterial growth is clearly more than a simple linear response to the supply of nutrients and it appears increasingly likely that it is under the control of specific proteins and other factors emanating both from bacteria and from eukaryotic cells. A telling example of this possibility occurs with the bacterium *Micrococcus luteus*. If cultures of this organism are starved and then fresh, nutrient-replete, media are supplied, the bacteria will stop growing after several divisions. This is believed to be due to the production of inhibitors of cell division. However, if media from growing *Micrococcus luteus* cultures are added to starved cells they continue to divide. These and other findings suggest that bacterial growth is dependent on specific growth factors produced by the bacteria, or possibly produced by cells with which the bacteria are in contact. A number of proteins and low molecular mass molecules have been found to promote bacterial growth in culture. Do factors from eukaryotic cells play any role in bacterial growth? This is a question which is attracting increasing attention. A number of molecules of eukaryotic origin, such as serotonin and catecholamines, have been reported to stimulate bacterial growth. Moreover, the last few years have seen a steady trickle of reports showing that various cytokines can promote the growth of a variety of bacteria.

Do cytokines play any role in the normal growth of bacteria, and could these molecules have a role in controlling the growth of bacterial pathogens or the bacteria constituting the commensal microflora which grow on mucosal surfaces? Answers to these key questions have not yet been achieved but a number of recent studies provide partial answers. The first report to support this idea was from Charles Dinarello's laboratory in Boston that interleukin (IL)-1 stimulated the growth of 'virulent' strains of *E. coli*. Dinarello is one of the major pioneers of cytokine, particularly IL-1, research. Comparing six virulent and four avirulent strains of *E. coli*, the former showed enhanced growth, in terms of colony-forming units (CFUs), in the presence of IL-1 concentrations as low as 10 ng/ml. In contrast, avirulent strains did not respond to IL-1. Other cytokines such as IL-4 and tumour necrosis factor (TNF) had no effect on the growth of these virulent strains. Of significance was the finding that the IL-1-induced bacterial growth could be almost totally abolished by a natural antagonist of IL-1 – IL-1ra (described in Chapters 3 and 8) – which suggested that there was an IL-1 receptor involved. Virulent, but not avirulent, strains of bacteria bound IL-1 in a cell density-dependent manner and binding was saturated at a concentration of 20 pg/ml. The binding of labelled IL-1 could be competed by an excess of unlabelled IL-1 and the kinetics of desorption suggested that bacteria contained 20–40 000 binding sites per bacterium, which is significantly greater than the numbers of IL-1 receptors present on human or murine cells. These findings are both interesting, and controversial, as another group reported that they could not reproduce such results; but Dinarello suggested that the failure to replicate the results from his laboratory was because freshly isolated virulent strains rapidly lost their ability to respond to IL-1. This suggests that an IL-1 binding and transduction system is induced by host factors and is rapidly lost in the absence of such signals. Another group reported that a virulent strain of *E. coli* responded by increasing its growth rate when incubated in the presence of two growth-promoting cytokines: interleukin (IL)-2, which is a growth factor for T lymphocytes or granulocyte—macrophage colony–stimulating factor (GM-CSF), a growth factor for neutrophils and monocytes (see Chapter 3). The effects of the IL-2, which is heat labile, could be blocked by heating the cytokine or by adding an IL-2-neutralizing antibody. The possibility that this strain of *E. coli* contained receptors for IL-2 or GM-CSF was explored indirectly by the demonstration that the bacteria depleted the medium of IL-2. It was established that this depletion was not due to proteolytic cleavage of the IL-2. The bacterial growth-promoting effect of IL-2 might explain the high levels of opportunistic infections found in patients receiving IL-2 for cancer or AIDS treatment. IL-2 has also been reported to bind to the yeast *Candida albicans*, which is a major pathogen causing mucocutaneous infections. The bound IL-2 was still biologically active, suggesting that the fungal receptor was different from that used by mammalian cells.

It has also been reported that certain Gram negative bacteria (*Salmonella typhimurium*, *Shigella flexneri* and *E. coli*) have receptors for the proinflammatory cytokine TNFα. Radioiodinated TNFα bound to these bacteria and it could be competed by unlabelled TNFα, but not by another cytokine with similar biological activity but different structure – TNFβ (lymphotoxin). Binding of TNFα could also be inhibited by exposing bacteria to the protease trypsin, or by heating bacteria to 52°C for a few minutes. This strongly supports the hypothesis that the receptor is a protein. Binding was not inhibited by monoclonal antibodies which react with the 55 kDa or 75 kDa cell surface receptors for TNFα (see Chapter 3) nor did these antibodies bind to Western blots of whole bacteria. Thus it appears that these various bacteria have a unique receptor for TNFα. The role of this receptor and its binding to TNF in stimulating bacterial growth was, unfortunately, not examined.

Tuberculosis, caused by *M. tuberculosis*, is still the most common chronic infectious disease. *M. tuberculosis* is well recognized for its slow growth in culture and for its ability to enter a quiescent (dormant) state in which it is relatively resistant to antibiotics. Dormancy is one of the major reasons for the spread of *M. tuberculosis* and its resistance to treatment. Do host factors have any influence on the growth of this organism, either free-living or within cells? One of the most interesting findings is that *M. tuberculosis* and *M. avium*, when cultured extracellularly, can be stimulated to grow by epidermal growth factor (EGF). This is a 6 kDa peptide member of a family of mitogenic peptides which includes transforming growth factor (TGF)α, amphiregulin, heparin-binding EGF and betacellulin. EGF promotes the division of several cell populations with the exception of haematopoietic cells. EGF caused significant enhancement of mycobacterial growth at concentrations as low as 50 ng/ml. However, EGF had no influence on the growth of bacteria growing inside macrophages. Radioiodinated EGF bound to bacteria and could be competed with cold EGF, and Scatchard analysis revealed that the receptor on *M. avium* had a disociation constant (K_d) of 2×10^{-10} M and that there were approximately 500 receptors per cell. The affinity of this bacterial EGF receptor is similar to that of the high-affinity 170 kDa EGF receptor found on mammalian cells, and isolation and cloning revealed a 37 kDa protein with significant homology to the glyceraldehyde 3-phosphate dehydrogenase (GAPD) of group A streptococci. GAPD is a glycolytic enzyme which catalyses the oxidative phosphorylation of glyceraldehyde 3-phosphate to produce 1,3-bisphosphoglycerate and reduced NAD^+. However, this enzyme seems to have a wide range of other functions which may or may not rely on its enzymatic activity. For example, in streptococci, GAPD acts as a receptor for plasmin, fibronectin, lysozyme and cytoskeletal proteins. It has also been found to phosphorylate proteins on human pharyngeal cells and may function as an intercellular signalling system in this respect.

Dormancy is a problem in treating tuberculosis as dormant bacteria are not killed by antibiotics and such bacteria are able to proliferate and reactivate the disease in patients after cessation of therapy. This raises the possibility that EGF could be used in combination with antibiotics to try and kill all the mycobacteria in the body.

More studies are obviously needed, but the available evidence suggests that a number of free-living bacteria have high-affinity receptors for several cytokines and that binding these cytokines can promote bacterial growth (Table 10.1). The nature of the receptors, with the exception of the identification of GAPD, is still unknown, and the intracellular signalling pathways utilized to promote cell division remain to be identified (see Chapter 3 for a description of intracellular signalling). Since proteoglycans in the extracellular matrices of mammalian cells bind a range of cytokines it has been suggested that surface carbohydrates on bacteria may function as 'receptors' for host cytokines. However, it is not clear how binding to cell surface carbohydrates would trigger bacteria to divide. So far no one has investigated whether the bacteria constituting the normal microflora contain cytokine receptors or respond to cytokines. This is a question which obviously needs to be answered.

In addition, a number of studies have shown that cytokines can influence the growth of intracellular bacteria such as *Mycobacteria* spp., *Brucella* spp. and *Legionella* spp. Studies of the growth of *M. tuberculosis* or *M. avium* in human or murine macrophages has shown that IL-1, IL-3, IL-6, IFNγ and transforming growth factor (TGF)β_1 can stimulate bacterial growth, while TNFα and IFNγ inhibit growth. Intracellular growth of *Brucella abortus* in murine macrophages was not influenced by IL-1α, IL-4, IL-6, TNFα or GM-CSF. However, IL-2 or IFNγ caused a reduction in the numbers of intracellular bacteria. TNFα also inhibited the

Table 10.1 Cytokines promoting growth of free-living bacteria

Cytokine	Bacteria whose growth is stimulated
Interleukin 1	Virulent strains of *E. coli*
Interleukin 2	Virulent strains of *E. coli*
Granulocye–macrophage CSF	Virulent strains of *E. coli*
Epidermal growth factor (EGF)	*M. tuberculosis* and *M. avium*
Tumour necrosis factor (TNF)α[a]	*E. coli, Sal. typhimurium, Sh. flexneri*

[a] TNF did not promote growth but bacteria had receptor for this cytokine.

growth of *Legionella pneumophila* in human peripheral blood monocytes. Inhibition of endogenous TNFα synthesis by pentoxyfylline caused an increase rate of proliferation of bacteria.

While it is still too early to see the complete story, the evidence to date suggests that proteins derived from host cells can interact with a range of bacteria and promote (or inhibit) their growth. The ability of bacteria to obtain growth factors from host cells fits in with the new paradigm of the interactive nature of cellular microbiology. However, it is perhaps counterintuitive to find that host defence cytokines such as IL-1, IL-2 and TNFα can bind and apparently promote the growth of bacteria. These cytokines are central to our antibacterial defences. The ability of infectious bacteria to use such cytokines for promoting growth, if a general finding, raises many questions about the true nature of bacteria–host interactions in infections. Again, one can learn valuable lessons from the behaviour of viruses. It is now well established that a number of viruses carry genes encoding for various growth factors such as EGF, TGFα and IL-6. The utilization of proinflammatory cytokines as growth promoters seems an excellent survival strategy for an infecting microorganism. Another strand in this tangled weave is the report, described in Chapters 3 and 9, that the quorum-sensing compounds, homoserine lactones (AHLs), can modulate immune responses. The AHLs are not growth factors but are sensors of bacterial concentration. Certain of the AHLs have been reported, albeit at high concentration, to inhibit the synthesis of proinflammatory cytokines and to block immune defence mechanisms. Certain of the bacteria using cytokines as growth promoters also utilize quorum sensing and it is possible that these bacteria have evolved to switch off host growth factor (cytokine) production when they reach a critical density (Figure 10.2).

Thus far, the discussion has focused on bacteria which are infectious. What about the very large numbers of bacterial species which constitute the normal or commensal microflora? What role do host factors play in their growth and maintenance? It is assumed that the commensal microflora utilize the nutrients available to the host, but little, if anything, is known about the utilization of host 'growth factors' for the maintenance of these bacteria. The nature of the signalling which must occur between the host and its commensal microflora is considered later.

EFFECT OF BACTERIA ON EUKARYOTIC CELL GROWTH AND SURVIVAL

Host factors can influence the growth of bacteria. Is this interaction two-way? Contamination of eukaryotic cell cultures with mycoplasmas is common and can slow down eukaryotic cell

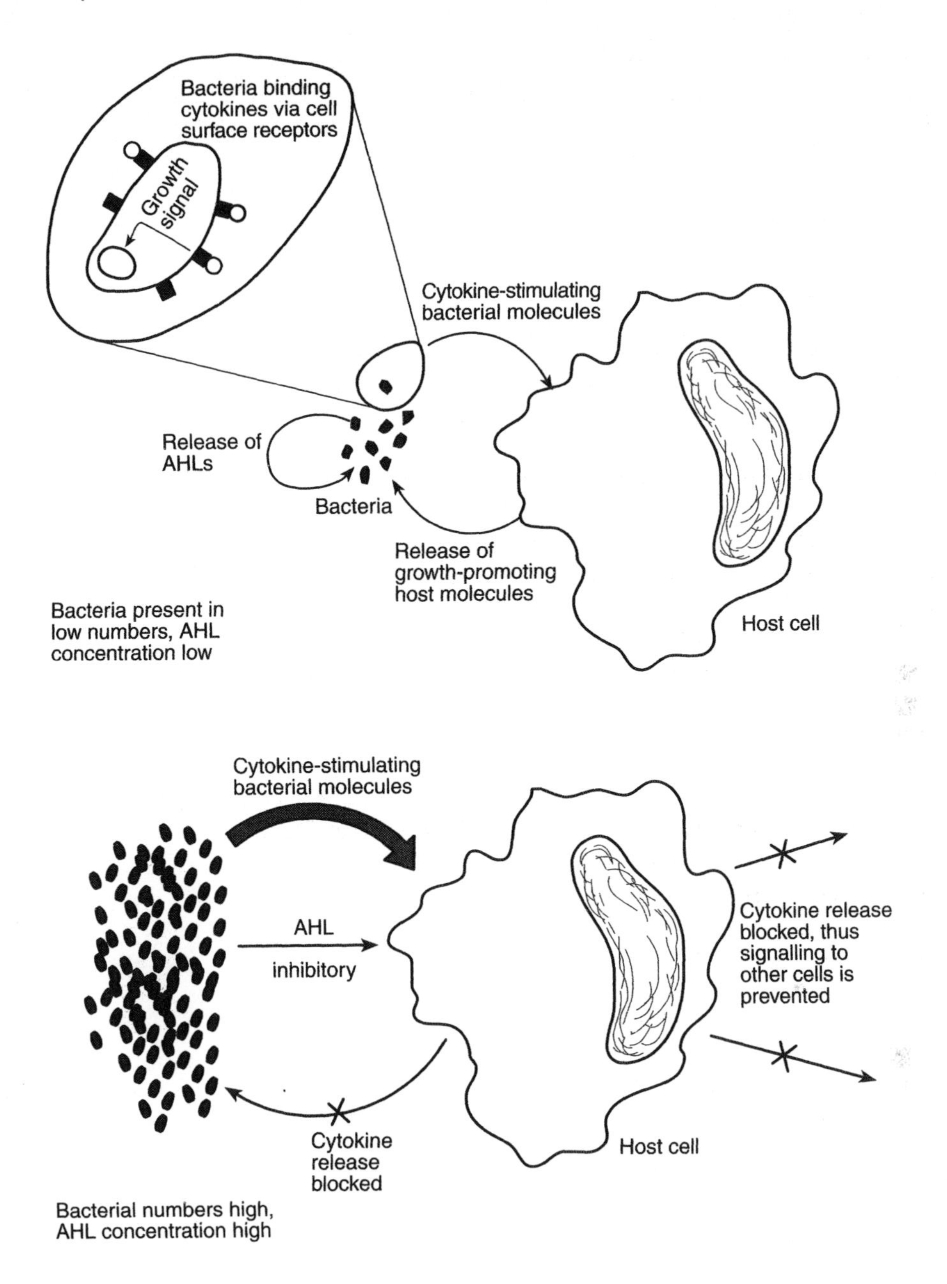

Figure 10.2 Possible growth control mechanism utilized by bacteria interacting with host cells. At low bacterial concentrations signals emanating from the bacteria stimulate nearby cells to produce pro-inflammatory cytokines which act as growth factors. As the bacteria increase in numbers the molecules inducing cytokine production also increase and could lead to production of inflammation and the killing of the bacteria. The finding that certain AHLs can inhibit cytokine synthesis suggests that these quorum-sensing compounds act both as stimulators of bacterial virulence gene expression and as inhibitors of host cytokine production, thus ensuring that bacteria reach suficient numbers to survive the defences of the host

growth, but the mechanism probably relates to competition for nutrients. Infections with planktonic microorganisms are often associated with inhibition of cell growth or the death of eukaryotic cells in culture. However, inhibition of cell growth and cell death are normally put down to nutrient competition, and the production of non-specific toxic agents, respectively. There is, however, growing evidence that bacteria can produce proteins which can exert control over the mechanism of the eukaryotic cell cycle. In addition to exerting control over the cell cycle, there are increasing numbers of reports that bacteria can produce substances which can exert control over apoptosis (see Chapter 2 for a description of apoptosis).

Bacterial control of the eukaryotic cell cycle

The eukaryotic cell cycle is a complex control system utilizing cyclins and cyclin-dependent kinases and phosphatases to drive cells through various cell cycle checkpoints and achieve daughter cell production, and has been described in Chapters 2 and 3. Various compounds have been discovered which can block the cell cycle in its various stages. For example, compounds which interfere with DNA synthesis (such as hydroxyurea) will block the cell cycle in early S phase, as it is in this phase of the cell cycle that the cell's DNA is replicated. In contrast, colchicine, an alkaloid isolated from the autumn crocus, blocks the polymerization of microtubules and arrests cells in the metaphase stage of mitosis (see Chapter 2). Other alkaloids such as vincristine and vinblastine with similar actions, and taxol, which stabilizes microtubules, are used in cancer chemotherapy. Sporadic reports appeared in the literature, starting in the 1970s, suggesting that crude fractions of bacteria, as opposed to the whole organisms, could affect the growth of cells in culture. Moreover, by the 1990s it was clear that several bacteria, which are normally found in association with mucosal surfaces, produced proteins able to directly modulate the growth of cells. Most of the proteins described have inhibitory effects, but some proteins have been discovered which actually promote cell growth (Table 10.2).

Bacterial inhibitors of cell proliferation

The first report of a bacterial putative cell cycle inhibitor concerned the capsular fraction of the oral Gram negative bacterium *Actinobacillus actinomycetemcomitans* which inhibited DNA synthesis by cultured human fibroblasts. Microscopic examination of the cells and flow cytometry of the cellular DNA suggested cell cycle arrest in G_2 prior to mitosis. This turned out to be due to an 8 kDa peptide called gapstatin. Analysis of the mechanism of action of gapstatin revealed that it had no direct effect on DNA or protein synthesis. Flow cytometry of the cellular DNA content revealed that gapstatin, like colchicine, caused cells to express $4n$ DNA content (see Chapter 2). However, when examined microscopically, gapstatin-treated cells failed to show the expression of metaphase chromosomes, indicating that the gapstatin was blocking cells in G_2. Using hydroxyurea synchronization (a method which blocks all cycling cells in culture at the beginning of S phase; removal of hydoxyurea results in all cycling cells being in the same phase of the cell cycle) cohorts of cells were followed through the cell cycle. Gapstatin completely blocked the appearance of mitotic figures in synchronized cell cultures. When gapstatin was added at various times after removal of the 'hydroxyurea block' it only inhibited entry of cells into mitosis when added early in the S phase. When added 4 hours after washing out the hydroxyurea, it did not block entry into

Table 10.2 Cell cycle-modulating bacterial (and protozoal) proteins

Bacterium	Protein	Mechanism
Escherichia coli	CDT	[I] cdc2 protein kinase dephosphorylation
Escherichia coli	CNF1	[I] acts on rho
Escherichia coli	CPE	[I] ?
Helicobacter pylori	100 kDa protein	[I] ?
Fusobacteriun nucleatum[a]	95 kDa protein	[I] inhibits cyclin production
Actinobacillus actinomycetemcomitans[a]	Gapstatin	[I] inhibits cyclin B production
Campylobacter rectus[a]	48 kDa protein	[I] ?
Prevotella intermedia[a]	50 kDa protein	[I] ?
Bordetella bronchiseptica	DNT	[I] inhibits cytokinesis
Salmonella typhimurium	STI	[I] ?
Pneumocystis carinii	?	[I] inhibits cdc2
Trypanosoma brucei	?	[I] block in G_0/G_1 mechanism unknown
Porphyromonas gingivalis[a]	FAF	[P] ?
Vibrio cholerae	Cholera toxin	[P]?
Pasteurella multocida	PMT	[P] ?
Bartonella henselae	?	[P]? apparently selective for vascular endothelial cells

CDT, cytolethal distending toxin; CNF, cytotoxic necrotizing factor; CPE, cytopathic effect; DNT, dermonecrotic toxin; STI, Sal. typhimurium inhibitor of T cell proliferation; FAF, fibroblast-activating factor; PMT, Pasteurella multocida toxin.
[I], inhibits cell proliferation; [P], promotes cell proliferation.
[a] Oral Gram-negative bacteria.
?, Mechanism of action not defined.

mitosis. One possible explanation for these findings was that gapstatin was blocking the synthesis of cyclins, and the cell content of cyclin A and cyclin B was measured in synchronized cultures. Gapstatin inhibited the expected rise in intracellular cyclin B levels while leaving the rise in cyclin A levels unaffected. Thus it appears that gapstatin blocks cell cycle progression by binding to a cell surface receptor and, in some way, inhibiting the synthesis of cyclin B and thereby blocking formation of the complex between cyclin B and cdc2 (also known as maturation promoting factor, MPF, and described in Chapter 2) which is required to drive cells into mitosis.

Act. actinomycetemcomitans is a member of the normal oral microflora, but is also an opportunistic pathogen and is implicated in periodontal disease, a very common form of chronic bacterially induced pathology. These diseases are caused by a variety of Gram-negative oral bacteria (periodontopathogens), which are part of the normal oral microflora and can overgrow and produce inflammation of the gingivae (gums). This inflammation is associated with the chronic destruction of the periodontal ligament and alveolar bone, which form the support structures for the teeth in the jaw bones. Two other periodontopathogens – *Campylobacter rectus* and *Prevotella intermedia* – also produce proteins (of 48 and 50 kDa respectively) which can inhibit cell proliferation. Another periodontopathogen – *Fusobacterium nucleatum* – produces a 95 kDa protein which, like gapstatin, has the ability to inhibit cyclin production. This 95 kDa heterodimeric protein, unlike gapstatin, blocks cells in mid-G_1 by a mechanism which appears dependent on the inhibition of the synthesis of a number of cyclins, A, B, D3 and E, but not cyclin D2 (see Chapters 2 and 3 for a description of cyclins and their role in the cell cycle).

Another bacterium which has shot to prominence in recent years is *Helicobacter pylori*. This organism colonizes the epithelial surface of the stomach and is responsible for gastric ulceration. *Hel. pylori* produces a 100 kDa protein which can inhibit cell proliferation. The mechanism of action of this protein is currently unknown, although it appears to have a different mechanism of action from gapstatin. *Sal. typhimurium* produces an 87 kDa protein which inhibits the proliferation of mitogen-activated murine T lymphocytes. This protein has been named *Sal. typhimurium*-derived inhibitor of T cell proliferation or STI for short and is suggested to be an immunosuppressive agent. *Pneumocystis carinii*, an opportunistic pathogen, is responsible for life-threatening pneumonia in AIDS patients. It has recently been reported that this organism is able to inhibit the cyclin-dependent kinase cdc2 without affecting levels of this protein. Cell cycle arrest is dependent on the binding of the organism (a fungus) to epithelial cells and the possibility that it involves some form of type III secretion mechanism needs to be considered.

Perhaps the most wide-ranging group of cell cycle-regulating proteins is produced by the common gut bacterium, *Escherichia coli*. As has been described in Chapter 1 there are now seven strains of *E. coli* which produce diarrhoeal diseases. A number of these strains produce pathology by the production of exotoxins. Two of these exotoxins have cell cycle-modulatory activity. Cytotoxic necrotizing factors (CNFs) were first described in 1983 in *E. coli* isolated human extraintestinal lesions. CNF1 is a heat-labile protein which causes giant cell formation in a variety of cell types and necrosis when injected into the skin of experimental animals. In recent years it has been shown that these toxins act on the 21 kDa Rho GTP-binding protein causing changes in cytoskeletal organization (see Chapters 2 and 7 for more details). These Rho-controlling toxins are also important in bacterial uptake into cells. One consequence of the cytoskeletal effects of the CNFs is that cytokinesis is blocked, causing the formation of giant cells. Of interest is the finding that the N-terminus of the CNFs exhibit around 30% identity to a potent mitogenic toxin from *Pasteurella multocida* – to be described in the next section. A separate factor has recently been described in pathogenic *E. coli* which causes HeLa cells to be blocked in G_2, but the nature of this activity has not been defined. The actin cytoskeleton and its control by GTP-binding proteins such as Rho, Rac and Cdc42 have been described in Chapters 2, 3 and 7 and the consequence of bacterial intereference with these proteins has been discussed in Chapters 5, 6 and 7.

A third cell cycle-regulatory group of proteins are the emerging family of toxins given the title cytolethal distending toxins (CDTs). These are found in unrelated bacterial species including *E. coli*, *Shigella dysenteriae*, *Campylobacter* spp. and *Haemophilus ducreyi*. CDT blocks HeLa cells in G_2 but, unlike gapstatin, does not inhibit the rise in cyclin B levels which occur during the progression from S to G_2. The progress from G_2 into mitosis is dependent on the complex between cyclin B and the protein kinase cdc2 (described in Chapter 2). The state of tyrosine phosphorylation of cdc2 is important for its activity. The initiation of mitosis is caused by the dephosphorylation of cdc2 by the phosphatase cdc25. CDT inhibits the dephosphorylation of cdc2 and it is this which prevents entry of cells into mitosis. The mechanism of action of the various bacterial proteins described on cell cycle progression is shown schematically in Figure 10.3.

Although the numbers of examples are still relatively small it is clear that bacteria which live in contact with epithelial surfaces, either as part of the nomal microflora or as pathogens, produce proteins which have the ability to inhibit cell cycle progression. The mechanism of action of three of these proteins has been partially elucidated and it seems clear that bacteria have evolved a group of proteins able to interfere with the cyclin/cyclin-dependent kinase/

phosphatase systems of eukaryotic cells. Other microbes – *Trypanosoma brucei* and *Pn. carinii* — also have the ability to arrest cell cycle progression. What value is a cell cycle inhibitory protein to a microorganism? The answer to this is not immediately apparent. However, if we refer back to Chapter 8 and the role of innate immunity in controlling infection, then one of the proposed mechanisms was the shedding of epithelial (mucosal) surfaces. Most epithelial surfaces shed the outermost layer of cells as part of a tissue replacement mechanism. This is most obvious with the skin, and indeed much of the dust in our houses is our own skin. It is also this outermost layer of cells on which bacteria live. Shedding is dependent on the division of cells and therefore if cell division was to be slowed bacteria would increase their colonization time on the epithelium. If this is the reason for the production of such eukaryotic cell cycle inhibitors, it remains to be determined how pervasive this activity is. The rapid replacement of epithelial cells is believed to be due to the physical and other demands on these cells and the fact that shedding removes bacteria is thought to be purely fortuitous. An alternative explanation is that shedding is designed first and foremost to remove bacteria. Only time will tell which hypothesis is correct.

Another possible function of bacterial cell cycle inhibitors is to inhibit myeloid and lymphoid cell proliferation which would occur in response to bacterial infections. This ability to inhibit lymphocye proliferation has been shown by a number of bacterial cell cycle inhibitors such as STI, described above. Thus the inhibition of cell cycle progression may be a novel immunosuppressive mechanism evolved by bacteria.

How did such cell cycle inhibitors proteins evolve? Bacteria are known to produce a large number of proteins and peptides which have the ability to inhibit the growth of other bacteria. These bacteriocins have some overlap with the antibiotic peptides. However, some bacteriocins have sophisticated mechanisms for blocking the bacterial cell cycle. For example, *E. coli* produces a 4 kDa peptide termed microcin B17 which acts as a DNA gyrase inhibitor in various Gram negative cells. Could inhibitors of the eukaryotic cell cycle have evolved from this bacterial antibacterial strategy? This raises the possibility that the cell cycle inhibitors produced by bacteria could also act to block bacterial proliferation and that such compounds are really antibiotics under a different guise. However, we have tested gapstatin for its ability to inhibit the growth of other bacteria and failed to find any inhibition. This may be why these bacterial proteins evolved to inhibit cyclin-dependent events, which are largely absent from the bacterial cell cycle.

One possible spin-off from the discovery that bacteria produce inhibitors of cell cycle progression is the possibility that such proteins could have clinical benefits as anticancer agents. This possibility will be discussed at the end of this chapter.

Bacterial stimulators of cell proliferation

Certain bacteria produce proteins whch promote the growth of eukaryotic cells. Unfortunately, much less is known about the capacity of bacteria to promote cell cycle progression and the possible benefit that such behaviour has for bacteria.

Four bacteria have so far been shown to synthesize and secrete mitogens. These are *Vibrio cholerae* which produces cholera toxin (CT), *P. multocida* producing *P. multocida* toxin (PMT), the oral bacterium *Porphyromonas gingivalis* which secretes a 24 kDa protein that has been termed fibroblast-activating factor (FAF) and *Bartonella henselae* (formerly *Rochalimaea henselae*), a rickettsia which produces a cell-selective mitogen. PMT, whose mechanism of action is described in Chapter 7, is able to stimulate the proliferation of both adherent and non-

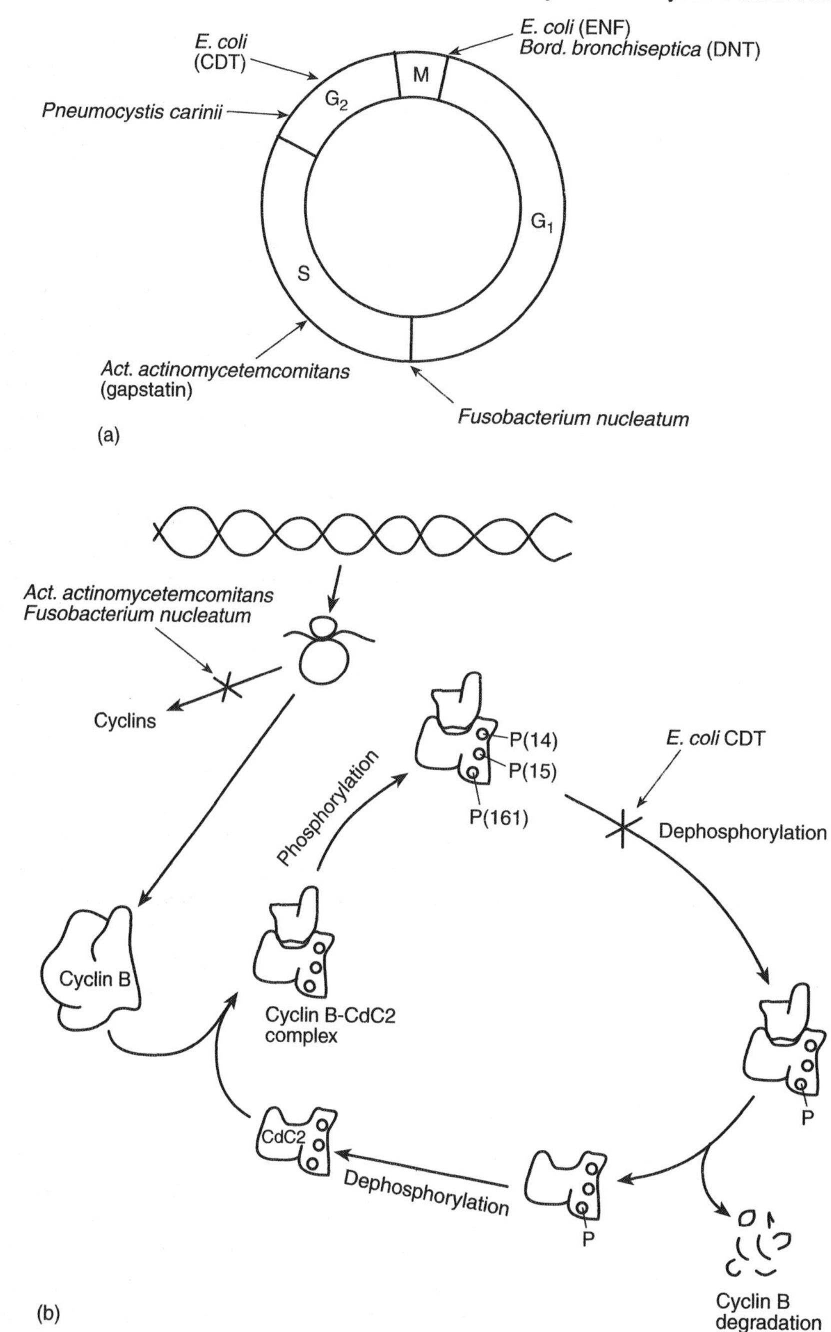

Figure 10.3 Schematic diagram showing the mechanisms of actions of the various bacterial proteins which inhibit eukaryotic cell cycle progression and whose cellular targets have been identified. (a) shows the points of the cell cycle at which the various bacteria known to inhibit the cell cycle and whose mechanism has been defined act. (b) shows the mechanism by which bacterial cell cycle inhibitory proteins work

adherent cells at concentrations as low as 1 picomole and is the most potent inducer of cell proliferation yet discovered. In the natural disease, PMT causes bone destruction – this toxin is an incredibly active promoter of bone resorption *in vitro*. The mechanism of action of PMT is not established but may be to do with the rapid promotion of proliferation of the two major cell populations in bone: the osteoblast (which synthesizes the extracellular matrix of bone) and the osteoclast (which breaks down the calcified bone matrix). FAF is a 24 kDa protein obtained from outer membrane vesicles which bud from *Por. gingivalis*. This protein is able to stimulate the DNA synthesis in cultured human gingival fibroblasts at low molar concentrations, independent of the synthesis of IL-1 (a potent fibroblast growth factor). *Por. gingivalis* is associated with the pathology of the periodontal diseases, in which the major pathology is bone destruction. *Bartonella henselae* causes cat scratch fever and bacillary angiomatosis (an angioma is a tumour composed of capillaries). Infection results in the production of new blood vessels and it is believed that this organism produces a protein, whose nature has not been elucidated, which is able to stimulate the selective proliferation of vascular endothelial cells.

Again, the question has to be asked – what evolutionary advantage would a bacterium gain from producing a factor which promoted cell proliferation? The answer to this question is not immediately apparent, and indeed is difficult to answer, as only a handful of bacteria have, as yet, been shown to produce such cell cycle-activating factors. This does not mean that other bacteria do not produce such stimulatory factors, only that they have not been looked for. One cause for concern is the enormous potency of PMT as a growth factor. This molecule is more potent than all comparable human growth factors, including platelet-derived growth factor (PDGF). PDGF is equivalent to the *sis* oncogene product, a protein produced by the simian sarcoma virus which produces sarcomas in monkeys (refer to Chapter 3 and its discussion of oncogenes). It is now recognized that many oncogene products – proteins involved in the cancer process – are either growth factors or growth factor receptors. The incredible potency of PMT as a growth factor raises the spectre that PMT and molecules like PMT, if they exist, could provide bacteria with the wherewithal to induce a malignant phenotype in host tissues. Bacteria have not been implicated in the pathology of cancer, but this is changing and there is the belief that *Hel. pylori* may have the capacity to promote the cellular changes leading to cancer. The reader should watch this space.

Bacterial control of apoptosis

As has been described in Chapter 2 homeostatic control in multicellular organisms requires that for every new cell produced an 'old' cell must die by the mechanism of apoptosis. Apoptosis also provides a defence mechanism in which damaged and potentially dangerous cells can be eliminated, thus rendering the organism free of their deleterious effects. Apoptosis is increasingly being seen as an important mechanism in the defence against infections. This is particularly true with virally infected cells, which often undergo spontaneous apoptosis or apoptosis induced by CD8 T cells (see Chapter 9). Apoptosis is controlled by a number of proteins including the caspases, which are proteinases, bcl-2, which inhibits apoptosis, and Bax, which can block the activity of bcl-2 (described in Chapter 2). One of these caspases, interleukin 1 converting enzyme (ICE, now known as caspase 1), was first discovered because of its ability to cleave the 31 kDa inactive pro-form of IL-1β to produce the 17 kDa active form and thus is a proinflammatory signal. A number of viruses encode proteins which can inhibit the caspase system and thus block apoptosis. Other viruses, such as HIV, appear to induce CD4 lymphocytes to undergo apoptosis. In recent years it has been demonstrated that a range of bacterial pathogens can induce apoptosis in a variety of host cell populations. No bacteria have so far been reported to inhibit apoptosis, but this is surely only a matter of time. In addition, it has been suggested that certain bacteria can utilize the apoptotic mechanism to induce inflammation. This, at first sight, appears to be counterintuitive as apoptosis is believed to have evolved to allow cells to be removed without the production of inflammation.

The macrophage is targetted by a number of bacteria which cause it to undergo apoptosis. Two pathogens which employ this method are *Shigella* spp. and *Salmonella* spp. *Sh. flexneri* produces an invasin (termed IpaB) which activates ICE and induces apoptosis. ICE knockouts are insensitive to the cytotoxic actions of *Shigella* spp. and in such animals infection with this bacterium does not produce inflammation. Inhibition of ICE by synthetic inhibitors also blocks the ability of *Shigella* to induce apoptosis. Conversely, microinjection of IpaB into macrophages is sufficient to induce apoptosis. *Salmonella typhi* produces a similar invasin (SipB) and likely acts by a similar mechanism. *Shigella flexneria*, the causative organism of bacillary dysentry, enters the submucosa of the colon via M cells and is taken up by macrophages and via IpaB induces the apoptotic mechanism and also stimulates the conversion of inactive pro-IL-1β to the active 17 kDa form. In addition, other cytokines, in particular IL-18, which also requires ICE for activation, could be formed if present in the cell in their pro-forms. IL-18 is an inducer of IFNγ synthesis and activates Th_1 lymphocytes. It is suggested that the production of these cytokines, via this apoptotic mechanism, results in the inflammatory pathology of bacillary dysentry.

A range of other bacteria including *Bacillus anthracis, Corynebacterium diphtheriae, Pseudomonas* spp., *Staphylococcus* spp., *Streptococcus* spp. and *Act. actinomycetemcomitans* induce apoptosis, mainly via the production of exotoxins. The action of bacterial superantigens has been described in Chapter 9. These exotoxins result in the apoptotic death of particular Vβ subsets of T cells.

THE COMMENSAL MICROFLORA: AN EXAMPLE OF CLOSE CELLULAR CONVERSATION?

Cellular microbiology has come to the fore in describing the mechanisms by which individual bacteria cause tissue pathology. However, while there may be a few dozen or so bacterial species which are recognized as being capable of causing tissue pathology, the average human body contains at least 1000 different species of bacteria which constitute the normal or commensal microflora (Figure 10.4). These organisms live on all the externally facing epithelial surfaces of the body, which includes the skin, oral cavity, oesophagus, upper respiratory tract, gastrointestinal tract and outer aspect of the urogenital tract (Figure 10.5). We collect our commensal microflora in the first hours to days after we are born as we change from being a germ-free (gnotobiotic) organism into a complex community of eukaryotic and prokaryotic cells. This must be one of the greatest shocks to the system as the epithelia of the neonate make contact with a multitude of diverse bacteria. It is probably only after weaning that the composition of the microflora becomes relatively stable. Surprisingly little is known of this important process of microbial colonization. There is limited evidence

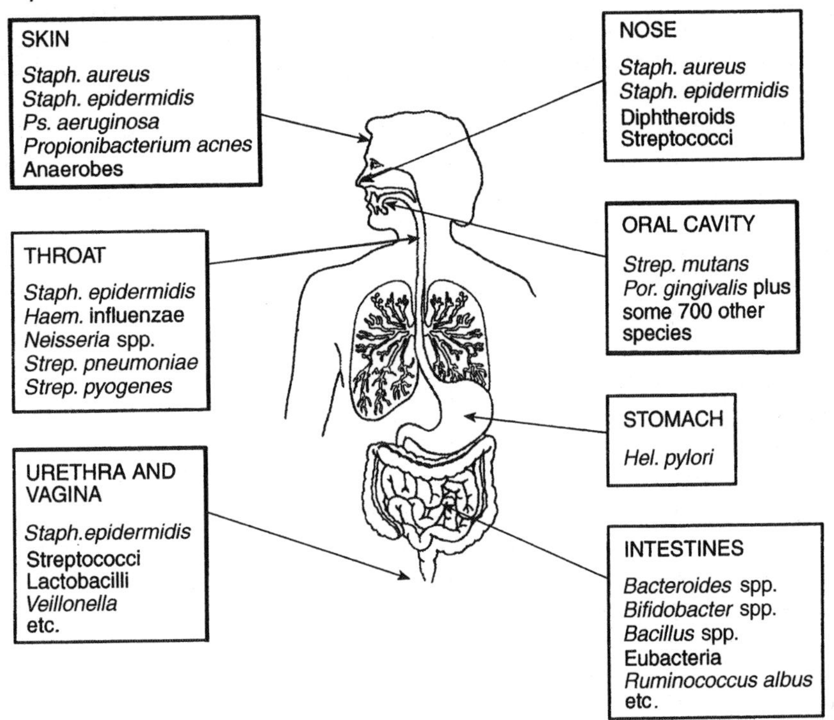

Figure 10.4 Diagram showing the sites in the human body where the commensal microflora are to be found and some of the organisms residing there

which suggests that the composition of the normal microflora is, to some extent, genetically determined. Thus the oral and nasal microflora of twins are more similar than that of comparable singleton children.

In terms of cell numbers we are 90% bacterial. Despite the complex systems of cellular immunity which have evolved to combat bacterial infections and the sophisticated machinery used by bacteria to defeat immune systems, the mucosal surfaces of the body can cope with living in close association with very large numbers of bacteria, many of which are perfectly capable of causing disease. For example, the majority of the members of the normal microflora are obligate anaerobic bacteria. Many of these bacteria can produce anaerobic and mixed anaerobic infections which can occur in practically any site in the body. Other bacteria which colonize man are opportunistic pathogens. Well-known examples of such organisms are *Staphylococcus aureus, Neisseria meningitidis, Haemophilus influenzae* and *Streptococcus pneumoniae*, which can produce a range of diseases many of which, like toxic shock, meningitis and pneumonia, can be lethal.

Thus, for most of us, and for most of the time, we live in harmony with our commensal microflora. This harmony has largely been taken for granted by microbiologists and other interested scientists. However, increased focus on the intimacy of the interactions which occur between bacteria and host eukaryotic cells suggested that the maintenance of a normal microflora requires some degree of conversation between the epithelial cells of the host mucosa and the bacteria which constitute the normal microflora.

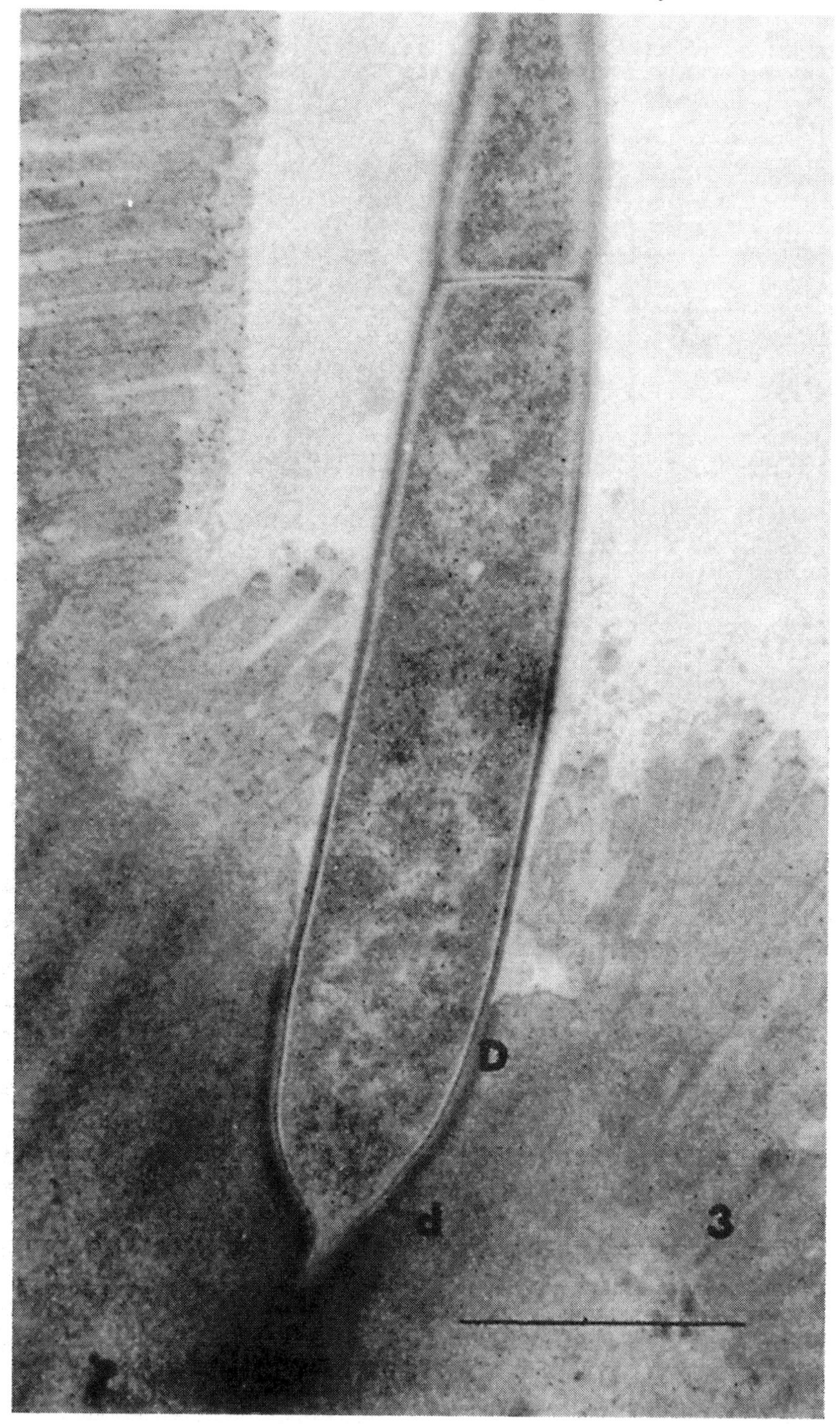

Figure 10.5 Photograph of commensal bacterium attached to an epithelial cell. Reproduced from Davis CP, Savage DC (1974) *Infect. Immun* **10**: 948–956 (with permission)

The commensal paradox

Modulins are molecules derived from bacteria which can induce host eukaryotic cells to produce cytokines (see Chapter 8). A large number of bacterial components are capable of inducing cytokine synthesis and some are extremely active in this respect. Thus, it is not simply lipopolysaccharide (LPS) which is the active cytokine inducer from bacteria. Many bacterial components are as active as the average LPS molecule and many are much more active. A good example are the bacterial exotoxins, a number of which are active at concentrations in the picomolar to attomolar concentration range. Even bacterial DNA can promote cytokine synthesis and cause pulmonary inflammation or septic shock in experimental animals.

Thus, many of the key components and exported molecules of bacteria are potent inducers of cytokine synthesis. The epithelial cells on which the commensal microflora exist are, like any other cell population, able to produce proinflammatory cytokines. Indeed, it is believed that this is one of their main functions. The obvious key signals which would warn the body that epithelia were under attack from bacteria are the proinflammatory cytokines.

Bacteria release large amounts of material, including components from their surfaces which would contain LPS and other cytokine-inducing structural components. The lifespan of bacteria is generally short and significant cell death occurs in any bacterial population, releasing a wide range of bacterial components, including DNA. The mucosal surfaces of the body, being in contact with large numbers of bacteria, must also be in contact with significant amounts of cytokine-inducing bacterial components. Logically, therefore, the epithelial surfaces of our bodies should be in a constant state of inflammation. However, equally obviously, this is not the case. This is the basis of what has been termed the *Commensal Paradox* (Figure 10.6).

A hypothesis to explain the commensal paradox

A hypothesis to explain this paradox is that the failure to release cytokines is because members of the commensal microflora produce and release proteins which modulate the production of proinflammatory cytokine by epithelial cells (Figure 10.7). The putative proteins, to which we have given the name microkines (a generic term to include similar viral and protozoal proteins), could either: (i) directly inhibit the formation of pro-inflammatory cytokines; (ii) induce the synthesis of cytokines with anti-inflammatory properties (e.g. interleukin (IL)-10 or transforming growth factor (TGF)β); (iii) do both; or (iv) act to modulate cytokine networks in the submucosal tissues by some, as yet, undefined mechanism (Figure 10.7). These proteins have been described in Chapters 8 and 9.

Evidence for the the microkine hypothesis

Two distinct areas of research provide support for the microkine hypothesis. The first, described in Chapter 3, is the discovery made over the past decade that viruses encode a large number of proteins which resemble cytokines or cytokine receptors and/or which can block the actions of cytokines (Table 10.3). Such proteins, termed virokines, appear to be a major part of the viral strategy to overcome the innate and acquired systems of immunity, which are largely controlled by cytokines.

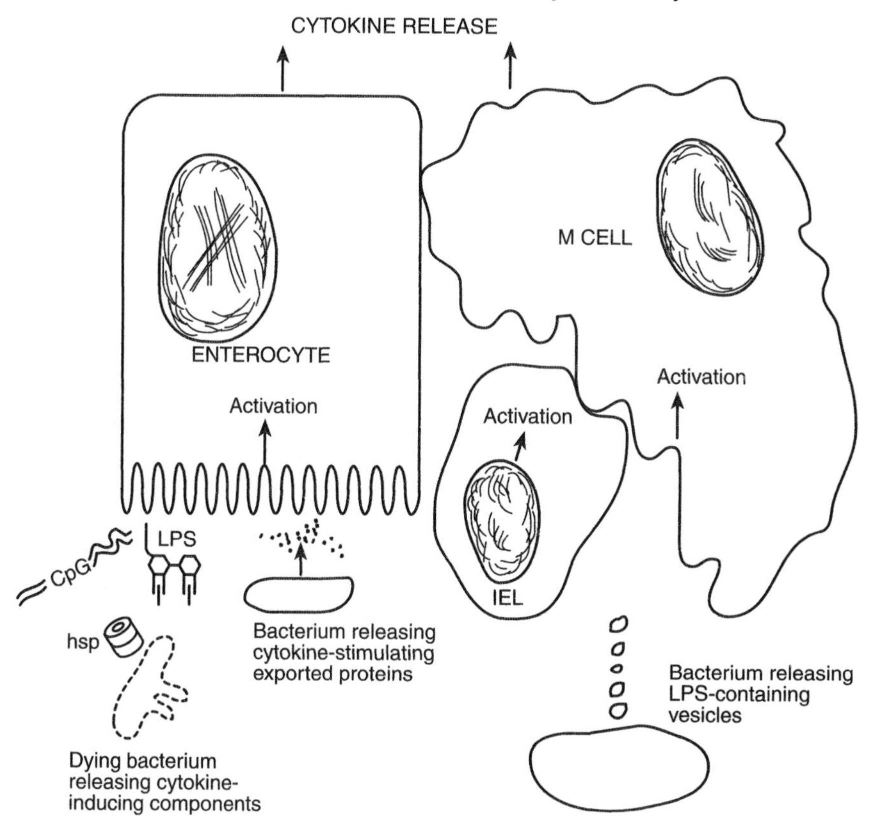

Figure 10.6 The commensal paradox. Diagrammatic illustration outlining the interactions occurring between commensal bacteria and the host epithelium which must lead to cytokine production. In this example the colonic microflora is being considered. Bacteria constantly release membrane components and exported proteins which have the capacity to induce cytokine synthesis. Many bacteria must die releasing LPS, CpG-containing DNA, heat shock proteins and many other cytokine-inducing molecules. These should interact with the epithelial barrier, which contains enterocytes, M cells and intraepithelial lymphocytes (IELs). These cells are capable of producing cytokines and should therefore induce a local inflammatory response

The second piece of evidence is that alteration in cytokine network signalling can produce profound host responses to the commensal microflora. The most interesting cytokine gene knockout is that of the IL-2 gene. This cytokine is a growth factor for Th_1 lymphocytes and the most obvious prediction is that IL-2 knockout animals would be immunosuppressed and susceptible to intracellular bacterial infections. In fact such animals died from a colitis which resembled the severe and chronic human disease, ulcerative colitis. In addition to colitis the IL-2-deficient mice also produced autoantibodies to constituents of the colon. As animals lacking the important T cell growth factor were likely to be immunocompromised they were examined for the presence of infectious organisms which could induce the pathology seen. None of the likely parasitic organisms could be found. The solution to the causation of the colitis came when IL-2-deficient animals were bred under germ-free conditions. Animals lacking both IL-2 and their commensal microflora did not develop colitis. This suggests that the colitis was in response to the normally non-inflammatory colonic microflora. A similar

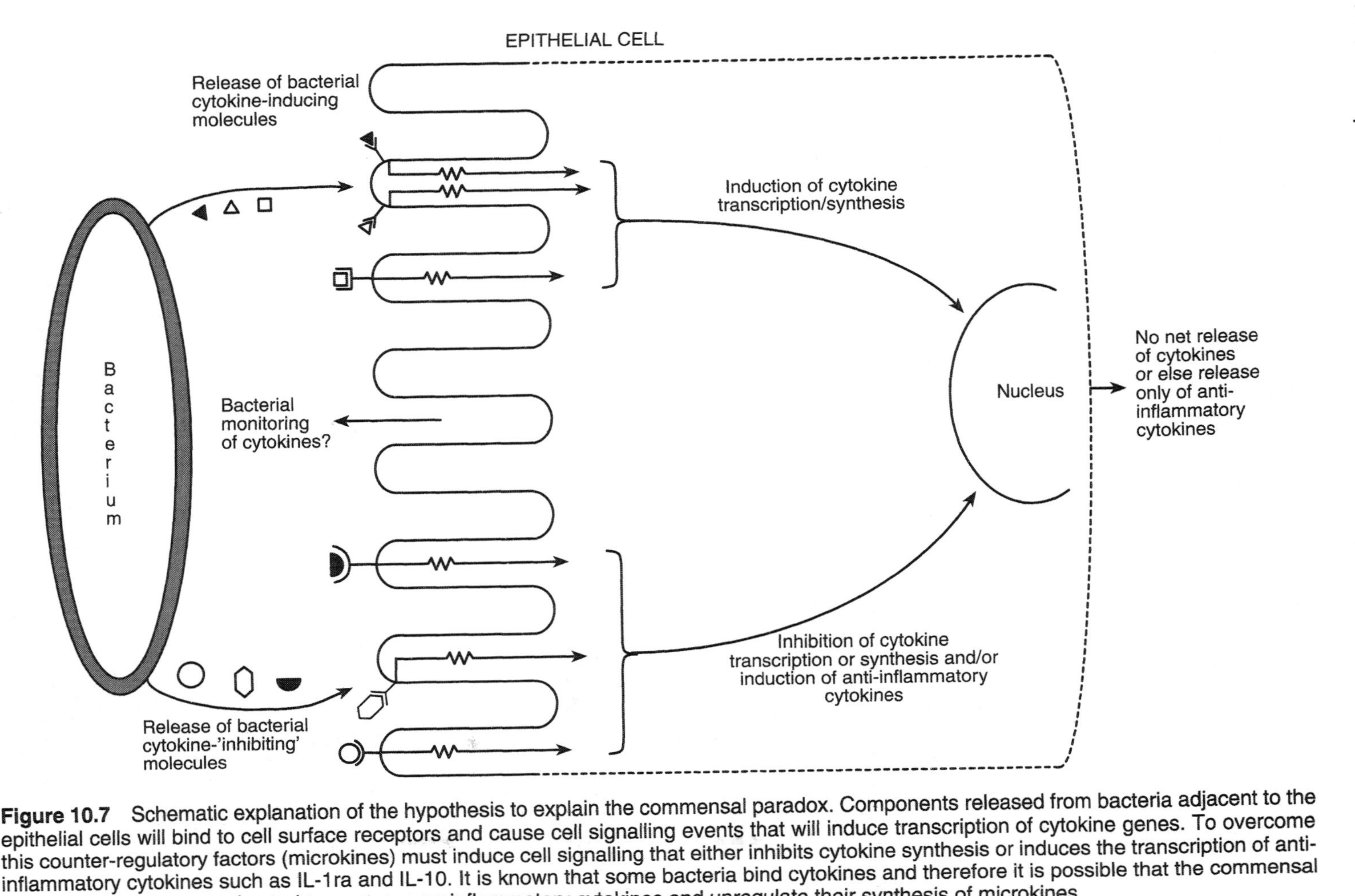

Figure 10.7 Schematic explanation of the hypothesis to explain the commensal paradox. Components released from bacteria adjacent to the epithelial cells will bind to cell surface receptors and cause cell signalling events that will induce transcription of cytokine genes. To overcome this counter-regulatory factors (microkines) must induce cell signalling that either inhibits cytokine synthesis or induces the transcription of anti-inflammatory cytokines such as IL-1ra and IL-10. It is known that some bacteria bind cytokines and therefore it is possible that the commensal bacteria can sense proinflammatory cytokines and upregulate their synthesis of microkines

Table 10.3 Cytokine-modulating genes and gene products produced by viruses

Virus	Viral protein	Host homologue	Function
Cowpox	crmA	Serpin type protease inhibitor	Inhibits ICE/blocks IL-1β synthesis
Baculovirus	p35	Protease inhibitor	Inhibits ICE/blocks IL-1β synthesis
Vaccinia	B15R	IL-1β receptor	Inhibits IL-1β activity
HVS	ORF78	Soluble IL-8 receptor	Inhibits IL-8 activity
HCMV	US28	Soluble IL-8 receptor	Blocks chemokine activity
Myxoma	T2	Soluble TNF receptor	Blocks TNF activity
Myxoma	T7	Soluble IFN_{γ} receptor	Blocks IFN_{γ} activity
Myxoma	SERP-1	Serine proteinase inhibitor	Inhibits ICE?
Myxoma	Myxoma growth factor	Epidermal growth factor-like proteins	Stimulates cell growth
SFV	T2	Soluble TNF receptor	Blocks TNF activity
EBV	BCRF1	IL-10	Anti-inflammatory
EHV2	IL-10-like protein	IL-10	Anti-inflammatory
Poxviruses		EGF, TGF_{α}	Cell growth promoters
KSHV	v-chemokines	MIP-I, MIP-II	?
KSHV	v-IL-6	IL-6	Inhibits apoptosis?
KSHV	v-IRF	IRF	?

EBV, Epstein–Barr virus; EHV2, equine herpesvirus type 2; HVS, herpesvirus saimiri; HCMV, human cytomegalovirus; KSHV, Kaposi's sarcoma-associated herpesvirus, also known as human herpes virus 6; IRF, interferon regulatory factor; MIP, macrophage inflammatory protein; SFV, Shope fibroma virus.

story has emerged with animals in which the gene for IL-10 has been knocked out. This could be due to a number of possibilities:

(i) In the absence of IL-2, the normal anti-inflammatory cytokine network which is maintained in the colon in the face of the proinflammatory cytokine-inducing bacterial components cannot be maintained and so the colon becomes inflamed. This explanation does not require significant conversation between the colonic bacteria and the epithelial cells.
(ii) In the absence of IL-2 the normal signalling between the colonic microflora and the colonic mucosa is altered such that signals emanating from the former can no longer be received as an anti-inflammatory message and inflammation ensues. As suggested in Figure 10.7 the bacteria of the commensal microflora may monitor key cytokines such as IL-2 and produce counterregulatory molecules in response. In the absence of IL-2 these microkines are no longer produced.
(iii) In the absence of IL-2 the two-way flow of information between the colonic mucosa and the colonic microflora is aberrant. The normal induction of anti-inflammatory microkines by the microflora is controlled by the mucosa. With changes in the cytokine networks in the host tissues the signals controlling the expression of these microkines are missing and so such vital anti-inflammatory proteins are not produced and inflammation ensues (Figure 10.8).

These explanations are, at the moment, simply guesses but are amenable to scientific testing. However, these knockout experiments show the importance of 'correct' tissue cytokine networks for the control of the host response to its own commensal microflora.

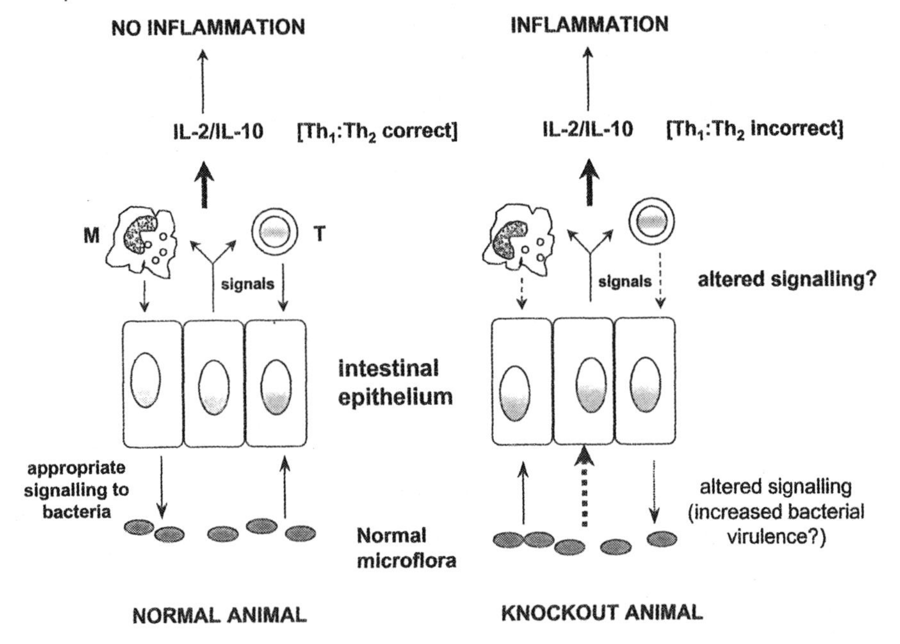

Figure 10.8 Schematic explanation of the IL-2 knockout results. The details of this diagram are explained in the text. Reproduced from Henderson *et al* (1998) Bacteria-Cytokine Interactions: In Health and Disease. Portland Press, London (with permission)

Viruses have evolved or pirated proteins with the ability to neutralize the activity of host cytokines. Do bacteria have the same capacity and are these proteins involved in the ability of multicellular eukaryotes to maintain large collections of bacteria within their bodies? It is only within the last five years that bacterial proteins with cytokine-modulating actions have begun to be described. During this period around 15 proteins from a variety of bacteria have been demonstrated to inhibit the synthesis or actions of host cytokines. Many of these proteins are able to block the synthesis of proinflammatory cytokines such as TNF and a smaller number can block IL-1 synthesis (Table 10.4). Many are bacterial exotoxins which, although they can block TNF synthesis, will also stimulate the synthesis of other proinflammatory cytokines such as IL-6. A few microkines can inhibit the production of certain types of cytokines. For example, an undefined protein factor from enteropathogenic strains of *E. coli* has been shown to inhibit the synthesis of various lymphokines (cytokines involved in the growth and activation of lymphoid cells), including the production of IL-2, but has no effect on the synthesis of proinflammatory cytokines. A similar finding was made with a purified 14 kDa protein from *Act. actinomycetemcomitans*. The possibility exists that proteins with this spectrum of activity may have a specific effect on the acquired immune system while leaving the innate systems intact. Other bacteriokines may be targetted at the innate systems of immunity.

We predict that members of the commensal microflora will be found to produce large numbers of proteins able to modulate mucosal cytokine networks and prevent the mucosa from becoming inflamed in response to released proinflammatory bacterial constituents. These proteins may have therapeutic potential.

Secreted proteinases are another mechanism utilized by bacteria to control host responses. Proteinases are well-recognized factors in the pathology of a number of important idiopathic

Table 10.4 Microkines: anticytokine proteins from bacteria

Protein	Cytokine whose synthesis is inhibited
Cholera toxin	TNF (not IL-6 or TGFβ)
Pertussis toxin	IL-1
Anthrax oedema toxin	TNF (stimulates IL-6)
Pseudomonas aeruginosa exotoxin	IL-1, TNF lymphotoxin, IFNγ
Botulinum toxin type D	TNF
Yersinia enterocolitica YopB	TNF (no effect on IL-1 or IL-6)
Proteus mirabilis 39 kDa protein	TNF
Sal. typhimurium-derived inhibitor of T cell proliferation (STI)	IL-2 (stimulates IFNγ)
Brucella-derived protein	TNF (not IL-1 or IL-6)
Enteropathogenic *E. coli* protein	IL-2, IL-4, IL-5, IFNγ (not IL-1, IL-6, IL-12 or RANTES)
14 kDa *Act. actinomycetemcomitans* protein	IL-2, IL-4, IL-5, IFNγ

diseases of man including emphysema, asthma, arthritis and cancer. A role for bacterial proteinases in nutrition and bacterial growth has been recognized for some time. There is now evidence that bacteria can modulate inflammatory mechanisms by their released proteinases. For example, bacterial proteinases can activate the kinin-generating cascade leading to the production of vasoactive peptides. On the other hand, bacterial proteinases can inhibit inflammation by inactivating the anaphylotoxin C5a. Bacterial proteinases can also cause the release of cytokine receptors from cells. The consequences of inactivating pro- or anti-inflammatory cytokines, activating proinflammatory cytokines and releasing soluble cytokine receptors (which would then act as cytokine antagonists) via bacterial proteinases are almost impossible to define. The proteinase Arg-1 gingipain (R1-proteinase) of the oral Gram negative bacterium *Porphyromonas gingivalis* is capable of proteolytically cleaving and inactivating cytokines such as IL-1 and IL-6. The protease is capable of self-cleavage and one of the cleavage products has itself the ability to stimulate cytokine synthesis. Thus, this one proteinase seems to carry with it the capacity to modulate cytokine networks. These findings suggest that bacterial proteinases could be therapeutic targets for the treatment of bacterial infections. From the available evidence it would appear that bacterial proteinases can either be classified as modulins or bacteriokines or, indeed, both.

CELLULAR MICROBIOLOGY AND IDIOPATHIC DISEASES

At the beginning of this century infectious diseases were the major killers worldwide. By the 1960s it was felt that bacterial infections were largely under control. By this time the major killers in the developed world (although not in the rest of the world, which contains the majority of the population) had become cardiovascular disease (which includes strokes and atherosclerosis) and cancer. Other major disease which caused severe morbidity and could be fatal included the arthritic diseases (rheumatoid arthritis, osteoarthritis and osteoporosis), gastric ulceration and asthma. A number of these conditions were becoming prevalent largely as a result of health care and nutritional measures which had caused a dramatic increase in average lifespan. One suggested explanation for the large increase in the prevalence of asthma is that it is a result of the decrease in the incidence of tuberculosis. When the population was rife with tuberculosis the majority of individuals had a constant boost to develop Th_1 responses (which are needed to combat intracellular parasites). However, when the population is free of tuberculosis Th_2 responses can become prominent and it is such responses that trigger mast cells and eosinophils believed to be responsible for atopic conditions which are related to asthma. By the 1960s biomedical research had become a major industry in North America and Europe and enormous amounts of funds were funnelled into researching the causes and pathology of cancer and heart disease and in later years arthritis and asthma. In consequence, the funds focused on understanding bacterial infections largely dried up.

Infectious diseases have a clear causation – the infectious agent. However, many of the diseases afflicting humans are idiopathic, that is, of unknown origin. This is largely true of cancers and cardiovascular disease and certainly is true of arthritis and asthma. There are obvious predisposing factors to certain of these diseases, the most obvious being smoking and diet. However, we still know little about the causation of the various conditions which are contained within these four categories of disease.

Could infectious agents be involved in any of these major idiopathic conditions? The underlying cellular mechanisms responsible for some of the pathology (the inflammation) of arthritis, asthma, cardiovascular disease and cancer are identical, as far as can be deduced, from what would be happening in infection. Thus to a greater or lesser extent the pathology of these conditions utilizes the defence mechanisms that the organism uses to combat infection. The first evidence that infectious agents could be responsible for some idiopathic conditions was the demonstration by Peyton Rous that chicken sarcomas (tumours of connective tissue) could be transmitted by cell-free filtrates of the original tumour. This established that malignancy could be transmitted by viruses (in this case the Rous sarcoma virus – RSV) but it was only in the 1970s that the molecular mechanism by which RSV transforms cells, via the v-Src oncogene, was discovered. These oncogenes, which are present in a range of viruses, have been found to encode either the cytokines known as growth factors or the receptors/post-receptor kinases for these growth factors. In more recent years there has been the suggestion that certain bacteria could be involved in the process of carcinogenesis. The most persuasive evidence, such as it is, is that *Hel. pylori*, now known to be responsible for gastric ulceration, is also a risk factor for the development of gastric cancer. The evidence for this hypothesis is, at present, circumstantial and is based on the finding that there is an association between colonization with this bacterium and cancer. Additional evidence supporting the possibility that bacteria could induce cancer comes from the finding described earlier in this chapter that a growing number of bacteria have the ability to control the cell cycle and apoptosis. As aberration of these cellular events is the foundation stone of carcinogenesis, the possibility that bacteria cause cancer cannot be ignored.

For much of the twentieth century the causation of arthritis was believed to be infectious. In the period encompassing the 1970s until the 1990s the prevailing paradigm was that the inflammatory condition, rheumatoid arthritis, was due to autoimmunity. The autoimmune hypothesis provided a mechanism but not an aetiology. Over the past 10–15 years attention has focused on the concept that in most autoimmune diseases the problem is cross-reactivity between some microbial constituent, to which the patient becomes immunized, and a self component, made by every cell and responsible for protein folding – the heat shock proteins – described in detail in Chapter 9. These heat shock proteins (Hsps) are also known as stress proteins and molecular chaperones. A key finding is the large degree of amino acid sequence conservation of these proteins between pro- and eukaryotic Hsps. A second major finding is that the Hsps are major immunogens during bacterial infections. This homology and the large immune response that we raise when confronted with bacterial Hsps may lead to the autoimmune cross-reactivity in patients with arthritis. This hypothesis is still the subject of extensive research. Bacterial superantigens have also been implicated as a causative agent of autoimmune disease (see Chapter 9). The presence of superantigens and the resultant disappearance of V_β T lymphocyte subsets can therefore be seen as pathological 'fingerprints' in blood or in lesional tissues. This is known as skewing of the V_β T cell population. In diseases such as rheumatoid arthritis, psoriasis and the periodontal diseases such skewing has been reported to exist, suggesting that some bacterial superantigen has been at work. The loss of such a large number of T cells could destabilize the control of immunity and in some way lead to a chronic inflammatory state.

Another common chronic inflammatory disease which can be fatal is ulceration of the stomach and duodenum. This condition was often attributed to a stressful lifestyle and was treated surgically and by use of histamine antagonists. *Hel. pylori* was found in stomach ulcers but the possibility that it could cause ulceration was discounted until eventually Koch's postulates were satisfied. One 'experiment' which supported the contention of the infectious nature of gastric ulcers was done by Barry Marshall, a clinician working in the laboratory of J. Warren, the discoverer of *Hel. pylori*. In the great tradition of the clinical scientist, Marshall had his stomach examined using an endoscope to confirm that it was healthy. He then drank a culture of *Hel. pylori* and two weeks later he had a second endoscopy which revealed that his stomach was inflamed, and a biopsy taken at the same time demonstrated the presence of viable bacteria. It is now accepted that *Hel. pylori* is the causative agent of stomach ulcers and the treatment currently recommended is a triple combination of bismuth with amoxicillin and metronidazole.

The discovery that a well-known idiopathic disease was caused by a bacterium has shaken the foundations of medicine and will make it easier in the future to test hypotheses that suggest that particular idiopathic diseases are due to infection. A good example of a 'disease' which may be caused by an infectious agent are cardiovascular diseases. These include well-known conditions such as hypertension, stroke, angina, myocardial infarction and atherosclerosis. The latter disease is a chronic inflammation of the blood vessels which leads to occlusion of vessels and loss of blood supply to tissues. If this occurs in the arteries supplying the heart it can cause angina and heart attacks. There is now evidence that certain bacteria may be implicated in the pathology of atherosclerosis. The intracellular bacterium *Chlamydia pneumoniae*, which causes about 10% of pneumonia cases in the USA, has recently been linked to the pathology of atherosclerosis. Patients with atherosclerosis generally have high levels of antibodies to this pathogen and the organism has been found in the diseased arteries of up to 60% of patients. In one study 80% of the arteries from atherosclerosis patients demonstrated the presence of *Chl. pneumoniae*, while only 4% of controls specimens were infected. It is still early days and the results need to be confirmed but it is nonetheless a very interesting connection. A second group of bacteria which have recently been implicated in the pathology of atherosclerosis are those implicated in the chronic dental conditions known as the periodontal diseases. In these lesions build-up of oral Gram negative commensal bacteria between the teeth and the gums results in chronic inflammation and damage to the jaw bone supporting the teeth. A number of studies have now indicated a higher incidence of cardiovascular disease in patients with periodontal conditions. The suggestion is that the entry of these oral bacteria into the bloodstream of individuals with 'gum' disease, consequent upon eating or brushing the teeth, is a chronic inflammatory stimulus which can promote the inflammatory atherosclerotic pathology in susceptible individuals. Again, much more evidence needs to be collected before this idea becomes a medical reality.

The latest disease proposed to be caused by bacteria is kidney stones. In this case the bacteria believed to be responsible have been termed nanobacteria because of their size (about one tenth the diameter of *E. coli*). These bacteria, under certain conditions, can build calcified shells around themselves and in one study, all of the human kidney stones examined contained such nanobacteria.

THE APPLICATION OF CELLULAR MICROBIOLOGY TO THE GENERATION OF NOVEL THERAPEUTICS

Advances in our understanding of bacteria–host interactions are likely to have a profound effect on our ability to control human diseases, including infectious diseases, idiopathic inflammatory conditions and cancer. The most straightforward example of this is the application of our new-found knowledge to the development of the next generation of antibacterial drugs. The current antibiotics in clinical use have largely come from the last generation of research and no new classes of antibiotics have been developed in the past 25 years. Antibiotic resistance continues to increase, and it has been recognized for some considerable time that new approaches to antibiotic design had to be taken. The development of new antibacterial strategies is likely to be the first of the spin-offs from our new understanding of bacterial–eukaryotic cell interactions.

Antibacterial agents

The discovery in the 1980s–1990s that all organisms produce antibacterial peptides has suggested that these natural products should be be utilized for treatment of infections, and currently clinical trials of a number of such peptides are under way. These peptides appear to have been in existence for many millions of years and do not appear to have engendered significant resistance among bacteria during that time. However, the fact that these molecules are peptides, and fairly large peptides at that, brings with it some problems. Peptides are expensive to produce and are difficult to administer as oral administration is likely to lead to proteolytic breakdown. They are also potentially immunogenic and could give rise to immunological responses such as anaphylaxis. These problems could be overcome by using peptides topically. In the future it is likely that the approach taken by the drug industry will be to make non-peptide mimetics of these antibacterial peptides. Another possible approach is to induce the production of antibacterial peptides in selected tissues. It may be possible to do this once we understand the nature of the signals which induce antibacterial peptide synthesis (see Chapter 8).

The development of novel antibacterial drugs in the future will depend upon the recognition of specific molecular targets in bacteria. With the rapid increase in sequence information obtained from bacterial genomics and the use of techniques such as *in vivo* expression technology (IVET) and signature tag mutagenesis (described in Chapter 4) to identify virulence mechanisms, it is certain that many novel therapeutic targets will be identified.

Identification of molecular targets is the first step in the development of pharmaceuticals, although the development of compounds which can interfere with the identified targets is not always straightforward. However, it is likely that in the next 10–20 years a large number of novel therapeutics to treat individual bacterial infections will become available.

Anti-inflammatory and immunomodulatory drugs

The inflammatory defences against infectious diseases are evolved mechanisms that are also utilized in most idiopathic diseases (described above). Such diseases probably afflict around one in 10 of the world's population. Thus there is a pressing need to be able to control the inflammatory response. The immune system of the host organism and the countermeasures utilized by infectious agents have co-evolved genetic systems designed to maximize the survival of the infecting organism and the host organism. In Chapters 8 and 9 and in earlier sections of this chapter the mechanisms utilized by bacteria, protozoans and viruses to limit the inflammatory defences of the host have been highlighted. It is likely that these immune evasion mechanisms developed by infectious agents could be developed as therapeutic agents for the treatment of idiopathic inflammatory diseases. Moveover, it is likely these immune evasion mechanisms can be tailored to individual diseases. For example, the bacterial mechanism which can inhibit complement-mediated host defences could be developed to treat diseases in which the major pathology is due to complement activation. Such diseases include rheumatoid arthritis and systemic lupus erythematosus. Bacteria produce a range of proteins which arrest the cell cycle. Some of these proteins have been discovered by testing their actions on lymphocytes. As has been described in Chapter 9, the acquired immune response requires the clonal proliferation of selected T and B lymphocytes. Bacterial proteins able to inhibit lymphocyte proliferations could have obvious immunosuppressive actions. Bacterial proteins inducing apoptosis of leukocytes could also be used as immunosuppressive agents. Bacterial exotoxins are powerful modulators of cell behaviour which could also be utilized to inhibit leukocyte functions.

Microkines – microbial proteins with cytokine-modulating activity – could have therapeutic benefit in a wide range of inflammatory diseases by blocking proinflammatory cytokine networks.

Anticancer therapy

As described earlier in this chapter, it is now established that bacteria synthesize proteins which have selective effects on the eukaryotic cell cycle. These proteins could have obvious clinical benefit in the treatment of cancer and of other conditions in which aberrant cell proliferation lies at the heart of the disease.

THE FUTURE OF CELLULAR MICROBIOLOGY

The term cellular microbiology was only coined within the past few years and yet the subject has already produced an outpouring of fascinating findings about the interactions between infectious agents, mainly bacteria, and the host. There are likely to be many major advances in the next decade and it is a brave individual who can try and predict the future. In terms of the application of technology, bacterial genomics will accelerate and more and more complete genomes will be available for analysis. These sequences, along with those of Archaea and Eukarya, will allow a more precise definition to be made of the evolutionary relationships between these three superkingdoms. They will also provide a fresh insight into the requirements, in terms of genes, for the creation of a cell. Genomics will generate information for those biochemists and cell biologists interested in the integration of cell biochemical pathways. The generation of gene sequences from different cells encoding the same proteins will provide a new perspective on the structural requirements for particular protein function.

Cellular microbiology is illuminating the intimacy of the interactions between bacteria and host eukaryotic cells. It should be borne in mind that the eukaryotic cell is a symbiont containing bacterial components including mitochondria, chloroplasts and the centriole. The story of the symbiosis between bacteria and eukaryotic cells is told in great detail, and with a feeling of great wonder, by Professor Margulis in her exceptional book *Symbiosis in Cell Evolution*. In her book the experiments of the cell biologist Dr Jeon are described. During the study of nucleocytoplasmic interactions of *Amoeba proteus* in the 1960s and early 1970s Jeon found that the cultured cells were infected with an unknown bacterium and their growth rate declined significantly. The amoeba that survived this infection were found to have become symbionts and contained on average 40 000 bacteria per cell. In the absence of the bacteria (induced by antibiotics or by nuclear transplantation) the amoeba were unable to undergo cytoplasmic division at the end of mitosis. Thus it appears that bacterial/eukaryotic symbiosis is not just a 'once in a lifetime' experience. Are such interactions between bacteria and eukaryotic cells occurring constantly and if they are what are the consequences for the evolution of life on our planet?

We have concentrated on the interactions between Bacteria and Eukarya and have ignored the Archaea. However, examples of this third superkingdom of life are known to exist as commensal organisms in ruminants and in man. Many questions need to be addressed concerning Archaeal/Eukaryal interactions. How much crosstalk is there between Archaea and Archaea? Do Archaeal species cause disease? What are the nature of the interactions between Archaea and Bacteria?

Little attention has been paid in this small book to the genetics of susceptibility to infectious diseases. For example, it is known that individuals who are heterozygous for sickle-cell anaemia or β thalassaemia have reduced susceptibility to *Plasmodium falciparum* malaria. This is believed to be due to heterozygotes being more effective at expressing malarial antigens on their red blood cells. In terms of susceptibility and response to bacterial infections there is growing evidence that polymorphisms in the promoter elements of cytokine genes are important. Some individuals will produce significantly more IL-1 or IL-6, for example, than others in response to a particular proinflammatory signal. Such elevations in the cytokine response can be protective. However, as cytokines are also the main causation of tissue pathology then their overproduction can also be deleterious. With the enormous numbers of these proteins and their intimate involvement in innate and acquired immune responses, unravelling their genetics in terms of susceptibility to bacterial infections will be a major but necessary task.

As described in Chapter 8, in inbred mouse strains susceptibility or resistance to infection with intracellular parasites has been shown to be controlled by the natural resistance-associated macrophage protein gene *Nramp*1 on chromosome 1. Nramp now appears to comprise a novel family of membrane proteins involved in control of divalent ions. Allelic variants in the human *Nramp1* locus appear to be associated with susceptibility to leprosy and tuberculosis and this is another area in which there are likely to be exciting advances in our understanding of susceptibility and resistance to infections.

Another major source of information on the genetics of susceptibility/resistance to infections will come from the human genome project, as homology searching of expressed sequence tags (ESTs) will disclose the numbers of genes with homology to genes such as *Nramp1*.

The use of 16S ribosomal RNA gene sequence analysis is opening a whole new world of microbial diversity which was unimaginable only a generation ago. The next 20 years will, we envisage, see a new picture emerging of the evolutionary history of life on earth and of the parts played by the three superkingdoms of life in this process. It will be an exciting time to be involved in science and we hope that some of the readers of this book will be encouraged to become cellular microbiologists and join in the chase.

Further Reading

Books

Cann AJ (1997) *Principles of Molecular Virology* (2nd edn). Academic Press, London.

Henderson B, Poole S, Wilson M (1998) *Bacteria/Cytokine Interactions in Health and Disease*. Portland Press, London.

Kuhn TS (1972) *The Structure of Scientific Revolutions* (2nd edn). University of Chicago Press, Chicago.

Margulis L (1993) *Symbiosis in Cell Evolution* (2nd edn). Freeman, New York.

Murray A, Hunt T (1993) *The Cell Cycle*. Freeman, New York.

Tannock GW (1995) *Normal Microflora: An Introduction to Microbes Inhabiting the Human Body*. Chapman & Hall, London.

Reviews

Baba T, Schneewind O (1998) Instruments of microbial warfare: bacteriocin synthesis, toxicity and immunity. *Trends Microbiol* 6: 66–71.

Henderson B, Poole S, Wilson M (1996) Bacterial modulins: a novel class of virulence factors which cause host tissue pathology by inducing cytokine synthesis. *Microbiol Rev* 60: 316–341.

Hersh D, Weiss J, Zychlinsky A (1998) How bacteria initiate inflammation: aspects of the emerging story. *Curr Opin Microbiol* 1: 43–48.

Kaprelyants AS, Kell DB (1996) Do bacteria need to communicate with each other for growth? *Trends Microbiol* 4: 237–242.

Knowles DJC (1997) New strategies for antibacterial drug design. *Trends Microbiol* 5: 379–383.

Zychlinsky A, Sansonetti PJ (1997) Apoptosis as a proinflammatory event: what can we learn from bacteria-induced cell death? *Trends Microbiol* 5: 201–204.

Papers

Higgins TE, Murphy AC, Staddon JM, Lax AJ, Rozengurtz E (1992) *Pasteurrella multocida* toxin is a potent inducer of anchorage-independent cell growth. *Proc Natl Acad Sci USA* 89: 4240–4244.

Pancholi V, Fischetti VA (1997) Regulation of the phosphorylation of human pharyngeal cell proteins by group A streptococcal surface dehydrogenase: signal transduction between streptococci and pharyngeal cells. *J Exp Med* 186: 1633–1643.

Sparwasser T, Miethke T, Lipford G, Borschert K, Hacker H, Heeg K, Wagner H (1997) Bacterial DNA causes septic shock. *Nature* 386: 336–337.

Bellamy R, Ruwende C, Corrah T *et al.* (1998) Variations in the NRAMP1 gene and susceptibility to tuberculosis in West Africans. *N Engl J Med* 338: 640–644.

Index